COMPUTATIONAL MODELING OF ISSUES
IN MATERIALS SCIENCE

COMPUTATIONAL MODELING
OF ISSUES
IN MATERIALS SCIENCE

PROCEEDINGS OF SYMPOSIUM D
ON
COMPUTATIONAL MODELING OF ISSUES
IN MATERIALS SCIENCE

OF THE 1997 ICAM/E-MRS SPRING CONFERENCE
STRASBOURG, FRANCE, JUNE 16-20, 1997

Edited by

H. DREYSSÉ
Université Louis Pasteur, Strasbourg, France

Y. KAWAZOE
Institute of Materials Research, Tohoku University, Sendai, Japan

L.T. WILLE
Florida Atlantic University, Florida, USA

C. DEMANGEAT
Université Louis Pasteur, Strasbourg, France

ELSEVIER
AMSTERDAM–LAUSANNE–NEW YORK–OXFORD–SHANNON–TOKYO

Published by:
Elsevier
Elsevier Science SA
Avenue de la Gare 50
1003 Lausanne
Switzerland

ISBN: 0-444-20514-4

Reprinted from

COMPUTATIONAL MATERIALS SCIENCE Vol. 10 (1-4)

The manuscripts for the Proceedings were received by the Publisher between
13 August and 15 September 1997.

Transferred to digital printing 2006

Printed and bound by CPI Antony Rowe, Eastbourne

Sponsors

This conference was held under the auspices of:

The Council of Europe
The Commission of European Communities

It is our pleasure to acknowledge with gratitude the financial assistance provided by:

Banque Populaire	(France)
Elsevier Science	(The Netherlands)
Office du Tourisme, Strasbourg	(France)
Université Louis Pasteur, Strasbourg	(France)
Institut de Physique et Chimie des Matériaux, Strasbourg	(France)
Ministère de l'Enseignement Supérieur et de la Recherche	(France)
HCM Network "Ab-initio calculation of complex processes in materials"	(EU)

ELSEVIER

Volume 10, Numbers 1–4, 1998

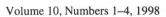

COMPUTATIONAL
MATERIALS
SCIENCE

Computational modeling of issues in materials science

Editors: Hugues Dreyssé, Yoshi Kawazoe, Luc T. Wille, Claude Demangeat

Contents

Contents

Electronic structure. Methodology

Complex materials and alloys

ELSEVIER

Computational Materials Science 10 (1998) xiii–xiv

COMPUTATIONAL
MATERIALS
SCIENCE

Preface

The following pages contain the Proceedings of Symposium D, entitled 'Computational Modeling of Issues in Materials Science', presented at the combined 1997 International Conference on Applied Materials/ European Materials Research Society Spring meeting (ICAM'97/E-MRS'97) held in Strasbourg (France), from 16 to 20 June 1997.

With 181 submitted abstracts and 12 sessions running for the entire duration of the meeting, this was the largest symposium at the conference. Attendants came from all five continents, with participants from as far away as South Africa and Australia. Also a large representation from Eastern Europe was present.

The symposium was kicked of with a plenary talk by Prof. Volker Heine (Cambridge University) entitled: 'The design of *computer experiments*'. This speaker emphasized the predictive power of computer calculations, the possibility of accessing regions in parameter space that cannot be attained by experiment, and the ability of simulations to provide virtually unlimited space and time resolution. He also stressed the need for computer experiments to provide 'understanding', which allows a massive amount of data to be collapsed in a few simple rules.

The remainder of the symposium consisted of a series of lively sessions with the following titles:

1. General simulation methods
2. Ab initio and tight-binding molecular dynamics
3. Surfaces and films
4. Complex materials and parallel computing
5. Fracture
6. Polymers
7. Micromagnetism
8. Magnetism
9. Electronic structure (specific materials)
10. Electronic structure (methodology)
11. Complex materials and alloys
12. Clusters.

There were 14 invited talks, 54 contributed papers, and 62 posters (divided over two sessions) actually presented at the symposium.

It became evident at the meeting that computational materials science has truly emerged as a field in itself. The range of phenomena studied and the variety of techniques used were truly astonishing and indicate that the subject has sufficiently matured that technologically relevant information can now routinely be extracted from computational modeling. Moreover, these models increasingly use atomistic information (frequently in the form 'ab initio' calculations) from which macroscopic parameters may be determined.

The contributions and discussions at this symposium clearly indicated that some major themes have emerged. Several papers showed that parallel computers will play a major role in the further development of the field: simulations with 100 million atoms, approaching macroscopic dimensions, have now become possible on the latest parallel hardware. Clearly, new algorithmic ideas will be needed and, in particular, scalable software will be essential for further progress. Thus, so-called order-N methods were the subject of lively debate.

Elsevier Science B.V.

Another clear trend was the emergence of the Car-Parrinello method as a workhorse for the most advanced simulations, simultaneously combining electronic structure and molecular dynamics optimization. Only a few years ago it was considered computationally very expensive and of such a sophistication that only experts could use it. The advent of faster hardware and the diffusion of computer codes have brought this technique within the reach of many research groups.

The local density approximation (LDA) of the density functional theory (DFT) has been at the center of much work in electronic structure calculations for several decades. Although it has been remarkably successful, a number of its limitations have been known for a while and several authors at the meeting discussed schemes for going beyond LDA.

Atomistic simulations frequently operate on microscopic time and length scales: nanometers and femtoseconds are the basic units in this realm. The macroscopic world however operates on a very different scale. While a correct description of many observations requires quantum mechanics for a true understanding, the relevant measurements are done on the macro-scale. How to consistently go from the micro- to the macro-scale (and also describe the intermediate meso-scale) remains one of the great unsolved puzzles in computational materials science and was the subject of much discussion at the symposium.

Science often progresses by analogies: techniques that are valuable in one field are frequently found to be also applicable in another. The interdisciplinary side of computational studies of matter was demonstrated in several talks, where authors borrowed methods from nuclear physics, fluid dynamics, and other subjects.

The organizers benefited greatly from the expert advice provided by the members of the International Organizing Committee. They include: V. Alessandrini (Paris, France), K. Binder (Mainz, Germany), J. Connolly (Lexington, USA), P.H. Dederichs (Jülich, Germany), M. Doyama (Tokyo, Japan), R.A. de Groot (Nijmegen, Netherlands), J. Hafner (Wien, Austria), V. Kumar (Kalpakkam, India), A.A. Lucas (Namur, Belgium), J.L. Moran Lopez (San Luis Potosi, Mexico) and A. Zangwill (Atlanta, USA).

Financial assistance for the symposium was provided by the Université Louis Pasteur (Strasbourg), the Institut de Physique et Chimie des Matériaux de Strasbourg, the Ministère de l'Enseignement Supérieur et de la Recherche for supporting travel and lodging expenses of Eastern European participants and by the European HCM network Ψ_k 'Ab initio (from electronic structure) calculation of complex processes in materials'. The support from these organizations was essential and is gratefully acknowledged.

All in all, this was a very productive symposium. New collaborations were started, many novel ideas were generated, and a large amount of information was disseminated. Because of the wide range of subjects, participants gained new appreciations for the scope of the field but also for the unity that has emerged. The meeting gave an excellent idea of the status of computational materials science anno 1997. It is hoped that the reader will find in these proceedings a more permanent record of these developments.

Hugues Dreyssé, Strasbourg
Yoshi Kawazoe, Sendai
Luc T. Wille, Boca Raton
Claude Demangeat, Strasbourg

ELSEVIER

Computational Materials Science 10 (1998) 1–9

COMPUTATIONAL
MATERIALS
SCIENCE

Large-scale simulations of brittle and ductile failure in fcc crystals

Farid F. Abraham [a,*], J.Q. Broughton [b]

[a] *IBM Research Division, Almaden Research Center, San Jose, CA 95120-6099, USA*
[b] *Naval Research Laboratory, Washington, DC 20375, USA*

Abstract

We are simulating the failure of three-dimensional notched fcc solids under mode one tension using molecular dynamics, simple interatomic potentials and system of tens of millions of atoms. We have discovered that crystal orientation with respect to the uniaxial loading is important; the solid fails by brittle cleavage for a notch with (1 1 0) faces and by ductile plasticity for a notch with (1 1 1) faces or (1 0 0) faces. We argue that the competition between bond-breaking and interplanar slippage is governed by the nonlinearity and anisotropy of the crystal elasticity near materials failure. If the speed of the (1 1 0) brittle crack velocity reaches approximately one-third of the Rayleigh sound speed, a "brittle-to-ductile" transition occurs and is consistent with the onset of a dynamic instability of brittle fracture. Such an instability was seen in our earlier two-dimensional fracture simulations of rare-gas films and appears to be a general feature of the dynamic brittle fracture process. We close with a simulation showing the consequences of a brittle crack colliding with a void. Copyright © 1998 Elsevier Science B.V.

We have been simulating rapid fracture of two- and three-dimensional crystals using molecular dynamics for system sizes up to 100 million atoms and time scales approaching nanoseconds [1–4], and have found that the rare-gas solid cleaves as a brittle solid for a notch with (1 1 0) faces. We note that the (1 1 0) face does not have the lowest surface energy and that according to the Griffith criterion this should not be a favorable choice for expecting brittle cleavage. Probably more surprising is that an fcc solid of rare-gas atoms could fracture brittlely. Experience tells us that fcc crystals are typically ductile, whether it be a rare-gas solid or a metal [5]. While material type seems unimportant, we have now discovered that crystal orientation with respect to the uniaxial loading is important. We find that the solid fails by brittle cleavage

for a notch with (1 1 0) faces and by ductile plasticity for a notch with (1 1 1) faces or (1 0 0) faces. We argue that the competition between bond-breaking and interplanar slippage is governed by the nonlinearity and anisotropy of the crystal elasticity near materials failure. While the elastic fcc solid is approximately isotropic for small uniaxial deformations, it is highly anisotropic near the stability limit, a feature that is not generally accounted for in fracture mechanics [6]. The nature of brittle failure is directly correlated with bulk anisotropic elasticity in a sensible way; in the stress–strain dependence, the crystal direction that becomes elastically unstable for the smallest strain (and stress) defines the cleavage plane. We learn that this gives a different prediction for the favored cleavage plane with respect to Griffith's criterion using equilibrium surface energies. We now present the detailed results.

The molecular dynamics simulation technique is based on the motion of a given number of atoms governed by their mutual interatomic interactions given by a continuous potential and requires the numerical integration of the equations of motion, $F = ma$. A simulation study is defined by a model created to incorporate the important relevant features of the physical system of interest. These features include the specified boundary and initial conditions and the choice of the interatomic force law. Most of our fracture simulations assume a rare-gas solid. The van der Waals bonding giving the cohesion of the rare-gas solid can be modeled accurately by a two-body Lennard-Jones potential and has served as a paradigm for studying classical many-body phenomena of atomistic systems in computational physics [7]. We choose to do our fracture studies with the simple rare-gas solid defining our *model* material [8].[1] However, we have extended our results to include other interatomic force laws, like the Morse pair potential and the embedded-atom many-body (EAM) potential, to describe the metal nickel. We have learned that our findings are generic to the fcc packing.

We discovered in our two-dimensional molecular dynamics simulations for fracture that the tip follows a path associated with the highest surface energy face. This is in contrast to the Giffith theory which associates the lowest energy surface as the favored cleavage direction [3]. In fact, it was suggested that our early results "have to be interpreted with care since they were not obtained for cracks on the cleavage plane of the triangular lattice [9]". However, we discovered that the two-dimensional bulk elastic modulus has a pronounced anisotropy for large strains and argued that this is the origin of this puzzling feature. Going to three dimensions, we also find a pronounced elastic anisotropy at large strains for the fcc packing of simple atomic systems. In Fig. 1, we present the stress–strain curves for the three-dimensional LJ lattice for uniaxial loading in the $\langle 1\,1\,1 \rangle$, $\langle 1\,0\,0 \rangle$ and $\langle 1\,1\,0 \rangle$ directions. We

---

[1] On p. 73 in commenting on Griffith's desire to have a simpler experimental material which would have an uncomplicated brittle fracture, Gordon writes, "In those days models were all very well in the wind tunnel for aerodynamic experiments but, damn it, who ever heard of a model material?

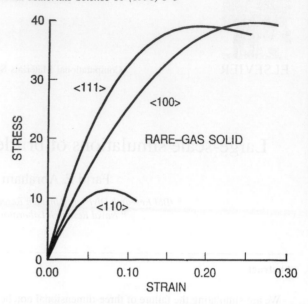

Fig. 1. The stress–strain curves for the three-dimensional fcc rare-gas lattice for uniaxial loading in the $\langle 1\,1\,1 \rangle$, $\langle 1\,0\,0 \rangle$ and $\langle 1\,1\,0 \rangle$ directions, for zero lateral stress.

calculated these curves for zero lateral stress boundary conditions. The elastic properties under large deformations are highly anisotropic. Most important, we note that instability occurs for a very low strain of 8% for uniaxial tension applied in the $\langle 1\,1\,0 \rangle$ direction, the corresponding value of the stress being 0.14 GPa. For the other two directions $\langle 1\,1\,1 \rangle$ and $\langle 1\,0\,0 \rangle$, the instability occurs at 19 and 30% strain, respectively, and a fourfold increase in stress. If you want to break a crystal by uniform expansion and you want to use the least effort, choose the $\langle 1\,1\,0 \rangle$ direction (of course assuming that plasticity does not intervene before brittle failure). In the failure of notched solids, imposed strains on the order of a few percent are required because of the magnified stress at the crack tip. This was the case in our first three-dimensional simulation with a $(1\,1\,0)$ notch where the slab cleaved brittlely at an applied strain of 1.5% [4]. We will now discuss simulations with all three notch faces and argue that the anisotropy of the bulk stress–strain governs whether the fcc solid fails by brittle cleavage or by ductile plasticity.

For the remainder of the study, we express quantities in reduced units; lengths are scaled by σ, energies by ϵ, where ϵ is the LJ well depth and σ is where

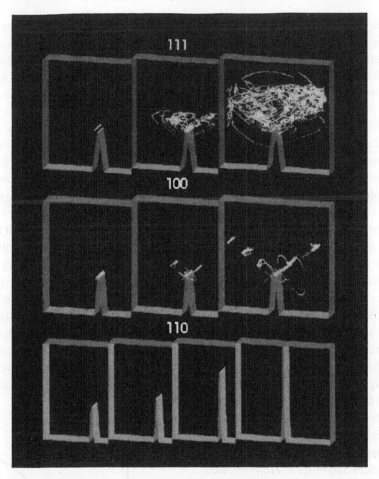

Fig. 2. Time sequence of the propagating crack is shown as overlapping landscapes of the growing surface due to brittle fracture or ductile failure for three notch surfaces; (1 1 1), (1 0 0), and (1 1 0). Only atoms with a potential energy less than 97% of the bulk value are displayed, resulting in the selected visualization of atoms neighboring surfaces and dislocations.

the potential goes through zero. The system is a three-dimensional slab with L_x layers by L_y layers by L_z layers for the three orthogonal sides. The notch is a slit beginning midway along L_x for $y = 0$, having a y extension of length $l_t = 120$ which extends through the entire thickness L_z. The exposed notch faces are in the y–z planes. For (1 1 0) faces, the crack is in the $\langle 1\bar{1}0 \rangle$ direction and $L_z = 108$; for (1 0 0) faces, the crack is in the $\langle 1 1 0 \rangle$ direction and $L_z = 144$; for (1 1 1) faces, the crack is in the $\langle \bar{1}10 \rangle$ direction and $L_z = 216$ atoms. For all faces, $L_x = 336$ and $L_y = 336$ layers. The total number of atoms in the slabs range from 10 to 12 million, and the physical dimensions of the slabs

are comparable. Periodic boundary conditions are imposed between the x–y faces at $z = 0$ and $z = L_z$. The slab is initialized at zero reduced temperature. An outward strain rate $\dot{\varepsilon}_x$ is imposed on the outer most columns of atoms defining the opposing vertical faces of the slab. A linear velocity gradient is established across the slab, and an increasing lateral strain with time occurs in the solid slab with an applied strain rate of $\dot{\varepsilon}_x = 0.00033$. If and when the solid fails brittlely at the notch tip, the imposed strain rate is set to zero.

In Fig. 2, snapshot pictures of the failure dynamics are presented for the three notched surfaces. In order to see into the interior of the solid, we show only those atoms that have a potential energy greater than −6.1, where the ideal bulk value is −6.3. This trick has been used very effectively in our two- and three-dimensional studies for displaying dislocations, micro-cracks and other imperfections in crystal packing and reduces the number of atoms seen by approximately two orders of magnitude. The visible atoms are associated with faces of the slab and initial notch, surfaces created by crack motion, local interplanar separation associated with the material's dynamic failure at the tip, and topological defects created in the otherwise perfect crystal. Our choice of displaying the visible atoms in a perspective view of the slab as white on a black background gives the appearance of looking through a clear plastic (Lucite) solid block with etched surfaces and etched interior regions defining the locations of dislocations. The sequences begin when the solid has been stretched by ∼1.5% and 45 (reduced time units) from the start of the pull. The (1 1 0) sequence, in reduced time units, is 45, 112, 146, 214 and is displayed as the bottom row of "off-front" perspective views for Fig. 2; the (1 0 0) sequence is 45, 79, 124 and is the middle row; the (1 1 1) sequence is 45, 79, 112 and is the top row of the figure. For all notches, the solid begins to fail at the notch tip at or before 10 000 time-steps, or reduced time of 45, when the solid has been stretched by ∼1.5%. However, their failures are different. For the (1 1 0) notch, the solid fails at the notch tip brittlely with planar cleavage of the bonds between the two (1 1 0) neighboring atomic sheets defined by the initial notch and continues until it passes through the entire length of the slab. For the (1 0 0) and (1 1 1) notches, the failure is ductile, by plastic deformation and the emission of dislocations traveling along (1 1 1) slip planes. In Fig. 2, the dislocations are seen as slip *strings* flying from the crack tip of the (1 0 0) and (1 1 1) notches (top and middle rows in the figure).

As set down by Rice and Thomson [10], ductile versus brittle behavior may be regarded as a competition between dislocation emission at an atomically sharp crack and cleavage decohesion. Most of their work has been directed to the understanding of the inherent ductile or brittle nature of a given material or the description of the brittle–ductile transition as

a result of temperature. Modes of dislocation nucleation from a crack tip depends on how the slip planes cut the crack edge for a particular crack face. The slip plane's orientation with respect to a particular crack plane may be parallel, inclined or oblique, and each crack plane has three unique slip planes. Predicting dislocation nucleation in each circumstance is difficult, especially within the framework of elasticity theory. However, using the approximation of a linear elastic crack tip field, Gao [11] has calculated the Schmid factors for fcc crystals with cracks on (1 1 1), (1 0 0) and (1 1 0) planes and the respective directions. He has concluded that the maximum Schmid factor is only slightly different between the three crack orientations because of the large number of slip systems (1 1 1) ⟨1 1 0⟩ in the fcc packing. This implies that a difference in failure (brittle versus ductile) cannot be easily explained as a significant difference of the critical shears for the various crack orientations. Furthermore, the Griffith argument based on surface gives us a means of estimating the favored path for brittle failure. For the LJ crystal, the surface energies

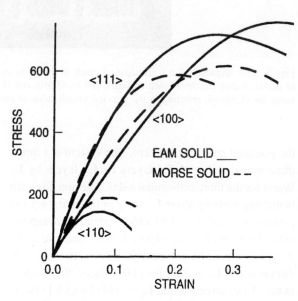

Fig. 3. The stress–strain curves for the three-dimensional fcc nickel lattice for uniaxial loading in the ⟨1 1 1⟩, ⟨1 0 0⟩ and ⟨1 1 0⟩ directions, for zero lateral stress. The Morse pair potential and the embedded-atom potentials are used to represent the nickel interatomic potential.

Fig. 4. Temporal sequence of the crack dynamics in a 100 million atom slab (ordered from left-to-right and top-to-bottom) is shown as landscapes of the crack surface growing due to brittle fracture (the first two images for times 1, 32), the initiation of the instability (time 43), and the subsequent growth of dislocations (the last three images for times 52, 68, 90) after the transition to ductility.

for the (1 1 1), (1 0 0) and (1 1 0) faces are in the ratios 1.0 : 1.05 : 1.12, respectively [12]. Griffith's criterion predicts the most favorable cleavage plane to be the closed-packed (1 1 1) surface. If ductile failure is to preempt brittle cleavage, its most likely occurrence should be for the highest surface energy crack; i.e., the (1 1 0) crack, everything else being equal. We observe quite the opposite.

How can we rationalize this observation? We believe that the globally expanded lattice should more nearly approximate the local atomic environment at the crack tip at rupture, more so than the simple cutting of bonds bridging two neighboring planes as in the Griffith picture. The *effective surface energies* now become proportional to the areas under the stress–strain curves for the respective crystal directions in Fig. 1.

Using these new surface energies with the Griffith picture suggests that cleavage failure would be most favorable for the (1 1 0) plane, followed by the (1 1 1) and (1 0 0) planes. The critical shear stress for dislocation nucleation has not been achieved before the tensile stress for failure has been reached for (1 1 0) cleavage. But the critical tensile stress for (1 0 0) and (1 1 1) cleavage is much larger (well over a factor of 8!, see Fig. 1), and the critical shear stress (which is comparable for all faces) preempts brittle failure at the notch for these faces.

"So much for Lennard-Jonesium, what about a *real* material?" is a question often heard in one's quest to be relevant. In Fig. 3, we present the same calculation as in Fig. 1, but for two potentials describing the metal nickel, the Morse pair potential [13][2] and the EAM potential [14]. The curves for the Morse potential essentially fall on top of the Lennard-Jones curve, when normalizing them to the interatomic well depth. We also note that the EAM solid gives the similar dependences as the simple pair potentials. We should state our concern that using ANY potential for a real metal through the entire range of strain to failure is highly questionable and should be scrutinized by doing first principles calculations. In any case, the anisotropic elastic behavior is maintained because it is a simple consequence of the fcc packing upon distortion. This suggests the possibility that nickel will behave like a rare-gas solid, of course on a different strength scale; i.e., our findings are generic to the fcc packing.

The (1 1 0) notch can be "driven" from brittle to ductile failure if a sufficient strain is applied. We have learned that the crack tip begins moving when the solid has been stretched by $\sim$1.5%. Instead of setting the imposed strain rate equal to zero at initiation of crack-tip motion, we now set the imposed strain rate to zero at δT_s time beyond motion; the "driving force" for the crack dynamics increases with increasing δT_s.

Fig. 4 shows our simulation of the 100 million atom slab ($L_z = 896$) and $\delta T_s = 16$ (reduced time), where the dynamic "brittle-to-ductile" transition was discovered and observed to occur at $\delta T_t = 43$. In the time interval 0–42, the crack motion is representative of

---

[2] We take $m = 1.5$.

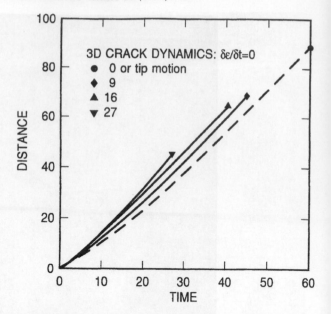

Fig. 5. The crack-tip distance versus reduced time is presented for various times δT_s where the strain rate is set equal to zero.

brittle fracture, with planar cleavage of the bonds between the two (1 1 0) neighboring atomic sheets defined by the initial notch. At a time of 43, the crack edge begins to blossom into a "flower of loop dislocations". The crack slows to a stop and continues to dissipate elastic energy through the continued creation and motion of dislocations.

We have determined the dynamic history of the crack motion up to time δT_t for different choices for δT_s in order to establish a relationship between the dynamics and the fracture transition [15]. In Fig. 5, we present the crack-tip distance versus reduced time for various times where the strain rate is set equal to zero. As demonstrated in Fig. 1, we might expect that there is a time δT_t in the dynamics where the brittle fracture crosses over to ductile failure, and we indeed observe that $\delta T_t = 45, 40, 27$ for $\delta T_s = 9, 16, 27$. Only for $\delta T_s = 0$ do we observe that the crack propagates brittlely throughout the entire length of the solid. A primary interest is to find a dynamic feature at the transition point that is common to all of the simulations. Such a feature was found in our two-dimensional simulations where the speed for the onset of the brittle fracture instability was $\sim$33% of the Rayleigh speed.

Fig. 6. Time sequence of a straight crack colliding with a disk shaped void.

In three dimensions, we now find that the transition from brittle fracture to plastic failure occurs when the tip speed v_t equals 0.36 of the Rayleigh sound speed c_R; $v_t/c_R = 0.36, 0.36, 0.37$ for $\delta T_s = 9, 16, 27$. This supports the following scenario for the dynamic failure process in the three-dimensional rare-gas solids. From zero velocity, the crack accelerates smoothly by brittle fracture along the (1 1 0) plane until it approaches 0.36 of the Rayleigh sound speed. At this point, a dynamic instability in the brittle fracture process occurs. The brittle nature of this failure process marginally exists for the perfect (1 1 0) planar cleavage, and the deviation from planar cleavage gives rise to plasticity, and a spontaneous proliferation of dislocations.

Our last example, illustrated in Fig. 6, shows the consequences of a brittle (1 1 0) crack colliding with a void. This simulation assumes that the fcc metal Ni is described by the Morse potential adopted by Milstein [13]. Instead of the crack edge "jumping across" the diameter of the void upon collision, locations along the crack edge stop upon contact with the circular boundary of the penny-void, while the remaining portion of the crack edge attempts to continue its forward brittle motion. It has the appearance of a "soft" straight object (a knife edge) hitting a rigid obstacle. However, in Fig. 6 we are showing only the crack and the void. By showing all of the atoms with a deficit potential energy (Fig. 7), we see a great deal of plasticity

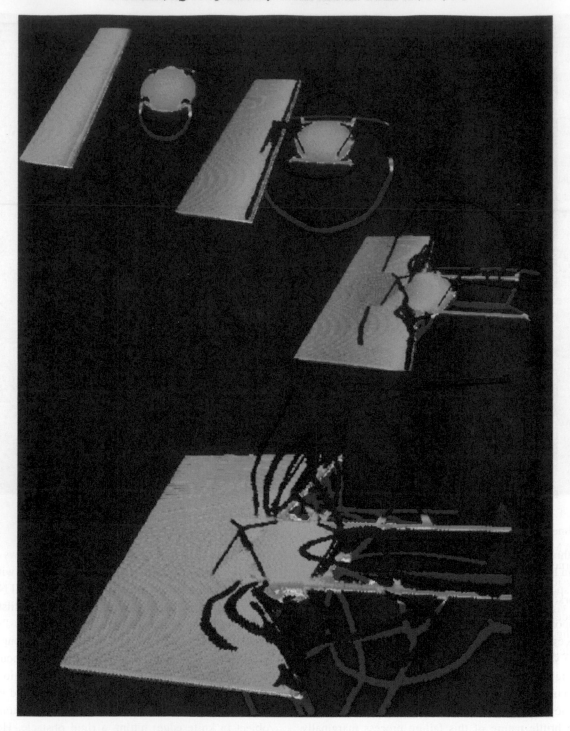

Fig. 7. Time sequence of a straight crack colliding with a disk shaped void (similar to Fig. 6), but showing the dislocations created in the process.

occurring upon collision (the appearance of dislocations beginning even before collision). This kills the brittle nature of the failure process and gives rise to a dominating ductility at contact.

Argon et al. [16] have discussed the important progress that has been made by recourse to hybrid continuum-atomistic approaches based on the use of a Peierls interplanar potential. They state, "Ultimately, a full understanding of the brittle–ductile transitions must come from atomistic models of the formation and outward propagation of the dislocation embryo at the crack tip". Certainly, our discovery by atomistic simulation of the character of materials failure on hyperelasticity for the simplest of solids should become a test-bed for future theories addressing the competition between brittle and ductile failure. In this system, we see brittleness and ductility coexisting and depending on the intrinsic elastic anisotropy arising from extreme deformation. This delicate competition between atomic separation and atomic slippage is exhibited in our *model* material and may exist in most fcc solids.

A multimedia version of this atomistic simulation of fracture is available via the World Wide Web: http://www.tc.cornell.edu/~farid/fracture/100million/

FFA is grateful for collaborative efforts with Dave Schneider, Bruce Land, Joe Skovira, Jerry Gerner and Marcy Rosenkrantz at the Cornell Theory Center, Cornell University, Ithaca, NY, who made this study possible. N. Govind, UCLA, and Xiaopeng Xu, IBM, worked on calculating the stress–strain relation for modeled solids. This research was conducted in part, using the resources of the Cornell Theory Center, which receives major funding from the National Science Foundation and New York State with additional support from the Advanced Research Projects Agency, the National Center for Research Resources at the National Institute of Health, IBM Corporation and members of the Corporate Research Institute.

References

[1] F.F. Abraham, D. Brodbeck, R. Rafey, W.E. Rudge, Phys. Rev. Lett. 73 (1994) 272.

[2] F.F. Abraham, D. Brodbeck, W.E. Rudge, X. Xu, J. Mech. Phys. Solids, to be published.

[3] F.F. Abraham, Phys. Rev. Lett. 77 (1996) 869.

[4] F.F. Abraham, D. Schneider, B. Land, J. Skovira, J. Gerner, M. Rosenkrantz, J. Mech. Phys. Solids, to be published; also in: D. Wirtz, T. Halsey (Eds.), Statistical Mechanics in Physics and Biology, MRS Symposium Proceedings, vol. 463, Materials Research Society, Pittsburgh, 1997.

[5] A. Kelly, N.H. Macmillan, Strong Solids, Clarendon Press, Oxford, 1986, p. 10.

[6] L.B. Freund, Dynamical Fracture Mechanics, Cambridge University Press, New York, 1990.

[7] M.L. Klein, J.A. Venables, Rare Gas Solids, vol. I, Academic Press, London, 1976.

[8] J.E. Gordon, The New Science of Strong Materials, Princeton University Press, Princeton, 1988.

[9] P. Gumbsch, Atomistic Modeling of Failure Mechanisms, in: H.O. Kirchner, L.P. Kubin, V. Pontikis (Eds.), Computer Simulation in Materials Science: Nano/Meso/Macroscopic Space and Time Scales, NATO ASI Series E, vol. 308, Kluwer Academic Publishers, Dordrecht, 1996, p. 227.

[10] J.R. Rice, R. Thomson, Phil. Mag. 29 (1974) 73.

[11] H. Gao, private communication.

[12] J.Q. Broughton, G. Gilmer, Acta Met. 31 (1983) 845.

[13] F. Milstein, J. Appl. Phys. 44 (1973) 3833.

[14] M. Daw, S. Foiles, M. Baskas, Mater Sci. Rep. 9 (1993) 251.

[15] F.F. Abraham, Europhysics Lett. 38 (1997) 103.

[16] A.S. Argon, G. Xu, M. Ortiz, Kinetics of dislocation emission from crack tips and the brittle to ductile transition of cleavage fracture, in: R. Blumberg Selinger, J. Mecholsky, Carlsson, E. Fuller, Jr. (Eds.), Fracture-Instability Dynamics, Scaling, and Ductile/Brittle Behavior, MRS Symposium Proceedings, vol. 409, Materials Research Society, Pittsburgh, 1996, p. 29.

ELSEVIER

Computational Materials Science 10 (1998) 10–15

COMPUTATIONAL
MATERIALS
SCIENCE

Computer simulations of martensitic transformations in NiAl alloys

R. Meyer *, P. Entel [1]

*Theoretische Tieftemperaturphysik, Gerhard-Mercator-Universität–Gesamthochschule Duisburg,
Lotharstraße 1, D-47048 Duisburg, Germany*

Abstract

$Ni_{1-x}Al_x$ alloys in the concentration range $34\% < x < 40\%$ exhibit a martensitic transformation from an austenitic phase with bcc structure to a close-packed structured martensitic phase. Above the transformation temperature electron microscopy shows the occurrence of tweed like structures which are accompanied by a considerable softening of the phonon energies at $\mathbf{q} = \frac{1}{6}[1\,1\,0] - TA_2$. We have done molecular dynamics simulations employing a semi-empirical model which allows us to study the transformation on an atomistic length scale. Our results show that local distortions of the crystal lattice, which come from the atomic disorder of the alloys, are responsible for the occurrence of tweed phenomena. Copyright © 1998 Elsevier Science B.V.

Keywords: Martensitic transformation; Molecular dynamics simulation; NiAl alloys

1. Introduction

Martensitic transformations are diffusionless structural phase transitions of first order with an invariant line [1]. These transformations play an important role in materials science since they have a strong influence on the formation of microstructure of a material. Despite more than a century of investigation there are still many unsolved problems in this field. Especially the atomistic processes driving the structural transformation are not well understood. A recent review about martensitic transformations and related phenomena has been given by Delaey [2].

Among the materials exhibiting martensitic transformations are $Ni_{1-x}Al_x$ alloys in the concentration range $34\% < x < 40\%$. Experimental details can be found in [3]. At high temperatures these alloys have a disordered B2 crystal structure, where one sublattice is occupied solely by Ni atoms, while the second sublattice contains both, Ni and Al, atoms. On cooling this austenite phase is transformed to a martensite phase with a close-packed structure called 7R. Well above the transformation temperature electron microscopy reveals a so-called tweed pattern indicating local distortions of the crystal structure along the $[\bar{1}\,1\,0]$ direction which is accompanied by a considerable softening of the $\frac{1}{6}[1\,1\,0] - TA_2$ phonon mode [4].

In order to get more insight in these premartensitic phenomena and the nucleation of the martensitic phase we have done molecular dynamics simulations of $Ni_{64}Al_{36}$ employing a semi-empirical model. These simulations show a transition from a high temperature B2 phase to a fct like structured phase at low temperatures which is quite similar to the experimentally

* Corresponding author. Tel.: +49 203 379 3323; fax: +49 203 379 2965; e-mail: ralf@thp.uni-duisburg.de.
[1] E-mail: entel@thp.uni-duisburg.de.

observed martensite phase. At high temperatures local distortions of the crystal lattice are present, which become more pronounced as the temperature is decreased. In addition to this the regions of homogeneous distortions tend to grow until they reach the size of the simulation cell which marks the onset of the structural transformation. Calculations of the static structure factor show that this behavior leads to the anomalies observed as tweed contrast in electron microscopy. In addition to this it can be seen from the simulations that the local distortions in the tweed phase give rise to the formation of twin boundaries of the martensite phase.

2. Details of the calculations

For our simulations we have developed a model for the description of $Ni_{1-x}Al_x$ alloys which is based on EAM (embedded-atom method) potentials introduced by Daw and Baskes [5]. While this semi-empirical method allows for a better description of the elastic properties of metals than ordinary pair potentials do, the calculations are still amenable to allow for the simulation of thousands of particles. The parameters entering in our model have been fitted to the experimental values of lattice constant, cohesive energy, elastic constants and selected phonon frequencies of Ni and Al. The interactions between Ni and Al atoms have been chosen to give a good representation of lattice constant, cohesive energy and elastic constants of ordered $Ni_{50}Al_{50}$. Due to limited space, a detailed description of the parameters of the model and the resulting properties cannot be given here and will be published elsewhere [6].

We have done molecular dynamics simulations, making use of the velocity form of the Verlet-Algorithm [7] with a time step of 1.5 fs. We have used periodic boundary conditions and the Nosé–Hoover thermostat method [8] together with the Parrinello–Rahman scheme [9] in order to get an isothermal–isobaric ensemble with a fluctuating simulation cell. All simulations were carried out at zero pressure. Starting from an initial configuration of 16^3 cubic bcc cells containing 2949 Al atoms ($x = 36\%$) which

are randomly distributed over one of the sublattices, we equilibrated our system at $T = 800$ K. Hereafter, the temperature was decreased subsequently in steps of 50 K down to $T = 600$ K, then in steps of 25 K until $T = 500$ K was reached and afterwards we cooled down to $T = 250$ K in steps of 10 K. Finally simulations have been carried out at temperatures of $T = 225, 200, 150,$ and 100 K. At each temperature 10 000 simulation steps were performed. Similar calculations have been done using an initial configuration of 40^3 elementary cells (128 000 atoms) with the same Al concentration. Due to the enormous computational demand of these calculations only a smaller set of temperatures could be handled so far. The configurations of the smaller system obtained at $T = 800, 400,$ and 100 K were used for additional runs of 2000 steps. During these runs the positions of the particles after each 20th step were recorded in order to calculate the static structure factor

$$I(\mathbf{k}) = \frac{1}{N} \sum_{i,j} f_i f_j \langle \exp(i\mathbf{k}(\mathbf{r}_i - \mathbf{r}_j)) \rangle, \qquad (1)$$

which can be measured by electron microscopy. Here N is the number of atoms and f_i is the scattering factor of atom i which has been taken to be $f_{Ni} = 12.85$ a.u. and $f_{Al} = 11.53$ a.u. [10].

3. Results and discussion

Visual inspection of the atomic configurations obtained from our simulations showed that at high temperatures the crystal structure of the systems is B2. On cooling a structural transformation occurred which led to a twinned fct structure similar to the experimentally observed one. In Fig. 1 we show the thermal expansion coefficient obtained by numerical derivation of the mean volume of the simulation cell. Both system sizes show a strong increase of the thermal expansion coefficient indicating the structural transformation. The large system shows a sharp peak around 340 K which is in agreement with the transformation temperature given by Shapiro et al. [4]. In contrast to this the thermal expansion coefficient of the smaller system exhibits strong fluctuations over a broad range

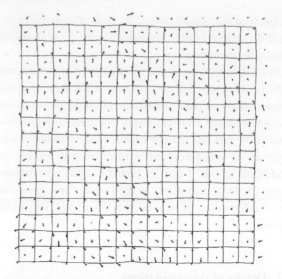

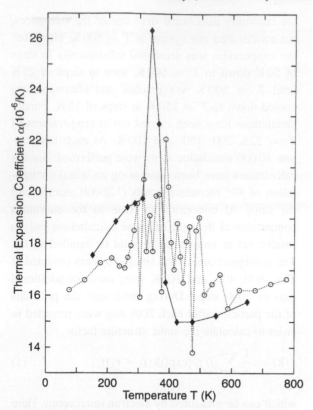

Fig. 1. Thermal expansion coefficient of $Ni_{64}Al_{36}$ obtained from molecular dynamics simulations by numerical derivation. Circles (diamonds) represent results from simulations of the system with 16^3 (40^3) elementary cells.

of temperatures between 200 and 600 K. This must be an artifact of finite size effects. In other words, the transformation process possesses a typical length scale which is larger than the system size in the case of a small system.

In order to get insight into the transformation process we have calculated the static structure factor and the displacement of the mean atomic positions from ideal positions of the B2 lattice. Results are given for the small system in Figs. 2–4 for $T = 800$, 400, and 100 K. When comparing these figures the different magnification of the displacements ($3\times$ in Fig. 2 and 3, $1.5\times$ in Fig. 4) has to be kept in mind. From Fig. 2 it can be seen that even at a temperature of 800 K there are strong static deviations of the atoms from ideal lattice positions. Since there are no defects present in our system, atomic disorder must be

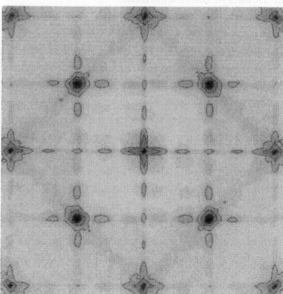

Fig. 2. Mean displacements ($\times 3$) from ideal lattice positions of the atoms of two $(0\,0\,1)$ layers of $Ni_{64}Al_{36}$ at $T = 800\,K$ (top); static structure factor of the system at this temperature in the plane $k_z = 0$ (bottom).

responsible for these deformations. The fact that the displacements of the atoms are nearly isotropically distributed is reflected by the contour lines of the static structure factor around the four $[1\,1\,0]$ peaks, which are nearly circular with only small streaks along the $[\bar{1}\,1\,0]$ directions. On further cooling the regions with strong atomic displacements grow until

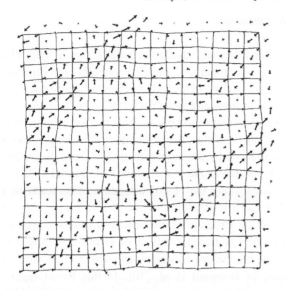

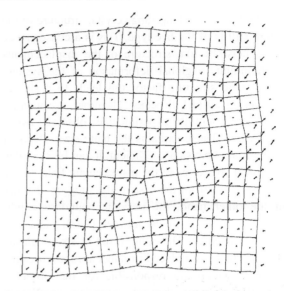

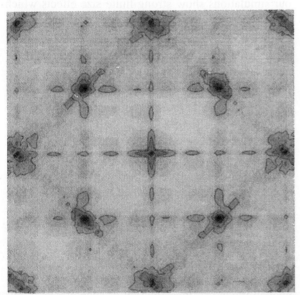

Fig. 3. Mean displacements ($\times 3$) from ideal lattice positions of the atoms of two ($0\,0\,1$) layers of $Ni_{64}Al_{36}$ at $T = 400\,K$ (top); static structure factor of the system at this temperature in the plane $\mathbf{k}_z = 0$ (bottom).

Fig. 4. Mean displacements ($\times 1.5$) from ideal lattice positions of the atoms of two ($0\,0\,1$) layers of $Ni_{64}Al_{36}$ at $T = 100\,K$ (top); static structure factor of the system at this temperature in the plane $\mathbf{k}_z = 0$ (bottom).

they touch each other. At this point the isotropy of the distortions is destroyed. At $T = 400\,K$ the displacements of the atoms shown in Fig. 3 are mainly confined in the $[\bar{1}\,1\,0]$ directions (the components of the displacements perpendicular to the plane drawn are rather small) which causes a strong streaking of

the static structure factor along the $[\bar{1}\,1\,0]$ directions. The form of the peaks in $I(\mathbf{k})$ is rather similar to the results obtained by Becquart et al. [10] at $T = 600\,K$ in their investigation of the tweed structure of NiAl alloys. On the other hand the static structure factor also shows that the long-range crystal structure of the system still is of the bcc type. This is a consequence of the network like structure of the displacements which inhibits a global deformation of the system. In the static structure factor this leads to a similar streaking at all [1 1 0] peaks. Therefore, we identify this structure with the experimentally observed tweed pattern. Fig. 3 also explains the finite size effects of the thermal expansion coefficient. It can be seen that the typical structures of the displacement field have a size of about half of the system size (approx. 20 nm). On the other hand, the periodic boundary conditions impose correlations between [1 1 0] planes on the same distance. This must result in strong elastic stresses which are reduced in rather abrupt relaxation processes which are responsible for the observed fluctuations of the thermal expansion coefficient. An inspection of the displacement field of the large system at the same temperature shows that the size of structures is about the same, therefore much less than the correlation length imposed by the boundary conditions and in rough agreement with experimental results [4].

Fig. 4 shows how this tweed structure finally leads to the formation of martensite in our simulations. In this figure the displacements of the atoms are all along one of the two possible $[\bar{1}\,1\,0]$ directions. The displacements along the perpendicular direction have vanished completely. Now the four [1 1 0] peaks in the static structure factor are no longer equivalent. While two of them show almost no streaking the other two have split into two peaks, which reflects the formation of two variants of martensite which can also be seen in the real space picture. A detailed analysis of the static structure factor revealed that the martensite phase has a tetragonal face centered structure with lattice parameters $a = 7.25$ a.u. and $c = 5.95$ a.u. which is in good accordance with the experimental values obtained by Martynov et al. [11]. Finally it can be seen from Figs. 3 and 4 that the lines of strongest displacements in the tweed phase give the twin boundaries between the two martensitic variants. This is so because at the lines of maximum displacement the direction of the shear distortion changes its sign.

4. Summary

Our molecular dynamics simulations of $Ni_{64}Al_{36}$ using a semi-empirical model correctly reproduce the experimentally observed martensitic transformation from an austenite phase with B2 structure to a fct structured martensite phase. At very high temperatures we find local distortions of the crystal lattice which are induced by the atomic disorder. These distortions grow in size as the temperature is decreased and tend to align along the [1 1 0] directions. This leads to a characteristic streaking of the [1 1 0] Bragg peaks in the static structure factor which is observed experimentally as tweed contrast in electron microscopy. These results are in agreement with former simulations of Becquart et al. [10]. In this tweed regime our simulations show strong finite size effects which demonstrate the necessity of large scale simulations. Further cooling leads to the formation of martensite in our simulations. The resulting structure consists of two twinned domains of different martensite variants. It can be seen from our simulations that the distortions present in the tweed regime, lead naturally to the formation of such twin boundaries. In contrast to Rubini and Ballone [12] who report a weakly first order transition in their simulations we find a broad transformation range in case of small systems which tends to get smaller with increasing system size. The reason for this might be the very small system size used in [12] which is smaller than the characteristic size of the tweed structures we find.

Another feature of the tweed phase observed experimentally is the strong softening of the $\frac{1}{6}[1\,1\,0] - TA_2$ phonon mode [4]. Investigation of such effects in our simulations is in progress but not finished yet. First results indicate strong temperature dependencies of the $[1\,1\,0] - TA_2$ phonon branch. These studies will probably also help to explore the driving force of the structural transition. We think that the instability of Ni in the bcc phase which has been found in simulations of $Fe_{80}Ni_{20}$ [13] might also play a role in the destabilization of the B2 phase of NiAl. Another interesting thing would be to see whether there is a thermal hysteresis when we increase the temperature in our simulations.

Acknowledgements

We thank the Höchstleistungsrechenzentrum (HLRZ) at Forschungszentrum Jülich, Germany, for CPU time on its Intel Paragon and Cray T3E parallel computers. Part of our work has been calculated there. This work has been supported by the Deutsche Forschungsgemeinschaft (SFB 166).

References

[1] M. Cohen, G.B. Olson and P.C. Clapp, in: Proc. ICOMAT'79 (MIT Press, Cambridge, MA, 1979) pp. 1–11.

[2] L. Delaey, Diffusionless transformations, in: Materials Science and Technology, ed. P. Haasen, Vol. 5 (VCH, Weinheim, 1991).

[3] D. Schryvers, J. Phys. (Paris) IV C2 (1995) 225.

[4] S.M. Shapiro, B.X. Yang, Y. Noda, L.E. Tanner and D. Schryvers, Phys. Rev. B 44 (1991) 9301.

[5] M.S. Daw and M.I. Baskes, Phys. Rev. Lett. 50 (1985) 1285; Phys. Rev. B 29 (1984) 6443.

[6] R. Meyer and P. Entel, in preparation.

[7] M.P. Allen and D.J. Tildesley, Computer Simulations of Liquids (Clarendon Press, Oxford, 1987).

[8] S. Nosé, Mol. Phys. 52 (1984) 255; W.G. Hoover, Phys. Rev. A 31 (1985) 1695.

[9] M. Parrinello and A. Rahman, Phys. Rev. Lett. 45 (1980) 1196; J. Appl. Phys. 52 (1981) 7182.

[10] C.S. Becquart, P.C. Clapp and J.A. Rifkin, Phys. Rev. B 48 (1993) 6.

[11] V.V. Martynov, K. Enami, L.G. Khandros, A.V. Tkachenko and S. Nenno, Scr. Metall. 17 (1983) 1167.

[12] S. Rubini and P. Ballone, Phys. Rev. B 48 (1993) 99.

[13] R. Meyer and P. Entel, Phys. Rev. B, submitted.

ELSEVIER

Computational Materials Science 10 (1998) 16–21

COMPUTATIONAL
MATERIALS
SCIENCE

Computer simulation of martensitic textures

A. Saxena [a,*], A.R. Bishop [a], S.R. Shenoy [b], T. Lookman [c]

[a] *Los Alamos National Laboratory, Los Alamos, NM 87545, USA*
[b] *ICTP, Trieste, Italy*
[c] *University of Western Ontario, Canada*

Abstract

We consider a Ginzburg–Landau model free energy $F(\epsilon, e_1, e_2)$ for a (2D) martensitic transition, that provides a unified understanding of varied twin/tweed textures. Here F is a triple well potential in the rectangular strain (ϵ) order parameter and quadratic e_1^2, e_2^2 in the compressional and shear strains, respectively. Random compositional fluctuations $\eta(r)$ (e.g. in an alloy) are gradient-coupled to ϵ, $\sim -\sum_r \epsilon(r)[(\Delta_x^2 - \Delta_y^2)\eta(r)]$ in a "local-stress" model. We find that the compatibility condition (linking tensor components $\epsilon(r)$ and $e_1(r)$, $e_2(r)$), together with local variations such as interfaces or $\eta(r)$ fluctuations, can drive the formation of global elastic textures, through long-range and anisotropic effective ϵ–ϵ interactions. We have carried out extensive relaxational computer simulations using the time-dependent Ginzburg–Landau (TDGL) equation that supports our analytic work and shows the spontaneous formation of parallel twins. and chequer-board tweed. The observed microstructure in NiAl and $Fe_x Pd_{1-x}$ alloys can be explained on the basis of our analysis and simulations. Copyright © 1998 Elsevier Science B.V.

Keywords: Twinning; Tweed; Time-dependent Ginzburg–Landau; Model A dynamics

1. Introduction

A variety of minerals, ferroelectrics, ferroelastics, magnetoelastic materials, ceramics, Jahn–Teller materials, and most notably the shape memory alloys (e.g. NiTi, FePd, $CuAuZn_2$) undergo a diffusionless, displacive (i.e. martensitic [1]), weakly first order structural transition and usually exhibit twinning below the transition temperature T_0 (thermally induced martensite) and even above T_0 under external loading (stress induced martensite). In addition, these materials display a rich, stress-sensitive, family of elastic domain-wall patterns, both above and below T_0 as observed in transmission electron microscopy (TEM) [2]. The role of surface (or interface) energy is very important in determining fine-scale features, e.g. twin width, near an austenite twinned-martensite interface (habit plane). For instance, the twins are stabilized by a long range, habit plane mediated, nonlocal elastic interaction [3] in these materials.

* Corresponding author.

These materials also exhibit transformation precursors that can occur up to hundreds of degrees above the transition temperature T_0. Many types of pretransitional structures (or "mesoscopic textures") have been observed in TEM including the so-called "tweed" (criss-cross pattern of twins) patterns [2]. The understanding of pattern formation in elastic materials is of much interest, with a variety of models and mechanisms invoked separately, for twins and tweed [3–7].

2. Model

The modulated phases can be understood quite generally within a Ginzburg–Landau framework if, in addition to the traditional elasticity terms, one adds appropriate nonlinear and nonlocal (strain gradient) terms to the elastic energy functional [4]. The model presented below obtains tweed without explicit disorder, and as a stable pattern, formed by energy-lowering cross-gradient terms. The additional terms are quadratic in the strain and fourth order in strain gradients (arising from compositional disorder in the alloy), with all symmetry allowed terms consistently retained. The model synthesizes a variety of properties specific to these materials. It contains two important features: (a) a cross-derivative gradient term that favors domain wall crossing, and (b) the idea of hierarchical (e.g. Cayley tree) splitting of the domain walls from microscopic scales at the habit plane to macroscopic scales inside the tweed.

In two dimensions we define the components of strain tensor (without "geometric nonlinearity") by

$$\varepsilon_{ij} = \frac{1}{2}\left(\frac{\partial u_i}{\partial x_j} + \frac{\partial u_j}{\partial x_i}\right), \quad i, j = x, y.$$

In symmetrized form we write the area (e_1), shear (e_2) and the rectangular (ϵ) strains as

$$e_1 = \frac{1}{\sqrt{2}}(\varepsilon_{xx} + \varepsilon_{yy}), \qquad e_2 = \varepsilon_{xy}, \qquad \epsilon = \frac{1}{\sqrt{2}}(\varepsilon_{xx} - \varepsilon_{yy}).$$

The above ideas are embodied in the following (dimensionless) elastic model Hamiltonian:

$$H = H_{\text{bulk}} + H_{\text{grad}} + H_\alpha + H_{\text{twin}}, \tag{1}$$

$$H_{\text{bulk}} = \sum_i [(\tau - 1)\epsilon_i^2 + \epsilon_i^2(\epsilon_i^2 - 1)^2] - \sum_i P_i \epsilon_i, \quad \tau = \frac{T - T_c}{T_0 - T_c}. \tag{2}$$

$$H_{\text{grad}} = \frac{a}{4}\sum_i [(\nabla_x \epsilon_i)^2 + (\nabla_y \epsilon_i)^2] + \frac{b}{8}\sum_i [(\nabla_x^2 \epsilon_i)^2 + (\nabla_y^2 \epsilon_i)^2], \tag{3}$$

$$H_\alpha = -\frac{\alpha}{8}\sum_i (\nabla_x^2 \epsilon_i)(\nabla_y^2 \epsilon_i), \tag{4}$$

$$H_{\text{twin}} = v \sum_{i \neq j} \frac{\epsilon_i \epsilon_j}{|r_i - r_j|}. \tag{5}$$

Here ϵ_i are dimensionless, scaled local rectangular (i.e. deviatoric) strains defined on the sites of a 2D square lattice; P_i and τ are dimensionless stress and scaled temperature in the ϕ^6 (i.e. triple well) potential, respectively. T_c denotes the temperature at which the shear modulus would soften completely, i.e. the elastic constants would satisfy $C_{11} = C_{12}$. Of the three elastic gradient coefficients (a, b, α), b and α are possibly modified by compositional fluctuations, and are necessarily positive. For a specific material these coefficients can be determined from the measured phonon dispersion data [4]. The gradient terms (H_{grad} and H_α) are evaluated using discrete derivatives on the lattice. For $P = 0$, H_{bulk} has three minima for $0 < \tau < \frac{4}{3}$, one minimum at $\epsilon = 0$ (pure austenite) for $\tau > \frac{4}{3}$, and two side minima (two pure martensitic variants) for $\tau < 0$. The range for stable tweed is $1 < \tau < \frac{4}{3}$. There are three degenerate minima at $\tau = 1$. H_{twin} represents the habit plane-mediated long-range elastic interaction (of strength ν) which stabilizes twins below T_o [3]. $|r_i - r_j|$ denotes the distance between sites i and j on the square lattice.

3. Compatibility condition

There is a connection between compressional, shear and rectangular strains, as they are different components of the strain tensor, i.e. are derivatives of the same underlying displacement field $\mathbf{u}(r)$. The order parameter $\epsilon(r)$ acts as a source term in the compatibility equation, inducing variations in $e_1(r)$, $e_2(r)$. The 2D compatibility constraint, that is satisfied at all times, is [5]

$$\Delta^2 e_1(r) - \sqrt{8}\Delta_x \Delta_y e_2(r) = (\Delta_x^2 - \Delta_y^2)\epsilon(r). \tag{6}$$

The validity of (6) as an identity can be seen, by rotating axes by $\pi/4$ to a preferred (primed) frame with displacements $u_x = (u'_x + u'_y)/\sqrt{2}$, $u_y = (u'_x - u'_y)\sqrt{2}$, and discrete derivatives $\Delta_x = (\Delta'_x + \Delta'_y)/\sqrt{2}$, $\Delta_y = (\Delta'_x - \Delta'_y)/\sqrt{2}$. Then (6) becomes

$$\Delta'^2 e_1 - \sqrt{2}(\Delta'^2_x - \Delta'^2_y)e_2 = 2\Delta'_x \Delta'_y \epsilon, \tag{7}$$

with $e_1 = (\Delta'_x u'_x + \Delta'_y u'_y)/\sqrt{2}$, $e_2 = (\Delta'_x u'_x - \Delta'_y u'_y)/2$, $\epsilon = (\Delta'_x u'_y + \Delta'_y u'_x)/\sqrt{2}$, in the rotated frame; (7) is then manifestly satisfied.

From the form (7) we see that for configurations for which $\Delta'_x \Delta'_y \epsilon = 0$, the compressional and shear strains can minimize their costs by taking on their trivial solutions of zero. Conversely, $Q = \Delta'_x \Delta'_y \epsilon \neq 0$ configurations, such as at domain-wall crossings in tweed; or parallel twins ending at habit planes, will induce nonzero compressional/shear strains, in general. The imposition of the compatibility constraint (implies $e_1 = e_1(\epsilon)$, $e_2 = e_2(\epsilon)$) leads to a dependence of the compressional and shear strain energy cost on the order parameter ϵ. This means that the free energy contains compatibility-induced effective ϵ–ϵ potentials in the bulk, and near habit planes. The simplified long-range term (Eq. (5)) is derived from these considerations.

4. Dynamical simulations

The order parameter here, rectangular strain henceforth denoted ϕ, is not conserved; therefore the dynamics is purely relaxational. We assume a Model A dynamics [8] for the evolution of ϕ after a quench. The equation is given by

$$\frac{\partial \phi}{\partial t} = -\Gamma \frac{\delta F}{\delta \phi} + \zeta, \tag{8}$$

where Γ is the relative strength of elastic energy to thermal energy and ζ is a noise term. In this section we denote strain by ϕ instead of ϵ and the free energy density by f instead of H. Note that the time evolution for elastic systems must be treated carefully. Specifically, the final state depends on the damping mechanism [9] and inertial effects (acoustic waves) must be taken into account [10]. The use of relaxation time approximation is justified since, instead of studying the time evolution per se, or details of the final state, our objective here is to demonstrate *qualitatively* the emergence of twinning near the habit plane below the transition temperature, and of the tweed above the transition. Both of these features are expected to be independent of the damping mechanism.

The free energy $F = \int f \, d^d \mathbf{r}$ and the noise ζ satisfy the following fluctuation–dissipation relationship:

$$\langle \zeta(\mathbf{r}, t) \zeta(\mathbf{r}', t') \rangle = 2\Gamma \delta(\mathbf{r} - \mathbf{r}') \delta(t - t').$$ (9)

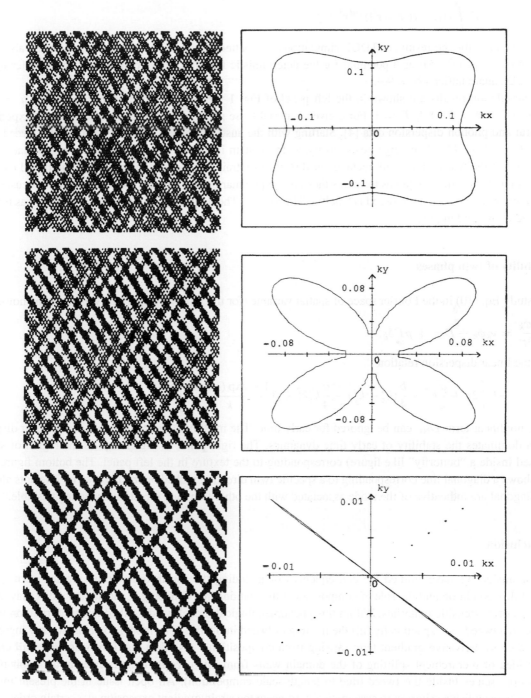

Fig. 1.

Using the free energy from Eqs. (1)–(5) we obtain in two dimensions ($d = 2$)

$$\frac{\partial \phi}{\partial t} = -\Gamma \left[(6\phi^5 - 8\phi^3 + 2\tau\phi - p) - \frac{a}{2}\nabla^2\phi + \frac{b}{4}\nabla^4\phi - \frac{\alpha}{4}\frac{\partial^2}{\partial x^2}\frac{\partial^2}{\partial y^2}\phi \right]$$
$$- \nu\Gamma \int \phi(\mathbf{r'}) R(|\mathbf{r} - \mathbf{r'}|) \, d^d\mathbf{r'} + \zeta. \tag{10}$$

Next, we describe the results of TDGL simulations (i.e. pattern formation) which are based on the above static Hamiltonian (Eqs. (1)–(5)) as a potential for the deterministic force term. We consider strain variables ϵ_i on an $N \times N$ 2D square lattice ($N = 96$).

The simulation results are shown in the left panel of Fig. 1 for representative parameter values: $a = 1, b = 0.01, \alpha = 2.4, \beta = 0.0001, \Gamma = 1$. For a given material these parameters are determined from the experimental structural and phonon dispersion data [4]. Starting with the austenite phase, twins ($T < T_0$) and a tweed pattern (for $\sim T_0 < T < T^{upper}$) emerge, respectively, as the system is cooled below T_0. The white and black regions correspond to the two regions with rectangular distortions (strain) in x and y directions. In the tweed case these are small local distortions whereas for twins they are the two martensitic variants. In Fig. 1 we have focused on the temperature range in which the tweed is "melting" into twins. The long-range interaction term H_{twin} was treated as discussed above and in [11].

5. Stability of twin phases

We study Eq. (10) in the Fourier space of spatial variable (for the following analysis, we ignore the noise):

$$\frac{\partial \phi_k}{\partial t} = \gamma_k \phi_k + D_{NL} + p\Gamma\delta_k, \tag{11}$$

where the linear dispersion relation

$$\gamma_k = -\Gamma \left[2\tau + \frac{a}{2}k^2 + \frac{b}{4}(k_x^4 + k_y^4) - \frac{\alpha}{4}k_x^2 k_y^2 + \beta\frac{1 - \exp(-r_d k)}{k} \right], \tag{12}$$

and the nonlinear term D_{NL} can be ignored for early time. The maximally unstable mode of the linear dispersion relation dominates the stability of early time dynamics. The right panel in Fig. 1 depicts the region of stability (enclosed inside a "butterfly" like figure) corresponding to the texture in the left panel. The bottom figure in this panel shows a diagonal line corresponding to a specific twin direction picked out by the system. The dots along the other diagonal are indicative of the mode associated with the other twinning direction becoming unstable.

6. Conclusion

In conclusion we have studied the consequences of including the fourth order strain gradient terms, elastic compatibility condition and the role of compositional fluctuations in a Ginzburg–Landau model that may apply to a subclass of martensitic materials. Within this phenomenological model we described texture formation such as: (i) twins, (ii) tweed, and specifically (iii) the melting of tweed into twins. The model contains two key ingredients, namely a cross-derivative gradient term (arising from compositional disorder) that favors domain wall crossing, and the idea of hierarchical splitting of the domain walls from microscopic length scales at the habit plane to macroscopic scales inside the tweed (due to length-scale competition). An open issue in the present modeling context is to establish a criterion to systematically truncate the strain gradient expansion at a certain order.

References

[1] C.M. Wayman, Introduction to the Theory of Martensitic Transformations (Macmillan, London, 1964).

[2] L.E. Tanner, A.R. Pelton and R. Gronsky, J. Phys. (Paris) 43 (1982) C4–169.

[3] G.R. Barsch, B. Horovitz and J.A. Krumhansl, Phys. Rev. Lett. 59 (1987) 1251; B. Horovitz, G.R. Barsch and J.A. Krumhansl, Phys. Rev. B 43 (1991) 1021.

[4] G.R. Barsch and J.A. Krumhansl, Phys. Rev. Lett. 53 (1984) 1069; Metallurg. Trans. A 19 (1988) 761; Proc. Int. Conf. on Martensitic Transformations (ICOMAT-92), eds. C.M. Wayman and J. Perkins (Monterey Institute of Advanced Studies, California, 1993) p. 53.

[5] S. Kartha, T. Kastàn, J.A. Krumhansl and J.P. Sethna, Phys. Rev. Lett. 67 (1991) 3630; J.P. Sethna, S. Kartha, T. Kastàn and J.A. Krumhansl, Phys. Scripta T 42 (1992) 214; S. Kartha, J.A. Krumhansl, J.P. Sethna and L.K. Wickham, Phys. Rev. B 52 (1995) 803.

[6] S. Semenovskaya and A.G. Khachaturyan, Phys. Rev. Lett. 67 (1991) 2223; Physica D 66 (1993) 205; S. Semenovskaya, Y. Zhu, M. Suenaga and A.G. Khachaturyan, Phys. Rev. B 47 (1993) 12 182.

[7] A.M. Bratkovsky, E.K.H. Salje and V. Heine, Phase Trans. 52 (1994) 77; 55 (1995) 79.

[8] P.C. Hohenberg and B.I. Halperin, Rev. Modern Phys. 49 (1977) 435.

[9] A.C.E. Reid and R.J. Gooding, Physica D 66 (1993) 180.

[10] F. Falk, J. Phys. C 20 (1987) 2501.

[11] C. Roland and R.C. Desai, Phys. Rev. B 42 (1990) 6658; C. Sagui and R.C. Desai, Phys. Rev. E 49 (1994) 2225; C. Sagui, A.M. Somoza and R.C. Desai, Phys. Rev. E 50 (1994) 4865.

ELSEVIER

Computational Materials Science 10 (1998) 22–27

COMPUTATIONAL
MATERIALS
SCIENCE

The microscopic theory of diffusion-controlled defect aggregation

E.A. Kotomin [a,b,*], V.N. Kuzovkov [a,c]

[a] *Institute of Solid State Physics, The University of Latvia, 8 Kengaraga Str., Riga LV-1063, Latvia*
[b] *Fachbereich Physik, Freie Universität Berlin, Arnimalee 14, 14195 Berlin, Germany*
[c] *Institut für Physikalische und Theoretische Chemie, Technische Universität Braunschweig,
D-38106 Braunschweig, Germany*

Abstract

The kinetics of diffusion-controlled aggregation of primary Frenkel defects (F and H centers) in irradiated CaF_2 crystals is theoretically studied. Microscopic theory is based on the discrete-lattice formalism for the single defect densities (concentrations) and the coupled joint densities of similar and dissimilar defects treated in terms of the Kirkwood superposition approximation. Conditions and dynamics of the efficient F center aggregation during crystal heating after irradiation are analyzed. Copyright © 1998 Elsevier Science B.V.

Keywords: Frenkel defects; CaF_2; F center; Aggregation; Metal colloid

1. Introduction

The primary radiation defects in ionic solids – the F centers (electron trapped by anion vacancy) and the H centers (interstitial halide atoms) – under intensive irradiation and at high enough temperatures are known to reveal *aggregation* which leads to the formation of alkali metal *colloids* and gas bubbles (see [1] and references therein). The intensive experimental studies of the conditions and efficiency of the metal colloid formation (such as the temperature interval, dose rate, etc.) continue nowadays for many alkali halides [2,3] and technologically important ceramics [4]. This problem is also interesting from the fundamental point of view, being an example of *pattern formation* and *self-*

organization in reaction–diffusion systems far from equilibrium [5]. In our case the process includes A, B particle random walks on a lattice (diffusion), particle interaction and bimolecular annihilation of dissimilar particles, $A + B \rightarrow 0$. Irradiation is modeled by a permanent particle source with a given dose rate (intensity).

Colloid growth study by means of the direct computer (Monte Carlo) simulations is a very difficult problem since the mobilities of the two kinds of defects involved – H and F centers – differ typically by 15 orders of magnitude and for monitoring slow F center aggregation, the time increment is dictated by mobility of fast H centers.

Existing analytical theories of the radiation-induced defect aggregation and colloid formation could be classified into three categories: macroscopic, mesoscopic, and microscopic [6,7]. We refer to as *microscopic* the first-principle theory which treats

---

* Corresponding author. Tel.: +37-1-2263085; fax: +37-1-7112583.

elementary processes at atomic scale and using no fitting or uncertain parameters but only several basic defect parameters like the diffusion energies and interaction energies. A simplified continuum approximation has been used up to now. More realistic, discrete-lattice microscopic theory was developed very recently [7] and applied to the F center aggregation in NaCl exposed to irradiation [8].

In this paper, we present briefly results obtained for microscopic simulations of the F center aggregation in CaF$_2$ crystals used as optical windows and potentially, in lithographic applications. We focus on the aggregation process *after* irradiation when a crystal is heated a given rate.

2. A model

Our physical model includes creation of the H and F centers (called hereafter defects A and B), with the dose rate p, AB pairs are not spatially correlated at birth and recombine when during their migration approach each other to within the nearest-neighbor (NN) distance. Therefore, their macroscopic concentrations always coincide, $n = n_A = n_B$. Isolated (single) defects hop with the activation energy E_λ is characterized by the diffusion coefficient $D_\lambda = D_0 \exp(-E_\lambda/k_B T)$, $\lambda = $ A, B. When several defects are closely spaced, the hop rate of a given defect to the nearest empty lattice site is determined by both the local defect configuration and the interaction between defects; this can change its *effective* diffusion coefficient D_λ^{eff} by many orders of magnitude compared to that for a single defect. It affects the effective reaction rate K of the A and B recombination; for the dilute particle system in the continuum approximation it is well known to be $K_0 = 4\pi r_0 (D_A + D_B)$, where r_0 is the recombination radius. Defect interaction is incorporated in the model via three types of the NN attraction energies for the two kinds of NN defects (in the spirit of the Ising model): E_{AA}, E_{AB} and E_{BB}. Namely, particle interactions make the kinetic equations essentially non-linear and thus able to manifest the self-organization (pattern formation) phenomena under the irradiation.

The experimental estimates of the activation energies for H and F centers in CaF$_2$ are 0.46 and 0.7 eV [9]. The attraction energies determining the H and F center attachment/detachment to/from similar-particle aggregates are less known. Calculations of the elastic interaction between F centers in KBr give the attraction energy about 0.02 eV [10]. This value will also be used in our calculations; for simplicity we assume that $E_{AA} = E_{BB}$ and $E_{AB} = 0$. In our calculations we simulated the experimental conditions of the low-energy electron irradiation [3] characterized by the dose rate about $p = 10^{17}$ cm^{-3} s^{-1}, which lasts for 30 min, with a subsequent heating at the rate of 1.3 K/min.

The *mathematical formalism* and the relevant computer code *Kinetica* will be described in detail elsewhere [11], it is a generalization of our previous microscopic many-point density approach [7] for the discrete-lattice case. This allows us to avoid the limitations of a continuum model, to increase the computation speed and thus to study the aggregation kinetics in a very wide time interval, exceeding many orders of magnitude. Theory is based on the Kirkwood superposition approximation for the three-particle densities. This formalism operates with a set of coupled kinetic equations for the lattice defect densities (total concentrations) $n_\lambda(t)$, $\lambda = 0$, A, B and the joint correlation functions $F_{\lambda\nu}(|r_\lambda - r_\nu|, t)$ where r_λ and r_ν are coordinates of two lattice sites and t the time. Since defect correlations are short-range, $F_{\lambda\nu}(r, t)$ strives for its asymptotic value of the unity (random particle distribution), as the relative coordinate increases, $r \to \infty$. That is, if some joint density $F_{\lambda\nu}(r, t)$ considerably exceeds the unity value, it means surplus of the defect pairs $\lambda\nu$ at a given relative distance compared to their random (Poisson) distribution, and vice versa. The simultaneous analysis of the joint correlation functions for similar (AA, BB) and dissimilar (AB) pairs, as well as for 0A, 0B pairs (empty site-defect) permit to study the *spatio-temporal* evolution of the strongly non-equilibrium system, in particular, crystals with radiation defects [7,11].

It is also convenient to characterize the aggregation process by monitoring the concentrations of *single* defects $n_\lambda^{(1)}$ (no other defects in NN sites) and *dimer* defects $n_\lambda^{(2)}$ (two similar defects are NN), which could be calculated from the joint densities in the standard *cluster approximation* [12]. Lastly, large defect aggregates could be characterized by the integral values of the mean number of particles N_A, N_B therein and their radii R_A and R_B.

3. Main results

Let us study the kinetics of defect concentration growth under CaF_2 irradiation at low temperatures when the F centers are definitely immobile but the H centers are moving either slowly (150 K) or already quite mobile (193 K). The first conclusion from the upper curve in Fig. 1 plotting a total concentration of H- (and F) defects is that at the end of irradiation (shown by an arrow) we are still in almost a linear regime of the defect accumulation. Under such a dose rate the concentration saturation could be expected only after 2–3 h of irradiation. Additional calculations show that neglect of the defect interaction would lead to one-to-two orders of magnitude reduction in defect concentrations at the end of irradiation (where a real saturation takes place), $n \approx 10^{18} \, \text{cm}^{-3} \, \text{s}^{-1}$, accompa-

nied by a very fast recombination when irradiation is switched off.

Fig. 1 shows that total concentrations of H centers almost coincide at the two temperatures, H concentration at 150 K irradiation exceeds that at 193 K irradiation only at $t > 70 \, \text{min}$ when irradiation is switched off and sample is heated up by ≈ 30 K. It is well seen that at a certain time t_0 concentration of single H centers, $n^{(1)}$, sharply descreases due to growth of dimer concentration, $n^{(2)}$, and the latter drops also due to growth of larger aggregates. The time t_0 decreses by three orders of magnitude when the irradiation temperature increases from 150 up to 193 K.

It should be reminded that F centers are complementary to the H centers and thus their total concentrations always coincide. At temperatures shown in Fig. 1 the F centers are immobile and their aggregation occurs only during sample heating *after* an irradiation. Let us consider this process.

Curves 1 in Fig. 2 show that the aggregate's size for the F centers and a mean number of defects therein begin to grow only at 250 K when the F centers become mobile. This temperature is in qualitative agreement with the experimental data on colloid growth in CaF_2 [3] as well as growth of M_A center concentration under pulsed electron irradiation of crystals doped with Na [13].

However, the electron irradiation experiments reveal small (1–2 nm in radius) aggregates of the F centers, whose size remains constant until heating up to 250 K. The fact that aggregate's size is independent of the irradiation temperature indicates for a process which takes place during irradiation and is characterized by a low activation energy (condsiderably less than that for the F center diffusion). To our mind, there are two candidates for this radiation-induced process. (i) Radiation-enhanced F center migration when the F center traps a hole, converts into an anion vacancy whose diffusion energy is about 0.4 eV, makes several hops and transforms (sooner or later) again into the F center after trapping an electron. (ii) Self-trapped excitons, making a diffusion motion with the activation energy not exceeding 0.4 eV, decay into pairs of the F, H centers not in regular lattice sites but preferentially nearby pre-existing F centers (e.g., due to drift

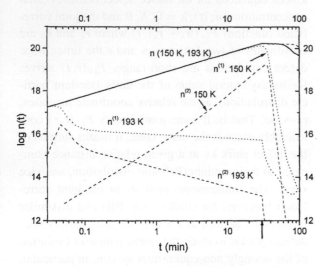

Fig. 1. Dynamics of the H center aggregation with the dose rate $p = 10^{17} \, \text{cm}^{-3} \, \text{s}^{-1}$ for irradiation at 193 and 150 K. The quantities n, $n^{(1)}$ and $n^{(2)}$ are total concentration, concentration of single and dimer centers, respectively. The end of irradiation is marked by an arrow.

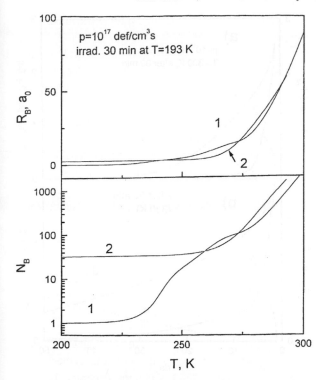

Fig. 2. Heating-induced growth of the mean radius of F aggregates, R_B (in units of the $F–F$ distance, $a_0 = 2.73$ Å), and a mean number of particles inside each aggregate, N_B. Curves 1 are for standard parameters described in the text, in curves 2 the activation energy for the radiation-enhanced F center diffusion is 0.4 eV.

in a lattice stress caused by defects). High mobility of the self-trapped excitons in CaF_2 is known from an efficient exciton energy transfer from host lattice to impurities [14]. In fact, in the second case F center aggregation induced by irradiation is again analogous to the radiation-enhanced F center diffusion with a reduced activation energy.

To simulate the radiation-enhanced process under question, we performed calculations with the F center activation energy of 0.4 eV under irradiation and 0.7 eV after irradiation. Curves 2 in Fig. 2 show remarkable agreement with the experiment – formation of small F aggregates with a radius of several nm which transform into larger aggregates at the temperatures when F centers are mobile. Evidently, it occurs via a slow motion of these small aggregates, as well as emission of F centers from small aggregates and

their attachment to larger aggregates. This is nothing but the so-called Ostwald ripening. Unfortunately, due to computational difficulties we cannot study the latest stage of the large aggregate growth, but its typical size of $100a_0 = 27$ nm is close to the experimental value (25 nm).

The analysis of the kinetics of the total F center concentration, as well as that for single F centers and F_2 dimers, plotted as a function of the temperature shows that at 200 K practically only single F center exists in the concentration close to 10^{21} cm^{-3} s^{-1} whereas concentration of dimers is smaller by three orders of magnitude. As the temperature reaches that of the F center mobility, 250 K, both concentrations drop by many orders of magnitude. However, the total F concentration decreases insignificantly thus indicating that most of the F centers are gathered now into large aggregates (colloids). This is another confirmation to what we concluded from Fig. 2.

Simulations of the radiation-enhanced F center mobility show that due to formation of small aggregates at low temperature, concentrations of F and F_2 centers are additionally reduced by 6 and 10 orders of magnitude, respectively. They decrease at 250 K, as in the previous case, when the F centers perform diffusion walks and thus could encount each other and aggregate.

Calculations of the *effective diffusion coefficient*, D_A^{eff}, show that the H centers (which are mobile under the irradiation temperature) are aggregated very quickly, in fractions of a minute. This reduces H average diffusion coefficient by two orders of magnitude as compared with the diffusion of free H centers in a regular lattice. In fact, at this stage the main contribution to the mobility comes from newly created H centers before they join aggregates. After the irradiation is finished, the mobility drops down additionally because no free H centers are created any longer. Upon heating, the effective diffusion coefficient begins to increase because of the separation of some of H centers from their aggregates and their recombination with the F centers. This recombination is an important factor for a considerable growth of the H center aggregates which takes place simultaneously with the aggregation of the F centers, at $T > 250$ K. Recombination results in disappearance in the region around small H aggregates of dispersed F centers. This creates an additional free space necessary for a further considerable growth of the H aggregates.

Unlike the H centers, the F centers remain immobile up to the time $t_0 = 60$ min when the temperature approaches F center mobility edge. However, if the radiation-enhanced F center diffusion takes place, mobility of the F centers during irradiation is close to that for the H centers and the F centers begin to aggregate already at low temperatures (see curves 2 in Fig. 2). However, growth of large F aggregates begins at any rate much later, at $T > 250$ K.

It is interesting to compare the above-said with the irradiation at high temperatures when *both* H and F centers are very mobile. To this end we performed calculations for the irradiation at room temperature, 300 K. In agreement with experiments, a concentration of electron centers (metal area) at the end of irradiation is nearly the same as at low-temperature irradiation. Unlike $T = 193$ K, now we observe growth of the F aggregates *during* the irradiation. However, the mean size of this aggregate remains small ($R_B < 10a_0$ after 30 min) due to the efficient $F–H$ recombination. This is again in agreement with experiments. When the irradiation is switched off, the F aggregates begin to grow. This process could be well illustrated by the analysis of the joint correlation functions characterizing the relative *spatial distribution of defects* (Fig. 3).

Large magnitudes of the joint correlation functions of similar defects, F_{AA}, F_{BB}, at short relative distances clearly demonstrate the strong aggregation of *both* H and F centers. At the end of irradiation (Fig. 3(a)) the relative distance of $r \approx 10a_0$, where F_{BB} (curve 2) approaches the asymptotic value of unity, agrees with the above-mentioned calculation of the effective radius R_B. The effective radius of the H aggregates is larger, about $30a_0$. After 25 min of heating up to 330 K (Fig. 3(b)) these radii increase significantly, up to $R_B \approx 30a_0$ and $R_A \approx 70a_0$, respectively.

The correlation function for the dissimilar defects, $F_{AB}(r)$ (curve 4), is anti-correlated to F_{AA}, F_{BB}. At the end of irradiation it increases from zero at $r \leq a_0$ up to unity at $r \approx 30a_0$, which gives us an estimate of the average distance *between* H- and F aggregates.

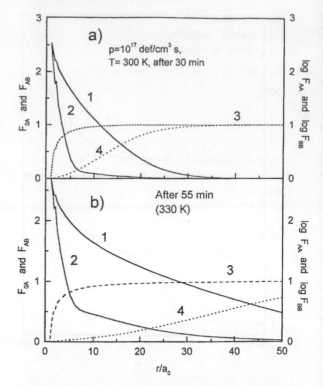

Fig. 3. The joint correlation functions vs. the dimesionless relative distance r between defects for irradiation at 300 K; (a) corresponds to the end of irradiation, whereas (b) after subsequent 25 min heating (up to 330 K). Curves 1 and 2 demonstrate the $H–H$ and $F–F$ correlations, curve 3 the "empty site-F center" correlation, and curve 4 the correlation of dissimilar defects ($F–H$). Note that F_{AA} and F_{BB} are plotted in the semi-logarithmic scale.

Lastly, the joint correlation function "empty site-F center", F_{0A}, curve 3, shows that these aggregates have small, dense cores (there is almost no empty sites in their centers) but they are quite loose on their periphery, $r \geq 10a_0$.

4. Conclusion

The presented microscopic theory of diffusion-controlled radiation defect aggregation reproduces main experimental results on the electron-irradiated CaF_2 and permits to understand the mechanism and kinetics of this process. First conclusion is that small metal colloids observed under irradiation at low temperatures ($T \approx 200$ K) can arise only due to radiation-enhanced F center diffusion. Study of its mechanism is of great interest. Another conclusion is that metal colloids have a dense core but could be rather loose at their periphery. Growth of the F aggregate occurs simultaneously with that for the H centers. Theory shows that the critical temperature at which begins an intensive F center aggregation is defined not only by the activation energy of the F center diffusion but significantly affected by the interaction energy between F centers and their aggregates; this determines the rate of a single F center attachment/detachment to the aggregate necessary for the transformation of small metal colloids into larger colloids (Ostwald ripening).

Acknowledgements

Authors are indebted to E. Matthias and M. Reichling for stimulating discussions. This research has been supported by the Volkswagen Foundation and the Commission of the European Communities (contract ERB CIPDCT 940008, amendment to ERB CHRX CT 930134, the European network SSASS).

References

[1] A.E. Hughes and S.C. Jain, Adv. Phys. 28 (1979) 717; P.W. Levy, J. Phys. Chem. Solids 52 (1991) 319; W.J. Soppe and J. Prij, Nucl. Technol. 107 (1994) 243.

[2] J.R.W. Weerkamp, J.C. Groote, J. Seinen and H.W. den Hartog, Phys. Rev. B 50 (1994) 9781.

[3] R. Bennewitz, M. Reichling et al., Radiation Effects and Defects in Solids 137 (1995) 19; E. Stenzel, N. Bouchaala, S. Gogol, T. Klotzbücher, M. Reichling and E. Matthias, Mater. Sci. Forum 239–241 (1997) 591; N. Bouchaala, Diploma Thesis, Freie Universität Berlin (1997).

[4] S.J. Zinkle, Nucl. Instr. Meth. B 91 (1994) 234.

[5] M. Cross and P.C. Holenberg, Rev. Mod. Phys. 65 (1993) 3.

[6] U. Jain, Ph.D. Thesis, Harwell (1977); U. Jain and A.B. Lidiard, Phil. Mag. 35 (1977) 245; A.B. Lidiard, Phil. Mag. A 39 (1979) 647.

[7] E.A. Kotomin and V.N. Kuzovkov, Modern Aspects of Diffusion-Controlled Reactions, Comprehensive Chemical Kinetics, Vol. 34 (Elsevier, Amsterdam, 1996); V.N. Kuzovkov and E.A. Kotomin, Rept. Progr. Phys. 51 (1988) 1479; E.A. Kotomin and V.N. Kuzovkov, Rept. Progr. Phys. 55 (1992) 2079.

[8] V.N. Kuzovkov and E.A. Kotomin, J. Phys.: Cond. Matter 7 (1996) L481.

[9] K. Atobe, J. Chem. Phys. 71 (1979) 2588.

[10] K. Bachmann and H. Peisl, J. Phys. Chem. Sol. 31 (1970) 1525.

[11] V.N. Kuzovkov, E.A. Kotomin and W. von Niessen, Phys. Rev. E (1997), submitted.

[12] H. Mamada and F. Takano, J. Phys. Soc. Japan 25 (1968) 675.

[13] V.M. Lisytsin and V.Yu. Yakovlev, Sov. Phys. – Proc. Higher Schools, Phys. 23 (3) (1980) 110.

[14] K.A. Kalder and A.F. Malysheva, Sov. Optics Spectr. 31 (1971) 252.

ELSEVIER

Computational Materials Science 10 (1998) 28–32

COMPUTATIONAL
MATERIALS
SCIENCE

Investigation of Be diffusion in InGaAs using Kick-out mechanism

J. Marcon *, S. Gautier, S. Koumetz, K. Ketata, M. Ketata

LCIA/INSA de Rouen, BP 08, F76131 Mont Saint Aignan, France

Abstract

Be diffusion during post-growth annealing has been investigated in InGaAs epitaxial layers. Kick-out mechanisms considering species charges, build-in electric field and Fermi-level effect have been studied. Several forms of Kick-out mechanism have been implemented in our simulation programs. Experimental concentration profiles obtained by SIMS have been compared systematically with the results of simulations. We have deduced that the Kick-out mechanism $Be_i^0 \Leftrightarrow Be^- + I_{III}^+$ is the dominating diffusion mechanism in InGaAs under our experimental conditions ($C_0 = 3 \times 10^{19}\,\mathrm{cm}^{-3}$). Copyright © 1998 Elsevier Science B.V.

1. Introduction

In order to improve InGaAs/InP device fabrication processes and effectively control the performances of heterojunction bipolar transistors (HBTs), accurate simulation of p-type dopant diffusion is required [1]. Indeed, the undesired diffusion of p-type dopant during high-temperature device fabrication often degrades the performances of compounds [1]. Consequently, the understanding and the control of p-type dopant diffusion in InGaAs epitaxial layers are necessary.

Investigations on Zn and Be diffusion mechanisms in the binary and ternary InP based compounds are still limited [2–4]. In this paper, we propose a Kick-out model taking into account species charges, build-in electric field and Fermi-level effect [5–8] to simulate the Be diffusion in InGaAs epitaxial layers. Using Kick-out mechanism and point defect nonequilibrium, we show that a good agreement between experimental and simulated profiles could be obtained.

2. The Kick-out mechanism

There has been a great deal of discussion in the literature concerning the fundamental type of exchange between interstitial and substitutional species in III–V materials [3,5]. Comparing the superlattice disordering obtained from Be and Zn in-diffusion and out-diffusion experiments, some authors conclude that the Kick-out mechanism governs

---

* Corresponding author. Tel.: (33) 2-35-52-84-08; fax: (33) 2-35-52-84-83; e-mail: marcon@ibis.insa-rouen.fr.

0927-0256/98/$19.00 Copyright © 1998 Elsevier Science B.V. All rights reserved
PII S0927-0256(97)00131-6

Be and Zn diffusion in GaAs and GaAlAs [5]. Supposing that the Be diffusion is similar in GaAs and InGaAs [2], in this study, only the Kick-out mechanism has been considered. The exchange between interstitial and substitutional species is given by the Kick-out mechanism:

$$A_i^{n+} \Leftrightarrow A_s^- + I_{III}^{r+} + (n - r + 1)h^+ \tag{1}$$

where A_i^{n+} and A_s^- are interstitial and substitutional impurity atoms, respectively, I_{III}^{r+} represents gallium and indium self-interstitials and h^+ is a hole; n and r are positive integers.

To explain the Zn or Be diffusion profiles in GaAs compounds, several forms of Kick-out mechanisms have been proposed ($n = 0$, 1 or 2 and $r = 0$, 1 or 2 [4–8]). The diffusion equations for the two mobile species are the following [5]:

$$\frac{\partial C_i}{\partial t} = \frac{\partial}{\partial x}\left(D_i\frac{\partial C_i}{\partial x} - n.D_i\frac{C_i}{p}\frac{\partial p}{\partial x}\right) - \frac{\partial C_s}{\partial t}, \tag{2}$$

$$\frac{\partial C}{\partial t} = \frac{\partial}{\partial x}\left(D_I\frac{\partial C_I}{\partial x} - r\frac{C_I}{p}\frac{\partial C_I}{\partial x}\right) + \frac{\partial C_s}{\partial t} - K_I(C_I - C_I^{eq}), \tag{3}$$

$$K = \frac{C_i}{C_s C_i p^{n-r-1}} = \frac{C_i^{eq}}{C_s^{eq}C_I^{eq}(p^{n-r+1})^{eq}}, \tag{4}$$

$$p = \frac{1}{2}\left(C_s + \sqrt{C_s^2 + 4n_i^2}\right), \tag{5}$$

$$C_I^{eq} = C_I^{eq}(n_i)\left(\frac{p}{n_i}\right)^r, \tag{6}$$

where C_i is the concentration of A_i^{n+}, C_I the concentration of I_{III}^{r+}, C_s the concentration of A_s^-, D_i and D_I are the diffusivities of A_s^- and I_{III}^{r+}, respectively; p is the hole concentration and n_i is the intrinsic carrier concentration; n and r are the interstitial and self-interstitial atom charges, respectively. Eq. (4) is obtained using the mass action law in Eq. (1). K_I is a constant taking into account the efficiency of self-interstitial elimination by the crystal bulk [5]. The superscript "eq" notes the equilibrium values. $C_I^{eq}(n_i)$ is the self-interstitial concentration in intrinsic conditions. The second terms in the parentheses on the right-hand side of Eqs. (2) and (3) are due to the build-in electric field. It is assumed that substitutional atoms have a negligible diffusion coefficient [5].

There is no a priori justification for the assumption that the interstitial Be atoms are neutral or positively charged. In practice, the best fit of experimental profiles provides the best justification for assuming the species charges (interstitials and self-interstitials) [5–8].

For the case of epitaxial layers, the boundary conditions are:

in the doped layer: $\quad C_s = C_s^{eq}, \quad C_i = C_i^{eq} \quad$ and $\quad C_I = C_I^{eq},$ \hfill (7)

in the undoped layer: $\quad C_s = 0, \quad C_i = 0 \quad$ and $\quad C_I = C_I^{eq}(n_i).$ \hfill (8)

The coefficients C_i^{eq}, $C_I^{eq}(n_i)$, D_i, D_I, k_+, k_- and K_I are used as parameters. Their evaluation from simulation data will be discussed in Section 5.

3. Numerical procedure: The method of lines

An effective method of solving large parabolic systems (diffusion reaction equations) is the numerical method of lines [9]. Using finite differences on space, these PDEs (partial differential equations) are converted into ODEs' (ordinary differential equations) systems for each grid point $x = x_j$ ($j = 1, \ldots, n - 1$).

For the ODE integrator, we used a Runge–Kutta method with adaptive stepsize control to achieve some predetermined accuracy in the solution with minimum computational calculation [10].

4. Application of models to Be diffusion in InGaAs

In order to determine the species charges, the systems of equations corresponding to the Kick-out mechanisms ($n = 0, 1$ and $r = 0, 1, 2$) have been computed. Typical simulated profiles have been presented in Figs. 1(a) and (b). The parameter values have been listed in Table 1 where D_I and K_I have been fixed to $D_I = 5 \times 10^{-13} \, \text{cm}^2 \, \text{s}^{-1}$ and $K_I = 0, 1 \, \text{s}^{-1}$ in order to observe a double front on diffusion profiles (kink effect). We categorize the various profiles into three basic groups: (i) type I profiles where the diffusion front are very steep which is explained by a very strong variation of effective diffusion coefficient; (ii) type II profiles with a less steep diffusion front; (iii) type III profiles with a slowly decreasing low-concentration portion which is explained by a specific form of mass action law (Eq. (4) with $n = 0$ and $r = 2$). The differences between these simulated profiles could be explained by the species charges (A_i^+ and A_i^0) and the build-in electric field occurring during the p-type dopant diffusion in III–V compounds.

The calculated diffusion profiles have been compared with the SIMS depth profiles. The best fit will determine the choice of the species charges.

Qualitative comparisons between simulated profiles with our experimental data (Fig. 2) [4] show clearly that only the mechanism $\text{Be}_i^0 \Leftrightarrow \text{Be}_s^- + I_{III}^+$ could fit correctly with our experimental profiles.

5. Comparison between experiments and numerical simulations

The mechanism $\text{Be}_i^0 \Leftrightarrow \text{Be}_s^- + I_{III}^+$ has been used to fit our experimental profiles. We have found that the p-type dopant diffusion depth is mainly determined by the $C_{\text{eff}} = D_i C_i^{\text{eq}} / C_s^{\text{eq}}$ values and the simulated profiles seem to

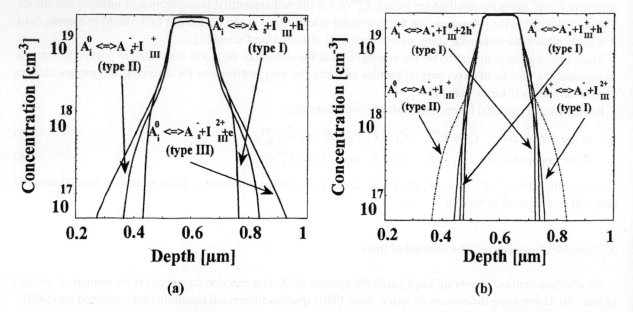

Fig. 1. Typical computed impurity profiles for different models: (a) neutral beryllium interstitials, (b) charged beryllium interstitials

Table 1
Simulation parameters for the simulated (Be) diffusion profiles

Model	Units	$T = 700°C$	$T = 800°C$
C_s^{eq}	cm^{-3}	3×10^{19}	3×10^{19}
C_i^{eq}	cm^{-3}	3×10^{17}	3×10^{17}
$C_I^{eq}(n_i)$	cm^{-3}	2.93×10^{15}	1.21×10^{16}
D_i	$cm^2 s^{-1}$	9.0×10^{-11}	1.5×10^{-9}
D_I	$cm^2 s^{-1}$	3×10^{-12}	1.4×10^{-10}
K_I	s^{-1}	0.6	2
n_i	cm^{-3}	4.02×10^{17}	7.70×10^{17}
t	s	65	65

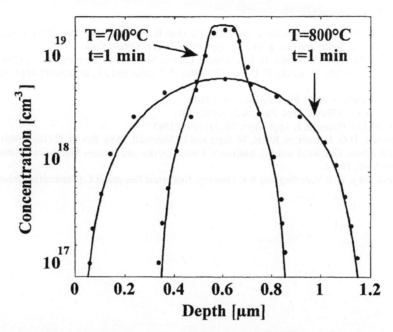

Fig. 2. The experimental [Be] diffusion profile (points) and $C_s + C_i$ calculated curves (full curves) corresponding to annealing temperatures of $T = 700°C$ and $T = 800°C$ ($C_0 = 3 \times 10^{19}$ cm^{-3}).

be approximatively insensitive to the variation of individual C_i^{eq} and D_i parameter values if their product is kept constant [3,5]. This means that the same experimental profile could be fitted with several C_i^{eq} and D_i individual values.

Consequently, our aim is not to propose a complete and consistent set of Be diffusion parameters but to deduce or extract some physical and absolute values (as D_{eff}). Accurate determination of the coefficients C_i^{eq}, D_i, $C_I^{eq}(n_i)$, D_I and K_I for III–V compounds would be the most significant development of future research.

6. Conclusion

In summary, Be diffusion profiles in InGaAs epitaxial layers have been simulated by using the Kick-out models $Be_i^0 \Leftrightarrow Be_s^- + I_{III}^+$.

The observed double profiles have been explained by a supersaturation of self-interstitial on the diffusion front.

Acknowledgements

This work has been supported by the CNET, Laboratories of Bagneux (contract No. 95 6B). The authors wish to thank C. Dubon-Chevallier for initiating this study, P. Launay for fruitful discussions, J.L. Benchimol for epitaxial growth, C. Besombes for RTA, F. Lefebvre for help in SIMS analysis and J.G. Caputo for numerical calculation.

References

[1] N. Jourdan, F. Alexandre, C. Dubon-Chevallier, J. Dangla and Y. Gao, IEEE Trans. Elec. Dev. 39 (4) (1992) 767.
[2] E.G. Scott, D. Wake, G.D.T. Spiller and G.J. Davies, J. Appl. Phys. 66 (11) (1989) 5344.
[3] J. Marcon, S. Gautier, S. Koumetz, K. Ketata, M. Ketata and P. Launay, Solid State Commun. 101 (3) (1997) 159.
[4] S. Koumetz, J. Marcon, K. Ketata, M. Ketata, C. Dubon-Chevallier, P. Launay and J.L. Benchimol Appl. Phys. Lett. 67 (15) (1995) 2161.
[5] S. Yu, T.Y. Tan and U. Gösele, J. Appl. Phys. 69 (6) (1991) 3547.
[6] M. Uematsu, K. Wada and U. Gösele, Appl. Phys. A 55 (1992) 301.
[7] J.C. Hu, M.D. Deal and J.D. Plummer, J. Appl. Phys. 78 (3) (1995) 1595.
[8] G. Bösker, N.A. Stolwijk, H.G. Hettwer, A. Rucki, W. Jäger and U. Södervall, Phys. Rev. B 52 (16) (1995) 11 927 .
[9] G.F. Carey, W.B Richardson, C.S. Reed and B.J. Mulvaney, Circuit Device and Process Simulation: Mathematical and Numerical Aspects (Wiley, New York, 1996).
[10] W.H. Press, S.A. Teukolsky, W.T. Vetterling and B.P. Flannery, Numerical Recipes in C (Cambridge University Press, Cambridge, 1992).

ELSEVIER

Computational Materials Science 10 (1998) 33–37

COMPUTATIONAL
MATERIALS
SCIENCE

Mesoscopic study of laser absorption by a transparent ceramic

R. Mendes Ribeiro [a,*], Marta M.D. Ramos [a], A.M. Stoneham [b]

[a] *Departamento Física, Universidade do Minho, Largo do Paço, 4709 Braga Codex, Portugal*
[b] *Physics Department, University College London, Gower Street, London WC1E 6BT, UK*

Abstract

The understanding of the processes occurring on the target in pulsed laser deposition (PLD) is crucial for a fast optimisation of the deposition parameters in order to obtain high quality thin films.

Phenomenon occurring in the target like the ejection of large particulates that deposit on the substrate or the formation of a rough cone shaped morphology that affect the deposition process cannot be understood in the framework of atomistic simulations, since the processes involve very large volumes. Integration of the heat equations does not seem to be the appropriate approach for the study of PLD, since it ignores the actual ways by which the energy is transferred to the target and transported through it.

Mesoscopic modelling provide solutions in an intermediate scale where both results from atomistic studies and methods characteristic of macroscopic modelling are used. We are developing a mesoscopic model for PLD. In this paper, we show the results for the evaporation of a transparent material in which only structural defects can absorb light. The preliminary results show that the generated electric fields play a dramatic role in the process. Copyright © 1998 Elsevier Science B.V.

Keywords: Mesoscopic modelling; Pulsed laser deposition; PLD

1. Introduction

The evaporation of materials using a nanosecond pulsed laser in the near UV (PLD, pulsed laser deposition) is a promising technique for thin film deposition of high technological materials [1]. Our work concerns the understanding of the role of the target microstructure in the process of pulsed laser ablation. Here we limited the problem to the UV regime and to pulse duration of 30 ns (which include the excimer lasers). The reason is that different wavelengths, pulse durations and other deposition regimes will lead to very different results and physical phenomena, and would

therefore compromise our objective of understanding PLD. We are interested in laser ablation for evaporation in order to produce thin films, and so we will study the process for fluences above laser ablation threshold, which usually is in the range of a few J/cm^2.

In PLD, there are specific problems that need to be solved or controlled. The main problem of this technique is that it is frequent that some aggregates of particles, typically of submicronic size, deposit on the film, invalidating some of the possible applications. A good quality target [2] and a careful choice of the deposition parameters minimises this problem, but frequently some other system is needed, like a chopper that cuts the low velocity particles [3].

Another important problem is the variation in the morphology of the target as the evaporation proceeds

* Corresponding author. Tel.: 351 53 604335; telex: 32135
RUTMIN P; fax: 351 53 678981; e-mail: rribeiro@ci.uminho.pt.

[4], which can lead to a change in the deposition rate and an increase in the number of large particulates deposited on the substrate [5].

Here, we shall look at the beginning of the process, studying the interaction of the laser beam with the target, the processes of absorption and energy transfer and the removal of the material.

Typically, these problems have been studied by two different approaches: an atomistic one, where an atom by atom evaporation process is considered; the mechanism can be as sophisticated as quantum molecular dynamics (MD) calculations [6], in a cluster framework or as a Monte Carlo (MC) calculation [7]. The other approach, more generally employed, is to integrate the diffusion equations of heat, leading to a temperature distribution in the target [8], or to make some energy balance calculations [9] in order to estimate the amount of material removed.

In both approaches, the problem is studied from a limited point of view that ignores the other: atomistic calculations cannot be performed on a large amount of material; on the other hand, heat or energy balance equations cover the description and understanding of the physical microscopic phenomena, like energy conversion, transfer and accumulation, as well as the actual mechanisms of particle and aggregate ejection, which are important for the global understanding of the process.

We believe that a better understanding of PLD will come from mesoscopic modelling, like the one we present here. The conceptual idea is that one should start from the results obtained in atomistic models, see what is relevant in them for our specific problem and use these results (not repeat the calculations) in a larger scale, including microscopic features that cannot be included in an atomistic model, like micron size surface morphologies and grains.

2. Description of the model

In this model, we consider four types of particles: atoms, neutral defects, ionised defects and electrons. We simulated a transparent target like MgO, that typically has a high density of absorbing defect centres

in places like dislocations [10] and where the surface atoms absorb radiation with energy of 5 eV [11] which we consider here as the photon energy. Defects are considered the only radiation absorbents, ionising and emitting electrons. At the beginning, we assumed that the target only has a small density of defects except in very tiny regions as the surface and some dislocations, where the density of defects is high.

The electrons are assumed nearly free (in the conduction band) and, if their density is high enough, they can absorb laser radiation collectively. Electrons are free to move and can transport energy to other target places. The only means of energy transfer to the lattice is by the recombination of the nearly free electrons with ionised defects. In this case, all the energy of the electron plus the recombination energy is given to the lattice. We do not consider any other heat diffusion process explicitly, which means that phonon diffusion is not considered, for instance.

The model can trace as a function of time the total energy of the electrons and the lattice, the density of electrons, neutral defects and ionised defects, as well as the number of evaporated species. The sample is divided in cubic elements in which calculations are performed. The number of atoms in each element is calculated by substracting from the initial value the number of evaporated atoms in each time step. The atoms have a binding energy. If their energy is higher than the binding energy, they are assumed to evaporate. We have the lattice energy for each element and we consider a Maxwell–Boltzmann energy distribution to know the number of atoms evaporated in each time step.

To calculate the number of electrons we take into account: the number of electrons in the previous time step; the number of generated electrons by defect ionisation; the number of electrons diffusing to the element considered; the number of electrons drifting to the element, considered driven by the electric fields generated when the number of electrons is not equal to the number of ionised defects, somewhere in the ensemble; the number of evaporated electrons from the element considered and the number of electrons in the element considered recombining with ionised defects.

The number of electrons generated by defect ioni-sation is proportional to the defect cross-section, the density of defects and the number of photons per unit area. Diffusion and drift are calculated using a fi-nite element method. For diffusion we adopt periodic boundary conditions; for the calculation of the elec-tric field we use periodic boundary conditions, but assume that the surface and the bulk are uncharged. Electrons can evaporate only from the elements at the surface, and are assumed to evaporate if their energy is higher than the work function, including the gener-ated electric fields. To know how many electrons have sufficient energy to evaporate, we assume a Maxwell–Boltzmann distribution of energies in each element.

The neutral and ionised defects are treated similarly to the electrons, except that they are not allowed to diffuse nor drift, and they are assumed to evaporate at the same rate as the atoms (they belong to the lattice) and so the actual density of ionised defects is different from the density of electrons; this generates an elec-tric field which changes the work function and makes electrons drift.

3. The numerical implementation

A plane set of 15 × 15 cubic elements 5 nm size is considered (Fig. 1). The atom density is equal to 5.35×10^{22} atoms/cm^3 everywhere. In the calculations described here, the density of defects varies from el-ement to element, in the following way: the surface elements (region C) are assumed to have a density of defects which is 1% that of the atoms. The bulk ele-ments have 0.01% defects (region F). The central col-umn (region E) is considered to be a grain boundary, and so the density of defects there is 100 times that of adjacent material, including the surface element (re-gion A). The two columns near the grain boundary (region D) are considered to have 10 times the density of defects of the material, in order to allow a softer de-crease in the density of defects. So, in region B, 10% of the atoms are defects, and in region A 100% are defects.

The initial density of electrons, ionised defects and the initial temperature are set to zero. Other parameters

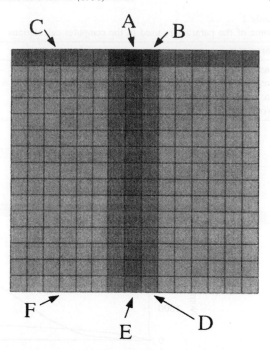

Fig. 1. Schematic of the ensemble; letters identify regions of different defect concentration (see text).

are listed in Table 1. The electron mobility is very low, as is expected for a material like MgO.

4. Results and discussion

The calculations were performed for the first 3 ns of the laser pulse, assuming a square profile of the pulse. The electrons that are generated by the ionisa-tion of defects are the first to evaporate (Fig. 2), be-cause their concentration increases fast and they start absorbing radiation strongly. The electron evaporation rate is limited by the positive electric field they leave behind (Fig. 3).

This positive electric field increases until the atoms start evaporating, taking with them the ionised de-fects. The electric field limits the evaporation rate of the electrons and keeps their density high in region A, where most of them are generated. The electron energy increases as they absorb the laser radiation and this energy is rapidly transferred to the lattice through the recombination with the ionised defects.

Table 1
Some of the parameters used in the computer calculations

Parameter	Value	Parameter	Value
Laser fluence	2.0 J/cm^2	Atom binding energy	15.625 eV
Laser wavelength	248 nm	Excitation energy	5.0 eV
Laser pulse duration	30 ns	Defect cross section	10^{-17} cm^2
Work function	3.25 eV	Electron mobility	8.78 × 10^{-4} m^2/V s

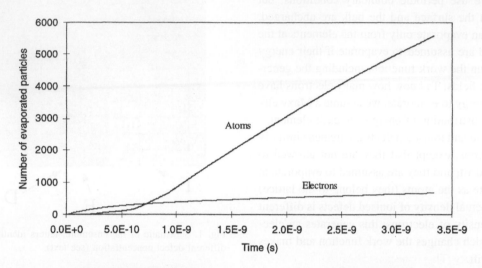

Fig. 2. Number of evaporated atoms and electrons as a function of time.

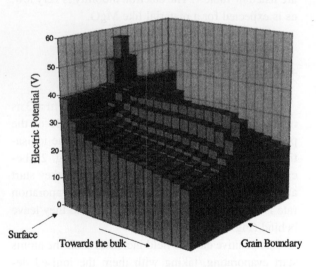

Fig. 3. The electric potential 1 ns after the beginning of the laser pulse.

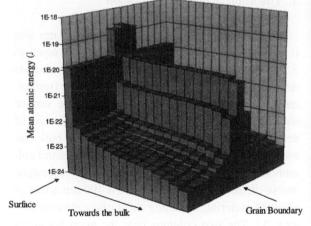

Fig. 4. Mean atomic energy in each element after 3 ns (log scale).

This recombination process is stronger where the defects are more abundant, and so the increase in the energy released is very localised to these places (Fig. 4).

In fact, the material starts to evaporate in the grain boundary and around it.

Using this assumption, we have studied the effect of the absorption of laser radiation by a target with a specific grain structure [12]. As the target evaporates, some surface roughness shows up as a consequence of preferential evaporation at the grain boundaries. The target with wide grain boundaries develops a cone morphology as observed in laser ablation experiments.

The electric potential plays an important role in this model. It keeps electrons in the target, allowing their density to increase and so they can absorb laser radiation more strongly. It also drives electrons generated in the bulk towards the surface. This effect is counterbalanced by diffusion of electrons, which tends to reduce the concentration of electrons where it is higher.

The generated electric field is very high and may nucleate dislocations, increasing the number of defects. This effect is not included in this model, but will be soon. We expect that this effect will lead to an increase in the heated volume and so to a wider evaporation.

We studied the influence of the work function on the qualitative behaviour of the model, and it does not affect significantly. This is because the electric potential generated by evaporation of electrons increases very soon to values that are much greater than the work function, and this determines the subsequent evolution.

5. Conclusions

A new type of mesoscopic modelling of pulsed laser ablation is introduced. A key point of this model is to consider a variety of physical processes taking place in the target, namely a specific absorption process (in this case, absorption by defects and electrons), as well as diffusion and drift by charge carriers, in a large scale. This model is readily scaled to larger volumes.

In transparent targets, the absorption of the laser radiation occurs mainly in places of high defect concentration, such as grain boundaries and dislocations, and material evaporation starts preferentially at those places. We believe that this preferential evaporation at grain boundaries (where the defect concentration is higher) may be a plausible explanation for target cone formation during laser ablation, and is a factor in particulate generation.

Acknowledgements

We wish to acknowledge the JNICT for its financial support under the project No. PBIC/C/FIS/2151/95.

References

[1] D.B. Chrisey and A. Inam, MRS Bulletin 2 (1992) 37; J.C. Miller, ed., Laser Ablation: Principles and Applications (Springer, Berlin, 1994).
[2] O. Auciello, A.I. Kingon, A.R. Krauss and D.J. Lichtenwalner, NATO ASI Series, Vol. E234.
[3] T. Venkatesan, X. Wu, R. Muenchausen and A. Pique, MRS Bulletin 2 (1992) 54.
[4] W. Kantek, B. Roas and L. Schultz, Thin Solid Films 191 (1990) 317.
[5] O. Auciello, A.I. Kingon, A.R. Krauss and D.J. Lichtenwalner, NATO ASI Series, Vol. E234.
[6] H. Gai and G.A. Voth, J. Appl. Phys. 71 (1992) 1415
[7] N. Itoh, J. Kanasaki, A. Okano and Y. Nakai, Ann. Rev. Mater. Sci. 25 (1995) 97.
[8] D. Bhattacharya, R.K. Singh and P.H. Holloway, J. Appl. Phys. 70 (1991) 5433.
[9] R.K. Singh and J. Narayan, Phys. Rev. B 41 (1990) 8843.
[10] J.T. Dickinson, S.C. Langford and J.J. Shin, Phys. Rev. Lett. 73 (1994) 2630.
[11] A.M. Stoneham and P.W. Tasker, Mat. Res. Soc. Symp. 40 (1985) 291.
[12] R.M. Ribeiro, M.M.D. Ramos and A.M. Stoneham, Appl. Surf. Sci. 109/110 (1997) 158.

ELSEVIER

Computational Materials Science 10 (1998) 38–41

COMPUTATIONAL
MATERIALS
SCIENCE

Modeling of InSb and InAs whiskers growth

I. Bolshakova *, T. Moskovets, I. Ostrovskii, A. Ostrovskaya, A. Klimenko

Lviv Polytechnic State University, 1, Kotlyarevskogo St., 290013, Lviv, Ukraine

Abstract

For InSb whiskers grown by the chemical transport reaction (CTR) method in the InSb-J_2 system the relation between whisker diameter (d), crystallization zone temperature (T_{cr}) and thermoprocessing time (t) is found in a certain range of these values. This relation is of practical importance when whiskers are grown for using as active areas of magnetic sensors. The dependence of the InSb and InAs whisker growth rate on the diameter is also considered, main kinetic parameters of the whisker crystallization are determined. In particular, the crystallization energy for InSb and InAs whiskers is equal to 63.9 and 147 kJ mol^{-1}, respectively, that confirms the whisker formation by CTR-method in accordance with the vapour–liquid–crystal mechanism. Copyright © 1998 Elsevier Science B.V.

1. Introduction

Practical needs of the device miniaturization promote the development of science towards production, research of properties and application of low-dimensional structures. These structures include whiskers with perfect crystal structure and surface which are obtained of desired dimensions and shapes without mechanical and chemical processing. These features are responsible for good operating characteristics of whisker-based devices. Thus, InSb and InAs whiskers have found a wide use for manufacturing magnetic microsensors with the active area volume of 10^{-5} mm^3. A wide application of such whiskers as sensitive elements in various devices may be only provided that a large quantity of whiskers are grown in a controlled mode. However, till now initial stages of InSb and InAs whisker growth were not studied

experimentally to a sufficient degree. Lack of data concerning the effect of various technological factors on the growth rate and dimensions still exists. Besides, general principles of the whisker growth have not been determined yet.

This work describes the simulation and analysis of kinetic parameters of InSb and InAs whisker growth performed to optimize the technology and ensure its reproducibility.

2. Experiment

InSb and InAs whiskers are grown by the CTR-method in the isolated iodide system [1]. The whisker growth is carried out in quartz ampoules vacuumized to 10^{-5} mm Hg. Dopants are Te, Sn, Ge. Crystals of various diameters and shapes can be obtained accordingly to growth conditions (the source zone temperature, crystallization zone temperature, dopant concentration, transport agent concentration,

* Corresponding author. Tel.: +38 0322 721632; fax: +38 0322 744300; e-mail: mms@inessa.lviv.ua.

processing time, etc.). In experimental batches available the InSb whisker diameter varies from 6 to 50 μm, and for InAs from 10 to 70 μm. The source zone temperature and the crystallization temperature were 950–1000 and 710–740 K for InSb whiskers, and 1100–1150 and 800–1050 K for InAs whiskers, respectively. The transport agent concentration falls within the range 0.5–1 mg cm^{-3}. Directions of the InSb whisker growth are $\langle 1\,1\,1 \rangle$, $\langle 2\,1\,1 \rangle$ and $\langle 1\,1\,0 \rangle$. Selecting technological conditions one can get the InAs material in the form of either whiskers with the growth direction $\langle 1\,1\,1 \rangle$, or crystal wafers.

3. Experimental results and discussion

3.1. Simulation of the whisker growth technology

The whisker growth performed by the CTR-method is a multi-factor process. The dependence of the crystal diameter upon technological conditions during the growth is of a practical importance. Parameters which have a significant effect on dimensions are the crystallization zone temperature (T_{cr}) and the thermoprocessing time (t). Using the mathematical planning of the experiment one can define the dependence of the whisker diameter on these parameters, $d = f(T_{cr}, t)$.

With computer processing of experiments on the InSb whisker growth we have chosen the mathematical model described by the following equation:

$$d = k_0 + k_1 \cdot T_{cr} + k_2 \cdot t + k_3 \cdot T_{cr} \cdot t, \qquad (1)$$

where k_0, k_1, k_2, k_3 – coefficients.

Taking into account the experimental data (Table 1) we can write Eq. (1) in the following form:

$$d = -979 + 1.364 T_{cr} - 26.39t$$
$$+ 3.73 \cdot 10^{-2} T_{cr} \cdot t. \qquad (2)$$

Using this dependence we further consider the InSb whisker growth process in two cases:
(1) $d = f(T)$ for $t = $ const, that corresponds to the whisker axial growth according to the vapour–liquid–crystal (VLC) mechanism;

Table 1
Dependence of the InSb whisker diameter on the crystallization zone temperature and the thermoprocessing time

No.	d (μm)	t (h)	T_{cr} (K)
1	6	6.33	720
2	15	26.00	720
3	50	26.00	735
4	30	6.33	735

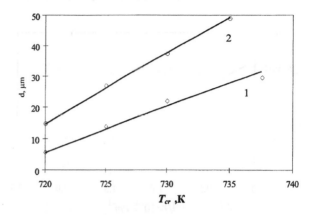

Fig. 1. Dependence of the InSb whisker diameter on the crystallization zone temperature for different thermoprocessing times: 1–6.33, 2–26 h.

(2) $d = f(t)$ for $T = $ const, that describes the whisker radial thickening according to the vapour–crystal (VC) mechanism.

In the first case we obtain whiskers with diameters of 6–30 μm within the temperature range 720–735 K at the constant thermoprocessing time. Fig. 1 shows the InSb whisker diameter as a function of T_{cr} for various t. This dependence demonstrates that the whisker diameter increases as T_{cr} increases. When T_{cr} decreases we obtain whiskers with the diameter $d < 6$ μm.

In the case when $d = f(t)$ for $T = $ const we can evaluate the InSb whisker radial growth rate using the formula (2). We have found this velocity to be equal to 0.45 μm h^{-1} for $T_{cr} = 715$ K. The radial velocity increases from 0.45 up to 1 μm h^{-1} as the crystallization zone temperature raises to 735 K.

3.2. Studies of the InSb whisker growth kinetics

Main kinetic parameters of InSb whiskers were studied for various T_{cr} in accordance with the method

Table 2

Kinetic parameters of the InSb whiskers for $T_{cr} = 720$ K and $t = 6.33$ h

No.	$V \cdot 10^6$ (cm s^{-1})	$l \cdot 10^4$ (cm)	$r \cdot 10^4$ (cm)	$1/r \cdot 10^{-3}$ (cm)$^{-1}$
1	2.06	468.7	3.5	2.8
2	1.84	418.7	2.7	3.6
3	1.5	343.5	2.0	5.0
4	1.21	275.0	1.5	6.7

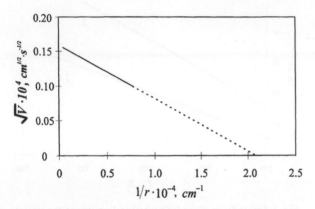

Fig. 2. Dependence of the growth rate on the n-type InSb whisker inverse radius.

described in [2,3]. The growth rate is determined as a ratio of a whisker length (l) to a growth time. The accuracy of measuring whisker dimensions with an optical microscope is $\pm 0.5 \mu$m. On the other hand, the growth velocity can be represented as the thermodynamic relation $V = b(\Delta\mu/kT)^n$, where b – kinetic crystallization coefficient, $\Delta\mu$ – effective chemical potential difference (for IbSb) in the vapour phase and whisker, k – Boltzmann constant, T – temperature, and n – an integer or fractional number. It follows from experimental results for the InSb-J$_2$ system (Table 2) and the dependence of V versus the inverse radius $1/r$ (Fig. 2) that $n = 2$, i.e. the growth rate in this system is defined as

$$V = b \cdot (\Delta\mu/kT)^2. \tag{3}$$

Such dependence confirms the quadratic relation for the whisker growth in the InSb-J$_2$ system. Extrapolating the straight line (Fig. 2) to $V = 0$ we derive the critical diameter (d_k). Then the effective supersaturation $\Delta\mu/kT$ is determined accordingly to the

Table 3

Temperature dependence of the effective supersaturation and the kinetic coefficient b

No.	T_{cr} (K)	$(\Delta_0/kT) \cdot 10^{-3}$	$b \cdot 10^2$ (cm s^{-1})
1	715	5.20	3.7
2	720	5.05	4.0
3	725	4.85	4.3
4	730	4.70	4.6

relation

$$\frac{\Delta\mu_0}{kT} = \frac{4\Omega\alpha_{vc}}{kT} \cdot \frac{1}{d_k}, \tag{4}$$

where $\Omega = 7 \cdot 10^{-23}$ cm^3 – specific volume, and $\alpha_{vc} = 580$ erg cm^{-2} – specific free energy of the vapour-condensed vapour interface [3].

Taking into account that $\Delta\mu = \Delta\mu_0 - 2\Omega\alpha_{vc}/r$ we extract the kinetic crystallization coefficient b from Eq. (3). Computations show that $b = 3.7 \cdot 10^{-2}$ cm s^{-1} at $T_{cr} = 720$ K, while the effective supersaturation is $5.2 \cdot 10^{-3}$ and $d_k = 0.9 \cdot 10^{-4}$ cm. For GaAs we have obtained b of the same order of magnitude ($b_a = 1.0 \cdot 10^{-2}$ cm s^{-1}; $b_b = 5.4 \cdot 10^{-2}$ cm s^{-1}) [3].

Results obtained for the temperature dependence of b during experiments with InSb whiskers are given in Table 3.

Taking into consideration that b is related to T_{cr} and crystallization energy ΔE as $b = b_0 e^{-\Delta E/kT}$ we can express ΔE (per 1 mol) using crystallization coefficient values b_1 and b_2 determined for crystallization zone temperatures T_1 and T_2:

$$\Delta E = \frac{N_0 k \ln b_1/b_2}{1/T_2 - 1/T_1}, \tag{5}$$

where N_0 – Avogadro constant. InSb whisker crystallization energy obtained from (5) is equal to 63.9 kJ mol^{-1}. Thus, determined in our work InSb and InAs whisker crystallization energies are comparable with those for Si whiskers (201 kJ mol^{-1}). Whisker crystallization energy varies also when crystals are doped during the growth. For instance, when introducing Pt in the gas phase while Si whiskers grow the crystallization energy decreases to 48 kJ mol^{-1}.

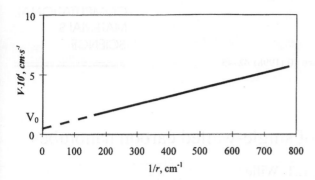

Fig. 3. Dependence of the InAs whisker growth rate on the inverse radius ($V_0 = 0, 5 \cdot 10^{-4}$ cm s^{-1}).

3.3. Studies of the InAs whisker growth kinetics

The typical dependence of the InAs whisker growth rate on the whisker radius (Fig. 3) is different from that obtained for InSb whiskers. As the crystal radius increases the growth rate decreases and gradually approaches a certain non-zero constant value. The similar result was obtained before for Si whiskers in [4], and the authors noted also the decrease of the growth rate as transversal dimensions of crystals increase. When InAs whiskers are of sufficient thickness, the whisker axial growth can be represented without considering diffuse flows from the crystal side face, with the Gibbs–Thompson effect being neglected. In accordance with results given in [5], for $d > 10\,\mu$m the InAs whisker growth rate can be described as

$$V = b \left[\frac{2(1 - \cos \Theta_k)}{\sin \Theta_k} \left(\frac{\Delta \mu_0}{kT} + \frac{2\Omega \alpha_{1g}}{kTr} \right) \right], \quad (6)$$

where Θ_k – contact wetting angle between a liquid drop and a crystal formed, and α_{1g} – specific free energy of the liquid–gas interface.

The experimental dependence $V = f(1/r)$ (Fig. 3) can be described as follows:

$$V = V_0 + \frac{\tan(\alpha)}{r}, \quad (7)$$

where V_0 – segment till the intersection point on the ordinate axis, and α – angle of slope for the straight line $V = f(1/r)$. Solving Eqs. (6) and (7), we find that $b = 1.1 \cdot 10^{-1}$ cm s^{-1} and $\Delta \mu_0 / kT = 8.6 \cdot 10^{-4}$. These b values are of one order of magnitude greater

than those for GaAs submicron whiskers [2] and InSb whiskers grown by the CTR-method in the InSb-J$_2$ system.

In addition, temperature dependence of b for InAs whiskers was studied within the T_{cr} range 800–1050 K. $\Delta E = 147$ kJ mol^{-1} was estimated using the formula (5).

4. Conclusions

Using basics of the mathematical planning of experiments and empirical results concerning the InSb whisker growth by the CTR-method in the InSb-J$_2$ system we have obtained the dependence (2). It enables us to reduce the number of technological experiments and control an unknown value d in a purposeful way within the range 6–50 μm. For InSb whiskers the axial (0.01–0.02 μm s^{-1}) and radial (0.45–1 μm s^{-1}) growth rates are evaluated. The axial growth rate is greater than the radial one by two orders of magnitude for InAs whiskers. Kinetic coefficients and crystallization energy are determined for InSb and InAs whiskers.

Results obtained are of fundamental and practical importance. The whisker growth occurs under nonequilibrium conditions. The growth rate characterizes the degree of non-equilibrium. It means that, InAs of smaller diameters grow under more non-equilibrium conditions in comparison with crystals of larger diameters. We have observed the inverse dependence for InSb whiskers. This fact should have an effect on physical properties of crystals of different diameters, and therefore, on operating characteristics of microsensors.

References

[1] V. Voronin, I. Maryamova, A. Ostrovskaya, Cryst. Prop. and Prepar. 36–38 (1991) 340.
[2] E. Givargizov, J. Crystal Growth 31 (1975) 20.
[3] A. Schetinin, A. Dunayev, O. Kozenkov, Izv. vuzov, Fizika. 3 (1982) 111.
[4] J. Weyher, Cryst. Growth. 43 (2) (1978) 235.
[5] B. Darinskii, O. Kozenkov, A. Schetinin, Izv. vuzov, Fizika. 12 (1986) 18.

ELSEVIER

Computational Materials Science 10 (1998) 42–45

COMPUTATIONAL
MATERIALS
SCIENCE

Simulations of the elastic response of single-walled carbon nanotubes

C.F. Cornwell, L.T. Wille *

Department of Physics, Florida Atlantic University, Boca Raton, FL 33431, USA

Abstract

Using the Tersoff–Brenner potential we have performed molecular dynamics simulations of nanotubes under axial strain, analyzing both compression and stretching forces. These large-scale simulations were carried out on a MasPar massively parallel computer. The elastic response is investigated and expressions for various elastic constants are derived from the simulations. Typical failure modes are also shown and discussed. Copyright © 1998 Elsevier Science B.V.

Keywords: Nanotubes; Elasticity; Parallel computing; Molecular dynamics

Carbon nanotubes have been the subject of intense research over the past several years. These unique materials possess properties that are of theoretical and practical interest to both the scientific and industrial community. Although fairly large quantities of these materials can now be readily produced, isolation of tubes with uniform characteristics has proven to be difficult. Therefore, researchers have turned to realistic computer simulations to gain some insight into the nature of these structures. Simulations are especially valuable as they allow complete control over the structures and their treatment, thus permitting a detailed microscopic study with potentially unlimited time resolution. Of course, the crucial ingredient in the simulations is the expression for forces and energies, which may be derived from a classical potential or from fully quantum mechanical calculations. The former has the advantage of simplicity and allows large systems to be studied in a time-efficient manner, while the latter is

considerably more accurate and reliable, at the cost of large amounts of computer time which limits the size of systems that can be investigated. In either case, the ultimate test of the simulations is a comparison with experiment. If calculated trends turn out to be confirmed by direct measurement, one gains confidence in the methodology and its predictions. These may then be used to indicate the most promising avenues for further experimental scrutiny.

Numerous calculations have been performed on nanotube structures concerning their energetics and structural properties (see [1,2] and references therein). The present paper is concerned with the response of single-walled carbon nanotubes to axial compression and expansion. The elastic properties of these tubes have been examined for small strains by a number of other authors, but our interest here is not only in the elastic response but also in the plastic deformation regime. A comparison between the discrete and continuum description of the plastic deformation regime has recently been reported by Yakobson et al. [1]. While these authors used conjugate-gradient methods

* Corresponding author. Tel. +1(561) 297-3379; fax: +1(561) 297-2662; e-mail: willel@acc.fau.edu.

the present work employs finite temperature molecular dynamics (MD) simulations to examine the plastic deformation regime. The scheme described by White et al. [3] is used to construct and classify the tubules. A total of 28 tubules of the form $T(n, n)$ was studied, with n ranging from 4 to 32.

The MD simulations were carried out on a MasPar MP-1 massively parallel computer using algorithms optimized for short-ranged many-body potentials [4]. The MD calculations used a time step of 15 fs with a standard Verlet algorithm. The simulations were carried out using the Tersoff–Brenner potential [5,6] which has been widely used in the study of fullerenes. It has been found to accurately describe the bond-energy and elastic properties of a wide range of hydrocarbons as well as the diamond and graphite phase of carbon. The potential is short-ranged and consists of two- and three-body contributions in order to mimic directional bonding.

Before subjecting the computer generated tubes to any axial strain, they were quenched for several thousand iterations by scaling the velocities until the final temperatures of the tubes were less than 0.005 K. Next, a quasi-static deformation of the length of the tube was accomplished by gradually scaling the tube's length and allowing it to relax and quench after each step. To eliminate the boundary effects at the tube ends the relative positions of the atoms in the last four rings at each end of the tube were held fixed [2]. The quenching process was typically taken out to 120 ps.

We started by linearly expanding several of the tubes and found that they could be stretched to approximately 140% of their relaxed length. One such tube of the form $T(18, 18)$, with a radius of 12.47 Å is shown in Fig. 1. To be noted is that the tube has retained its overall cylindrical symmetry, but that there is a considerable flaring-out at the ends of the tube due to an accommodation of strain-energy. When expanded beyond 140% of their unstretched length the tubes would tear, recoil, and break into fragments. This was true for tubes over a wide range of diameters. Other simulations using tight-binding MD and the Tersoff–Brenner empirical potential reported tube failure at 120% and 140%, respectively [7,1].

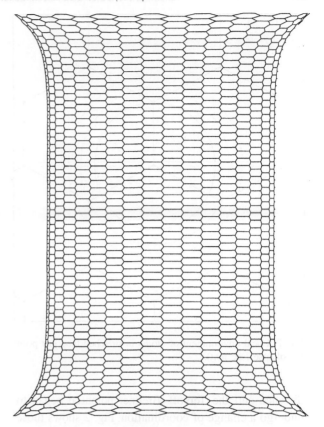

Fig. 1. Tube of the form $T(18, 18)$ stretched to the critical point, just prior to catastrophic failure.

Fig. 2 shows the local environment of an atom in the center of the tube as viewed perpendicular to the tube axis immediately prior to failure. The ellipses in the figure represent the intersection of two surfaces. One of the surfaces used to define the ellipses is a cylinder centered on the tube axis with a radius equal to that of the central part of the tube. The other surfaces are spheres centered on the central atom in the figure with radii equal to the equilibrium distance (solid line) and the inner (dotted line) and outer (dashed line) cutoff distances of the Tersoff–Brenner potential. Initially the relaxed tubes have their nearest neighbors located on the inner ellipse at the potential's equilibrium distance. However, as the tube is stretched, the atoms are pulled from their equilibrium positions towards the inner cutoff distance. In all cases, the tubes failed when the two second-neighbor atoms moved outside the inner cutoff

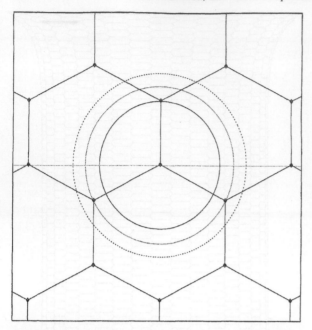

Fig. 2. Local environment of atoms in stretched tube just prior to collapse (see text).

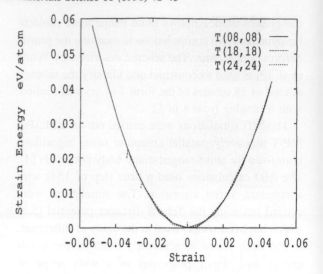

Fig. 3. Strain energy vs. strain for several tubes.

distance and the Tersoff–Brenner potential began to undergo a change of coordination. Fig. 2 shows those two atoms just about to cross that critical distance. Note that the first-neighbor atom is still very closely to its equilibrium distance from the central atom, due to the fact that the strain is applied along the tube axis. In the potential construction the cutoff distance was chosen to include only nearest neighbor interactions and although it is not completely arbitrary there is a great deal of flexibility in choosing this distance. Therefore the fact that the tubes fail at 140% elongation is a result of how the potential is constructed and not of any physical property of the system. Nevertheless, the actual deformation and failure mechanisms should be accurately described by this potential.

Next a series of compression studies was performed. For small strains, in the elastic regime, the response was similar for both the stretched and compressed tubes. The atoms did not undergo large displacements from their equilibrium positions and the strain energy increased quadratically with strain (Fig. 3). Here the strain (ϵ) is a dimensionless quantity defined as $(L - L_0)/L_0$, where L_0 is the length of the relaxed

tube and L is the length of the strained tube. Upon increasing the linear compression we found that the compressed tubes would maintain their cylindrical symmetry until some critical value of strain was reached at which point they would buckle or bend. The response in the elastic regime was similar for all tubes but the critical strain was found to vary with tube radius. The atoms did not undergo large displacements from their equilibrium positions and no bond breaking or forming was observed during this process in contrast to the stretched tubes. The final configuration of the tubes depended on the rate of increase in strain and the rate of energy dissipation. However, the tubes generally ended up in one of the buckled or kinked modes described by Yakobson et al. [1] and Xie et al. [8]. The failure of the tube resulted in a release and redistribution of strain energy in the system as can be seen in Fig. 4 in which the light areas represent the atoms with the higher strain energy. As an example, tube $T(18, 18)$ had an average strain energy of -7.368 eV/atom with a standard deviation of 0.0001 eV at its equilibrium length. Just prior to failure the average strain energy was -7.346 eV/atom with a standard deviation of 0.02 eV. Patterns in the strain energy showed up as rings around the tube at the ends as seen in Fig. 4(a). Strain energy is released in the failure process. The buckled tube had

(a)

(b)

Fig. 4. Snapshots of plastically deformed tube at the end of compression cycle. The tube is of the form $T(18, 18)$, with a radius of 12.47 Å. (a) Just prior to plastic deformation the strain build-up modulates the cylinder wall. (b) After the tube has permanently deformed, strain has been released but residual build-up can still be seen along the folds.

an average strain energy of -7.348 eV/atom with a standard deviation of 0.128 eV. The overall energy of the system was lower and the remaining strain energy was concentrated along the folds in the buckled and bent tubes as seen in Fig. 4(b).

An analysis of the elastic region in Fig. 3 is particularly enlightening. First one notes the symmetric parabolic shape of the energy (E) vs. strain (ϵ) curve, suggesting that Hooke's law is valid for small strains, be it for compressed or stretched tubes. Fitting an expression of the form: $E(\epsilon) = 1/2 \, k \times \epsilon^2$ to this parabola, we find a spring constant $k = 40$ eV/atom. Second, we calculated Young's modulus for the various tubes using the procedure outlined elsewhere [2]. This leads to an expression of the form:

$$Y(\text{GPa}) = 4296(\text{GPa Å})/r \, (\text{Å}) + 8.24(\text{GPa}). \qquad (1)$$

For typical tube radii ranging from 5 to 10 Å this gives Young's moduli between 400 and 800 GPa. These are on the same order of experimentally measured values [9].

Summarizing, classical MD simulations of single-walled carbon nanotubes indicate that these tubes respond elastically for small axial strains, both in compression and under stretching forces. A spring constant and Young's modulus can be determined to describe this regime. Tubes that are stretched beyond a critical value (uniformly, 140% of the relaxed length) collapse suddenly and lead to fragmentation. However, it is argued here that the precise location of this critical point may be an artefact of the potential rather than a fundamental quantity. Nevertheless, it is believed that the scenario leading up to the collapse is a plausible one. Under compression, tubes deform plastically once a certain critical threshold is reached. This critical strain depends on tube radius. The deformed tubes may be generically classified and exhibit some residual stress. While the observed trends are plausible and consistent, it would be very interesting if quantum mechanical calculations could be performed to confirm some of the observations. Ultimately, the greatest value of a classical potential is its easy and flexibility in use, which allows it to point the way to more accurate studies as well as possible experiments. It is hoped that the present paper will stimulate such further work.

References

[1] B.I. Yakobson, C.J. Brabec and J. Bernholc, Phys. Rev. Lett. 76 (1996) 2511.
[2] C.F. Cornwell and L.T. Wille, Solid State Commun. 101 (1997) 555.
[3] C.T. White, D.H. Robertson and J.W. Mintmire, Phys. Rev. B 47 (1993) 5485.
[4] L.T. Wille, C.F. Cornwell and W.C. Morrey, Mat. Res. Soc. Symp. Proc. 408 (1996) 125.
[5] J. Tersoff, Phys. Rev. Lett. 61 (1988) 2879.
[6] D.W. Brenner, Phys. Rev. B 42 (1990) 9458.
[7] R. Yu, M. Zhan, D. Cheng, S. Yang, Z. Liu and L. Zheng, J. Phys. Chem. 99 (1995) 1818.
[8] S.S. Xie, W.Z. Li, L.X. Qian, B.H. Chang, C.S. Fu, R.A. Zhao, W. Y. Zhou and G. Wang, Phys. Rev. B 54 (1996) 16 436.
[9] M.M. Treacy, T.W. Ebbesen and J.M. Gibson, Nature 381 (1996) 678.

ELSEVIER

Computational Materials Science 10 (1998) 46–50

COMPUTATIONAL
MATERIALS
SCIENCE

A new model of DLA under high magnetic field

Hiroshi Mizuseki *, Kazumi Tanaka, Keiko Kikuchi, Kaoru Ohno, Yoshiyuki Kawazoe

Institute for Materials Research, Tohoku University, Sendai 980-77, Japan

Abstract

A new Monte Carlo model is introduced to describe diffusion-limited aggregation (DLA) with extra forces arising from Lorentz's and/or Coulomb forces. Furthermore, we simulate a behavior of multiparticle diffusive aggregation to examine the resultant pattern of crystal in electrochemical deposition. Different patterns grown under various external forces are produced by Monte Carlo simulations. In the present model, the basic movement of particles is a random walk, with different transition probabilities in different directions, which characterizes stochastically the effect of extra forces. In case of assuming a high magnetic field, pattern formations which are qualitatively different from the standard DLA model are observed and they are successfully compared with preexisting experiments (Mogi et al., 1991). The present numerical results of electrochemical deposition show that the generated patterns strongly depend on the force acting on ions and their concentration (Sawada et al., 1986). Copyright © 1998 Elsevier Science B.V.

Keywords: Fractal dimension; Diffusion-limited aggregation (DLA); Magnetic field; Crystal growth; Random walk; Electrochemical deposition; Monte Carlo method

1. Introduction

The phenomena of crystal growth in diffusive system have recently attracted considerable attention [1,2]. Pattern-formation examples in diffusive system include electrochemical deposition, crystal growth, viscous fingering, dielectric breakdown, chemical dissolution and bacterial colonies. An approximation to these phenomena is provided by the Laplacian growth model that can be simulated by the diffusion-limited aggregation (DLA) [1]. These patterns generated by DLA are analyzed by computational and experimental methods. Several extensions of the DLA model have been proposed to take into account electrochemical deposition [3], magnetic field [4], concentration [5], drift [6], convection [7] and sticking probability [5]. A variety of computer simulations have been carried out to investigate the relationship between the patterns of the crystal and the movement behavior of the particles. Consequently, the morphology of the crystal strongly depends on the dynamics of the crystal growth process.

As a new type of perturbation, Mogi et al. at our Institute applied a high magnetic field for the dendritic growth of silver [8], lead [9] and zinc [10] in solutions and obtained new specific patterns of metal leaves. Under a high magnetic field, the branches bend slightly and the shape of the envelope becomes circular. These results imply that the particle diffusion in the medium is strongly affected by the magnetic field and this decisive difference in the particle movement

* Corresponding author. Tel.: +81 22 215 2054; fax: +81 22 215 2052; e-mail: mizuseki@imr.tohoku.ac.jp.

changes the final pattern from the DLA result. Since the direction of bending is reversed, according to the reversal of magnetic field, the observed bending of the crystal is attributed to the cyclotron motion of the Ag ions. These experimental works have also shown that the effect of magnetic field is universal, i.e., independent of the details of the chemical properties of the solution. Therefore, we expect to reproduce this universality observed in the experiments via computer simulation. The first aim of this paper is to examine various patterns of crystal growths under an external force involving the stochastic treatment of the cyclotron motion.

Sawada et al. [11,12] show that the patterns in electrochemical deposition depend on the concentration of the solution and applied voltage of electrode. This experiment shows that different pattern formations under various conditions such as an ion concentration and an applied voltage in the Zn from a thin layer of $ZnSO_4$ solution appeared. It is worthwhile to examine the subject more closely. Therefore, we have to inquire, to some extent, into the force that is affected by other ions. In this paper, we present a new model for the electrodeposition process. The multiparticle DLA simulation is carried out to simulate this behavior of particle [11,12], because the particle is affected by Coulomb forces from the electrode as well as from the other ions.

The present paper is organized as follows: A new growth model is introduced in the next section, which is affected by external force. A large computer simulation is performed and the obtained numerical results are given in Section 3. They are compared with the experimental observations also. Sections 4 and 5 are devoted, respectively, to discussion and conclusion.

2. Model and numerical method

Our present model is basically a Monte Carlo simulation on a square lattice. The inclusion of external forces shows that the pattern formation is governed not only by the diffusion field, but also by the magnetic field, gradient of concentration, electric field, and so on. The effect of the magnetic field is included by simply changing the probability of the movement perpendicular to the Monte Carlo movement. We assume two environments in crystal growth: high magnetic field and gradient of ion concentration. Multiparticle aggregations have been simulated at various concentration f in the range of 0.05–0.4 and crystal sizes from 500×500 to 800×800. In the present simulation, when the particle arrives to contact with the crystal, it sticks permanently, and a sticking probability between the ion and the crystal is 1. Furthermore, we have not allowed that two ions occupy the same square sites. In this simulation, the area far from the electrode was held at initial concentration f. During simulations, f is fixed. This is essentially the model decribed previously [4] where it was used to study the effect of external force in crystal growth.

3. Results

3.1. High magnetic field

Numerical simulations were performed based on the models described in the preceding section. Typical growth patterns obtained by the present research are indicated in Figs. 1 and 2 together with the experimental results [8]. The numbers on the axes show the mesh points and therefore the sizes of the model clusters. By the standard DLA growth model (Fig. 1), the well-known pattern with the fractal dimension of 1.67 is obtained for the square lattice case.

Fig. 2(b) shows the experimentally observed crystal pattern under high magnetic field [8]. This figure bears resemblance to the present simulation from the view point of the bending of the pattern. The numerically obtained pattern in Fig. 2(a) is basically similar to the pattern of the experimental result. The numbers and the thickness of the branches in Figs. 2(a) and (b) are still different because of the small number of particles in simulation compared with experiment. If we can perform quantitatively larger scale simulations which are not possible at the moment, the number of branches increase and accordingly the branches will become thinner.

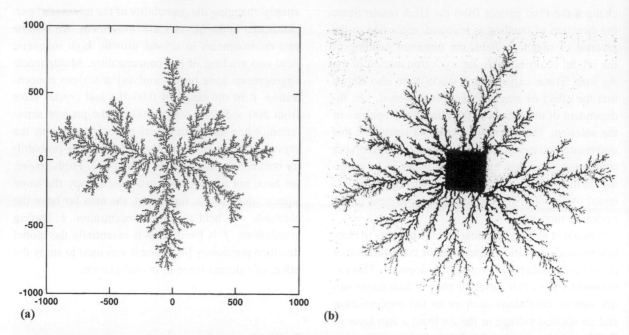

Fig. 1. (a) Simulated pattern of the standard DLA model on square lattice. (b) Observed pattern of silver leaves without magnetic field. The square at the center of the metal leaf is a piece of copper metal [8].

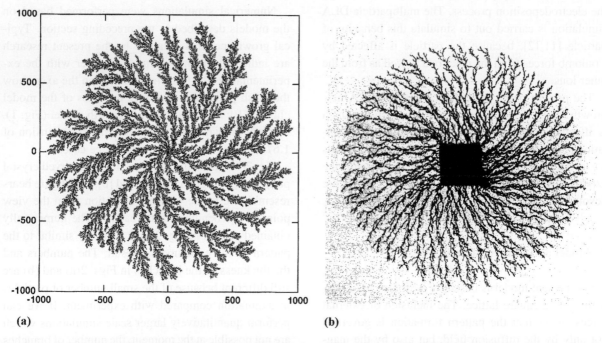

Fig. 2. (a) Simulated crystal growth with the effect of Lorentz force and gradient of electrolyte concentration, respectively. The conditions used are $a = 1.0$, $\theta = 10°$, and ionization tendency parameter $b = 0.035$. The notation of parameters was referred to [4]. (b) Observed pattern of silver leaves under 8T. The magnetic field was applied perpendicular to the plate [8].

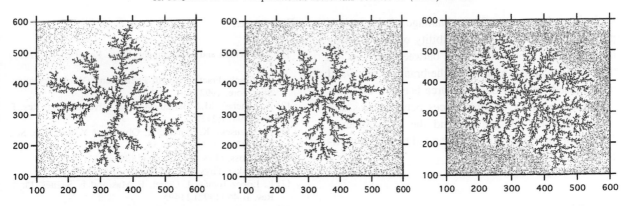

Fig. 3. Multiparticle diffusive aggregation to point electrode from a concentration (a) $f = 0.05$, (b) $f = 0.1$ and (c) $f = 0.2$.

3.2. Electrochemical deposition

Typical growth patterns obtained by the present research are indicated in Fig. 3. Figs. 3(a)–(c) show the results of aggregation from a point for different f. Naturally, in low concentration, the obtained pattern is very similar to the standard DLA pattern. On the other hand, in high concentration, the number of branches increases and accordingly the space between the branches becomes small. In the case of random structures such as present study, we need the so-called density–density correlation function $c(r)$ to detect the fractality of the crystal pattern [2]. This equation is the expection value of the existence that two points separated by a distance r belong to the structure. In low concentration, Fig. 4 shows that the density–density correlation function within the clusters decays according to a power law. Furthermore, Fig. 4 shows that $c(r)$ increases at higher concentration.

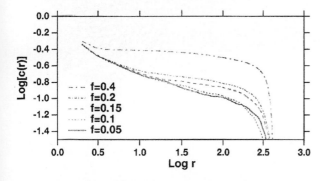

Fig. 4. Double logarithmic plot of the density–density correlation function $c(r)$ for multiparticle aggregation.

4. Discussion

The success in understanding the transfer processes by external force in crystal growth has led to widespread interests from the technological point of view. An origin of curve of the branch in the metal leaf [8–10] is a problem that should not be ignored. If the convection exists in the system, the branch of the metal leaf should be bent. If the pattern of the convection effect is investigated, these results [7] show that the branches have spiral form. The observed metal leaves are not spiral, and the curvature is uniform. Therefore, we propose the effect of Lorentz force in this crystal growth process.

Sawada et al. [12] found several qualitatively different growth forms in electrochemical deposition of Zn from a thin layer of $ZnSO_4$ solution. The self-similarity of the crystal in electrochemical deposition disappears when applied voltage reaches a certain strength. Within the present model, we can reproduce that the pattern of the resultant crystal depends on the ion concentration and the applied voltage.

5. Conclusions

Using a Monte Carlo simulation, the pattern of metal leaf growths under high magnetic field and electrochemical deposition are investigated. The results are presented by comparing the patterns between the standard and perturbed DLA models arising from various environments. The advantage of this multi-particle simulation is the possibility of investigating the effects of Coulomb forces between the ions. In this simulation, the number of the branches of crystal is increased by the Coulomb force between the ions. Furthermore, the Coulomb forces between ions and electrode affect the pattern and the self-similarity. This information can not be obtained by single particle simulations.

Acknowledgements

The authors would like to express sincere thanks to the crew of the Supercomputing Center of Institute for Materials Research, Tohoku University, for their continuous support of the HITAC S-3800/380 super-computing system.

References

[1] T.A. Witten and L.M. Sander, Phys. Rev. Lett. 47 (1981) 1400.
[2] T. Vicsek, Fractal Growth Phenomena (World Scientific, Singapore, 1992).
[3] A. Sánchez, M.J. Bernal and J.M. Riveiro, Phys. Rev. E 50 (1994) R2427.
[4] H. Mizuseki, K. Tanaka, K. Ohno and Y. Kawazoe, Sci. Rep. RITU A 43 (1997) 55.
[5] R.F. Voss, Phys. Rev. B 30 (1984) 334.
[6] P. Meakin, Phys. Rev. B 28 (1983) 5221.
[7] L. López-Tomás, J. Claret, F. Mas and F. Sagués, Phys. Rev. B 46 (1992) 11495.
[8] I. Mogi, S. Okubo and Y. Nakagawa, J. Phys. Soc. Jpn. 60 (1991) 3200.
[9] I. Mogi, S. Okubo and Y. Nakagawa, J. Cryst. Growth 128 (1993) 258.
[10] I. Mogi, M. Kamiko, S. Okubo and G. Kido, Physica B 201 (1994) 606.
[11] M. Matsushita, M. Sano, Y. Hayakawa, H. Honjo and Y. Sawada, Phys. Rev. Lett. 53 (1984) 286.
[12] Y. Sawada, A. Dougherty and J.P. Gollub, Phys. Rev. Lett. 56 (1986) 1260.

ELSEVIER

Computational Materials Science 10 (1998) 51–56

COMPUTATIONAL
MATERIALS
SCIENCE

Ab initio supported model simulations of ferroelectric perovskites

M. Sepliarsky [a,b], R.L. Migoni [a,b,*], M.G. Stachiotti [a,b]

[a] *Instituto de Física Rosario (CONICET-UNR), Bv. 27 de Febrero 210 Bis, 2000 Rosario, Argentina*
[b] *Facultad de Ciencias Exactas, Ingeniería y Agrimensura (UNR), 27 de Febrero 210 Bis, 2000 Rosario, Argentina*

Abstract

While ferroelectric $KNbO_3$ and $BaTiO_3$ present the cubic–tetragonal–orthorhombic–rhombohedral phase sequence with decreasing T, their related compounds $KTaO_3$ and $SrTiO_3$ are incipient ferroelectrics which exhibit the ferroelectric soft mode but remain paraelectric up to $0\,K$. Despite the large amount of research attracted by the above phenomenology since long ago, quite a few realistic model simulations are available. Many lattice dynamical calculations have been performed for each isolated material, but the only unified view of their ferroelectric behavior has been provided by the nonlinear oxygen polarizability model. Although initially based on phonon data for the cubic phase, recent ab initio calculations confirm that the model is basically correct with regards to the energetics involved in the various ferroelectric distortions in $KNbO_3$. Our molecular dynamics simulations show the crossover from a soft mode to an order–disorder dynamics in the cubic phase of $KNbO_3$, the appearance of its various ferroelectric phases, and the soft-mode behavior of $KTaO_3$. Copyright © 1998 Elsevier Science B.V.

Keywords: Ferroelectric transitions; Order–disorder; Soft-mode

1. Introduction

Recent first principles calculations based on the density functional theory in the local density approximation have raised again the question on the microscopic mechanism of the phase transition in ferroelectric ABO_3 perovskites. While a soft-mode theory had been initially proposed as the origin of the transition, already early experiments on $BaTiO_3$ and $KNbO_3$ have shown characteristics of order–disorder phenomena [1]. Both materials show the same sequence of transitions with decreasing temperature from the cubic paraelectric phase. They first become ferroelectric tetragonal with polarization along a [1 0 0] direction, then orthorhombic with [1 1 0] polarization and finally rhombohedral with [1 1 1] polarization. The results of numerous attempts to observe a ferroelectric soft mode in these compounds had to be interpreted in terms of an overdamped oscillator with such a high damping that it could hardly be distinguished from a relaxational motion [2–4]. Thus, an eight-site order–disorder model had been proposed with eight equivalent [1 1 1] off-center energy minima for the transition metal B ion [1,5]. According to this model, all sites are visited with equal probability by the B ion in the cubic phase; four sites with equal projection along one of the cubic axes become preferentially populated in the tetragonal phase; among these only two nearest-neighbor sites remain most probable in the orthorhombic phase, and finally only

* Corresponding author. Address: Instituto de Fisica Rosario, Bv. 27 de Febrero 210 Bis, 2000 Rosario, Argentina. Tel.: (0054-41)-853200/853222; fax: (0054-41)-821772; e-mail: migoni@ifir.ifir.edu.ar.

one site is occupied in the rhombohedral phase. The ab initio calculations of the total energy vs. B ion displacements in the cubic phase at the experimental 0 K extrapolated volume of $BaTiO_3$ [6] and $KNbO_3$ [7] confirm the presence of absolute minima at the [1 1 1] off-center positions and higher energy minima for [1 0 0] displacements. These minima are very sensible to volume.

On the other hand, the related perovskites $SrTiO_3$ and $KTaO_3$ do show a quite well-defined ferroelectric soft mode which, however, does not become unstable down to 0 K just due to quantum fluctuations [8,9]. Ab initio calculations for $KTaO_3$ show extremely shallow [7] or even absent [10] [1 1 1]-off-center minima for the soft-mode displacement pattern of the ions. The situation is more complex in $SrTiO_3$ due to competition with the antiferrodistortive instability.

The lattice dynamics and ferroelectric soft-mode behavior of the above-mentioned four compounds has been described accurately by the nonlinear oxygen polarizability (NOP) model within the self-consistent phonon approximation (SPA) [8,9,11,12]. This model locates at oxygen the polarization effects produced by variations of the O–B distance, which are largely due to hybridization between oxygen p and transition metal d orbitals [6,7]. Thus a shell model is considered with a fourth-order term in the oxygen shell-relative-to-core displacement component along the O–B bond. This interaction stabilizes the model, which is otherwise unstable against the ferroelectric mode displacements. However, the SPA does not allow for the observed order–disorder behavior. With an exact numerical treatment of the model we have recently shown for $KNbO_3$ and $KTaO_3$ that the behavior of total energy vs. B ion displacements on a (1 1 0)-plane is in good agreement with the ab initio results and is consistent with the sequence of transitions observed in $KNbO_3$ [13].

We therefore examine the dynamical behavior of the model with the aim to verify the soft-mode to order–disorder crossover proposed for $KNbO_3$ as well as the soft-mode behavior in $KTaO_3$. To this purpose we perform first a constant volume molecular dynamics simulation in the cubic phase and analyze the dynamical structure factor $S(\mathbf{q} = 0, \omega)$ as a function of

temperature. A more detailed discussion is given elsewhere [14]. In addition, we present here preliminary results on constant pressure molecular dynamics simulations of $KNbO_3$, in a range of temperatures which includes its four phases.

2. Procedure

A description of the model and its parameters for $KTaO_3$ and $KNbO_3$ have been previously given [13]. The equations of motion are:

$$\mathcal{M}\ddot{\mathbf{u}} = (\mathcal{S} + \mathcal{C}^{ZZ})\mathbf{u} + (\mathcal{S} + \mathcal{C}^{ZY})\mathbf{w}, \tag{1}$$

where $\mathbf{u}$ and $\mathbf{w}$ denote core and shell-relative-to-core displacements, respectively. The matrix $\mathcal{S}$ represents the short-range shell–shell interactions, and $\mathcal{C}^{ZZ}$ and $\mathcal{C}^{ZY}$ represent the Coulomb ion–ion and ion–shell interactions, respectively, Y denoting shell and Z total ionic charges. In spite of the linear appearance of the above equation, $\mathbf{w}$ is a nonlinear function of $\mathbf{u}$ through the adiabatic condition, which requires that the electronic shell configuration be a potential minimum at every instantaneous core configuration:

$$\frac{\partial V}{\partial \mathbf{w}} = (\mathcal{S} + \mathcal{C}^{YZ})\mathbf{u} + (\mathcal{S} + \mathcal{K} + \mathcal{C}^{YY})\mathbf{w}$$
$$+ \frac{1}{3!}k_{OB,B}\mathbf{f}(\mathbf{w}) = 0. \tag{2}$$

Here the matrix $\mathcal{C}^{YZ}$ represents the Coulomb shell–shell interaction and $k_{OB,B}$ is the quartic core–shell coupling constant at oxygen. $\mathbf{f}(\mathbf{w})$ is a vector with components $f_\alpha(l\kappa) = w_\alpha^3(l\,O_\alpha)\delta_{\kappa,O_\alpha}$, where α, κ and l denote cartesian, ionic site and cell indices, respectively, and O_α denotes an oxygen whose B neighbor ions lie in α-direction. $\mathbf{w}$ is solved iteratively using the steepest descent procedure at each time step of the dynamics.

For the analysis of the dynamical behavior in the paraelectric phase we perform molecular dynamics simulations in the (N, V, T) ensemble and employ a Beeman algorithm [15]. We consider $4 \times 4 \times 4$ unit cells with periodic boundary conditions. Once thermalization and stability is achieved, we take 12 000 time steps for the evaluation of the physical quantities.

The temperature of the system is adjusted by letting a Nosé–Hoover thermostat [16] act on each core co-ordinate. This is achieved by modifying the previous equations of motion (1) as follows:

$$m_i \ddot{u}_i = F_i - \eta_i m_i \dot{u}_i, \tag{3}$$

where F_i represents the right-hand side of Eq. (1), and the "friction" coefficient η_i of the thermostat evolves according to

$$\dot{\eta}_i = (m_i \dot{u}_i{}^2 - k_B T)/(k_B T \tau^2). \tag{4}$$

This procedure was necessary to drive the system into thermal equilibrium. Since the anharmonic term of the NOP model couples a limited number of modes to each other (including the ferroelectric mode), many modes do not interchange energy. As a consequence, the temperature of the coupled modes may differ from the mean temperature of the system, depending on the initial conditions of the simulation. This led us to a wrong evaluation of temperature and possibly also to distortions of the dynamical structure factor in our preliminary calculations [17].

We evaluate the dynamical structure factor:

$$S(\mathbf{Q}, \omega) = \int_{-\infty}^{+\infty} dt\, e^{i\omega t} \sum_{l\kappa} \sum_{l'\kappa'} e^{i\mathbf{Q}\cdot(\mathbf{R}_\kappa^l - \mathbf{R}_{\kappa'}^{l'})}$$
$$\times \langle \mathbf{Q}.\mathbf{u}_\kappa^l(t)\mathbf{Q}.\mathbf{u}_{\kappa'}^{l'}(0)\rangle, \tag{5}$$

where the wave vector $\mathbf{Q} = \mathbf{G} + \mathbf{q}$, with $\mathbf{G}$ a reciprocal lattice vector and $\mathbf{q}$ within the first Brillouin zone. We take $\mathbf{q} = 0$ for the analysis of the long-wave modes. The choice of $\mathbf{G}$ affects the relative intensities of peaks corresponding to different modes.

In order to study the capability of the model to describe the ferroelectric phases of $KNbO_3$, we perform further simulations in the (N, P, T) ensemble, thus allowing for strains. To this pur-pose we determine short-range Buckingam po-tentials from the harmonic force constants [11]. To attain equilibrium at the experimental $0\,K$ ex-trapolated lattice constant $(4.000\,\text{Å})$ we readjusted slightly the Nb–O potential. For the numerical

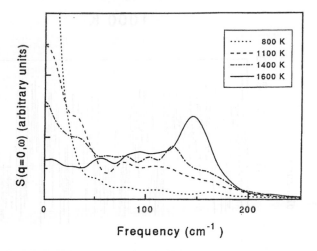

Fig. 1. Dynamical structure factor $S(\mathbf{q} = 0, \omega)$ of $KNbO_3$ at several temperatures in the paraelectric phase, for $\mathbf{G} = (2, 0, 0)2\pi/a$, in the low-frequency range corresponding to the ferroelectric mode.

calculations we use the code DL-POLY[1] with the addition of the anharmonic core–shell interaction at oxygen.

3. Results and discussion

For $KNbO_3$ at 730 K, $S(\mathbf{q} = 0, \omega)$ shows well-defined peaks in quite good agreement with the ex-perimental data and the SPA for the TO long-wave phonons, except for the lowest TO mode, i.e. the fer-roelectric one, which does not appear at all.

We search for evidence of the ferroelectric mode at increasingly higher temperatures. As shown in Fig. 1, no well-defined structure appears below the melting point of the material (1323 K). On the other hand, we observe a quasi-elastic component of $S(\mathbf{q} = 0, \omega)$ (central peak) which sharpens and grows in inten-sity with decreasing temperature. As we will further discuss below, this is the signature of a relaxational dynamics characteristic of an order–disorder transi-tion mechanism. Nevertheless, with the aim to ob-serve if a crossover to a soft-mode dynamics can be

[1] DL-POLY is a package of molecular simulation routines written by W. Smith, Daresbury T.R. Forester, and UK. Ruther-ford Appleton Laboratory, Daresbury.

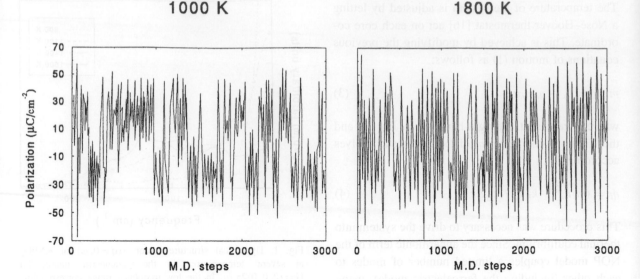

Fig. 2. Time evolution of a single cell polarization in KNbO$_3$ at representative temperatures of the order–disorder (1000 K) and the soft-mode (1800 K) regimes.

detected, we increase the temperature of our simulations above the actual melting point. A wide structure in the 70–140 cm^{-1} frequency interval at 1100 K develops a peak at ∼125 cm^{-1} as the temperature is increased to 1400 K. This peak, which corresponds to the ferroelectric mode, shifts to a higher frequency and gains intensity as the temperature is further increased to 1600 K.

This dynamical behavior constitutes a theoretical evidence of a crossover from a displacive to an order–disorder mechanism of the ferroelectric phase transition in KNbO$_3$, as has been proposed to interpret various experiments [4]. The soft-mode peak appears at temperatures above the range of available experimental data.

The characteristics of the particle dynamics in the just remarked two regimes are visualized in Fig. 2 through the time evolution of the polarization in a single unit cell. At 1800 K only fast oscillations around zero polarization are observed, while at 1000 K, fast oscillations around finite polarization values coexist with slower changes in the polarization sign. The critical motion, therefore, possesses two components with different time scales. While one component is associated with oscillations about quasi-equilibrium

positions and give the phonon feature, the other corresponds to a relaxational motion between equiprobable positions. The mean frequency of the polarization reversals at 1000 K gives an inverse relaxation time ∼27 cm^{-1}, in close agreement with experiments [4]. The polarization reversals become less frequent upon cooling, which is related to the narrowing of the central component of $S(\mathbf{q} = 0, \omega)$. The time evolution of the supercell polarization shows a similar behavior with temperature. Polarization reversals were observed down to 350 K, a permanent polarization appearing in [1 1 1] direction at lower temperatures. The supression of the intermediate [1 0 0] and [1 1 0] polarized phases is due to the constraint of the rigid cubic lattice [5].

For KTaO$_3$, $S(\mathbf{q} = 0, \omega)$ displays a well-defined ferroelectric mode peak in a wide temperature range, as shown in Fig. 3. The temperature dependence of the frequencies at which the peaks are centered is in good agreement with inelastic neutron scattering data and with the SPA calculations [9]. By contrast, the temperature dependence of the peak width is opposite to the experimental observation [18]. This could be an evidence for a very flat energy minimum for the ferroelectric mode, instead of a double-well structure

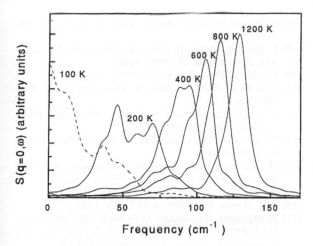

Fig. 3. Dynamical structure factor $S(\mathbf{q}=0, \omega)$ of KTaO$_3$ at several temperatures.

[10]. Then anharmonic perturbation theory would apply, which predicts a decrease of damping with temperature. The very large peak widths observed at the lower temperatures, and the central peak appearing at 100 K, can be considered spurious effects of the classical simulation, where the quantum zero point energy is not considered.

We proceed now to the simulations of KNbO$_3$ through its various phases. We start from 100 K and increase temperature by steps of 50 up to 1200 K. At each temperature we perform runs of 32 000 time steps after equilibration. The various phases are detected by analyzing the time evolution of each component of the total polarization, and its mean value over the whole run. We found the tetragonal and orthorhombic deformations to be essential for the stability of the corresponding phases. To this purpose we readjusted the parameters K_{OB} and $K_{OB,B}$ so that the experimental deformations were reproduced. Thus we end up with $K_{OB} = 66.00 \text{ eV/Å}^2$ and $K_{OB,B} = 300.0 \text{ eV/Å}^4$.

We obtain the mean values of polarization components at the various temperatures represented in Fig. 4. Up to ∼300 K we observe a [1 1 1] polarization, which corresponds to the rhombohedral phase, from ∼300 to ∼600 K one polarization component vanishes, thus indicating the orthorhombic phase, and from ∼600 to ∼730 K the tetragonal phase is distinguished through the vanishing of two polarization

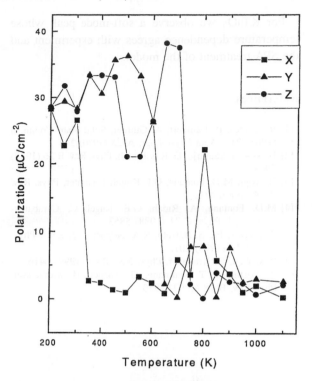

Fig. 4. Mean value of each polarization component for the supercell as a function of temperature in KNbO$_3$.

components. Between ∼730 and ∼830 K it is not clear if the system is in the tetragonal or the cubic phase, which is seen clearly above ∼830 K. These results need to be improved through longer simulation runs, smaller temperature steps and refinement of the potentials to achieve a quantitative description of the material.

4. Conclusions

The NOP model shows clearly a soft-mode to order–disorder crossover in the dynamics of the paraelectric–ferroelectric transition of KNbO$_3$, as suggested from experiments. The order–disorder behavior extends up to nearly the melting point, thus not allowing the direct experimental observation of the ferroelectric mode. This is distinguished in the simulation only beyond the melting point. The model allows also to describe the three ferroelectric phases of KNbO$_3$.

For KTaO$_3$ we observe a soft-mode peak whose temperature dependence agrees with experiment and the SPA treatment of the model.

References

[1] R. Comes, R. Lambert, A. Guinier, Solid State Comm. 6 (1968) 715; Acta Crystallogr. A 26 (1970) 244.

[2] H. Vogt, J. Sanjurjo, G. Rossbroich, Phys. Rev. B 26 (1982) 5904.

[3] H. Vogt, M.D. Fontana, G.E Kugel, P. Günter, Phys. Rev. B 34 (1986) 410.

[4] M.D. Fontana, A. Ridah, G.E. Kugel, C. Carabatos-Nedelec, J. Phys. C 21 (1988) 5853.

[5] S. Chaves, F.C.S. Barreto, R.A. Nogueira, B. Zêks, Phys. Rev. B 13 (1976) 207.

[6] R. Cohen, H. Krakauer, Phys. Rev. B 42 (1990) 6416.

[7] A. Postnikov, T. Neumann, G. Borstel, M. Methfessel, Phys Rev. B 48 (1993) 5910.

[8] R. Migoni, H. Bilz, D. Bäuerle, Phys. Rev. Lett. 37 (1976) 1155.

[9] C. Perry, R. Currat, H. Buhay, R. Migoni, W. Stirling, J. Axe, Phys. Rev. B 39 (1989) 8666.

[10] D.J. Singh, Phys. Rev. B 53 (1996) 176.

[11] G.E. Kugel, M.D. Fontana, W. Kress, Phys. Rev. B 35 (1987) 813.

[12] D. Khatib, R.L. Migoni, G.E. Kugel, L. Godefroy, J. Phys.: Condens. Matter 1 (1989) 9811.

[13] M. Sepliarsky, M.G. Stachiotti, R. Migoni, Phys. Rev. B 52 (1995) 4044.

[14] M. Sepliarsky, M.G. Stachiotti, R.L. Migoni, Phys. Rev. B 56 (1997) 566.

[15] D. Beeman, J. Comput. Phys. 20 (1976) 130.

[16] W.G. Hoover, Phys. Rev. A 31 (1985) 1695.

[17] M. Sepliarsky, R.L. Migoni, M.G. Stachiotti, Ferroelectrics 183 (1996) 105.

[18] H. Vogt, H. Uwe, Phys. Rev. B 29 (1984) 1030.

ELSEVIER

Computational Materials Science 10 (1998) 57–62

COMPUTATIONAL
MATERIALS
SCIENCE

Rotational dynamics in orientationally disordered K ClO$_4$

F. Affouard [a], Ph. Depondt [b],*

[a] *Laboratoire de Dynamique et Structure des Matériaux Moléculaires, CNRS ura 801,*
Université Lille 1, 59655 Villeneuve d'Ascq Cedex, France
[b] *Laboratoire des Milieux Désordonnés et Hétérogènes, CNRS ura 800, Université Pierre et Marie Curie,*
Casier 136, 75252 Paris Cedex 05, France

Abstract

A molecular dynamics simulation of a $8 \times 8 \times 8$ cells potassium perchlorate crystal in the orientationally disordered phase was carried out using a CM-5 parallel computer in order to study the individual and two-molecule rotational dynamics of the ClO$_4$ ions. Symmetry adapted rotator functions and the rotation matrix are used to overcome the difficulty of dealing with orientational variables. The crystal structure is fcc, with most probable T$_d$ symmetry molecular orientations, in agreement with experiments. The rotational motion turns out to be diffusive, but molecules which are far from the most probable orientations have greater rotational velocity. The relative orientations of two neighboring molecules last longer than that of a single molecule, showing coupled motion. Copyright © 1998 Elsevier Science B.V.

Keywords: Molecular dynamics simulation; Orientational disorder; Rotational dynamics

1. Introduction

Some crystals, constituted of molecules of so-called 'globular' shape, display a phase in which the molecular centers of mass are ordered in a usually fcc lattice while the orientations are dynamically disordered [1]. Detailed studies of the rotational dynamics are relatively rare (e.g. [2–5]) because orientational variables such as Euler angles are, by no means, easy to handle.

Potassium perchlorate K$^+$ClO$_4^-$ displays above approximately 310°C a cubic phase [6,7] in which the perchlorate ions are orientationally disordered, with maxima of the orientational probability density function in T$_d$ symmetry orientations, i.e. when the four three-fold axes of the tetrahedral ions are parallel to the four crystal three-fold axes [8] (Fig. 1).

A constant-pressure molecular dynamics simulation [9], of a sample of 2048 K ClO$_4$ units ($8 \times 8 \times 8$ cells) was carried out with rigid perchlorate molecules. The atom–atom interaction potentials are of the type: $\exp(-r^{-6})$, for short range interactions, and an Ewald sum is done for Coulomb interactions [10]. The structures of both the low and high temperature phases are well reproduced [11,12]. The simulation was done at $T = 500$ K.

2. Orientational variables

We use here two sets of tools: symmetry adapted rotator functions which have the convenient property

---

* Corresponding author. Tel.: (33) 1 44 27 42 33; fax: (33) 1 44 27 38 82; e-mail: dep@drp.jussieu.fr.

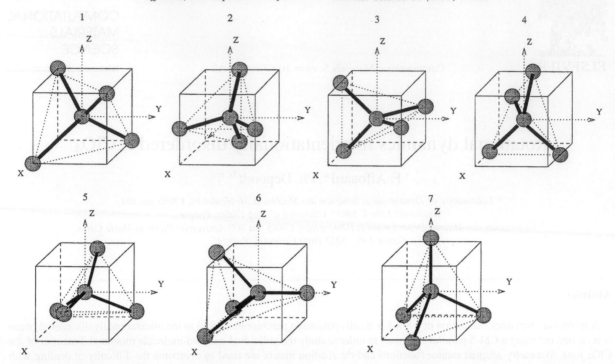

Fig. 1. Orientations for which functions $\mathcal{R}_l^m$, $l \in [1, 7]$, have extrema. Situation 1, called T_d, yields $\mathcal{R}_1^m = 1$: the four molecular three-fold axes are parallel to the crystal three-fold axes; two equivalent but physically distinct such orientations occur via a 90° rotation around a crystal four-fold axis. In situations 2, 3, 4 (functions $\mathcal{R}_2^m$, $\mathcal{R}_3^m$, $\mathcal{R}_4^m$) one molecular binary axis is parallel to a crystal four-fold axis, X for $l = 2$, Y for $l = 3$ and Z for $l = 3$, thus producing six D_{2d} orientations via a 90° rotation around the common symmetry axis. In situations 5, 6, 7 (functions $\mathcal{R}_5^m$, $\mathcal{R}_6^m$, $\mathcal{R}_7^m$), one-molecular three-fold axis is parallel to a crystal four-fold axis, X for $l = 5$, Y for $l = 6$ and Z for $l = 7$, the orientations of the three other molecular three-fold axes being irrelevant. In a crystal with cubic symmetry, functions $\mathcal{R}_{2-4}^m$ on the one hand, and functions $\mathcal{R}_{5-7}^m$ on the other are equivalent through symmetry.

that their extrema are in high symmetry orientations, and on the other hand, the instantaneous rotation matrix which yields for each molecule the rotational velocity ω and rotational axis $\mathbf{u}$.

Symmetry adapted rotator functions are defined in [4]: those that are of interest to us are given in Table 1 and Fig. 1. Function $\mathcal{R}_1^m$ is of special interest since its extrema are for T_d orientations.

The orientational matrix $\mathbf{M}_m(t)$ of molecule m at time t, obtained from the simulation, expresses in a reference frame attached to the crystal axes and the origin of which is the molecular center of mass, the coordinates $\mathbf{r}$ of a vector, given its coordinates $\mathbf{r}_0$ in the molecular frame: $\mathbf{r} = \mathbf{M}_m(t)\mathbf{r}_0$. We compute $\mathcal{R}_l^m(t)$, $l \in [1, 7]$, from the expressions in Table 1 and the single-molecule orientational correlation functions

$$\mathcal{C}_{ll'}^{I}(t) = \langle \mathcal{R}_l^m(t') \mathcal{R}_{l'}^m(t' + t) \rangle.$$

The two-molecule orientational correlation functions writes

$$\mathcal{C}_{ll'}^{II}(t) = \langle \mathcal{R}_l^m(t') \mathcal{R}_{l'}^{m'}(t' + t) \rangle,$$

where molecules m and m' are first neighbors in the fcc lattice.

The molecular rotational matrix from t to $t + \delta t$, $\mathbf{R}(t, \delta t)$, is defined by (subscript m omitted):

$$\mathbf{M}(t + \delta t) = \mathbf{R}(t, \delta t)\mathbf{M}(t).$$

The instantaneous rotation axis $\mathbf{u}$ is a unit vector that is unaffected by $\mathbf{R}(t, \delta t)$, thus

$$\mathbf{R}(t, \delta t)\mathbf{u} = \mathbf{u}.$$

The rotation angle $\omega \delta t$, for $\delta t = 0.05$ ps, is obtained by applying $\mathbf{R}(t, \delta t)$ to a vector that is perpendicular

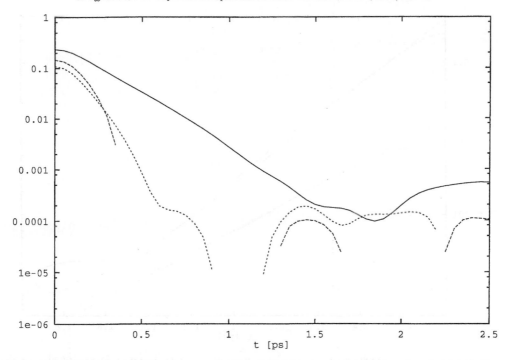

Fig. 2. Log-normal plot versus time of $C_{l,l}^{I}$ for $l = 1$ (continuous line), $l = 2, 3, 4$ (dashed line), $l = 5, 6, 7$ (dotted line).

Table 1
Expressions of the rotator functions $\mathcal{R}_l^m$ used in this papers

l	$\mathcal{R}_l^m$
1	$(3\sqrt{3}/4) \sum_{i=1,4} x_i y_i z_i$
2	$(3\sqrt{3}/8) \sum_{i=1,4} x_i (y_i^2 - z_i^2)$
3	$(3\sqrt{3}/8) \sum_{i=1,4} y_i (z_i^2 - x_i^2)$
4	$(3\sqrt{3}/8) \sum_{i=1,4} z_i (x_i^2 - y_i^2)$
5	$(3\sqrt{5}/40) \sum_{i=1,4} x_i (5x_i^2 - 3)$
6	$(3\sqrt{5}/40) \sum_{i=1,4} y_i (5y_i^2 - 3)$
7	$(3\sqrt{5}/40) \sum_{i=1,4} z_i (5z_i^2 - 3)$

Where x_i, y_i, z_i are the coordinates of the unit vector that is parallel to the ith molecular Cl–O bond of molecule m (subscript m being omitted for simplicity) in a reference frame attached to the crystal axes and the origin of which is the molecular center of mass.

to **u**. The rms ω_0 of ω is given by thermodynamics: $\frac{1}{2}I\omega_0^2 = \frac{1}{2}I\langle\omega^2\rangle = \frac{3}{2}k_bT$ where I is the moment of inertia of the perchlorate ion, k_b Boltzmann's constant, and T is the temperature. For a perchlorate ion at $T = 500\,\mathrm{K}$, $\omega_0 = 3.09\,\mathrm{rd/ps}$ (i.e. approximately $177°/\mathrm{ps}$).

3. Results

3.1. Single-molecule orientational correlation function

Fig. 2 shows the time dependence of $C_{ll}^I(t)$ for $l = 1$–7. The correlation function C_{11}^I has a larger initial value than the others since the rotator function $\mathcal{R}_1^m$ has extrema ± 1 when the molecule is in T_d orientation. The correlation function $C_{11}^I(t)$ follows a decay law $e^{-t/\tau}$ with a correlation time $\tau \simeq 0.2\,\mathrm{ps}$. For times greater than $1.5\,\mathrm{ps}$ only noise is obtained; no oscillation is present.

The functions C_{22}^I, C_{33}^I, C_{44}^I produce identical results, as they should. They decrease to 0 much faster than $C_{11}^I(t)$. The functions C_{55}^I, C_{66}^I, C_{77}^I are also identical since they are equivalent, with approximate exponential decay ($\tau \simeq 0.1\,\mathrm{ps}$).

The correlation functions $C_{ll'}^I(t)$ with $l' \neq l$ were also computed showing no visible correlation.

Rotational jump models of the type often used to interpret experimental results [2,3,5] seem to be

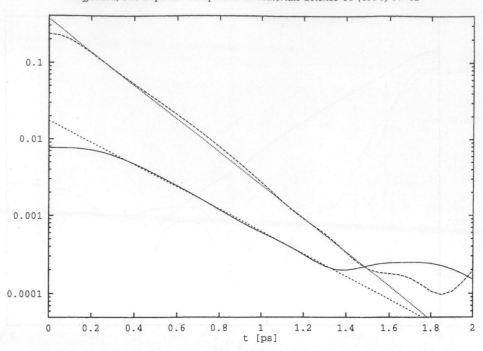

Fig. 3. Log-normal plot versus time of $|C_{l,l}^{II}|$ for $l = 1$ (continuous line), $C_{l,l}^{I}$, $l = 1$ (dashed line), $0.175 e^{-t/0.30}$ and $0.37 e^{-t/0.20}$ for comparison.

disqualified for potassium perchlorate since had the molecules undergone jumps, e.g. from one T_d orientation to the other, C_{11}^{I} would have shown some periodicity due to the repetition of the same simple motion. We may also discard rotational motions about a high symmetry axis, of the type suggested in [8]: e.g. a rotation from a T_d orientation to the other about a crystal four-fold axis passes necessarily through a D_{2d} orientation (cf. Fig. 1) and should yield a contribution for $l = 1$ with $l' = 2, 3, 4$, depending on whether the rotation axis is x, y or z, since the functions $\mathcal{R}_l^m$, $l = 2, 3, 4$, display maxima for D_{2d} orientations.

Rotational diffusion seems therefore to be the type of motion most compatible with our results.

3.2. Two-molecule orientational correlation function

The results for $C_{11}^{II}(t)$ are shown in Fig. 3 along with $C_{11}^{I}(t)$, the single molecule equivalent, for comparison ($C_{11}^{II}(t)$ is negative, which confirms the anti-

ferromagnetic correlation between the orientations of first neighbors [11]: when one molecule has an orientation of T_d symmetry, its neighbors tend to have the other T_d orientation). The time dependence between $t = 0.4$ and 1.3 ps is an exponential decay law with $\tau \simeq 0.30$ ps. The decay time to $\frac{1}{10}$th of the initial value is almost 1 ps instead of 0.55 ps for $C_{11}^{I}(t)$, showing that two-molecule dynamics is significantly slower than single-molecule motion.

In all situations for which $l' = l \neq 1$ and $l' \neq l$ no correlation was observed: this means that the above result holds for molecules in or close to T_d orientations only.

3.3. Rotation matrix

The distribution of angular velocities is found to follow Maxwell's distribution, and the distribution of the rotational axis $\mathbf{u}$ is isotropic, showing no favored rotational trajectory such as a rotation about a symmetry axis or a simple combination

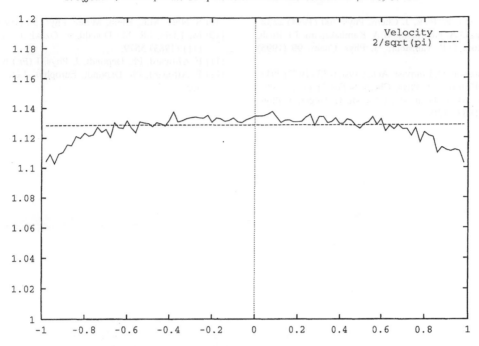

Fig. 4. Distribution of rotational velocity (in units ω_m), as a function of $\mathcal{R}_1^m$.

thereof. This confirms the lack of structure in $l \neq l'$ rotators.

Fig. 4 shows the distribution of angular velocities as a function of $\mathcal{R}_1^m$ compared with the average velocity $\tilde{\omega} = 1.1284\,\omega_0$

For $|\mathcal{R}_1^m| < 1$ (i.e. far from T_d orientations), the rotational velocity is larger than for $\mathcal{R}_1^m \simeq \pm 1$ (i.e. close to T_d).

4. Conclusion

We thus schematically consider two populations of molecules:

1. Molecules in T_d orientations which undergo correlated gear-type motion with their neighbors. The rotational velocity is slower than average.
2. Molecules not in T_d orientations with quicker individual rotational motion.

In the simple case of a one-dimensional random walk problem, the coupling of two 'walkers', discarding the details of the actual coupling, results in

effective motion slower than in the case of two independant walkers, since they will sometimes attempt to go in opposite directions and thus cancel out each other's movement. We suggest that this could be the explanation for the slower motion of T_d-oriented molecules in Fig. 4, the main result in this paper.

Acknowledgements

The authors gladly acknowledge extensive use of the CM-5 computer at the Centre National de Calcul Parallèle pour les Sciences de la Terre (Institut de Physique du Globe de Paris), on which the simulations were carried out.

References

[1] J.N. Sherwood, The Plastically Crystalline State, Wiley, New York, 1979.
[2] Ch. Steenbergen, L.A. de Graaf, Physica B 96 (1979) 1.
[3] R. Gerling, A. Hüller, J. Chem. Phys. 78 (1983) 446.

[4] W. Breymann, R. Pick, J. Chem. Phys. 100 (1994) 2232.

[5] P. Gehring, D.A. Neumann, W.A. Kamitakahara, J.J. Rush, Ph.E. Eaton, D.P. VanMeurs, J. Phys. Chem. 99 (1995) 4429.

[6] G.B. Johansson, O. Linqvist, Acta Cryst. B 33 (1977) 2918.

[7] C.W.F.T. Pistorius, J. Phys. Chem. Solids 31 (1970) 385.

[8] B. Denise, M. Debeau, Ph. Depondt, G. Heger, J. Phys. (Fr.) 49 (1988) 1203.

[9] S. Nosé, M.L. Klein, Molec. Phys. 50 (1983) 1055.

[10] M. Klein, I.R. Mc Donald, Y. Ozaki, J. Chem. Phys. 79 (11) (1983) 5579.

[11] F. Affouard, Ph. Depondt, J. Phys. I (Fr.) 6 (1996) 1.

[12] F. Affouard, Ph. Depondt, Europhys. Lett. 33 (5) (1996) 365.

ELSEVIER

COMPUTATIONAL
MATERIALS
SCIENCE

Computational Materials Science 10 (1998) 63–66

First-principles simulations of organic compounds: Solid CO_2 under pressure

François Gygi [1]

Institut Romand de Recherche Numérique en Physique des Matériaux (IRRMA), CH-1015 Lausanne, Switzerland

Abstract

We present ab initio simulations of organic crystals carried out within Density Functional Theory using plane waves and norm-conserving pseudopotentials. Applications to the study of two crystalline phases of carbon dioxide under pressure are discussed and compared to recent experimental results. The use of gradient corrected exchange-correlation functionals is found to be important for a good description of intermolecular interactions. Copyright © 1998 Elsevier Science B.V.

Keywords: Carbon dioxide; Organic crystals; Ab initio simulations

1. Introduction

Ab initio molecular dynamics simulations (for a review, see e.g. [1]) of organic molecules are currently the subject of intense research. The application of traditional ab initio simulation methods to organic compounds is made difficult by the presence in those compounds of first-row elements, which require a high numerical accuracy, and by the small energy scales typically involved in organic chemical reactions. The weak intermolecular interactions characterizing organic compounds (hydrogen bonds, electrostatic or van der Waals interactions) also emphasize the need for accurate density functionals, such as the recently proposed Generalized Gradient functionals [2].

Even the simplest organic crystals have only been recently described in a satisfactory manner using first-

principles methods. The widely used local density approximation (LDA) of Density Functional Theory is generally considered to be insufficient for a proper account of the the cohesive energy of most solids. In spite of this, the LDA often yields accurate results for geometries (bond lengths and cell dimensions), in particular for covalent crystals. The situation is much less satisfactory in weakly bonded crystals, in which the subtle nature of the bonding is not adequately modelled by the LDA.

In the important example of ice, intermolecular interactions are dominated by hydrogen bonds, and an improved treatment of exchange and correlation was found to be essential for a quantitative description of bonding. After a number of years of numerical experimentation on various gradient-corrected density functionals, some functionals that give a satisfactory description of hydrogen bonds have now been identified [3–5].

Some other organic crystals, such as benzene, are characterized by even weaker intermolecular

[1] Tel.: +41 21 693 5327; fax: +41 21 693 6655; e-mail: gygi@irrma.epfl.ch.

interactions, such as electrostatic multipolar and van der Waals interactions. A correct description of such systems within Density Functional Theory is even more challenging, and the very existence of density functionals applicable to these situations is far from established. In recent calculations on crystalline benzene, Meijer and Sprik [6] have demonstrated that the BLYP functional (see e.g. [7]), which correctly accounts for hydrogen bonds in water, is unable to represent correctly the cohesive properties of the crystal. Van der Waals interactions were found to contribute significantly to the cohesive energy of benzene. The cohesive energy calculated in [6] by Meijer and Sprik was *negative*, indicating that the crystalline form of benzene is unstable, in clear contradiction with experiment. All density functionals in use to date fail to represent adequately the long-range features of the van der Waals interaction.

In this paper, we consider the case of crystalline carbon dioxide, which has recently been the subject of refined experimental studies. Until the end of the 1980s, the cubic (Pa3) phase (see Fig. 1) was the only known solid phase of CO_2. In 1988, on the basis of empirical Monte-Carlo simulations, Etters and Kuchta [8,9] predicted the existence of an orthorhombic (Cmca) phase (Fig. 1) above 11 GPa. This new phase was subsequently observed experimentally by

Aoki et al. [10], who also noted the coexistence of the Pa3 and the Cmca phase at 11.8 GPa. The large quadrupolar moment of the CO_2 molecule is expected to play an important role in the long-range part of intermolecular interactions. Quadrupole–quadrupole interactions have been invoked to explain the stability of the Pa3 phase at unusually high temperatures and pressures, as compared to e.g. N_2 and CO, which exhibit other crystalline phases [8]. Ab initio calculations of the high pressure phases of CO_2 constitute a stringent test of density functionals which will help choosing the most accurate functionals for calculations of organic molecular crystals. First-principles calculations are also the tool of choice for the exploration of other crystalline phases that might be present at high pressures.

2. Computational approach

We have carried out ab initio calculations of solid carbon dioxide for both the Pa3 and the Cmca phases at various volumes and cell dimensions. We use norm-conserving pseudopotentials [11] and represent electronic wave functions using plane waves with an energy cutoff of 70 Ry. Test calculations performed on isolated molecules have confirmed the adequacy

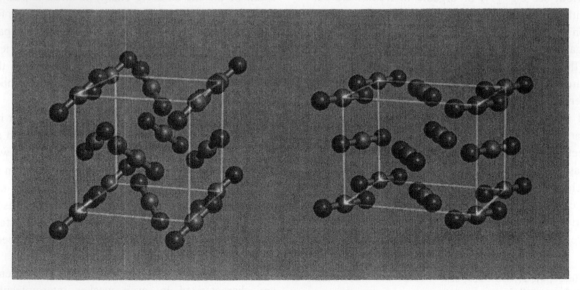

Fig. 1. Unit cells of the cubic Pa3 phase (left), and orthorhombic Cmca phase (right) of CO_2.

of these parameters to the description of CO_2. We use both the LDA and the Generalized Gradient Approximation of Perdew, Burke and Ernzerhof (PBE) [4] for the description of exchange and correlation. The stress tensor was calculated using a modification of the technique described by Focher [12]. Equilibrium cell dimensions under pressure were calculated using a second order damped molecular dynamics approach. All atomic positions were fully relaxed in all calculations.

3. Pa3 phase

A cubic cell of 12 atoms (4 molecules) was used to represent the Pa3 unit cell. Electronic wave functions were calculated at the Γ point only of the cubic Brillouin zone, which is justified by the weak interaction expected to take place between molecules. Calculations including 96 atoms (32 molecules) were carried out to check the accuracy of the Brillouin zone sampling. Atomic coordinates were first relaxed in a cubic cell of fixed lattice parameter (5.29 Å), resulting in a pressure of -2.08 GPa. The negative sign of the pressure indicates that the crystal is under tensile stress at that volume. The cell was subsequently relaxed under zero applied pressure. After some oscillations, the lattice constant reached an equilibrium value of 5.15 Å. The same relaxation procedure was repeated at a pressure of 7.64 GPa, corresponding to the experimental conditions of the observation of the Pa3 phase in [10]. The zero-pressure experimental lattice parameter (5.54 Å) was taken from [8]. Our zero-pressure LDA result underestimates the lattice constant of the Pa3 cell by $\simeq 7\%$. This result confirms the general tendency of the LDA to underestimate lattice constants in most crystals, although the amplitude of the error in the present case is larger than is usual in covalent crystals. The PBE value (5.65 Å) is much closer to the experimental value and overestimates it by a small amount (2%). This is again in agreement with the trend observed for various generalized gradient functionals when applied to solid state calculations [13]. The same conclusions hold at a pressure of 7.64 GPa, where the LDA lattice constant (4.84 Å) underestimates the ex-

perimental value (5.06 Å, from [10]) by 4%. The corresponding PBE value (5.13 Å) is again in much closer agreement with experiment and overestimates it by only 1.4%. We note that for both the LDA and the PBE approximations, the relative error in the lattice constant decreases under pressure. This can be attributed to the increase of the electronic density in the intermolecular regions as molecules get in closer contact. The approximations made in density functionals are most critical at low electronic densities, so that DFT errors are expected to decrease as pressure increases.

4. Cmca phase

Similarly, 12 atoms (4 molecules) were used to represent the Cmca orthorhombic phase of CO_2. The cell parameters were relaxed under 11.8 GPa, which corresponds to the conditions of observation of the Cmca phase in [10]. Relaxed LDA cell parameters at 11.8 GPa are $a = 4.47$, $b = 4.10$, $c = 5.91$ Å, to be compared to the experimental values of $a = 4.33$, $b = 4.66$, $c = 5.96$ Å. Using the PBE approximation, we find $a = 4.76$, $b = 4.21$, $c = 6.01$ Å, in closer agreement with experiment. We note that, for both the LDA and PBE calculations, the ordering of a and b is reversed with respect to experimental values. This behaviour is unexpected, in particular in the case of LDA, for which a systematic underestimation of lattice parameters is the rule. If confirmed by more complete calculations, this result raises the interesting possibility of having to reinterpret the powder diffraction patterns of [10] with a and b reversed. Calculations of diffraction intensities are under way to explore this hypothesis. In the experimental results of [10], the CO_2 molecular axes are tilted away from the c crystallographic axis by $\phi = 52°$. Our calculated value of 54° (for both LDA and PBE) is in excellent agreement with the experimental one.

5. Pa3-Cmca transition

The enthalpies of the Pa3 and Cmca crystals at zero temperature were extracted from the above results in order to estimate the transition pressure. LDA results yield a transition pressure of 9.3 GPa, whereas the PBE value is 2.2 GPa. It should be noted that these numbers are most sensitive to the parameters of the calculation, and might change, in particular if a better Brillouin zone sampling is used. Experimentally, the exact transition pressure is not known accurately, due to a large hysteresis [8]. In 1988, Olijnyk et al. [14] reported a transition from Pa3 to a then unidentified phase around 10 GPa. Liu et al. [15] observed indications of phase transitions as low as 2.3 GPa. Aoki et al. [10] report a transition pressure of 9.9 GPa, associated with a small volume change of $0.16 \pm 0.9\%$. Calculated transition pressures show qualitative agreement with experiment. However, the large uncertainties in the calculated and measured values preclude a more detailed discussion of the transition pressure.

6. Conclusions

Calculated properties of cubic and orthorhombic carbon dioxide show remarkable agreement with experimental results, in spite of the difficulties that could be expected from the use of Density Functional Theory for weakly interacting molecules. While the LDA exhibits the well-known tendency to predict small unit cell volumes, the PBE approximation of exchange and correlation essentially corrects this error and restores an excellent agreement with experiment. The importance of quadrupole–quadrupole intermolecular interactions – which are treated exactly within our calculations – might explain this good agreement. The calculated ordering of lattice parameters in the Cmca

phase disagree with experimental values, which suggests a possible reinterpretation of experimental data. While computed transition pressures show qualitative agreement with experimentally measured values, more work along these lines is needed in order to attain a higher precision. Our results confirm the excellent behaviour of the recently introduced PBE functional already observed by other authors, and make it a very interesting candidate among density functionals for organic molecular crystals simulations.

References

[1] G. Galli and A. Pasquarello, in: Computer Simulation in Chemical Physics, eds. M.P. Allen and D.J. Tildesley, NATO ASI Series C, Vol. 397 (Kluwer, Dordrecht, 1993) p. 261.
[2] J.P. Perdew, in: Electronic Structure of Solids '91, eds. P. Ziesche and H. Eschrig (Akademie Verlag, Berlin, 1991) p. 11.
[3] M. Sprik, J. Hütter and M. Parrinello, J. Chem. Phys. 105 (1996) 1142.
[4] J.P. Perdew, K. Burke and M. Ernzerhof, Phys. Rev. Lett. 77 (1996) 3865.
[5] D.R. Hamann, Phys. Rev. B 55 (1997) R10157.
[6] E.J. Meijer and M. Sprik, J. Chem. Phys. 105 (1996) 8684.
[7] B.G. Johnson, P.M.W. Gill and J.A. Pople, J. Chem. Phys. 98 (1993) 5612.
[8] B. Kuchta and R.D. Etters, Phys. Rev. B 38 (1988) 6265.
[9] R.D. Etters and B. Kuchta, J. Chem. Phys. 90 (1989) 4537.
[10] K. Aoki et al. Science 263 (1994) 356.
[11] D.R. Hamann, Phys. Rev. B 40 (1989) 2980.
[12] P. Focher, G.L. Chiarotti and M. Bernasconi, Europhys. Lett. 26 (1994) 345.
[13] A. Dal Corso, A. Pasquarello and R. Car, Phys. Rev. 53 (1996) 1180.
[14] H. Olijnyk, H. Däufer and H.-J. Jodl, J. Chem. Phys. 88 (1988) 4204.
[15] L. Liu, Nature (London) 303 (1983) 508.

ELSEVIER

Computational Materials Science 10 (1998) 67–74

COMPUTATIONAL
MATERIALS
SCIENCE

On the effect of quench rate on the structure of amorphous carbon

V. Rosato [a,*], M. Celino [a], L. Colombo [b]

[a] *ENEA, Ente per le Nuove Tecnologie, l'Energia e l'Ambiente, High Performance Computing & Networking Project,
C.R. Casaccia, PO Box 2400, 00100 Roma AD, Italy*
[b] *Istituto Nazionale per la Fisica della Materia (INFM) and Dipartimento di Scienza dei Materiali,
Università degli Studi di Milano, Via Emanueli 15, 20126 Milano, Italy*

Abstract

We have simulated, via tight binding molecular dynamics (TBMD), the process of the quench from a melt of an atomic scale system of carbon. We have correlated the local properties of the resulting structure to the quench rate used to bring the liquid phase beyond the glass transition temperature. Results have been analyzed also in terms of the hamiltonian model used to describe the simulated system. In this respect, amorphous structures generated via tight binding and ab initio molecular dynamics have been compared. Results indicate that quench rates as slow as 10^{14} K/s produce the onset of an increasingly high fraction of threefold coordinated sites in the structure. Moreover, it has been put in evidence the tendency of the tight binding approach to favor threefold coordinated sites with respect to fourfold coordinated, even in the fast quench rates domain. Copyright © 1998 Elsevier Science B.V.

1. Introduction

The allotropy of carbon leads this element to form intrinsically different disordered (amorphous) structures (a-C hereafter) whose properties may be, in principle, correlated to several parameters, e.g.: the mass density and the preparation route followed to induce the glass transition (i.e. the quench rate in the case where the classical technique of rapidly quenching a melt is used). A two-fold behavior has been clearly evidenced by comparing simulation and experimental data: low density ($\rho \leq 2\,\mathrm{g/cm^3}$) samples present a predominant fraction of sp^2 coordinated atoms while in high density ones ($\rho \geq 2\mathrm{g/cm^3}$), in

turn, the majority of atoms assumes an sp^3 coordination [1,2]. While in the former case the situation is less controversial, it has been recently argued [3] that the coordination of simulated high density a-C systems largely depends on the preparation technique (i.e. on the quench rate used to bring the system below the liquid-to-amorphous transition [3,4]). These results seem to indicate that the quench rate could control the fraction of sp^2/sp^3 bonded sites in high density ($\rho > 3\,\mathrm{g/cm^3}$) samples [3]. The results described in [3] have been obtained by first-principle molecular dynamics, and therefore they are limited by the huge computational effort. In fact, the quench rate domain there adopted can be hardly compared to typical experiments. We believe that this feature can affect the study of structural trends in a-C as function of sample preparation, especially when the

---

* Corresponding author. Tel.: +39 6 3048 4825; fax: +39 6 3048 4729; e-mail: rosato@casaccia.enea.it.

small system size (just 64 atoms) can produce sizable statistical errors.

In the present work, we made use of a semi-empirical scheme, namely tight binding molecular dynamics (TBMD), which allowed for large-scale calculations. TBMD is the simplest scheme within a quantum mechanical framework for describing the energetics and allotropy of the C–C bond. It provides an intuitive conceptual outline for the physical understanding of amorphous materials, still keeping low the computational workload.

The main aim of this work is to assess the effect of the quench rate on the microscopic scale structure of the amorphous phase produced by the melt-quench technique. Furthermore, since we are aware of the semi-empirical character of the tight binding model, we comment about the role played by the system (i.e. the type of hamiltonian used) by comparing TBMD data to first-principle results.

2. Computation results

We have simulated a carbon model system containing $N = 216$ atoms (initially arranged in the diamond structure), described by means of a tight binding representation as in [5]. All simulations have been performed at constant volume so as to keep the system density constant. This requirement has been imposed by the need to compare our results with data at the same number density, as the structural properties of a-C systems are a sensitive function of this quantity [2,4,6]. The time-step was as small as 0.7 fs for an optimal integration of equations of motion by velocity-Verlet algorithm.

The system has been firstly heated up to $T = 6500$ K at constant volume, with a thermal cycle of 2100 time-steps. The large temperature has been chosen in order to prepare a reliable liquid structure within typical simulation times, by making use of

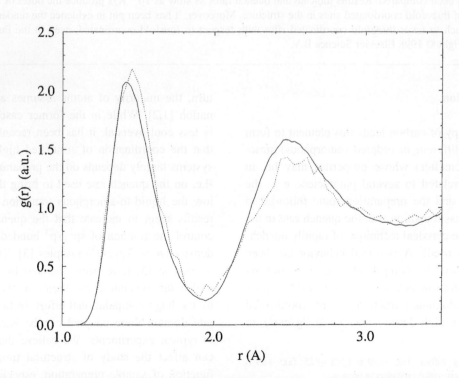

Fig. 1. $g(r)$ of the liquid at $T = 5000$ K. Full line: TBMD. Dashed line: first-principle results from [3,7].

Table 1
Local atomic coordination of the TBMD liquid and the CP liquid at equal density ($\rho = 3\,\text{g/cm}^3$) and $T = 5000\,\text{K}$ (the n_α values are in percent)

	n_1	n_2	n_3	n_4
TB liquid	2.3	14.4	73.6	9.7
CP liquid	3.1	10.9	62.5	23.4

the fast diffusion dynamics allowed in such temperature range. The system has been then relaxed at $T = 5000\,\text{K}$ and aged, at that temperature, for 1100 time-steps. Fig. 1 reports the comparison between the $g(r)$ evaluated upon equilibration of the liquid structure in the TBMD formalism with that pertaining to the liquid structure obtained by ab initio molecular dynamics in the Car–Parrinello formalism (CP hereafter) (liquid sample of 64 atoms, at a similar temperature and density [3,7,8]). The agreement is good, standing for a reliable TBMD model of l-C, as far as global structural properties are concerned.

A further property used to compare the two different liquid structures, related to the local atomic coordination, is the fraction of sites having a given coordination n_α.

As it is clear from Fig. 1 and Table 1, the two liquids show large similarities, although the sp^3-coordination of the CP liquid results to be much larger than that experienced in the TBMD liquid. This results from

a natural tendency of the present TBMD hamiltonian to favor a graphitic-like coordination with respect to a tetrahedral one, as it has been discussed in [2,4,9]. However, the TBMD liquid at $T = 5000\,\text{K}$ displays a diffusion coefficient $D = 7.14 \times 10^{-5}\,\text{cm}^2/\text{s}$ which is in agreement with that measured on the corresponding CP liquid ($D = 7.1 \times 10^{-5}\,\text{cm}^2/\text{s}$ [3]).

The TBMD liquid has been subsequently quenched down to $T = 300\,\text{K}$ with four different quench rates q_x ($x = 1, 4$) spanning more than two decades (e.g. $q_1 = 8.8 \times 10^{16}\,\text{K/s}$, $q_2 = 8.8 \times 10^{15}\,\text{K/s}$, $q_3 = 1.7 \times 10^{15}\,\text{K/s}$, $q_4 = 4.4 \times 10^{14}\,\text{K/s}$). The system's behavior has been monitored both in the as-quenched configurations and after a further relaxation period of 2500 time-steps spent at constant temperature, where the systems totally relax their structures.

The properties of the structures resulting upon the different quench rates have been compared to each other and to those resulting upon application of the same quench rates to the CP liquid structure. Table 2 reports the comparison among all the fractions of n-fold coordinated sites in the two classes of systems. The as-quenched U_q and the relaxed U_r energies are reported. In the same table, we have also reported the results of the site-coordination distribution issued by CP quench dynamics at different quench rates [3,7].

Table 2
Relevant thermodynamic and structural data for the considered amorphous structures

		Quench rate (K/s)	U_q (eV/atom)	U_r (eV/atom)	n_2	n_3	n_4
TBMD dynamics	From ab initio MD liquid	$q_1 = 8.8 \times 10^{16}$	−7.470	−7.639	1.6	69.0	29.3
		$q_6 = 2.6 \times 10^{16}$	−7.598	−7.602	4.6	50.7	44.6
		$q_2 = 8.8 \times 10^{15}$	−7.629	−7.636	4.7	50.0	46.3
		$q_3 = 1.7 \times 10^{15}$	−7.623	−7.628	4.7	63.5	32.8
		$q_4 = 4.4 \times 10^{14}$	−7.907	−7.915	—	100.0	—
	From TBMD liquid	$q_1 = 8.8 \times 10^{16}$	−7.418	−7.622	5.2	70.4	23.6
		$q_2 = 8.8 \times 10^{15}$	−7.683	−7.699	2.8	71.3	25.9
		$q_3 = 1.7 \times 10^{15}$	−7.795	−7.794	1.4	70.4	28.2
		$q_4 = 4.4 \times 10^{14}$	−7.877	−7.876	3.7	94.4	1.8
CP dynamics [7]	From ab initio MD liquid	$q_5 = 6.5 \times 10^{19}$			—	35.0	65.0
		$q_6 = 2.6 \times 10^{16}$			—	34.0	66.0
		$q_7 = 9.4 \times 10^{15}$			—	40.0	58.0

The rows comprised the "TBMD dynamics" contain the results of the quench of liquid structure obtained via a TBMD formalism, starting from ab initio (CP) liquid (The liquid structure hereby used have been kindly provided by M. Bernasconi [11] or from TBMD liquid. The rows comprised in the "CP dynamics" simply reports the data as collected from [3,7]. U_q and U_r are the energy per particle of the as-quenched and the aged amorphous structures, respectively. The n_α values are in percent.

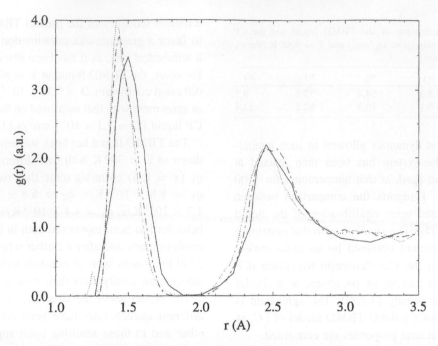

Fig. 2. $g(r)$ of amorphous systems generated, with different quench rates, from the TBMD liquid compared to the experimental correlation [3]. Full line = experimental result, dotted line = q_1, dashed line = q_3.

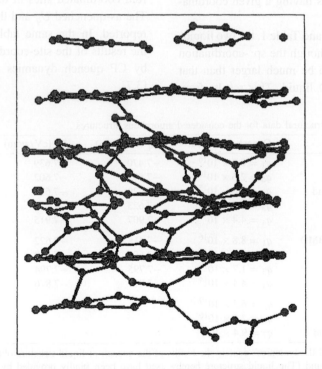

Fig. 3. Atomic-scale structure of the a-C issued upon quenching of the TBMD liquid with the quench rate q_4.

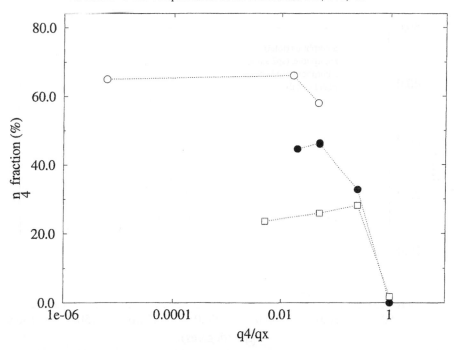

Fig. 4. Fraction of fourfold coordinated sites in the different amorphous configurations obtained by quenching the ab initio CP liquid configurations with TBMD (full circles) and ab initio CP dynamics (empty circles [3]). The same for the quench of TBMD liquid configurations performed by TBMD (empty squares).

In Fig. 2 we reported the radial distribution functions $g(r)$ for the a-C systems issued upon the different quench rates $q_1 - q_3$ performed starting from a l-C structure produced by TBMD. The slowest quench-rate q_4, in turn, produced a structure with a fraction of sp^2-sites as large as 100% (see Table 2). This is consistent with an atomic-scale structure which exhibits the typical graphite sheets, although highly distorted (Fig. 3).

The same procedure has been used to quench l-C structures produced in the CP formalism [3]. This results in the formation of several amorphous structures (issued by identical quench-rates as those used for TBMD liquid) whose local coordination has been compared in Fig. 4. In this figure, the fraction of sp^3-coordinated sites has been plotted as a function of the applied quench-rate (normalized to the slowest one, i.e. q_4). If we focus on the results obtained by quenching the CP liquid (by using both CP and TBMD) (circles in Fig. 4) a natural trend toward systems graphitization is evident. However, it must be

noticed that the quench of the TBMD l-C produces amorphous structures with a much smaller sp^3 fraction. This could be due to a poor reproduction of the liquid state by TBMD and by a natural tendency of this representation to underestimate the system tendency to keep sp^3-like coordination at high densities. This artifact could be related to the fitting procedure of the TB hamiltonian, which is performed by accounting for the energetic reproduction of graphite sheets. No information on the energetics of C–C interactions among different graphite sheets is thus contained in the formalism; for this reason, the sp^2 coordination is favored for energetic and entropic arguments. The raw reproduction of the energy of the sp^2 sites does not allow to counterbalance the entropic tendency which favors the creation of planar graphitic sheets in the system.

A further quantity which has been evaluated is the distribution of bond angles present in the amorphous structures. This quantity has been averaged over the "aging" period (see above) lasting 2500 time-steps. In

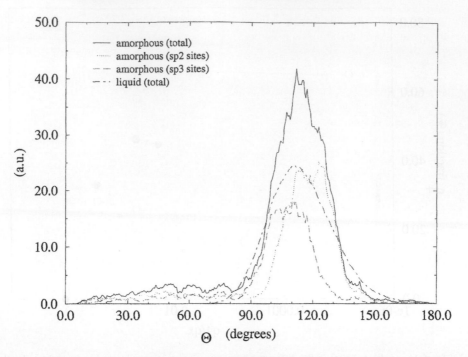

Fig. 5. Total bond-angle distributions for the TBMD liquid at $T = 5000\,\mathrm{K}$ and for the relaxed amorphous structure issued upon the quench rate $q_3 = 1.7 \times 10^{15}$ at $T = 300\,\mathrm{K}$. The two different contributions to the total bond-angle distribution of the amorphous structure, coming from the sp^2- and the sp^3-coordinated sites, have been also reported.

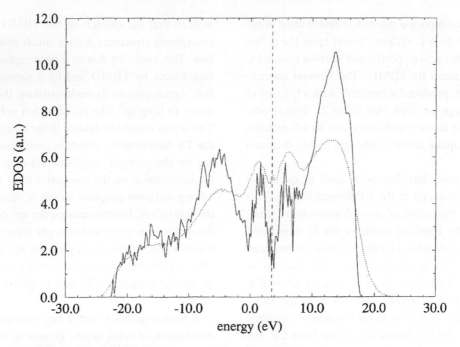

Fig. 6. EDOS for the liquid at $T = 5000\,\mathrm{K}$ (dotted line) and for the relaxed amorphous structure issued upon the quench rate $q_3 = 1.7 \times 10^{15}$ (full line) at $T = 300\,\mathrm{K}$. The dashed line represents the HOMO energy.

Fig. 5 we have reported the bond-angle distributions evaluated on both the liquid and the amorphous structure, for the a-C structures obtained with the quench rate q_3. The bond-angle distribution of the liquid structure shows a single-peak behavior, with average values located at $\theta = 110°$. For the amorphous structure, a much broader distribution has been found, with an apparent single-peak behavior (average located at $\theta = 113°$). However, the main peak has been found to be composed by two different peaks, the first arising from the sp^2-bonded sites (centered at $\theta \sim 120°$), the second coming from sp^3-bonded sites (centered at $\theta \sim 110°$). The spread in the bond-angle distribution is consistent with the presence of local strains which determine the occurrence of distorted configurations with respect to fully relaxed sp^2 or tetrahedral structures [7]. Results for the amorphous agree with the previous estimate of Frauenheim et al. [2] and slightly exceed those reported by Wang [10] for a-C structures at nearly the same density. The discrepancy could have been produced by the residual presence of strain in the relaxed a-C structure of [10].

The TB formalism allows to perform a detailed investigation of the electronic density of states (EDOS) of the systems by cumulating, during the simulation time, the energy of the system eigenvalues. The EDOS has been evaluated on both the liquid and the amorphous structure (Fig. 6). The liquid state shows the known metallic behavior while, in the amorphous state, a pseudo-gap is present at the Fermi energy.

3. Conclusions

Aim of this work has been to characterize amorphous structures produced by quenching, using TBMD with different quench rates, carbon liquid structures prepared by using both TBMD and CP formalism.

The obtained results indicated that:

(1) The average structural quantities for liquid and amorphous structures (namely the radial distribution function) are insensitive of the used formalism (TB or CP).

(2) The local arrangement characterizing the amorphous structures (i.e. the fraction of sp^2- and sp^3- coordinated sites) largely depends on the used quench rate. The large sp^3-coordination (typical of CP liquids) is kept for quench rates as fast as 10^{15} K/s. Slower quench rates, in turn, produce a progressive onset of regions of increasingly large content of sp^2- coordinated sites. This tendency is also visible in the quenching of TB liquids, although the latter are characterized by a smaller sp^2 fraction. The outcome structure presents a clear graphitic structure, with irregular hexagonal planes weakly bonded to each other (Fig. 3).

(3) The bond-angle distribution of the amorphous structure presents a broader structure than that of the liquid phase. This is related to the presence of distorted regions associated to large strains. Two different contributions (characterized by slightly different average values) are present. These are produced the bond-angle distributions typical of sp^2- and sp^3-sites.

Acknowledgements

The authors kindly acknowledge M. Bernasconi (Milan) for illuminating discussions and for having provided raw data of CP liquid and amorphous structures. One of us (LC) acknowledges computational support by the Swiss Center for Scientific Computing (CSCS/SCSC) Manno (CH), and partial financial support by the INFM Advanced Research Project CLASS.

References

[1] K.W. Gilkes, P.H. Gaskell, J. Robertson, Phys. Rev. B 51 (1995) 12303.
[2] Th. Frauenheim, P. Blaudeck, U. Stephan, G. Jungnickel, Phys. Rev. B 48 (1993) 4823.
[3] N.A. Marks, D.R. McKenzie, B.A. Pailthorpe, M. Bernasconi, M. Parrinello, Phys. Rev. B 54 (1996) 9703.
[4] S. Nosé, F. Yonezawa, Solid State Commun. 56 (1985) 1005.
[5] L. Colombo, in: D. Stauffer (Ed.), Annual Review of Computational Physics IV, World Scientific, Singapore, 1996, p. 147.
[6] G. Galli, R.M. Martin, R. Car, M. Parrinello, Phys. Rev. Lett. 62 (1989) 555.

[7] N.A. Marks, D.R. McKenzie, B.A. Pailthorpe, M. Bernasconi, M. Parrinello, Phys. Rev. Lett. 76 (1996) 768.

[8] G. Galli, R.M. Martin, R. Car, M. Parrinello, Science 250 (1990) 1547.

[9] V. Rosato, J. Lascovich, A. Santoni, L. Colombo, in preparation.

[10] C.Z. Wang, S.Y. Qiu, K.M. Ho, Comput. Mat. Sci. 7 (1997) 315.

ELSEVIER

Computational Materials Science 10 (1998) 75–79

COMPUTATIONAL
MATERIALS
SCIENCE

Tight-binding molecular dynamics simulations of the vibration properties of α-titanium

Jacob L. Gavartin *, David J. Bacon

Department of Materials Science and Engineering, The University of Liverpool, Liverpool L69 3BX, UK

Abstract

Phonon thermal anomalies in α-titanium are studied by means of tight-binding microcanonical molecular dynamics simulations. The frequencies of the zone centre [0 0 0 1]LO and TO phonons, and the [0 0 0 1]TA phonon at $q = \frac{1}{3}$ are determined at different temperatures via the power spectrum of the autocorrelation function associated with the corresponding projections of atomic velocities. It is shown that even at very low temperatures the effects of anharmonicity in vibrational properties are strong and dependent on both wave propagation direction and frequency. In particular, the frequencies of the TO and TA phonons are shown to decrease while the frequency of the [0 0 0 1]LO phonon increases with crystal temperature, in agreement with experiment. Copyright © 1998 Elsevier Science B.V.

Keywords: Titanium; Phonons; Molecular dynamics; Tight-binding

1. Introduction

The group IVB transition metals (titanium, zirconium and hafnium) possess a number of rather remarkable features in their properties attributed to strong anomalies in their phonon spectra. These include crystalline polymorphism [1], a high temperature saturation of the electrical conductivity [2], rapid increase of the constant-pressure heat capacity at high temperature [3], and anisotropy in diffusion and thermal properties. A vibrational anomaly in the low temperature hexagonal phase (α-phase) of both titanium and zirconium manifests itself through the relatively large decrease of the frequencies of all modes, except the

[0 0 0 1]LO branch, with increasing temperature [4,5]. The zone centre [0 0 0 1]LO phonon, in contrast, softens significantly with decrease of temperature and at room temperature exhibits a dip.

There are two effects causing thermal frequency shifts that need to be distinguished: (i) shifts due to lattice thermal expansion, and (ii) explicit frequency shifts due to the anharmonicity of the modes. The thermal lattice expansion causes a generally small albeit not uniform shift of the whole vibrational spectrum towards lower frequencies and can be estimated using the thermodynamic Grüneisen parameter, or microscopically, within the quasi-harmonic type approximations. However, the frequency shifts observed for all but the [0 0 0 1]LO phonon branches in α-Ti [5] and α-Zr [4] are much larger (on average seven times larger in Zr) than those estimated by the Grüneisen parameter. Moreover, the

* Corresponding author. Tel.: +44 0 151 794 4662; fax: +44 0 151 794 4675. On leave from the Chemical Physics Institute, The University of Latvia, Riga, Latvia.

0927-0256/98/$19.00 Copyright © 1998 Elsevier Science B.V. All rights reserved
PII S0927-0256(97)00186-9

[0 0 0 1]LO phonon is found to behave inversely, as noted above.

The experimental observations have also demonstrated that the frequency shifts are more pronounced for the long-wave phonons, suggesting that the long-range interatomic forces mainly determined by the electronic response on atomic motion are responsible. Thus, a study of phonon anomalies in these transition metals requires a calculational scheme that can account for effects of electron–phonon interaction and, at the same time, can be implemented as an O(N) method and used for long timescale MD simulations to study the anharmonic effects. A semiempirical tight-binding (TB) approach seems to be a compromise meeting these demands. In the present paper we adopt this approach for a study of α-Ti zone-centre phonons – [0 0 0 1]LO, and the TO$_\parallel$ (polarisation [2 $\bar{1}$ $\bar{1}$ 0]) and TO$_\perp$ (polarisation [0 1 $\bar{1}$ 0]), and a [0 0 0 1]TA mode at the $q = \frac{1}{3}$, the latter being important in a discussion of α-phase stability. Although the method would allow for it, we do not consider in this work the effects of thermal lattice expansion and electronic temperature, so phonon frequency dependence on temperature can be studied explicitly. Furthermore, we present simulation results for relatively low temperatures (50–400 K), where the harmonic approximation is generally expected to be satisfactory, and demonstrate that this is not the case for α-Ti.

2. Calculation procedure

We employ the orthogonal TB microcanonical molecular dynamics approach within the Goodwin–Skinner–Pettifor parametrisation scheme [6] implemented in the OXON code [7]. The details of the fitting of the titanium TB parameters (d-electrons only) are given in [8]. We mention here only that the parameters were fitted to reproduce structural, cohesion and elastic constants of the α-phase and were not specifically fitted to phonon properties. For the effective diagonalisation of the TB Hamiltonian a real-space recursion method implemented in the bond-order potential formalism [7] has been used. Thirteen moments in the expansion of the electron

density of states were used to ensure sufficient accuracy in evaluation of energy and forces.

For an analysis of mode frequencies we adopt the following strategy. First we define the projected velocity $\mathbf{v}_q(t)$ associated with an arbitrary displacement vector $\mathbf{a}_q$ as

$$\mathbf{v}_q(t) = \sum_{i=1}^{3N} a_{iq} \mathbf{v}_i(t), \tag{1}$$

where q indicates some generalised coordinate, index i numerates both atoms and their Cartesian coordinates, N is the number of atoms in the system, $\mathbf{v}_i(t)$ is the velocity of the coordinate i (x, y, or z component of some atom), and the summation is taken over all $3N$ coordinates of the system. Then, a vibrational density $Z_q(\omega)$ corresponding to the projected velocity $\mathbf{v}_q(t)$ is given by a Fourier transform of the corresponding velocity autocorrelation function

$$z_q(t) = \langle \mathbf{v}_q(0)\mathbf{v}_q(t)\rangle_0,$$

where the averaging is assumed over all the initial velocities $\mathbf{v}_q(0)$. We note that if $3N$ vector $\mathbf{a}_q$ corresponds to a polarisation of a particular normal mode Q, and the crystal is purely harmonic, then the corresponding vibrational density of states becomes simply a delta function: $Z_Q(\omega) = \delta(\omega - \omega_Q)$. Furthermore, under a constant volume-constraint the mode frequency ω_Q will not depend on temperature. The anharmonicity of the mode Q will be seen as both a shift of the peak position ω_Q and peak broadening due to phonon–phonon scattering.

The (NVE) molecular dynamics simulations were performed for a periodic system of 96 Ti atoms which contained six basal planes. The phonons in question correspond to a relative displacement of the basal planes. For instance, the zone centre TA phonon can be seen as a cooperative motion involving even and odd planes vibrating in antiphase in the direction parallel to themselves. Choosing the y-axis to be collinear with the [0 1 $\bar{1}$ 0] crystallographic direction, the corresponding coefficients a_{iQ} in Eq. (1) for $Q = TA_\perp$ are equal to 1 if i indicates a y component for an atom belonging to any odd plane, to -1 if i is a y component for an atom belonging to any even plane, and to 0 otherwise. Expansion coefficients a_{iQ} for other phonons can be found similarly.

3. Results and discussion

The vibrational spectra associated with the $[0\,0\,0\,1]$TA $(q = \frac{1}{3})$ phonon at different temperatures are shown in Fig. 1. We note that all the curves are normalised on their peak value, so the relative peak intensities are not meaningful. It is seen that the power spectrum of the autocorrelation function associated with the projected velocity produces a very sharp peak with position shifting with temperature quite markedly. The same applies to the zone-centre phonons although the corresponding peaks are generally broader and sometimes splitted. The peak positions as a function of temperature for all phonons studied are depicted in Fig. 2. It is seen that in the temperature interval up to 400 K the frequencies of the $[0\,0\,0\,1]$TA and zone-centre TO phonons decrease by about 13.5% and 3.5%, respectively. The frequency of

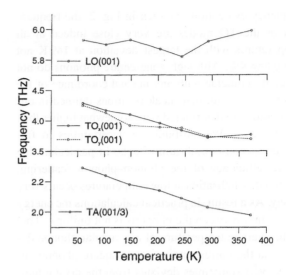

Fig. 2. The position of the peak of the vibrational density of states as a function of temperature for the zone-centre $[0\,0\,0\,1]$LO and basal plane TO phonons (polarisation: $x = [2\,\bar{1}\,\bar{1}\,0]$, $y = [0\,1\,\bar{1}\,0]$), and the $[0\,0\,0\,1]$TA $(q = \frac{1}{3})$ phonon.

the $[0\,0\,0\,1]$LO phonon exhibits a dip at about 250 K and increases with temperature above this point although not as dramatically as for the other phonons examined. It has been argued [5] that the inverse thermal behaviour of the $[0\,0\,0\,1]$LO phonon in α-Ti and Zr is associated with the electron entropy factor. Although this question requires further investigation, we note that in our modelling electron temperature was kept constant (the same for MD runs at all temperatures) and artificially high, $kT = 0.1$ eV, so the entropy effect on frequency shifts was neglected.

In spite of the fact that the TB parameters [8] used in the modelling were not specifically fit to the phonon properties, the frequencies obtained are found to agree well with experiment [5]. For a comparison, we quote the results of neutron scattering measurements on α-Ti available at the temperatures 295, 773 and 1054 K: $[0\,0\,0\,1]$TA $(q = 0.3)$ – 2.03, 1.83, and 1.68 THz respectively; TO – 4.1, 3.77, 3.53 THz, and $[0\,0\,0\,1]$LO phonon – 5.54, 6.14 and 6.33 THz.

We note that the basal plane phonons $TO_{\parallel}$ and $TO_{\perp}$ must degenerate at the zone centre, so we use their independent calculation for a test of accuracy in the

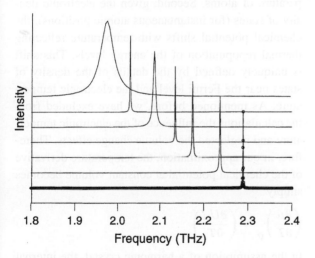

Fig. 1. The vibrational density of states associated with the $[0\,0\,0\,1]$TA $(q = \frac{1}{3})$ phonon calculated as a power spectrum of the corresponding projected velocity autocorrelation function. Different curves in ascending order correspond to the average temparatures: 59, 102, 148, 203, 243, 292, 371 K, and were obtained from MD runs of length 39.8, 23.8, 15.8, 25.26, 7.9, 12.39, 2.6 ps, respectively. All curves are normalised on their peak value.

frequency evaluation. As seen in Fig. 2, the frequencies of the TO modes are very close indeed at all temperatures with the largest deviation at 149 K not exceeding 4%. Although zone-centre phonons do not generally coincide with any normal coordinate of the finite size system, their peak positions defined in our calculation are determined with somewhat higher than 4% accuracy. We attribute the uncertainties in frequency to non-equilibrium effects. In particular, due to low efficiency of the phonon–phonon scattering, phonon thermalisation at low temperatures occurs very slowly. As a result, in practical calculations the energy distribution between the phonons is not uniform. That is, the average temperature of a particular phonon defined via the corresponding mean square phonon velocity $\langle \mathbf{v}_Q^2 \rangle$ sometimes deviates from the crystal temperature. This applies especially to the high energy phonons, since there is a very small number of them present in the system at low temperatures, and the long wavelength phonons, which are suppressed by the size of the system. Thus, although frequencies can be determined with a very high resolution, the corresponding temperature is not always well-defined. Further investigation of this problem is presently in progress. Another consequence of long phonon lifetime is the fact that the corresponding velocity autocorrelation functions exhibit a very weak decay. This causes a well-known effect of the artificial broadening in the respective frequency spectra for the MD runs of finite time. Rigorous analysis of the peak broadening at low temperature requires either very long MD simulations or some extrapolation procedures for the velocity autocorrelation function, but this goes beyond the scope of this paper.

In order to study the influence of the electron–phonon interaction on the crystal anharmonicity, we have calculated the average electronic chemical potential as a function of temperature (Fig. 3). The almost linear dependency obtained can be rationalised as follows. We recall that a chemical potential can be thought of as an electronic Gibbs free energy calculated per electron:

$$\mu = u_{el} - s_{el}T + \frac{P_{el}}{n},$$

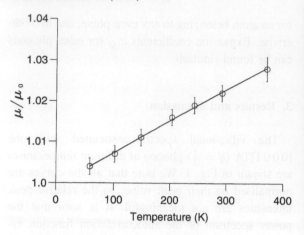

Fig. 3. The average electronic chemical potential as a function of temperature. The values are normalised on the Fermi-level value at $T = 0\,K$.

where u_{el}, s_{el} are the corresponding energy and entropy per electron, P_{el} is the electronic pressure, n is the electron concentration, and T is the temperature of the crystal. Thus, the chemical potential depends on temperature in two ways. First, being explicitly dependent on atomic coordinates, its average value changes with the level of atomic disorder, that is, with the temperature of atoms. Second, given the electronic density of states (for instantaneous atomic positions), the chemical potential shifts with temperature reflecting thermal repopulation of the energy levels. This shift is uniquely defined by the details of the density of states near the Fermi level and the electronic temperature. As mentioned before, we have excluded from the calculations the influence of the electronic temperature and neglected the volume change effects. Therefore, in our approximation, the temperature derivative of the chemical potential at constant volume becomes simply

$$\left(\frac{\partial \mu}{\partial T} \right)_V \approx \left(\frac{\partial U_{el}}{\partial T} \right)_V.$$

In the assumption of a harmonic crystal, the internal energy, U_{el}, is a quadratic form of atomic displacements. Furthermore, the mean square displacements $\langle Q^2 \rangle$ associated with normal modes must be proportional to the system's temperature. Therefore, neglecting the electronic entropy, the chemical potential in the harmonic approximation must be a linear function of temperature. It is interesting to note that the temperature dependence of the chemical potential does not deviate significantly from the linear law, as seen in Fig. 3. We conclude, therefore, that the anharmonicity seen in the chemical potential is not significant below 400 K, indicating that on the average, α-Ti at these temperatures behaves approximately harmonically. This is also confirmed by our calculations of the total atomic mean square displacement as a function of temperature.

Finally, the combination of the TB molecular dynamics with the method based on the analysis of the power spectrum of the projected velocities associated with particular modes, proved to be a sufficiently accurate and efficient tool in the study of phonon anomalies. Further analysis is required on the effects of the electronic temperature and thermal broadening of the vibration spectra. Work on the modelling of phase transformations and the effects of the presence of intrinsic defects in Ti is also currently under way.

Acknowledgements

The work was funded by the UK Engineering and Physical Science Research Council. JLG also acknowledges the Latvian Council of Science for support under the project no. 93.0270, and the Commission of the European Communities for support under contract ERB CIPDCT 940008. We are very grateful to A. Horsfield, D. Nguyen-Man, D. Pettifor and A. Girshick for software support and many stimulating discussions.

References

[1] V.V. Aksenenkov, V.D. Blank, B.A. Kul'nickiy, E.I. Estrin, Phys. Met. Metall. 69 (1990) 143.

[2] J.H Mooij, Phys. Stat. Solids A 17 (1973) 521.

[3] Samsonov (Ed.), Handbook of the Physicochemical Properties of the Elements, Oldbourne, London, 1968.

[4] C. Stassis, J. Zarestky, D. Arch, O.D. McMasters, B.N. Harmon, Phys. Rev. B 18 (1978) 2632.

[5] C. Stassis, D. Arch, B.N. Harmon, N. Wakabayashi, Phys. Rev. B 19 (1979) 181.

[6] L. Goodwin, A.J. Skinner, D.G. Pettifor, Europhys. Lett. 9 (1989) 701.

[7] A.P. Horsfield, A.M. Bratkovsky, M. Fearn, D.G. Pettifor, M. Aoki, Phys. Rev. B53 (1996) 12 694; A.P. Horsfield, A.M. Bratkovsky, D.G. Pettifor, M. Aoki, Phys. Rev. B 53 (1996) 1656.

[8] A. Girschik, A.M. Bratkovsky, D.G. Pettifor, V. Vitek, Phil. Mag. A (1997), to be published.

ELSEVIER

Computational Materials Science 10 (1998) 80–87

COMPUTATIONAL
MATERIALS
SCIENCE

Molecular dynamics in semiconductor physics

Sigeo Ihara *, Satoshi Itoh

Central Research Laboratory, Hitachi Ltd., Kokubunji, Tokyo 185, Japan

Abstract

Molecular dynamics, which is a powerful method for studying both structural and dynamic properties of condensed matter, is employed as a "microscope for the motion of atoms". Information about the motion of individual atoms that is difficult to obtain experimentally can be obtained using this method. Parallel computers with tera-flop speed and tera-byte memory will make it possible to perform hundreds of million particle simulations using empirical potential. Moreover, the macroscopic phenomena analyzed by continuum theory can be obtained by coarse-graining the atomic-level information provided by this approach. When considering the problem of data analysis, examples in semiconductor physics such as the implantation process are provided. Copyright © 1998 Elsevier Science B.V.

Keywords: Molecular dynamics; Parallel computer; Semiconductor; Cluster; Ion implantation

1. Introduction

The on-going scaling down of device technology will result in MOS devices with gate length below 1000 Å by the turn of this century. Recent experimental work shows that prototype devices of deep-submicron size can function as a single transistor [1]. One of the important problems in device physics, then, is how deep the sub-micron MOSFET will go. However, there are many limitations posing roadblocks to further reduction in scale [2]: fundamental dielectric problems, quantum phenomena and reliability limitations. The problems in the process for MOSFET include controlling impurity diffusion, developing a suitable method of ion implantation, ensuring the stability of the surface structure, and addressing the coverage problem.

Finding new materials with a high dielectric constant or a low dielectric constant is considered crucial from the point of view of materials physics. Ion implantation with low energy as well as high energy is very important from the point of view of device fabrication, and will be discussed in detail in a later section. Implantation is also important in other areas such as designing devices, since it is possible to obtain suitable or optimized device characteristics by changing the impurity concentration in the channel region. Reliability poses another important problem after fabricating deep-submicron devices. The next-generation of devices such as DRAM (dynamic random access memory) will be affected by the configuration of a small number of atomic defects in the crystal absorbing or releasing conduction electrons. The effect of the motion of atoms will be especially conspicuous in such high-density storage systems, so the reliability of chips and devices should also be analyzed from a molecular or atomic level.

---

* Corresponding author. Tel.: +81-423-23-1111, fax: +81-423-27-7742.

To understand such phenomena at the molecular/ atomic scale, simulations are an inevitable tool since they act as an efficient "microscope" in time and space. The reasons for doing atomic level simulation using various methods include: (1) to support the results of experiments, (2) to provide deeper analysis of the theory, (3) to predict the possibility of unknown phenomenon or matter, and (4) to discover proper conditions for processing. Another important benefit is that presenting simulation results using computer-generated graphics provides a precise image for the phenomena that experimental analysis cannot provide.

In the past, an atomic-level approach has been regarded only as an academic tool, but this situation is changing drastically. Semiconductor engineers are beginning to view an atomic-level approach as inevitable or even as the last hope. Shrinking device sizes and the increase in computational speed achieved by parallel computers make it practical to perform atomic level (or atomistic) simulations in the range of sub-0.1 micron MOSFET channel regions.

2. Simulation methods

Molecular dynamics [3,4] is perhaps the most powerful method for studying both structural and dynamic properties among all the various atomistic simulation methods. The simulation procedure of molecular dynamics is as follows. First, we determine the materials target and the simulation system, the boundary conditions, the number of atoms inside the system, and the frame of reference. Second, we determine the species of the atoms, the atomic structure whose coordinates are obtained by estimation or from a database.

Third, we determine the inter-atomic forces. If the method uses a quantum mechanical treatment, the inter-atomic forces between atoms are determined by solving the equation for many electrons. Because electrons are the origin of the forces between atoms, electronic states are calculated for the specific coordinates of the atoms. The empirical inter-atomic potential or forces of atoms, which is a function of the distance of atoms and the angle between them,

are sometimes employed to simplify the complex calculation of quantum mechanics.

Fourth, by assuming that the nuclei obey Newton's equations of motion, we calculate the new velocities and the coordinates of all atoms, i.e., the evolution of the atomic system (nuclei) from the forces acting on each atom. These procedures are very cumbersome, so we must rely on massively parallel computers. By repeating these procedures, we determine the velocities and coordinates of all atoms for each time step in the period of interest.

Finally, by averaging the number of atoms and time steps, we obtain the physical quantities, including thermodynamic properties such as latent heat and dynamic properties such as the diffusivity. Thus, molecular dynamics is sometimes called a method of computational statistical mechanics.

Based on these steps, molecular dynamics methods have been employed as a "microscope for the motion of atoms". Information about the atomic scale motions of individual atoms that are not detected by experiments can be obtained by this method. We can, in principle, link the macroscopic character of the materials with smearing or the coarse-graining of the evolution of individual atoms. Thus, hydrodynamics information can be derived from microscopic (and particle simulation) methods. Molecular dynamics methods are also employed to confirm the stability of hypothetical atomic structures obtained (purely) by imagination. Moreover, atomic-scale simulation methods are quite suitable for parallel computation. For example, molecular dynamics codes can often be parallelized using simple domain decomposition methods, since in many cases, the force calculation is well localized in space.

As a typical example, we consider the application of molecular dynamics to the ion implantation process in semiconductors [4–11]. This approach is useful for investigating the collision mechanism, because the time evolution of many particles (i.e., positions and velocities) are obtained simultaneously without assuming a specific collision mechanism. This is in contrast to the well-known Monte Carlo simulation method for implantation used by such systems as TRIM [12] and MARLOWE [13], where the binary collision mechanism is assumed to occur sequentially, which holds only in a high energy regime. However, for the collision cascade triggered by a single ion implantation in a low energy regime, or cluster implantation where many particles are involved naturally, its applicability is not clear, since in these regimes, the collision of one particle with the other occurs simultaneously.

3. Application to semiconductor physics

3.1. Recent trends

For sub-0.1μm MOSFET [14], it is said that ultra-shallow doping with a depth of less than 30 nm is required. In order to create low-leakage junctions for future high-speed and scaled-down device structures, low-energy implantation is necessary for preparing very shallow junctions constituting the source-channel and drain-channel regions of an MOS device. This extends the ion implantation technique towards lower energies of keV order. Unfortunately, standard ion-implantation technology, such as B^+ or BF_2^+ implantation, is confronted with several difficulties in this situation, including the inherent channeling effect, anomalous diffusion due to extraneous damage, and lateral diffusion under the mask [2].

A molecular or cluster soft landing technique is a promising solution [15] to reducing the kinetic energy per particle, since it reduces the effective impact energy more per constituting atom. The approach is bounded only by the technical limit of lowering the accelerating voltage in the ion-implanter. (For general account of implantation, see for example [16].)

Seeking low-energy and high-dosage boron ion implantation technology, Goto et al. [17] and Takeuchi et al. [18] recently used a decaborane, $B_{10}H_{14}$ molecule (with a plus charge) as an implanted ionized particle. Since $B_{10}H_{14}$ contains 10 boron atoms, it can be considered to be implanted with about a one-tenth lower effective acceleration energy and a 10-times higher effective beam current compared with those using B^+. However, neither the impact mechanics of cluster collisions on the Si(1 0 0) surface, the

influence of the cluster size, nor the number of induced defects are clear, despite a large amount of past work on surface bombardment of clusters [4,9,10].

The effect of cluster impacts on surfaces has previously been studied using molecular dynamics by many researchers [4]. The effect of silicon clusters on Si(1 0 0) surfaces [11], and Ar cluster bombardment on Si(1 0 0) [9,10] are especially relevant examples. Insepov and Yamada [9,10] calculated Ar_n ($n = 55 \sim 200$) clusters ionized with energy of 10–100 eV/cluster in two dimensions. To investigate the film deposition process, Gilmer and Roland [11] performed three-dimensional simulation of the impacts of 50 silicon clusters with energy of 0.5 keV on a Si (1 0 0) surface. However, their interest was crystal growth using molecular beam clusters, which has no direct relevance to ion implantation. To the authors' knowledge, the effect of silicon cluster impacts on Si(1 0 0) with energy of about 5 keV, which is a characteristic of the range of energies being used to implant dopants for semiconductor devices, has not been pursued. Our aim is to study in three-dimensional space using molecular dynamics the change in the implantation process resulting from use of clusters of atoms rather than a single atom (or ion) in the energy range of 5 keV.

3.2. Simulation system

The computational cell contains 108 000 Si atoms, with the base being a square 16 nm on an edge and the height being 8.3 nm. For the initial position of atoms in the substrate, we prepared the truncated ideal surface of silicon, i.e., we choose a perfect diamond structure as an initial position, with the (1 0 0) surface at the top. The Maxwellian velocity distribution was chosen as the starting velocities of particles. The periodic boundary conditions were applied for all directions, but the atoms at the bottom two layers were held fixed in their initial position, and all other atoms were allowed to move freely. However, the velocities of atoms are not controlled to monitor the temperature distribution of the system. In our calculation, the EVN ensemble molecular dynamics method with time reversible leap-frog time integration was used.

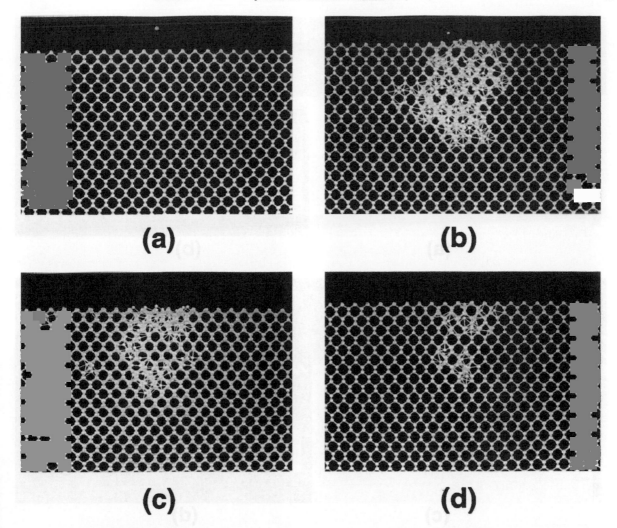

Fig. 1. Si–Si implantation with 1 keV. Initial state (a), $t = 119\,\mathrm{fs}$ (b), $t = 810\,\mathrm{fs}$ (c), and 2860 fs (d). Ball and stick model representation for three dimensional silicon cascade.

By thermalization using molecular dynamics, we obtained the reconstructed surface with dimers, which is the same as the previously obtained dimerized surface using the Stillinger–Weber potential [19].

The number of atoms in the incident cluster N_c was set to 1, 8, 20, and 90. Hereafter, we refer to a 'cluster' as a cluster with more than one atom to distinguish it from the single atom (ion) case. For each cluster or single atom, the incident kinetic energy is 5.7 keV. In most of the cases, each cluster only has translational velocity perpendicular to the surface: Our modeling implies that all particles in a cluster have the same velocity and only translational motion perpendicular to the surface.

3.3. Results for single particle impact

Before going into detail on cluster implantation, single ion implantation is discussed to clarify the changes in the physics of the impact. Diaz de la Rubia and Gilmer [5] studied the single ion (Si) bombardment of 5 keV on Si(1 0 0) surfaces using molecular dynamics employing Stillinger–Weber potential. They demonstrated the primary knock-on

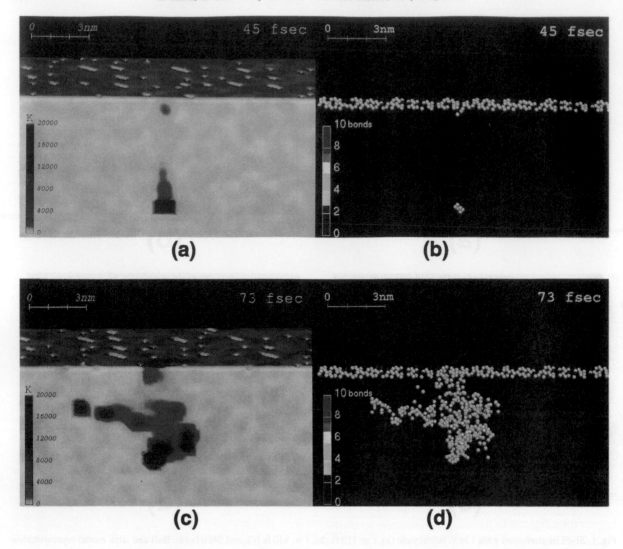

Fig. 2. Single particle implantation. Channeling process, temperature distribution (a) and defective particle image (b). Perfect collision process, temperature distribution (c) and defective particle image (d).

atom mechanism using single Si implantation. Although the amorphous region created by knock-on contains several hundred displaced atoms in the early stage, knock-on atoms result in few isolated Frenkel pairs, which finally collapse into clusters of interstitial or vacancies. They further suggested that these defect clusters are primarily responsible for the enhanced diffusion of dopant atoms commonly observed in low energy ion implanted

silicon. These views are consistent with experimental results.

Our results for single ion implantation, which are consistent with previous studies, are shown in Figs. 1 and 2. Fig. 1 shows the time evolution of the region of the collision cascade induced by a single particle. The annealing effect that results from the increasing temperature of the crystal, i.e. from the energy transferred from the incident particle, allows atoms to move

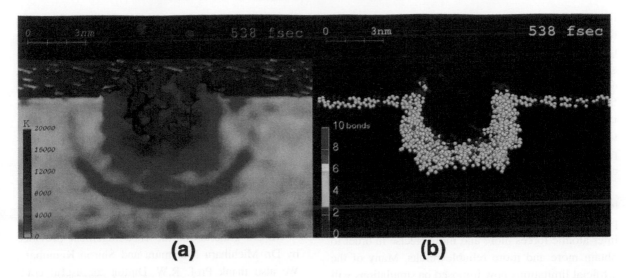

Fig. 3. Cluster implantation. Cluster with 90 silicon particle, (a) temperature distribution and (b) defective particle image.

to stable positions, thus the damaged region recovers with time. The temperature distribution obtained by smearing the particle velocity shows that the energy transfer is very small for the channeling case, Fig. 2(a). The particle image depicted for the defect, defined as an atom not having the standard four coordinates, shows that there is no damage, Fig. 2(b). While the energy of the incident particle is transferred completely to the substrate, indicating that the impact is localized and not symmetrical, there are many high temperature regions in the early period after the collision, Fig. 2(c). In this case, many defects are induced, but they finally will recrystallize with only a small number of Frenkel as indicated in Fig. 1(d).

3.4. Results for cluster impact case

The most prominent difference of cluster impacts from that of a single ion is the global influence of the impact on the surface. The damage of a cluster ion is isotropic compared with the single ion implantation case, where the collision is highly anisotropic. This follows because in the latter, there are channeling effects and the knock-on process in which the primary knock on atoms induce a large number of successive highly random collisions. In contrast, a cluster impact implies that the cross sectional area of the cluster is larger, so that the impact of a group of atoms with almost the same velocities will cause collective motion of particles. As we can see, the cluster impact causes large scale or collective phenomena such as shock waves. This is clearly seen in Fig. 3 of the temperature distribution of impact at 538 fs after the collision: the high temperature region moves with velocity of 8.43 km/s, which corresponds to the longitudinal sound velocity of silicon, without creating defects (see Fig. 3(b)). Our results also show that cluster size changes the nature of the impact; with a cluster size of more than 20 atoms, we can observe the shock wave. Our result shows that the impact energy is transferred partly by the shock wave or macroscopic vibrations, leading to suppression of the channeling effect. Thus, cluster implantation may not induce defects in deep regions, which would make it suitable for device design.

4. Future

Using a network environment, it is possible to perform materials simulations much faster than current technology allows. In this case, databases, electrical libraries, and massively parallel machines are connected by a network, which makes it easy to start up the simulations. The huge computational power in such a network also enables us to perform molecular dynamics simulations from a more fundamental and firm quantum mechanical basis. Furthermore, it makes it feasible to handle larger and larger system sizes. In these situations, storing the large amounts of data is a problem, so reproduction of data on demand will become more important.

Progress in hardware and algorithms is the main factor driving improvements in high performance computing. In the future, it will be possible to make the inter-atomic forces more and more precise in order to obtain more and more reliable results. Many of the artificial limitations now imposed on simulations will also be removed.

Another new direction is the development of a middleware framework that will enable the simulation scale to be increased by combining two different simulation programs with automatic data transfer. Such a framework will provide a practical use for molecular dynamics since it will be crucial to use the entire model hierarchy, ranging from quantum mechanics to hydrodynamics. From this point of view, molecular dynamics is the method that links the quantum and hydrodynamics regions. One possible example is the combination of molecular dynamics with a conventional process simulator that uses a continuum method. By combining these two methods, a wider region can be simulated with higher precision.

Another example is the combination of a particle Monte Carlo simulation for high energy particle collisions with molecular dynamics. By combining these two alternative methods, we can calculate high energy ion implantation much more precisely than those obtained by independent use of both methods. One of the other examples is the combination of the lattice gas Monte Carlo simulation for diffusion with molecular dynamics in order to simulate diffusion phenomena at low temperatures. By combining these two methods, we can calculate diffusion phenomena that cannot be obtained by molecular dynamics alone. Finally, parallel computers also make it possible to execute, for example, the ion implantation simulation with different

initial conditions independently on each processing unit. The goal here would be to obtain statistical data that is comparable with experimental results.

In conclusion, the frontier of future electronics, computers, multimedia, discovery of science and technology, CAD (Computer Aided Design) tools for materials, device/process, and biological sciences, will be opened through the use of atomic-level simulations such as molecular dynamics.

Acknowledgements

We are grateful for the encouragement provided by Dr. Michiharu Nakamura and Shiroo Kamohara. We also thank Prof. R.W. Dutton for useful comments on the early stages of our work. We also thank Prof. Richard Schlichting for his careful reading of the manuscript.

References

[1] R.W. Dutton, E.C. Kan, S. Onga, T. Okada, Extended Abstract of the 1996 International Conference on Solid State Devices and Materials, Yokohama, 1996, p. 7.
[2] G. Kamarions, P. Felix, J. Phys. D 29 (1996) 487.
[3] W.G. Hoover, Molecular Dynamics, Springer, Berlin, 1986; Wm.G. Hoover, Computational Statistical Mechanics, Elsevier, Amsterdam, 1991.
[4] R. Smith (Ed.), Atomic and Ion Collisions in Solids and at Surfaces, Cambridge University Press, Cambridge, 1997, and references cited there in.
[5] T.D. de la Rubia, G. Gilmer, Phys. Rev. Lett. 74 (1995) 2507.
[6] L.A. Marques, M.-J. Caturla, T.D. de la Rubia, G.H. Gilmer, J. Appl. Phys. 80 (1996) 6160.
[7] M. Jaraiz, G.H. Gilmer, D.M. Stock, T.D. de la Rubia, Nucl. Instr. and Meth. B 102 (1995) 180.
[8] M. Jaraiz, G.H. Gilmer, J.M. Poate, T.D. de la Rubia, Appl. Phys. Lett. 68 (1996) 409.
[9] Z. Insepov, I. Yamada, Nucl. Instr. and Meth. B 99 (1995) 248.
[10] Z. Insepov, I. Yamada, Nucl. Instr. and Meth. B 112 (1996) 16.
[11] G.H. Gilmer, C. Roland, Radiation Effects and Defects in Solids 130 (1994) 321.
[12] J.F. Ziegler, J.P. Biersack, U. Littmark, The stopping and range of ions in solids, in: J.F. Ziegler (Ed.), The Stopping and Range of Ions in Mater, vol. 2, Pergamon, New York, 1985.
[13] K.M. Klein, C. Park, Al F. Tash, IEEE Trans. Electron Device 39 (1992) 1614.

[14] S. Matsumoto, Extended Abstract of the 1996 International conference on Solid State Devices and Materials, Yokohama, 1996, p. 121.

[15] A.W. Kleyn, Science 275 (1997) 1440.

[16] E. Rimini, Ion Implantation: Basic to Device Fabrication, Kluwer Academic Publisher, Boston, 1995.

[17] K. Goto, J. Matsuo, T. Sugii, H. Minakato, I. Yamada, T. Hisatsugu, Techincal Digest of International Electron Device Meeting, 1996, San Diego, CA, IEEE, New York, 1996, p. 435.

[18] D. Takeuchi, N. Shimada, J. Matsuo, I. Yamada, Nucl. Instrum. Methods Phys. Res. B 121 (1997) 345.

[19] F.H. Stillinger, T.A. Weber, Phys. Rev. B 31 (1985) 5262.

COMPUTATIONAL
MATERIALS
SCIENCE

Computational Materials Science 10 (1998) 88–93

Quantum effects on phase transitions in high-pressure ice

Magali Benoit [a,*], Dominik Marx [b], Michele Parrinello [b]

[a] *Laboratoire des Verres-cc069, Université Montpellier II, Place E. Bataillon 34095 Montpellier, Cedex 05, France*
[b] *Max-Planck Institut für Festkörperforschung, Heisenbergstr. 1, 70569 Stuttgart, Germany*

Abstract

The H_2O ice phases VIII, VII, and X as well as their phase transformations are studied theoretically at 100 K as a function of pressure up to about 100 GPa. A combination of ab initio electronic structure calculations within the framework of density functional theory and the path integral representation of the nuclei is used. This allows the effects of thermal and quantum mechanical fluctuations on the properties of ice at high compression to be assessed separately and also in conjunction. Pronounced quantum effects are uncovered and different mechanisms are found to be at work at the antiferroelectric to paraelectric transition and the symmetrization transition. Copyright © 1998 Elsevier Science B.V.

Keywords: High-pressure ice; Phase transition; Quantum effects; Ab initio calculations

1. Introduction

The pressure–temperature $(P–T)$ phase diagram of H_2O ice is extremely rich and complex [1]. The various solid forms of ice have been, and still are, the subject of intense activities in experimental physics [1–8]. In the last decade, advances in high-pressure techniques permitted the in situ investigation of solid materials under extreme conditions such as very high pressures (up to the Megabar regime, i.e. ~ 100 GPa) using diamond anvil cells [2–8]. Above about 2 GPa, the phase diagram of ice is made of two molecular high density solid forms termed VII and VIII, which exist at low pressure above and below about 270 K and have very closed structures. In ice VII, the oxygen atoms form a body centered cubic (bcc) lattice, and the structure consists of two interpenetrating lattices of

cubic ice I_c, the dipole moments of the H_2O molecules being orientationally disordered [9–11]. In ice VIII, the molecules in the two sublattices have opposite dipole moments, leading to an antiferroelectric structure which has a slight tetragonal distortion from the cubic ice VII structure [3,10,12]; see Fig. 1 for a sketch of phases VIII. The interesting $P–T$ transition line between these two phases has been experimentally investigated [4] and shows a significant H/D isotopic effect on the transition pressure in the low temperature limit (about 60 GPa vs. 70 GPa), which essentially vanishes when approaching room temperature. Depending on the pressure and temperature domains, disorder at the VII–VIII transition line is believed to result from the competition between orientational and translational defect creation and propagation. At higher pressure, theoretical work predicted the pressure-induced symmetrization of the hydrogen bond, i.e. the breaking apart or dissociation of the H_2O molecules [13–15].

---

* Corresponding author. Tel.: +33 4 67 14 46 77; fax: +33 4 67 14 34 98; e-mail: magali@ldv.univ-montp2.fr.

The resulting structure termed ice X is expected to be of the cuprite type having a bcc oxygen lattice [2] with the protons located midway between two neighboring oxygen atoms. There have been a number of unconfirmed claims in the literature to have detected ice X, but only recent infrared experiments have provided evidence for a transition towards a symmetrized structure at about 60 GPa [6] or about 70 GPa [7].

Here, we report first preliminary results from the compression of ice VIII at 100 K up to about 100 GPa using a first principles simulation technique, ab initio path integrals, that treats both electrons and nuclei quantum mechanically [16]. Our method is able to describe realistically the hydrogen bonding including protonic defects, and the effects of thermal as well as quantum mechanical fluctuations such as tunneling and zero-point motion. It was already successfully used to study small molecules [17] and in particular charged water complexes in the gas phase [18].

2. Method

The ab initio path integral method [16] combines a Born–Oppenheimer first principles approach to electronic structure calculations with a quantum mechanical treatment of the nuclei at finite temperatures; we can present only a sketch of the basic idea and refer to the cited literature for further information. This method does not rely on a priori parametrized model potentials and reduces to the well-known Car–Parrinello ab initio molecular dynamics technique [19] if the nuclei are approximated as classical point particles. The electronic structure and thus the interactions between the nuclei are calculated within the framework of density functional theory in the Kohn–Sham formulation [20]. In particular, we supplement the local density approximation [21] with a gradient correction for the exchange energy [22]. Here, we expand the electronic wave function (of the cubic supercell with 16 H_2O molecules subject to periodic boundary conditions and fixed volume, see Fig. 1 for a sketch of half of the simulation box) in plane waves (only at the Γ-point of the Brillouin zone with a kinetic energy cutoff of 70 Ry) and use pseudo-potentials

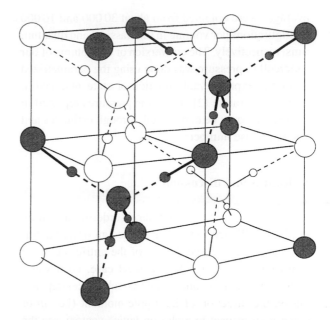

Fig. 1. Structure of antiferroelectrically ordered ice VIII. Only half of the simulation cell containing 16 H_2O molecules is drawn, with oxygen atoms marked by bigger circles and hydrogen atoms by smaller circles. The two sublattices formed by water molecules of opposite dipole orientations are, respectively, black and white. The paraelectric phase VII results if the dipoles are disordered; the small tetragonal distortion of ice VIII is not shown.

(of the Troullier–Martins type [23]) to represent the chemically inert core electrons. This type of approach is proven to describe hydrogen bonding reliably even in the presence of defects [18,24].

The quantum mechanics of the nuclei is described within the framework of Feynman path integral simulations [25]. The canonical path integral (neglecting the effects of nuclear exchange of identical particles) was discretized using $P = 16$ cyclically connected replicas. The sampling efficiency was increased by using staging together with Nosé–Hoover thermostat chains of length 3 coupled separately to each nuclear degree of freedom [26]. The fictitious electronic mass was set to 500 a.u., using a time step of 4.5 a.u. in simulations with classical nuclei (termed "classical") and 9.0 a.u. in simulations with quantum nuclei (termed "quantum"); in the latter case the fictitious nuclear masses were chosen to be four times as heavy as the physical masses. After careful equilibration, statistical averages were obtained from about 30 000 and 10 000 configurations per volume in the classical and quantum runs, respectively. The conversion of the volume of our supercell to pressure was done using the parametrized fit of the experimental equation of state obtained at room temperature [2]. Thus, the pressures reported in the text can only be considered as rough estimates and not as precise numbers.

3. Results and discussion

In order to follow the dipolar disordering transition of phase VIII, see Fig. 1, we defined an order parameter that measures the degree of the antiferroelectric character of ice VIII. In the ideal case, each water molecule belongs to one of the two sublattices depending on the direction of its dipole moment (i.e. *up* or *down* with respect to a chosen lattice vector), see the "black" and "white" water molecules in Fig. 1. Our order parameter is obtained by computing the dipole moment of every water molecule and adding these dipoles in each sublattice as

$$|\mathbf{D}| = \left\langle \frac{1}{2} \left| \frac{1}{N_{up}} \sum_{J \in up} \mathbf{d}_J - \frac{1}{N_{down}} \sum_{J \in down} \mathbf{d}_J \right| \right\rangle, \quad (1)$$

where $\langle \cdots \rangle$ denotes the statistical average in the quantum canonical ensemble. For simplicity, the molecular dipole moments are approximated by the purely geometric dipole vectors $\mathbf{d}_J = q_O \mathbf{R}_{O_J} + \sum_i q_H \mathbf{R}_{H_i}$ of a single molecule belonging to sublattices *up* or *down*. The vector $\mathbf{R}_{O_J}$ is the position vector of the Jth oxygen atom and $\mathbf{R}_{H_i}$ are the position vectors of its hydrogen neighbors. The charges are taken to be equal to $-2q$ for q_O and $+1q$ for q_H and are not calculated from the charge density. Following this definition, we find that the order parameter $|\mathbf{D}|$ is equal to $\approx 1.2q$ Å in the ideal ice VIII structure. In an infinitely large "perfect" paraelectric disordered structure, the value of $|\mathbf{D}|$ drops to zero.

The evolution of $|\mathbf{D}|$ along the trajectory is shown in Fig. 2 for three typical quantum simulations at 100 K with different volumes. At low pressure and after equilibration (Fig. 2(a)), the order parameter remains close to about $1.2q$ Å, which is the value of $|\mathbf{D}|$

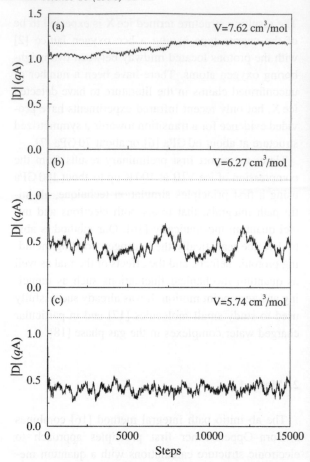

Fig. 2. Evolution of the antiferroelectric order parameter $|\mathbf{D}|$ defined in (1) along the trajectories from simulations with quantum nuclei for different volumes at 100 K; the equilibration phase is also shown and the dotted lines mark the value $1.2q$ Å. (a) $V = 7.62$ cm^3/mol (≈ 31 GPa), (b) $V = 6.27$ cm^3/mol (≈ 66 GPa), (c) $V = 5.74$ cm^3/mol (≈ 90 GPa); the fitted experimental equation of state [2] is used to convert V to the reported pressure.

that characterizes ice VIII. Upon further compression (Figs. 2(b) and (c)), the order parameter is strongly suppressed and fluctuates much more, in particular for the intermediate volume case in (b). Its mean value then saturates at about $0.4q$ Å , which seems to be the value in the paraelectric phase within our limited system size. The evolution of the order parameter as compression increases indicates that the antiferroelectric order disappears somewhere around roughly 50 GPa; note again that pressures are here obtained from the fit of the experimental equation of state at

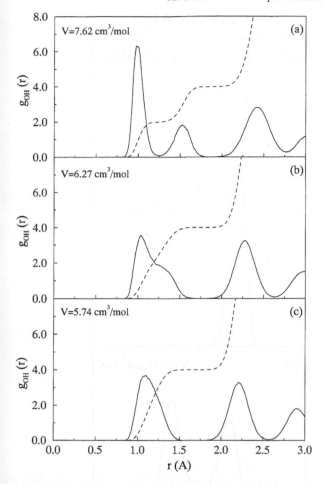

Fig. 3. Pair correlation functions $g_{OH}(r)$ (solid lines) from simulations with quantum nuclei for different volumes at 100 K; the dashed lines correspond to the integrated coordination numbers. (a) $V = 7.62\,cm^3/mol$ (≈ 31 GPa), (b) $V = 6.27\,cm^3/mol$ (≈ 66 GPa), (c) $V = 5.74\,cm^3/mol$ (≈ 90 GPa); the fitted experimental equation of state [2] is used to convert V to the reported pressure.

room temperature [2]. At the transition, the dipole moments probably become orientationally disordered in a structure similar to that of ice VII.

The structure of the ice sample at the three considered volumes is analyzed in Fig. 3 in terms of the OH radial distribution functions $g_{OH}(r)$ for the quantum simulations. At low pressure (Fig. 3(a)), the distribution function exhibits a clear first and second maximum corresponding to the covalent O–H bond and the hydrogen O···H bond with coordination numbers equal to 2 and 4, respectively. This is

the typical $g_{OH}(r)$ pattern of a molecular ice phase. As compression increases (Fig. 3(b)), the two maxima approach each other and finally merge to a single peak with a coordination number of 4 at highest compression (Fig. 3(c)). In the later case, the covalent bond and the hydrogen bond can no longer be distinguished from each other as is evident from Fig. 3(c). We therefore find a transition from a molecular phase towards a symmetrized phase, ice X, in which the H_2O molecules are broken apart and the protons reside midway between their two neighboring oxygen atoms.

Next we investigated the distribution of the proton positions between its two neighboring oxygen atoms in order to understand how disorder appears in the structure at the transition. In typical molecular phases, the protons sit between two oxygen atoms connected by a short covalent bond to one of them and by a weak hydrogen bond to the other. We computed the difference between the covalent bond length and the hydrogen bond length for each proton

$$\delta = R_{O_1H} - R_{O_2H}, \tag{2}$$

where R_{O_1H} and R_{O_2H} denote the OH distances between a given hydrogen atom and its two closest oxygen neighbors O_1 and O_2; symmetrization occurs evidently for $R_{O_1H} \equiv R_{O_1H}$, i.e. at $\delta \equiv 0$. The resulting distribution function averaged over all protons and many sampled configurations measures the probability of finding a proton at a certain relative position along its bonds to the two neighboring oxygen atoms; i.e. the two-sublattice symmetry of the antiferroelectric phase VIII is taken into account. Examples of such plots as obtained from simulations with quantum and classical nuclei are depicted in Fig. 4. At low pressure in both the quantum and classical cases (Figs. 4(a) and (d)) there is a single maximum located around $\delta \approx -0.5$ Å . Thus, every proton can be associated with a single oxygen atom and there is clearly two-sublattice antiferroelectric order present, i.e. ice VIII. At intermediate pressure, there are two maxima that are symmetric with respect to $\delta = 0$ (Figs. 4(b) and (f) for the quantum and classical cases, respectively). This phase is still molecular in nature, but does not possess dipolar antiferroelectric order, i.e. ice VII. By comparing the distributions for the same volume,

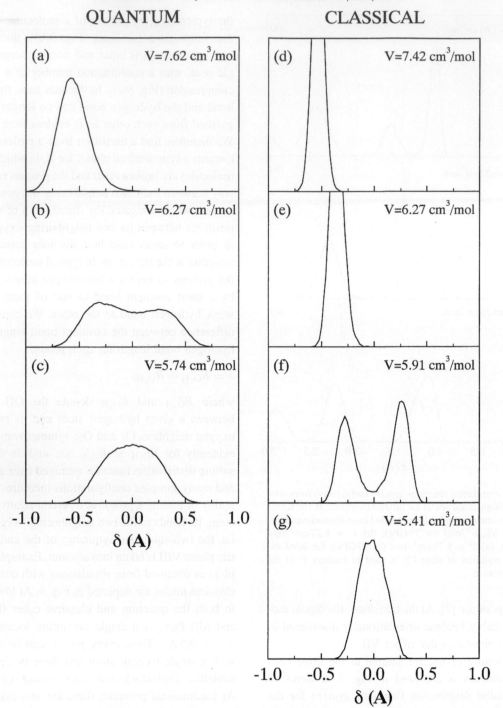

Fig. 4. Distribution function of the relative proton positions between the two neighboring oxygen atoms defined in (2) in the quantum case (left panels) and the corresponding classical cases (right panels) for different volumes at 100 K; the integral of all distributions is normalized to unity. Quantum: (a) $V = 7.62\,cm^3/mol$ ($\approx 31\,GPa$), (b) $V = 6.27\,cm^3/mol$ ($\approx 66\,GPa$), (c) $V = 5.74\,cm^3/mol$ ($\approx 90\,GPa$); Classical: (d) $V = 7.42\,cm^3/mol$ ($\approx 35\,GPa$), (e) $V = 6.27\,cm^3/mol$ ($\approx 66\,GPa$), (f) $V = 5.91\,cm^3/mol$ ($\approx 81\,GPa$), (g) $V = 5.41\,cm^3/mol$ ($\approx 110\,GPa$); the fitted experimental equation of state [2] is used to convert V to the reported pressure.

but with quantum and classical nuclei, in Figs. 2(b) and (e), one can clearly infer the crucial contribution that proton tunneling plays. At the highest pressures shown, one observes a single maximum centered around $\delta = 0$ Å in both cases (Figs. 4(c) and (g)). Thus, the most probable position of the protons is midway between its two oxygen neighbors, i.e. ice X. The classical picture is similar to the quantum one except that the merging of the two symmetric peaks into a single one appears at higher pressure with a much sharper distribution around $\delta = 0$ Å (Fig. 4(g)). Furthermore, phase VII is still present classically where phase X is observed quantum mechanically, thus zero-point motion effects seem to be the predominant quantum effect on the symmetrization transition.

4. Summary

First results of fully quantum mechanical atomistic simulations of compressed H_2O ice at 100 K up to pressures of about 100 GPa are presented. We find that the bcc oxygen structure remains stable up to our highest compression. Furthermore, we find evidence for a dipolar order–disorder transition from the antiferroelectric phase VIII to the paraelectric phase VII, followed by the symmetrization transition to phase X. Both transitions are accompanied by significant quantum effects, though of different origin.

Acknowledgements

The simulations were carried out on the IBM SP2 parallel computer at CNUSC (Centre National Universitaire Sud de Calcul, Montpellier, France) and at MPI Stuttgart.

References

[1] N.H. Fletcher, The Chemical Physics of Ice (Cambridge University Press, Cambridge, 1970); P.V. Hobbs, Ice Physics (Clarendon Press, Oxford, 1974).

[2] R.J. Hemley, A.P. Jephcoat, H.K. Mao, C.S. Zha, L.W. Finger and D.E. Cox, Nature (London) 330 (1987) 737.

[3] R.J. Nelmes, J.S. Loveday, R.M. Wilson, J.M. Besson, Ph. Pruzan, S. Klotz, G. Hamel and S. Hull, Phys. Rev. Lett. 71 (1993) 1192; J.M. Besson, Ph. Pruzan, S. Klotz, G. Hamel, B. Silvi, R.J. Nelmes, J.S. Loveday, R.M. Wilson, and S. Hull, Phys. Rev. B 49 (1994) 12540.

[4] Ph. Pruzan, J.C. Chervin and B. Canny, J. Chem. Phys. 99 (1993) 9842; Ph. Pruzan, J. Mol. Struct. 322 (1994) 279.

[5] K. Aoki, H. Yamawaki and M. Sakashita, Science 268 (1995) 1322; Phys. Rev. Lett. 76 (1996) 784.

[6] A.F. Goncharov, V.V. Struzhkin, M.S. Somayazulu, R.J. Hemley and H.K. Mao, Science 273 (1996) 218.

[7] K. Aoki, H. Yamawaki, M. Sakashita and H. Fujihisa, Phys. Rev. B 54 (1996) 15673.

[8] J.M. Besson, S. Klotz, G. Hamel, W.G. Marshall, R.J. Nelmes and J.S. Loveday, Phys. Rev. Lett. 78 (1997) 3141.

[9] K. Yamamoto, Jpn. J. Appl. Phys. 21 (1982) 567.

[10] W.F. Kuhs, J.L. Finney, C. Vettier and D.V. Bliss, J. Chem. Phys. 81 (1984) 3612.

[11] J.D. Jorgensen and T.G. Worlton, J. Chem. Phys. 83 (1985) 329.

[12] J.D. Jorgensen, R.A. Beyerlein, N. Watanabe and T.G. Worlton, J. Chem. Phys. 81 (1984) 3211.

[13] W.B. Holzapfel, J. Chem. Phys. 56 (1972) 712.

[14] F.H. Stillinger and K.S. Schweizer, J. Phys. Chem. 87 (1983) 4281; K.S. Schweizer and F.H. Stillinger, J. Chem. Phys. 80 (1984) 1230.

[15] C. Lee, D. Vanderbilt, K. Laasonen, R. Car and M. Parrinello, Phys. Rev. Lett. 69 (1992) 462; Phys. Rev. B 47 (1993) 4863.

[16] D. Marx and M. Parrinello, Z. Phys. B (Rapid Note) 95 (1994) 143; J. Chem. Phys. 104 (1996) 4037; M.E. Tuckerman, D. Marx, M.L. Klein and M. Parrinello, J. Chem. Phys. 104 (1996) 5579.

[17] D. Marx and M. Parrinello, Nature (London) 375 (1995) 216; Science 271 (1996) 179.

[18] M.E. Tuckerman, D. Marx, M.L. Klein and M. Parrinello, Science 275 (1997) 817.

[19] R. Car and M. Parrinello, Phys. Rev. Lett. 55 (1985) 2471.

[20] R.O. Jones and O. Gunnarsson, Rev. Modern Phys. 61 (1989) 689.

[21] J.P. Perdew and A. Zunger, Phys. Rev. B 23 (1981) 5048.

[22] A.D. Becke, Phys. Rev. A 38 (1988) 3098.

[23] N. Troullier and J.L. Martins, Phys. Rev. B 43 (1991) 1993.

[24] K. Laasonen, M. Sprik, M. Parrinello and R. Car, J. Chem. Phys. 99 (1993) 9080; E.S. Fois, M. Sprik and M. Parrinello, Chem. Phys. Lett. 223 (1994) 411; M. Tuckerman, K. Laasonen, M. Sprik and M. Parrinello, J. Chem. Phys. 103 (1995) 150; M. Sprik, J. Hutter and M. Parrinello, J. Chem. Phys. 105 (1996) 1142.

[25] D. Chandler, in: Liquids, Freezing and Glass Transition, eds. J.P. Hansen, D. Levesque and J. Zinn-Justin, (Elsevier, Amsterdam, 1991); D.M. Ceperley, Rev. Modern Phys. 67 (1995) 279.

[26] M.E. Tuckerman, B.J. Berne, G.J. Martyna and M.L. Klein, J. Chem. Phys. 99 (1993) 2796.

ELSEVIER

Computational Materials Science 10 (1998) 94–98

COMPUTATIONAL
MATERIALS
SCIENCE

First principle calculation of oxygen adsorption on a (1 0 0) silicon surface – First stages of the thermal oxidation

A. Estève [a,*], M. Djafari Rouhani [a,b], D. Estève [a]

[a] *Laboratoire d'Analyse et d'Architecture des Systèmes, CNRS, 7 Ave. du Colonel Roche, 31077 Toulouse, France*
[b] *Laboratoire de Physique des Solides, Univ. Paul Sabatier, 118, route de Narbonne, 31062 Toulouse, France*

Abstract

We present a full ab initio calculation of the oxygen atom in interaction with a (1 0 0) silicon surface. This calculation is based on the density functional theory (DFT) via the Harris functional. Preliminary results reported concern the calculation leading to the oxygen preferential setting on the surface. The total energy, angles and bond lengths are evaluated. We find two preferential sites for the oxygen which are linked in the oxidation process. One is identified as a molecular site and occurs first. The other is close to be a SiO_2 crystalline site. The effect of the oxygen density at the surface is also discussed. Copyright © 1998 Elsevier Science B.V.

1. Introduction

Because of its relevance to device technology, the formation of silicon dioxide layers by thermal oxidation is of great interest. During the last decade, the decreasing oxide thickness has increased the role of defects induced at the interface and their influence on the operating characteristics of the device. Despite an important technological and basic research effort, a successful modelling of silicon oxidation process of the first 30 Å thickness has never been achieved (on a macroscopic [1,2] or a microscopic basis). Coupled atomic mechanisms are certainly involved in this process. Even the reactivity of oxygen at the silicon surfaces remains controversial and many mechanisms are proposed [3–5] supported by an impressive number of

* Corresponding author. Tel.: 05 61 33 69 85; fax: 05 61 33 62 08; e-mail: aesteve@laas.fr.

characterization techniques (STM, UPS [6], infrared absorption spectroscopy using ^{18}O as a tracer [7], etc.). This is complicated by the passivation and cleaning processes. They introduce species at the surface that can change macroscopic properties (growth velocity, roughness, etc.) through microscopic mechanisms. Thus, we have to work at the atomic level.

Our final aim is to determine the microscopic mechanisms involved during the first stage of silicon oxidation. The mechanisms will, in the future, be introduced in a Monte Carlo simulation.

The present paper gives ab initio results concerning the atomic oxygen adsorption on the (1 0 0) silicon surface. The effect of surface coverage and the eventual interaction between oxygen atoms is investigated in detail. We will first present the calculation model and the type of configurations (number of atoms, layers, etc.) we used. In the second part, we will discuss the results for each configuration.

2. Model

Our calculation is based on the minimization of the total energy to find the optimal structure of the system. The total energy is calculated using the density functional theory (DFT) [8] via the Harris functional [9]. The DFT proved that the total energy is a unique functional of the electron density. The Harris functional is known to produce a fast ab initio total energy calculation. It allows accurate geometry optimizations in a small CPU time. It is true that the energies given by this functional are largely overestimated. These overestimated values are due to the crude treatment of the charge density in the Harris formalism. The energies given in this paper take the SiO total energy as a reference to give a more physical meaning to the calculated total energies. In the future, we may start from these geometries to perform calculations with higher accuracy in the energy evaluation by using pseudopotentials and a plane wave basis for the valence states.

The one k point ($k = 0$ in the brillouin zone) approximation used in this code has forced us to use large enough supercell. We first made series of calculations using 16 silicon atoms (eight layers of two atoms). The first two layers are free to relax, the others are given fixed positions. We also made calculation with 40 silicon atoms (five layers of eight atoms). Only the top layer can relax in this calculation. To create an infinite silicon surface for those configurations, we put a void space (20 Å) at the top of the silicon layers and we used periodic boundary conditions in the three axis. Sometimes, it has been necessary to simulate annealing which is possible in this Harris code. Actually, the calculation made at 0 K can easily fix the geometry configuration in an energy valley corresponding to a secondary minimum. The annealing allows the structure to clear the energy barrier.

3. Results

We have used different number of oxygen atoms to show the oxygen atom density effect on the silicon surface. But first, let us talk about the calculation

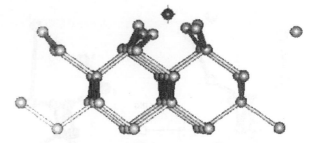

Fig. 1. [1 1 0] section view. One oxygen atom on a (1 0 0) silicon surface. The calculation supercell contains 40 silicon atoms.

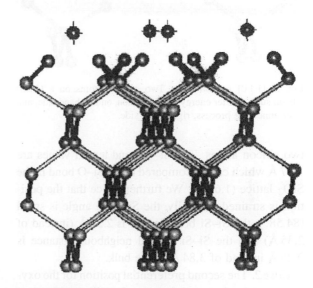

Fig. 2. [1 1 0] section view. One oxygen atom on a (1 0 0) silicon surface. The calculation supercell (16 silicon atoms) is four times reproduced.

results made with one oxygen for eight silicon atoms in the surface (low coverage).

As we can see in Fig. 1, the preferential position of the oxygen atom on the silicon surface is a bridge position (between two silicon atoms). The Si–O bond length values are 1.63 Å and the Si–O–Si angle is 137.8°. Elsewhere on the surface, we observe the silicon 2 × 1 reconstruction.

With one oxygen atom every two surface silicon atoms, we find two important cases:

Case 1. First, a bridge position shown in Fig. 2 which is the absolute minimum (1.87 eV for the adsorption energy) we found for the one oxygen configuration type. The oxygen is positioned between the

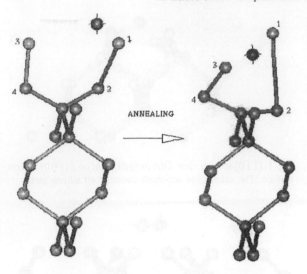

Fig. 3. [1 1 0] section view. Two oxygen atoms on a (1 0 0) silicon surface after energy minimization, on left-hand side, and after annealing process, right-hand side.

two silicon atoms. The Si–O bond length values are 1.67 Å which can be compared to the Si–O bond in the SiO$_2$ lattice (1.61 Å). We further notice that the position is strained. Actually, the Si–O–Si angle is small (84.58°), the Si–Si bond length is 2.28 Å (instead of 2.35 Å) and the Si–Si second neighbour distance is 2.25 Å instead of 3.84 Å in the bulk.

Case 2. The second preferential position for the oxygen atom is almost at the top of a silicon atom (see Fig. 3, left-hand side). The Si–O bond length is then 1.50 Å which is exactly the experimental value for the molecular SiO. A Si–Si bond is then broken (Fig. 4, left-hand side and Fig. 5, see Si1–Si2 and O–Si1). The molecular SiO then adsorbs on the surface with an energy of 0.63 eV, we can speak about physisorption in comparison with the 1.87 eV of the bridge position. This suggests that some SiO is formed and evaporated during the initial stage of oxidation that is documented experimentally for SiO$_2$ thickness inferior to 30 Å [10–12].

Now it is important to remember that the calculation is made at 0 K. Obviously, the initial position of the oxygen atom at the beginning of the minimization process will be very crucial. In Case 1, the initial site is symmetrical (middle position). Thus the oxygen interacts simultaneously with the two silicon

atoms. When the departure site is asymmetrical, one finds a very attractive process between the oxygen and the closest silicon atom. In fact, the results related to the above two cases are linked in time. Case 2 precedes Case 1 if we look at the total energies and the optimized geometries. To demonstrate this, we have simulated annealing using Case 2 configuration with the hope to converge to the first one. The result is shown in Fig. 3, right-hand side. As we said, we have a molecular SiO formation (Case 2) linked to a broken Si–Si bond (Fig. 5 and Fig. 4, first plot): we obtain a vacancy formation. With the annealing process, the SiO molecule rotates (see Fig. 3, right-hand side), adsorbs on a neighbouring site while the void space (at the vacancy site) is accentuated by the new local surface reconstruction (see the silicon atom Si2 moving to the bulk in Fig. 3, right-hand side). We notice that the Case 2 molecular bond length of 1.50 Å shifts to the crystal value (Si1–O: 1.68 Å). We have also the formation of a second Si–O crystalline bond 1.66 Å (Si3–O). The angle itself shifts more and more towards the crystal value and is stabilized at 161°. The grouping is then strongly directed and can be regarded as a nucleus site for the oxidation process. The adsorption energy becomes 1.42 eV. Then we can conclude that there is a geometrical and energetical convergence from Case 2 to Case 1.

With two oxygen atoms for two silicon atoms, we obtain the same process. We see the formation of two SiO molecules with bond lengths of 1.49 Å. Two Si–Si bonds are then broken.

With three oxygen atoms for two surface silicon atoms, one oxygen occupies the bridge site (see Fig. 2). The Si–O–Si angle is crystalline with an angle of 135°. The other two oxygens occupy the molecular sites (as shown in Fig. 3) on top of each silicon atom. The O–Si–O–Si–O chain is [−1 1 0] oriented.

With four oxygen atoms, two oxygen atoms form two SiO molecules (see Fig. 6, left-hand side, Si1–O1, Si3–O4). At the same time, two Si–Si bonds between the first and the second layers are broken (Si1–Si2, Si3–Si4). In a second stage, the molecules adsorb on the surface and one oxygen comes to the bridge position (Fig. 6, right-hand side, Si1–O3–Si3). The process is exactly the same as that of Case 2 followed

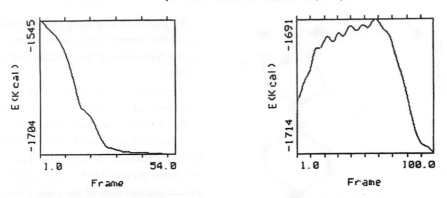

Fig. 4. Total energy of the minimization plus annealing procedure of Fig. 2.

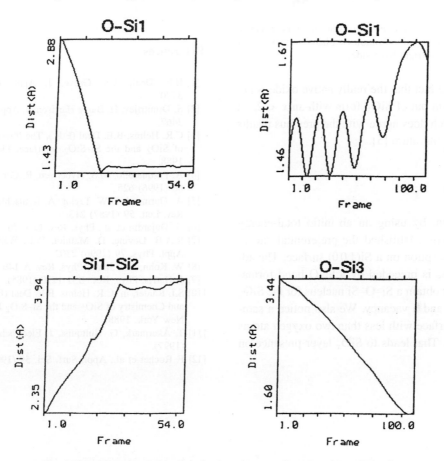

Fig. 5. Bond lengths of the minimization plus annealing procedure of Fig. 2.

by an annealing process. The angle is 135°, equal to that of a SiO₂ crystal. The three oxygens (O1,O2,O4) occupy molecular sites with bond lengths of 1.48, 1.49 and 1.53 Å, respectively (Fig. 6). One oxygen (O1) seems to go to the back bond (Si1–Si2). Here we have probably a frozen configuration because of the 0 K temperature. A dynamical calculation should lead to a broken back bond configuration. This could

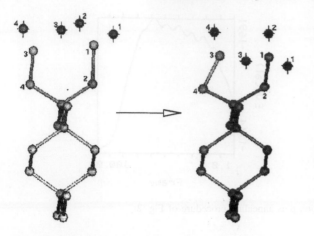

Fig. 6. [1 1 0] section view. Four oxygen atoms on a (1 0 0) silicon surface, during minimization (left-hand side) and after energy minimization (right-hand side).

demonstrate the fact that the really native oxide is not of the SiO_2 form but of SiO_x form with an $x < 2$, let us say 1.5, which does agree with the previous results reported in the literature [5].

4. Conclusion

In conclusion, by using an ab initio total-energy method, we have established the preferential site of the oxygen adsorption on a Si(1 0 0) surface. The adsorption process is preceded by a Si–O bond formation. We finally obtain a Si–O–Si nucleus for the SiO_2 crystal growth and a vacancy. We also notice a saturation of the surface with less than two oxygen atoms for one silicon. That leads to SiO_x layer presence on the surface.

We therefore can keep two mechanisms for future use in the Monte Carlo simulations:
- SiO formation at the surface (with rotation of this SiO);
- deposition of SiO which has to be considered as a nucleus site for the oxidation growth.

We are now investigating more precise energy calculation using CASTEP (CAmbridge Serial Total Energy Package), a code also based on the density functional theory (DFT). This calculation will give us a more precise idea of the energy barriers that will be injected in our Monte Carlo code.

References

[1] B.E. Deal, A.S. Grove, J. Appl. Phys. 36 (1965) 3770.
[2] S. Dimitrijev, H. Barry Harrison, J. Appl. Phys. 80 (1996) 2467.
[3] C.R. Helms, B.E. Deal (Eds.), The Physics and Chemistry of SiO_2 and the Si–SiO_2 Interface, Plenum, New York, 1988.
[4] A. Pasquarello, M.S. Hybertsen, R. Car, Appl. Phys. Lett. 68 (1996) 625.
[5] A. Ourmazd, D.W. Taylor, A. Rentschler, J. Bevk, Phys. Rev. Lett. 59 (1987) 213.
[6] G. Dujardin et al., Phys. Rev. Lett. 76 (1996) 3782.
[7] R.A.B. Devine, D. Mathiot, W.L. Warren, B. Aspar, J. Appl. Phys. 79 (1996) 2302.
[8] W. Kohn, L.J. Sham, Phys. Rev. A 140 (1965) 1133.
[9] J. Harris, Phys. Rev. B31 1770 (1985).
[10] S.I. Raider, in: C.R. Helms, B.E. Deal (Eds.), The Physics and Chemistry of SiO_2 and the Si–SiO_2 Interface, Plenum, New York, 1988.
[11] T. Åkermark, G. Hultquist, J. Electrochem. Soc. 144 (4) (1997).
[12] F. Rochet et al., Appl. Surf. Sci. 59 (1992) 117.

ELSEVIER

Computational Materials Science 10 (1998) 99–104

COMPUTATIONAL
MATERIALS
SCIENCE

Hydrogen chemisorption on the Si(1 1 1)$\sqrt{3} \times \sqrt{3}$R30°-Al, -Ga, -B surfaces: An ab initio HF/DFT molecular orbital modelling using atomic clusters

Sanwu Wang, M.W. Radny *, P.V. Smith

Department of Physics, The University of Newcastle, Callaghan, NSW 2308, Australia

Abstract

A comprehensive study of atomic hydrogen chemisorption on the Si(1 1 1)$\sqrt{3} \times \sqrt{3}$R30°-Al, -Ga and -B cluster modelled surfaces is presented using Hartree–Fock/density functional theory methods. Extrapolation of the results to the extended (1 1 1) silicon surface is also discussed. It is found that the chemisorption of hydrogen on the Al and Ga terminated surfaces induces a transition from the $\sqrt{3} \times \sqrt{3}$ structure to a local 1×1 : H-like reconstruction with a stable SiAl (or SiGa) sites. The subsurface boron induced $\sqrt{3} \times \sqrt{3}$ reconstruction is also lifted by hydrogen chemisorption but, in this case, boron adatoms are likely to be segregated on the surface, predominantly as BH or/and BH_2. Copyright © 1998 Elsevier Science B.V.

Keywords: Ab initio quantum chemical methods; Aluminium; Boron; Chemisorption; Gallium; Hydrogen; Silicon; Surface chemical interaction; Surface structure

Experiment has shown that the adsorption of Al, Ga, In and B on the Si(1 1 1) surface at 0.33 ML coverage leads to a $\sqrt{3} \times \sqrt{3}$R30° reconstruction in which the adatoms sit in the three-fold T_4 chemisorption sites atop the second layer Si atoms (Fig. 1) [1]. For the chemisorption of boron the experimental results have also suggested a second, more stable coexisting structure – the substitutional subsurface topology in which the boron atoms occupy the second layer S_5 positions, directly below the T_4 sites (Fig. 1) [2]. In contrast to the on-top Al/Ga/In/B-T_4 topologies the Boron-S_5 structure is stabilised by significant charge transfer to the subsurface boron from the neighbouring silicon atoms. It is the thin electrically active layer (the

so-called δ-layer) produced in this way which makes the B/Si system of practical importance because of its potential device applications [3]. For both the T_4 and S_5 structures, one might expect that the Si(1 1 1) surface would be passivated by the chemisorption of the trivalent group-III elements. Experiment suggests, however, that the chemisorption of atomic hydrogen onto these surfaces significantly changes their structures. Typically, as for the case of the Al terminated Si(1 1 1) surface [4], the exposure of 50 ML induces a structural transformation at room temperature from the $\sqrt{3} \times \sqrt{3}$ topology to 1×1 phase containing Al metallic clusters. The behaviour of boron is again in marked contrast to this picture. At low exposures (< 200 ML), the $\sqrt{3} \times \sqrt{3}$ periodicity of the B/Si(1 1 1) surface is still preserved, while at exposures > 200 ML, the 1×1 phase accompanied by some disordering of the surface

* Corresponding author. Tel.: + 61 49 21 5447; fax: +61 49 21 6907; e-mail: phmwr@cc.newcastle.edu.au.

has been observed [5]. In this paper we discuss the results of first principles Hartree–Fock (HF)/density functional theory (DFT) calculations of the atomic hydrogen treated Si(1 1 1)$\sqrt{3} \times \sqrt{3}$R30°-B, -Al and -Ga cluster modelled surfaces. Comparing our results with the experimental data permits us to distinguish the following hydrogenation processes of the Si(1 1 1)$\sqrt{3} \times \sqrt{3}$R30°-adatom surfaces: (a) the formation of a stable SiH + SiAl(Ga) layer, and a less stable SiH + AlH (or GaH) structure, and (b) the segregation of subsurface boron in the form of BH_2, and the formation of a stable BH_2 + SiH layer.

A basic MSi_5H_9 cluster, with M = Al, Ga or B, has been used to simulate the bonding at the T_4 and S_5 sites of the Si(1 1 1)$\sqrt{3} \times \sqrt{3}$R30°-adatom surface (Fig. 1). To validate the use of such small clusters, calculations employing clusters of progressively increasing size have also been performed. These additional clusters were MSi_8H_{11} (with three additional Si atoms

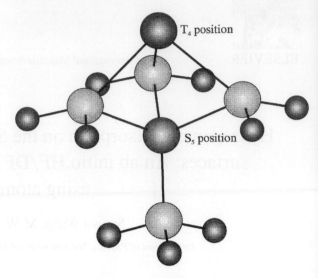

Fig. 1. The MSi_5H_9 cluster used to simulate the T_4 and S_5 adatom configurations of the $\sqrt{3} \times \sqrt{3}$-M surfaces, where M = Al, Ga and B. The dark spheres denote adatoms in the on-top T_4 and substitutional subsurface S_5 sites.

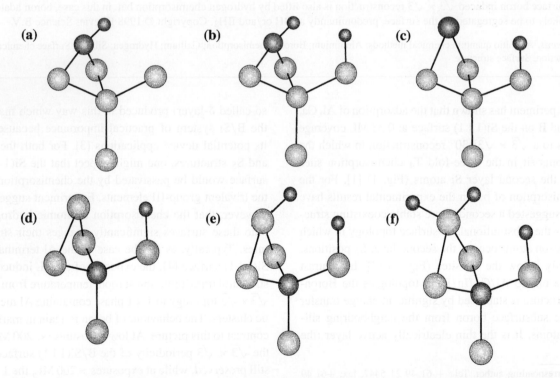

Fig. 2. The minimum energy bridge-site cluster geometries for (a) B-T_4/H; (b) B-T_4/2H; (c) Al-, Ga-T_4/H; (d) B-S_5/H; (e) B-S_5/2H; (f) B-S_5/3H.

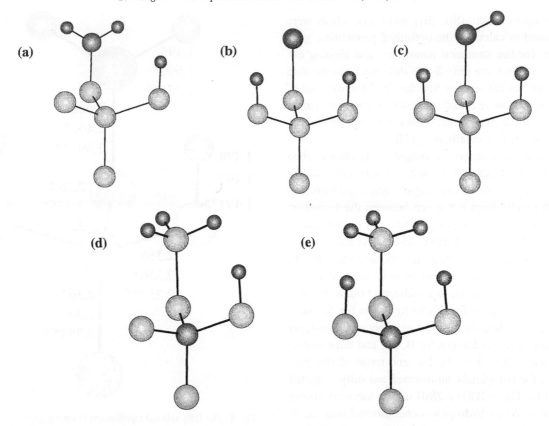

Fig. 3. The equilibrium on-top structures for: (a) B-T_4/3H; (b) Al-, Ga-T_4/2H; (c) Al-, Ga-T_4/3H; (d) B-S_5/4H; (e) B-S_5/5H.

on the third layer), $MSi_{11}H_{21}$ (six extra Si atoms on the second and third layers), and $MSi_{14}H_{27}$ (nine extra Si atoms including three more Si atoms on the first layer). The hydrogen atoms at the extremities of the clusters have been used to saturate the bulk dangling bonds. The calculations have been performed using the Becke3LYP option in the Gaussian94 program [6]. Becke3LYP is a hybrid method for exchange and correlation which incorporates an exact HF exchange energy [7] together with a semicmpirical combination of a local [8] and nonlocal [9] spin density electron correlation functional.

The total energy of each cluster has been optimised with respect to the position of the atoms within the smaller cluster, plus the chemisorbing hydrogens, but excluding the embedding H atoms. The sum of the total energies of the residual clusters, formed by removing the adatoms (Al, Ga, B or Si), or hydrogenated

species (AlH_x, GaH_x, SiH_x and BH_x), minus the energy of the original chemisorbed cluster, gives the binding energy of the desorbed adatom or species to the substrate. The double-ξ basis set 6–31g* with one set of d polarisation functions has been used to optimise the geometry of the clusters [10]. The binding energies have been calculated using the 6–31 + g(3df, 2p) basis set which supplements the 6–31 g basis with diffuse and polarisation functions: three sets of d functions and one set of f functions for heavy atoms, and two sets of p functions for hydrogens [11].

The minimum energy geometries of the Si_5H_9–Al and –Ga adatom clusters representing the T_4 structures were determined to be 1.14 and 0.65 eV, respectively, lower in energy than the corresponding S_5 adatom topologies. For boron, however, the total energy of the S_5 topology is 2.42 eV lower in energy than the related Si_5H_9B cluster geometry of the T_4 structure.

The larger clusters (Si$_8$, Si$_{11}$ and Si$_{14}$), which were also used to calculate the optimised geometries, gave values for the structural parameters and binding energies which were only 2–5% different from the data obtained for the smaller Si$_5$ clusters. Moreover, all of these results were found to be in very good agreement with both the experimental data and other theoretical calculations (for details see [12]).

As hydrogen atoms are progressively chemisorbed onto the stable adatom-T$_4$ or S$_5$ clusters, our total energy calculations reveal a systematic reconstruction which results from a balance between the formation and saturation of dangling bonds, and the release of strain within the bonds of the cluster. As more hydrogen atoms are chemisorbed near the adatom of the T$_4$ structure, the adatom moves from its original T$_4$ site (no hydrogens, Fig. 1), to an adjacent bridge site (one and two hydrogens for B, one hydrogen for Al(Ga) – Fig. 2), and then on-top of a neighbouring first-layer Si atom (three hydrogens for B, two and three hydrogens for Al(Ga) Fig. 3). The end result of this process is the completely undistorted and fully saturated Si–Al (or Ga or B)H$_2$ + 2SiH cluster topology shown in Fig. 4. When hydrogen is chemisorbed near the Si adatom of the S$_5$ structure (with boron at the S$_5$ subsurface position), the sequence is similar (Figs. 2(d)–(f), 3(d) and (e)) and results in an Si–SiH$_3$ configuration together with two SiH species being formed on the cluster.

The relative stability of the hydrogenated B/Si(1 1 1) T$_4$ and S$_5$ cluster geometries depends directly on the number of hydrogen atoms chemisorbed on the cluster. This dependence is seen in Fig. 5 where the difference between the total energies of the Boron-T$_4$ and Boron-S$_5$ structures, $\Delta E_T = E_T(T_4) - E_T(S_5)$, is plotted as a function of the number of chemisorbed hydrogen atoms (curve DeltaET). For one or two hydrogen atoms chemisorbed near an adatom, configurations with the boron atom in the S$_5$ position seem to be more energetically favourable than those with the boron atom in the T$_4$ position ($\Delta E_T > 0$). However, the T$_4$ structures are found to be more stable ($\Delta E_T < 0$) than the corresponding S$_5$ structures when three or four hydrogen atoms are chemisorbed near the adatom site. Extrapolating these cluster

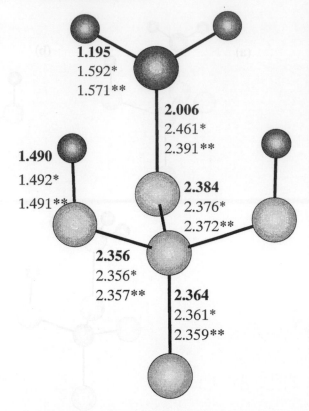

Fig. 4. The fully relaxed equilibrium cluster geometry obtained when four hydrogen atoms are chemisorbed near the adatom site of the T$_4$ $\sqrt{3} \times \sqrt{3}$-Al, -Ga and -B surfaces. The bondlengths, which are given in angstroms, correspond to B (bold), Al (*), and Ga (**).

results to the hydrogenated Si(1 1 1)$\sqrt{3} \times \sqrt{3}$R30°-B surface suggests that boron will most likely occupy a subsurface S$_5$ position at low hydrogen coverage, up to two hydrogen atoms per three substrate silicon atoms (i.e. ≤ 0.67 ML), but appear as an on-top adatom directly above one of the first-layer silicon atoms for hydrogen exposures greater than 0.67 ML. The binding energies of the BH$_x$ complexes for the T$_4$ structure are always higher than the corresponding energies of the SiH$_x$ species for the Boron-S$_5$ structure. Moreover, while the binding energies of the various species are found to progressively decrease with increasing number of hydrogens for both the T$_4$ and S$_5$ sites, the binding energies of the hydrogen bonded adatoms (Si for S$_5$ and B for T$_4$) in the most energetically stable structures are all quite high and

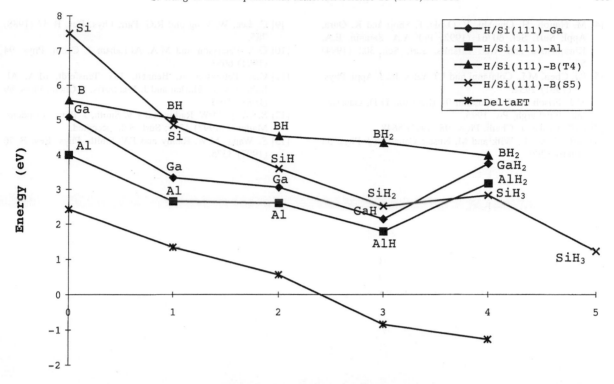

Fig. 5. The Becke3LYP binding energies of single (Al, Ga, B and Si), and hydrogenated (AlH_x, GaH_x, BH_x and SiH_x) adatoms on the T_4- and S_5-$\sqrt{3} \times \sqrt{3}$ structures. The difference between the total energies of the Boron-T_4 and Boron-S_5 structures ($\Delta E_T = E_T(T_4) - E_T(S_5)$), as a function of the number of chemisorbed hydrogen atoms, is also given.

lie in the range 3.5–5.0 eV (see Fig. 5). These data are consistent with the available experimental data for the H/Si(1 1 1)-B system [5] and are discussed elswhere [13]. For the Al and Ga adatom clusters, the smallest binding energies correspond to the desorption of AlH or GaH from the three hydrogen chemisorption structures (Fig. 4). The SiAl(Ga)–SiH bridge site and SiAl(Ga)–2SiH on-top configurations (Figs. 2(c) and 3(b)), exhibit very similar adatom binding energies 2.60 and 2.65 eV for Al, and 3.30 and 3.03 eV for Ga, respectively. These results suggest that hydrogenation of the Si(1 1 1)$\sqrt{3} \times \sqrt{3}$ R30–Al(Ga) surfaces will result in the formation of a stable SiAl(Ga) layer in which the metal atoms are directly bonded to the substrate and hydrogen atoms are sited directly above the first layer silicon atoms in a local 1×1 configuration. The larger binding energies for the Ga adatom suggests that a higher temperature may be necessary to observe such a transition on the Ga surface than on the Si(1 1 1) : Al surface.

References

[1] R.I.G. Uhrberg and G.V. Hansson, Crit. Rev. in Solid State and Mater. Sci. 17 (1991) 133; J.P. LaFemina, Surf. Sci. Rep. 16 (1992) 133; S. Kono, Surf. Rev. and Lett. 1 (1994) 359; H. Nagayoshi, Surf. Rev. and Lett. 1 (1994) 369; J. Nogami, Surf. Rev. and Lett. 1 (1994) 395; H.N. Waltenburg and J.T. Yates, Jr., Chem. Rev. 95 (1995) 1589.

[2] R.L. Headrick, I.K. Robinson, E. Vlieg and L.C. Feldman, Phys. Rev. Lett. 63 (1989) 1253; P. Bedrossian, R.D. Meade, K. Mortensen, D.M. Chen, J.A. Golovchenko and D. Vanderbilt, Phys. Rev. Lett. 63 (1989) 1257; I.-W. Lyo, E. Kaxiras and Ph. Avouris, Phys. Rev. Lett. 63 (1989) 1261.

[3] R.L. Headrick, A.F.J. Levi, H.S. Luftman, J. Kovalchick and L.C. Feldman, Phys. Rev. B 43 (1991) 14 711; R.L. Headrick, B.E. Weir, A.F.J. Levi, B. Freer, J. Bevk and L.C. Feldman, J. Vac. Sci. Technol. A 9 (1991) 2296.

[4] M. Naitoh, H. Ohnishi, Y. Ozaki, F. Shoji and K. Oura, Appl. Surf. Sci. 60/61 (1992) 190; A.A. Saranin, E.A. Khramtsova and V.G. Lifshits, Surf. Sci. 302 (1994) 57.

[5] P.J. Chen, M.L. Colaianni and J.T. Yates, Jr., J. Appl. Phys. 70 (1991) 2954.

[6] M.J. Frisch et al., Gaussian 94 (Revision D.1), Gaussian Inc., Pittsburgh, PA, 1995.

[7] A.D. Becke, J. Chem. Phys. 98 (1993) 5648.

[8] S.H. Vosko, L. Wilk and M. Nusair, Canadian J. Phys. 58 (1980) 1200.

[9] C. Lee, W. Yang and R.G. Parr, Phys. Rev. B 37 (1988) 785.

[10] G.A. Petersson and M.A. Al-Laham J. Chem. Phys. 94 (1991) 6081.

[11] G.A. Petersson, A. Bennett, T.G. Tensfeldt, M.A. Al-Laham, W.A. Shirlen and J. Mantzaris, J. Chem. Phys. 89 (1988) 2193.

[12] S. Wang, M.W. Radny and P.V. Smith, J. Phys.: Condens. Matter 9 (1997) 4535; Surf. Sci., accepted.

[13] S. Wang, M.W. Radny and P.V. Smith, Phys. Rev. B 56 (1997) 3575.

ELSEVIER

Computational Materials Science 10 (1998) 105–110

COMPUTATIONAL
MATERIALS
SCIENCE

Molecular dynamics description of silver adatom diffusion on Ag(1 0 0) and Ag(1 1 1) surfaces

N.I. Papanicolaou *, G.A. Evangelakis, G.C. Kallinteris

Department of Physics, Solid State Division, University of Ioannina, PO Box 1186, 45110 Ioannina, Greece

Abstract

The self-diffusion processes of single adatoms on Ag(1 0 0) and Ag(1 1 1) surfaces have been studied using molecular-dynamics simulations and a many-body potential derived in the framework of the second-moment approximation to the tight-binding model. Our results for the (1 0 0) surface indicate that, although the migration energy for hopping is lower than that of the exchange mechanism, the exchange diffusion is higher than hopping diffusion for temperatures above 600 K. The migration energy for the hopping mechanism is in very good agreement with the experiment and the results of ab initio calculations. We also find that for the Ag(1 1 1) face the dominant mechanism is the hopping, which exhibit Arrhenius behaviour with two distinct temperature ranges, corresponding to two different migration energies. The diffusion in the high temperature region is mainly due to correlated jumps requiring an activation energy which is in excellent agreement with the experimental data. In addition the temperature dependence of the mean-square-displacements and the relaxations of both surface atoms and adatoms are presented and compared with previous studies. Copyright © 1998 Elsevier Science B.V.

Keywords: Adatoms; Silver; Molecular dynamics simulation; Surface diffusion; Vibrations and relaxations of adsorbed and surface atoms

1. Introduction

The diffusion of single adatoms on metal surfaces is very important in understanding many physical phenomena related to surfaces, such as epitaxial thin film and crystal growth, heterogeneous catalysis, etc. Only a few experimental tools are suitable to investigate self-diffusion on surfaces: field ion microscopy (FIM), scanning tunnelling microscopy (STM), low-energy ion scattering (LEIS) and helium-beam scattering.

The self-diffusion of Ag on Ag(1 0 0) has been recently measured using the LEIS method [1], from

where the migration energy has been determined to a value of (0.40 ± 0.05) eV. In parallel, the system Ag/Ag(1 1 1) has been studied by means of STM [2] giving a corresponding migration energy of (97 ± 10) meV. From the theoretical point of view, the Ag/Ag(1 0 0) system has been studied by several groups using semi-empirical models, such as the embedded atom method (EAM) [3,4], the corrected effective medium (CEM) theory [5–7], the effective medium theory (EMT) [8], or first-principles techniques [9,10]. Furthermore, theoretical investigations have been performed for the Ag/Ag(1 1 1) system using the EAM [3,4], CEM [11], EMT [2,8,12] and ab initio calculations [9,13]. It is known that

* Corresponding author. Tel.: +30 651 98562; fax: +30 651 45631; e-mail: nikpap@cc.uoi.gr.

first-principles techniques provide accurate description of the atomic interactions, but they often require a lot of computational time; they are restricted therefore, to calculate only statically the energy barriers for diffusion. On the other side, the semi-empirical methods, although less precise, are able to simulate the dynamical character of diffusion, since they can deal with large systems for long simulation times. With the exception of Refs. [4,5,11], all the above studies refer to static calculations of self-diffusion parameters.

In this work we report detailed molecular-dynamics (MD) simulations of the self-diffusion processes of Ag/Ag(1 0 0) and Ag/Ag(1 1 1), using a many-body potential scheme within the second-moment approximation of the tight-binding (TB) theory. We deduce also the mean-square-displacements (MSD) and relaxations of both the surface atoms and adatoms in the normal to the surface direction as a function of temperature, as well as the adatom-vacancy formation energy. Our results are compared with available measurements and previous computations.

2. Computational details

The interactions of silver atoms are described by a many-body potential based on the TB method in the second-moment approximation for the attractive part and a Born–Mayer type for the repulsive contributions [14,15]. The functional form and the four adjustable parameters used in the present study are given in [16]. Although this potential scheme has been fitted to bulk properties, it has been employed successfully to predict various surface phenomena such as vibrations, relaxations and adatom diffusion of noble metals [17–20].

The MD simulations were performed for the (1 0 0) and (1 1 1) surfaces using a slab consisting of 40 and 36 layers, each layer containing 64 and 80 atoms, respectively. Periodic boundary conditions were applied in space and two free surfaces of each slab were simulated by fixing the supercell size (slab plus vacuum), in the normal to the surface direction, at a value twice the slab size. The simulations were carried out in the micro-canonical ensemble, using the Verlet algorithm

and a time step of 5 fs for the integration of the equations of motion. One adatom was placed on the free surfaces of each slab. In order to obtain reliable statistics for the calculation of diffusion coefficient, we performed simulations of 0.5–4 ns (depending on the temperature) for (1 0 0) face (600–850 K) and 0.5 ns for the (1 1 1) face (200–900 K). In the case of (1 0 0) face, where more than one diffusion mechanisms are present, we determined the frequency f of events for each diffusion process and we calculated the diffusion coefficient from the expression $D = f d^2/2z$, where d is the jump distance and z the dimensionality of the diffusion space (2 in our case). The diffusion coefficient for the (1 1 1) face was calculated using the Kubo integral of the velocity auto-correlation function [18]. Details for the calculations of the MSDs, relaxations, formation energies and static energy barriers are reported in [18].

3. Results and discussion

3.1. Mean-square-displacements and thermal relaxations

In Fig. 1 we present the temperature dependence of the MSDs in the normal to the (1 0 0) and (1 1 1) surfaces for both the surface atoms and adatoms. We can see that the vibrational amplitudes of the adatoms and the surface atoms are roughly the same up to 800 K in the case of (1 0 0) surface and up to 900 K for the (1 1 1) surface. Above these temperatures, the vibrations of the adatoms become greater than those of the surface atoms exhibiting anharmonic behaviour. Furthermore, it is clear that both surface atoms and adatoms on (1 0 0) face present higher MSDs than those on (1 1 1) face, indicating stronger coupling for the second surface. Our MSDs for the (1 1 1) surface agree well with measurements of this quantity using medium energy ion scattering [21].

In Fig. 2 we show the temperature dependence of the relaxations of both surface atoms and adatoms, in the perpendicular direction. We find that the (1 1 1) surface (Fig. 2(a)) presents a small contraction with respect to the bulk interlayer spacing at room temperature

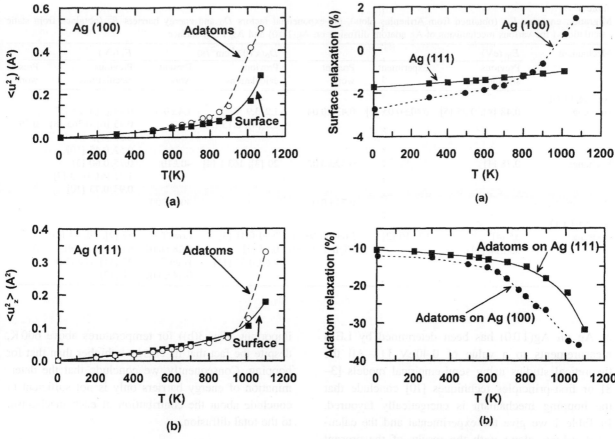

Fig. 1. Mean-square-displacements in the normal to (a) (1 0 0) and (b) (1 1 1) surface as a function of temperature; squares: silver surface atoms, circles: silver adatoms.

Fig. 2. (a) Relative change in the surface to bulk interlayer spacings with temperature for the Ag(1 0 0) (circles) and Ag(1 1 1) (squares) faces. (b) The same as (a), but for the adatoms.

(−1.6%) in agreement with the experimental results (−2.5%) [21]. Our findings are also very close to first-principles calculations (−1.3%) [9] and EAM simulations (−1.4%) [22]. Nevertheless, our model fails to reproduce the experimental observation of a dramatic expansion above 670 K (attaining 10% at 1150 K) [21]. Similar failure has been found within the EAM model [22], due probably to semi-empirical character of models. The (1 0 0) surface presents a more important contraction at low temperature (−2.7%) in agreement with ab initio computations (−1.9%) [9]. At high temperatures we observe a nonlinear expansion, denoting enhanced surface anharmonicity. Concerning the relaxations of the adatoms (Fig. 2(b)) we observe a contraction with strong temperature dependence for

both faces, which is more important in the (1 0 0) face.

3.2. Adatom self-diffusion on Ag(1 0 0)

The adatom self-diffusion on the fcc (1 0 0) metal face takes place via two main diffusion mechanisms that have been observed experimentally: (a) the bridge hopping mechanism, where the adatom hops from a 4-fold hollow site to an adjacent 4-fold hollow site over the 2-fold bridge site between the two sites; (b) the exchange mechanism, where the adatom replaces a surface atom, which moves to the 4-fold hollow site diagonal to the adatom's original 4-fold site. Recently, the migration energy for self-diffusion

Table 1
Migration energies E_m (obtained from Arrhenius plots), pre-exponential factors D_0 and energy barriers E_s (obtained from static calculations) for various mechanisms of Ag adatom diffusion on Ag(1 0 0) and Ag(1 1 1) surface

Mechanism	E_m (eV)			$D_0 (\times 10^{-3}\,cm^2/s)$		E_s (eV)	
	Previous simulations	Experiment	Present work	Previous calculations	Present work	Previous calculations	Present work
Ag/Ag(1 0 0) Hopping	0.48 [4], 0.25 [5]	0.40±0.05 [1]	0.43±0.04	1.2, 3.9 [3] 3.1 [4], 0.15 [5]	1.4±0.6	0.48 [3], 0.24 [5] 0.45 [6], 0.365 [8] 0.50 [9], 0.52, 0.45 [10]	0.46
Exchange	0.78 [4]		0.59±0.02	20 [3], 163.1 [4]	40±10	0.75, 0.60 [3] 1.01 [6], 0.73 [7] 0.93, 0.73 [10]	
Double exchange			0.74±0.06		300±200		
Ag/Ag(1 1 1) Hopping	55 [4] 55±11 [11]	97±10 [2]	Low T: 69±1 High T: 98±6	0.50, 0.41 [3] 0.15 [4] 0.69, 0.62 [11]	Low T: 0.28±0.01 High T: 0.54±0.01	67 [2], 59, 44 [3], 64 [8], 140 [9], 23 [11], 120 [12], 81 [13]	63

of Ag on Ag(1 0 0) has been determined by LEIS measurements to a value of 0.40 eV [1]. All the theoretical studies using semi-empirical models [3–8] or first-principles techniques [10] conclude that the hopping mechanism is energetically favoured. In Table 1 we give the experimental and the calculated values, along with the results of the present study for the migration energies E_m, the static barriers E_s and the pre-exponential factors D_0 for the various mechanisms. Our value, $E_m = 0.43$ eV, for the bridge mechanism is in excellent agreement with experiment [1] and MD-EAM simulations (0.48 eV) [4], while our static barrier, $E_s = 0.46$ eV, agrees well with the ab initio studies (0.50 eV [9], 0.45 eV-GGA, 0.52 eV-LDA [10]). The value we find for the exchange mechanism, $E_m = 0.59$ eV, is lower than that obtained in [4] (0.78 eV MD-EAM), due probably to the different potential model used. With the term *double exchange* (Table 1), we mean the more complicated process, whereas two surface atoms are involved [17].

In Fig. 3(a) we present the Arrhenius plots of hopping and exchange frequencies, while diagrams referring to the diffusion coefficients are shown in Fig. 3(b). It should be pointed out that, from the diffusion point of view, the exchange mechanism is more important (Fig. 3(b)) for temperatures above 600 K, despite the fact that its energy is higher than that for hopping. Consequently, we conclude that the determination of energy barriers only is not sufficient to conclude about the contribution of each mechanism to the total diffusion.

3.3. Adatom self-diffusion on Ag(1 1 1)

In Table 1 we give self-diffusion data for hopping on Ag(1 1 1), while in Fig. 4 we provide the corresponding Arrhenius diagram of the diffusion coefficient. As we can see in Fig. 4(a), it is possible to distinguish two different temperature ranges, with different migration energies: the low temperature range (200–500 K) with an energy of 69 meV and the high temperature range (500–900 K) requiring 98 meV. A detailed analysis of trajectories revealed that in the low temperature region, most of the diffusion events are simple hopping from a hcp to fcc stacking nearest neighbour sites or vice versa. For this mechanism, the static barrier was determined to be 63 meV, in very good agreement with the above migration energy. In the high temperature region we observed a lot of correlated and longer jumps, which have been also observed

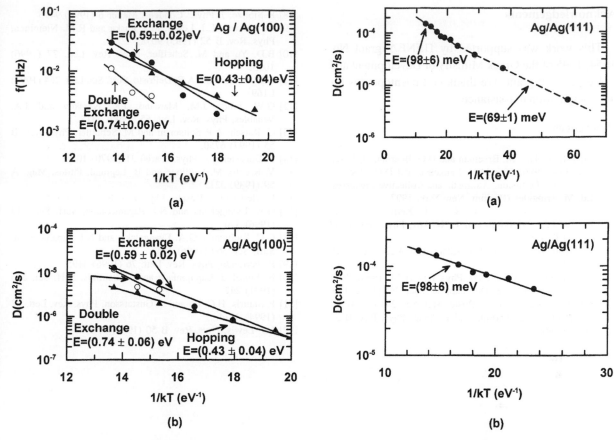

Fig. 3. Arrhenius plots of (a) hopping and exchange frequencies, (b) diffusion coefficients for Ag/Ag(1 0 0). The solid lines are least-square fits. Filled triangles and circles refer to jump and exchange mechanism, respectively; open circles to double exchange.

Fig. 4. (a) Arrhenius diagram of the diffusion coefficient of Ag adatoms on Ag(1 1 1) surface for the low (200–500 K) and high temperature (500–900 K) regions. (b) Detailed Arrhenius plot for the high temperature range.

by the MD-EAM [4], MD-CEM method [11] and also in the case of Cu/Cu(1 1 1) [18]. Our migration energies agree very well with recent STM studies (97 meV) [2] and ab initio computations (81 meV) [13].

Finally, in Fig. 5 we present the Arrhenius diagram of the concentration of the spontaneously created adatoms on Ag(1 0 0) and Ag(1 1 1), from which we deduce the formation energies for the adatom-vacancy pair to a value of (0.91 ± 0.03) eV and (2.45 ± 0.08) eV, respectively. The statically calculated formation energies, for the dissociated pair, 0.83 and 1.21 eV, respectively, can be compared with 0.69 and 1.03 eV from previous EMT theoretical studies [8].

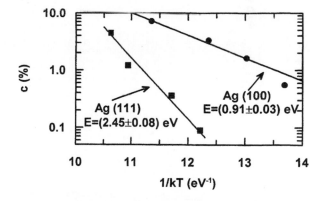

Fig. 5. Arrhenius plot of the concentration of the spontaneously created adatoms on Ag(1 0 0) (circles) and Ag(1 1 1) (squares) faces.

Acknowledgements

This work was supported by ΠΕΝΕΔ grant No. 1654/1995 of the Greek Ministry of Development and a Ψ_K network grant. We thank G. Leventopoulos for his computational assistance.

References

[1] M.H. Langelaar, M. Breeman and D.O. Boerma, Surf. Sci. 352–354 (1996) 597; M.H. Langelaar and D.O. Boerma, in: Surface Diffusion: Atomistic and Collective Processes, Ed. M. Tringides (Plenum, New York, 1997).

[2] H. Brune, K. Bromann, H. Röder, K. Kern, J. Jacobsen, P. Stoltze, K. Jacobsen and J. Nørskov, Phys. Rev. B 52 (1995) R14 380.

[3] C.L. Liu, J.M. Cohen, J.B. Adams and A.F. Voter, Surf. Sci. 253 (1991) 334.

[4] G. Boisvert and L.J. Lewis, Phys. Rev. B 54 (1996) 2880.

[5] D.E. Sanders and A.E. DePristo, Surf. Sci. 260 (1992) 116.

[6] L.S. Perkins and A.E. DePristo, Surf. Sci. 294 (1993) 67.

[7] T.J. Raeker, L.S. Perkins and L. Yang, Phys. Rev. B 54 (1996) 5908.

[8] P. Stoltze, J. Phys.: Condens. Matter 6 (1994) 9495.

[9] G. Boisvert, L.J. Lewis, M.J. Puska and R.M. Nieminen, Phys. Rev. B 52 (1995) 9078.

[10] B.D. Yu and M. Scheffler, Phys. Rev. Lett. 77 (1996) 1095.

[11] D.E. Sanders and A.E. DePristo, Surf. Sci. Lett. 264 (1992) L169.

[12] G.W. Jones, J.M. Marcano, J.K. Nørskov and J.A. Venables, Phys. Rev. Lett. 65 (1990) 3317.

[13] C. Ratsch, A.P. Seitsonen and M. Scheffler, Phys. Rev. B 55 (1997) 6750.

[14] F. Ducastelle, J. Phys. (Paris) 31 (1970) 1055.

[15] V. Rosato, M. Guillope and B. Legrand, Philos. Mag. A 59 (1989) 321.

[16] F. Cleri and V. Rosato, Phys. Rev. B 48 (1993) 22.

[17] G.A. Evangelakis and N.I. Papanicolaou, Surf. Sci. 347 (1996) 376.

[18] G.C. Kallinteris, G.A. Evangelakis and N.I. Papanicolaou, Surf. Sci. 369 (1996) 185.

[19] R. Ferrando, Phys. Rev. Lett. 76 (1996) 4195.

[20] B. Loisel, J. Lapujoulade and V. Pontikis, Surf. Sci. 256 (1991) 242.

[21] P. Statiris, H.C. Lu and T. Gustafsson, Phys. Rev. Lett. 72 (1994) 3574.

[22] L.J. Lewis, Phys. Rev. B 50 (1994) 17693.

ELSEVIER

Computational Materials Science 10 (1998) 111–115

COMPUTATIONAL
MATERIALS
SCIENCE

Modelling of boron nitride: Atomic scale simulations on thin film growth

Karsten Albe *, Wolfhard Möller

Research Center Rossendorf, Institute of Ion Beam Physics and Materials Research, PO Box 51 01 19, D-01314 Dresden, Germany

Abstract

Molecular-dynamics simulations on ion-beam deposition of boron nitride are presented. A realistic Tersoff-like potential energy functional for boron nitride, which was specially fitted to ab initio-data, has been used. The impact of energetic boron and nitrogen atoms on a c-BN target is simulated with energies ranging from 10 to 600 eV. The structural analysis of the grown films shows that a loose, dominantly sp^2-bonded structure arises at high ion flux. In no case the formation of a sp^3-bonded phase is observed, but the obtained films partially reveal textured basal planes as found in experiment. Two different growth regimes are identified for ion energies above and below 100 eV. Copyright © 1998 Elsevier Science B.V.

Keywords: Molecular dynamics; Boron nitride; Thin film growth

1. Introduction

Thin films of cubic boron nitride (c-BN) are of considerable practical interest in materials science due to their outstanding physical properties [1], such as extreme hardness, chemical inertness, high melting temperature, wide band gap and low dielectric constant, which are important for many commercial applications in modern microelectronic devices and protective coating materials. In recent years remarkable progress has been achieved in growing thin films with a significant content of c-BN crystallites, but the quality of the films deposited by several ion-beam assisted CVD and PVD methods [2] is still insufficient and the knowledge on the basic growth mechanisms is humble.

Boron nitride exhibits sp^2- and sp^3-bonded crystalline modifications similar to carbon. The hexagonal form (h-BN) is slightly different to graphite and the

zinc-blende modification (c-BN) is the analog to cubic diamond [1,3]. There are numerous similarities of both materials, but the deposition processes as well as the resulting film structures are obviously different. Because the chemical behavior of the constituents in both systems is quite distinct, analogies between carbon and boron nitride must be treated very carefully.

The physical properties of ion-beam deposited boron nitride, such as composition, structure, density, hardness, etc., depend strongly on deposition conditions. Experimentally, a competing growth of sp^2- and sp^3-bonded phases is found, which leads to a specific film structure with c-BN crystallites growing on an sp^2-bonded interlayer with vertically oriented planes [4]. Furthermore the c-BN growth regime only takes place in a specific window of deposition parameters close to the resputter limit. It depends rather complex on the three parameters substrate temperature, ion energy and flux ratio of ions to neutral atoms. The highest content of the dense modification

---

* Corresponding author. E-mail: K.Albe@fz-rossendorf.de.

is effectuated at about 300–500 eV ion energy, 500–600 K substrate temperature and a critical value of ion flux per deposited atom of 1–2 [5].

Two phenomenological growth models for c-BN are commonly discussed, which are known as "subplantation model" and "stress model". The former was introduced by Lifshitz et al. [6] for low energy subsurface processes in tetrahedral carbon and extended in a quantitative manner by Robertson [7]. The basic assumption is a balance between increment of local density under ion bombardment and relaxation processes due to the so-called thermal spikes, where defect diffusion and condensation is the driving force for relaxation and phase formation. The second model was proposed by McKenzie et al. [8]. Here the underlying idea is that formation of the dense BN modification occurs if the stress-induced hydrostatic pressure reaches the transition boundary between the sp^2- and sp^3-bonded phases in the phase diagram. Indeed, each of these models gives insights to selected aspects of BN thin film growth, but none can explain all of the commonly observed features of BN thin film growth not even, qualitatively. Therefore it is a need of more detailed investigations. Here atomistic computer simulations can be an appropriate tool in order to develop the understanding of relevant processes.

In this study we present molecular-dynamics simulations on the growth of boron nitride, which are possible now by use of a recently developed many-body potential for boron nitride. We study at the microscopic level the energy dependence of the structure of the growing film using a monoenergetic beam of neutral nitrogen and boron atoms.

2. Methods

2.1. Potential energy functional

Computer simulations of processes which lie within the scope of this work, i.e. thin film growth, require molecular-dynamics simulations on time scales in the range of hundredths of picoseconds and system sizes of many thousand atoms. Here, classical interatomic potentials have tremendous practical advances, be-cause the energy and consequently the force fields are calculated simply as functions of atomic coordinates. In this study an empirical potential energy functional for boron nitride has been used, which was developed to capture many of essential features of chemical bonding in systems consisting of boron and nitrogen [9]. It is a Tersoff-like potential, which describes structure and energy of BN polymorphs and clusters as well as pure nitrogen and boron bondings. Of course the electronic features of BN are without the scope of this potential, but to some extent they are "implicitly" taken into account by fitting the energy function to a wide range of ab initio-data.

The interatomic potential is a pair-like expression

$$\Phi = \frac{1}{2} \sum_{i \neq j} f_C(r_{ij})[f_R(r_{ij}) - b_{ij} f_A(r_{ij})]. \tag{1}$$

The distance between atom i and atom j is denoted by r_{ij} and the sum runs over all atomic sites. The repulsive and the attractive terms are chosen similar to a Morse potential as proposed by Brenner [10]:

$$f_R(r) = \frac{D_0}{S-1} \exp(-\beta\sqrt{2S}(r - r_0))$$
$$f_A(r) = \frac{SD_0}{S-1} \exp(-\beta\sqrt{2/S}(r - r_0)). \tag{2}$$

Here D_0 and r_0 are the dimer energy and separation, respectively, S and β fitting constants. The cutoff function f_C limits the interaction shell on the next neighbors inside the radius R:

$$f_C(r) = \begin{cases} 1, & r \leq R - D, \\ \frac{1}{2} - \frac{1}{2}\sin\{\pi(r - R)/(2D)\}, & \\ & |R - r| \leq D, \\ 0, & r \geq R + D \end{cases} \tag{3}$$

The expression b_{ij} determines the strength of f_A as a function of the number of interacting neighbors and bonding angles θ_{ijk},

$$b_{ij} = (1 + \gamma^n \chi_{ij}^n)^{-1/2n},$$
$$\chi_{ij} = \sum_{k \neq i,j} f_C(r_{ik}) g(\theta_{ijk}) \exp[\lambda_3^3(r_{ij} - r_{ik})^3],$$
$$g(\theta_{ijk}) = 1 + \frac{c^2}{d^2} - \frac{c^2}{[d^2 + (h - \cos\theta_{ijk})^2]}. \tag{4}$$

All constants n, c, d and h depend on the index pair ij. The corresponding parameter set is shown in Table 1.

Table 1

	BN-interaction	NN-interaction	BB-interaction
n	0.364153367	0.6184432	3.9929061
γ	0.000011134	0.019251	0.0000016
S_0	1.0769	1.0769	1.0769
β (Å^{-1})	2.043057	1.92787	1.5244506
D_0 (eV)	6.36	9.91	3.08
r_0 (Å)	1.33	1.11	1.59
c	1092.9287	17.7959	0.52629
d	12.38	5.9484	0.001587
h	−0.5413	0	0.5
λ_3 (Å^{-1})	1.9925	0	0
R (Å)	2.0	2.0	2.0
D (Å)	0.1	0.1	0.1

For short interatomic distances this potential is not sufficiently repulsive. Therefore the pair-like term has been splined to the repulsive ZBL potential [16].

2.2. Molecular dynamics

Molecular-dynamics simulations of growth processes under conditions similar to experiment remain a challenging problem for standard computer power, even if the use of empirical potentials can substantially increase the efficiency of such calculations. At present computer simulation studies have the capability for modelling detail processes. Therefore our purpose is to isolate the part that interatomic collisions and relaxations play during the layer growth.

The molecular dynamics simulations were performed using a simulation cell with periodic boundary conditions applied at the $x - z$ and $y - z$ edge planes of the system, where z denotes the direction of the surface normal. Link list technique as well as Verlet neighbor lists were applied and time integration was done by Beeman's Predictor–Corrector scheme. The alternating impact of energetic nitrogen and boron atoms on a c-BN target with dimer reconstructed boron terminated (1 0 0) surface has been simulated for different energies from 10 to 600 eV. This is equivalent to thin film deposition by mass selected ion-beam deposition [11], which allows growth of the cubic modification without use of any additional inert gas.

The target consists of 1216 atoms with 32 atoms per layer, which is sufficiently large with respect to

the penetration depth of boron and nitrogen in c-BN. All atoms were treated as neutral, because the true charge state is important only for the first collisions and does not affect the growth mechanism [12]. Inelastic processes due to electronic stopping have been neglected. For each energy a total of 300 projectiles were shot perpendicular to the surface at a random $x-y$ position. This corresponds to nearly 10 layers of an ideal c-BN or an atomic increase of 24% with respect to the initial target size. After each impact an analysis of the bond order and type was performed inside $r_{cut} = 1.9$ Å. The time intervals for particle impact have been chosen in dependence on energy from 500 fs at 10 eV to 2 ps at 600 eV. Because the resulting particle flux is several magnitudes higher than in a real experiment all atoms were cooled down to 700 K at each time-step with Berendsen's method [13] using $\Delta \tau = 50 - 100$ fs. Only the last 10 impinging atoms were followed with full dynamics employing a method of Luedtke and Landman [14]. This treatment avoids heating and melting of the target. The integration time step was 0.3 fs for all simulations. At the end all atoms were left free and the structure was relaxed for 25 ps. The resulting structures were analyzed twice, directly after the last impact and after the subsequent relaxation. Finally the bond order and environment of each atom was investigated and the fraction of sp^1-, sp^2- and sp^3-bonded atoms in the film was calculated.

3. Results and discussion

Fig. 1 shows the result of the bond order analysis. All curves exhibit a significant reversal point at about 100 eV ion energy. Here the sp^3-content of the film is nearly 75%. This means that all additional 300 atoms are purely sp^2-bonded, while the number of sp^3-bonds remains unaffected. Only at thermal ion energies (10 eV) a small increase of the fourfold coordinated atoms in the grown film is found.

Above the energy threshold of 100 eV the sp^2-fraction increases significantly. Here subsurface processes are dominant. The higher the energy the more atomic displacements and lattice distortions occur. The plot of the sp^2-fraction as function of energy has a

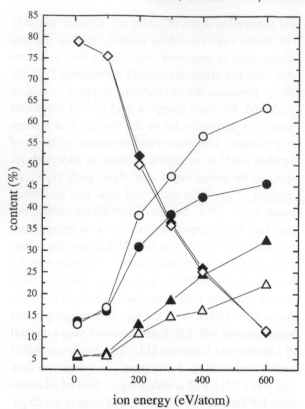

Fig. 1. Bond order analysis of a c-BN target with 1280 atoms after subsequent impact of 150 boron and 150 nitrogen atoms. Target has been cooled to 700 K after each impact. Diamonds (◇) correspond to the sp^3-, circles (○) to the sp^2-, and triangles (△) to the sp-content. Closed symbols denote the arrangement after the last impact, open symbols the arrangement after additional relaxation.

curved shape proportional to $\sqrt{E}$, which corresponds to the nuclear stopping power [15].

In all simulations the number of sputtered atoms and particles penetrating the target is lower than 1%. Generally, an amorphous-like, loose packed structure arises, which is built by two- and threefold coordinated atoms, but no evidence for the growth of a dense phase is found. Odd numbered rings, which are typical for amorphous carbon are not observable, because N–N and B–B connections are energetically unfavored. This proves the reliability of the energy function. At all energies a considerable amount of linear bonds occurs, which are necessary for producing the amorphous-like network. During the atomic relaxation the films are

Fig. 2. c-BN target after subsequent impact of 150 boron and 150 nitrogen atoms with 200 eV. The c-BN structure is partially destroyed. A small interlayer consisting of ordered, vertically oriented rings is visible.

densified and the number of sp-bonds decreases, while the sp^3-content is nearly unaffected by the subsequent relaxation. In Fig. 2 a snapshot of the target structure after the impact of 300 particles with 200 eV is depicted. Here the most remarkable fact is that on the unaffected c-BN region a small interlayer consisting of vertically oriented sixfold BN rings is observable, which is found in most simulations.

Finally, we can conclude: High flux ion bombardment on c-BN leads to a lowering of the atomic coordination in the film at energies above 100 eV. Although the applied potential energetically prefers the fourfold coordination, for entropy reasons the system tends to sp- and sp^2-bondings at conditions far from thermodynamic equilibrium. This is in agreement to Robertson's interpretation that a large number of rearrangements always leads to the formation of a low-dense phase [7]. As a consequence textured sp^2-bonded basal planes are a preferred structural arrangement. This observation corresponds to the experimental finding of textured basal planes in BN films [4] and sp^2-bonded top layers on the dense phase [5].

Hofsäss et al. [5] have recently reported that the lower energy threshold for the dense phase formation on c-BN is 100 eV. The simulations show that this is the lower limit for surface penetration. In contradiction to the experimental findings the simulations do not reveal any sp^3 formation above this threshold. Therefore we assume that the c-BN phase formation is an indirect, subsequent process, which occurs on much longer time scales, but needs the presence of well structured basal planes.

Acknowledgements

This work has been supported by a project of "Sächsisches Ministerium für Wissenschaft und Kunst".

References

[1] J. Edgar, in: J.H. Edgar (Ed.), Properties of Group III Nitrides, Inspec, 1994.
[2] S. Reinke, M. Kuhr, W. Kulisch, R. Kassing, Diamond Rel. Mat. 4 (1995) 272 .
[3] K. Albe, Phys. Rev. B 55 (1997) 6203.
[4] D.L.D.J. Kester, K.S. Ailey, D.J. Lichten-walner, R. Davis, J. Vac. Sci. Technol. A 12 (1994) 272.
[5] H. Hofsäss, H. Feldermann, M. Sebastian, C. Ronning, Phys. Rev. B 55 (1997) 13 230.
[6] Y. Lifshitz, S. Kasi, J. Rabalais, W. Eckstein, Phys. Rev. B 41 (1990) 10 468.
[7] J. Robertson, Diamond Relat. Mat. 5 (1996) 519.
[8] D. McKenzie et al., Diamond. Rel. Mat. 2 (1993) 970.
[9] K. Albe, W. Möller, K.-H. Heinig, Rad. Eff. Def. Sol. 141 (1997) 85.
[10] D. Brenner, Phys. Rev. B 42 (1990) 9458 .
[11] H. Hofsäss, C. Ronning, U. Griesmeier, M. Gross, Appl. Phys. Lett. 67 (1995) 46.
[12] M. Lu et al., Appl. Phys. Lett. 64 (1994) 1514 .
[13] H. Berendsen et al., J. Chem. Phys. 31 (1984) 3864.
[14] W. Luedtke, U. Landman, Phys. Rev. B 40 (1989) 11 733.
[15] W. Wilson, L. Hagmark, J. Biersack, Phys. Rev. B 15 (1977) 2458 .
[16] J.F. Ziegler, J.P. Biersack, U. Littmark, The Stopping and Range of Ions in Solids, Pergamon Press, Oxford, 1985.

ELSEVIER

Computational Materials Science 10 (1998) 116–126

COMPUTATIONAL
MATERIALS
SCIENCE

A Monte Carlo simulation of silicon nitride thin film microstructure in ultraviolet localized-chemical vapor deposition

J. Flicstein *, S. Pata, J.M. Le Solliec, L.S. How Kee Chun, J.F. Palmier, J.L. Courant

France Télécom CNET Laboratoire de Bagneux, PO Box 107, 92225 Bagneux, France

Abstract

Microstructural changes of surfaces and bulk of a SiN : H were investigated at the atomic level by a simulator. The simulator is based on a solid-on-solid type model for ultraviolet localized-chemical vapor deposition. The calculations consider the well-defined photolysis products adsorbed at atomic sites. Incorporation of main species is enabled by a Monte Carlo–Metropolis simulation technique. Photodeposition rates are obtained using bond dissociation energies. In this manner, the dependence of root-mean-square deviation of surface roughness and bulk porosity on operating conditions can be predicted. Photonucleation and photodeposition with a UV low pressure mercury lamp at low pressure and temperature were simulated onto indium phosphide substrate. Copyright © 1998 Elsevier Science B.V.

1. Introduction

Research in localized deposition to produce amorphous thin films onto III–V substrates is motivated by both low temperature deposit and high structural quality advantages enabled by the potential of ultraviolet induced chemical vapor deposition (UV CVD). So, applications of UV CVD with III–V materials are an attractive way of reducing deposition temperatures, to below the pyrolysis level of precursors, which are too high for such thermal sensitive materials. Thus, in addition to having the benefits of a versatile technique, for the manufacture of thin solid films for optoelectronic devices, it is equally possible to attain both, a high structural quality with a high interface quality [1]. However, in applications, this versatility can be perceived as a problem, since the optimization of the array of experimental parameters over a wide operating range can be extremely time-consuming and costly. Always there is a possibility that another set of conditions will give improved results. In the past models have been developed mostly for CVD [2] but also for UV CVD [3]. The models have included elementary gas-phase reaction. But the treatment of heterogeneous surface processes has been ignored. We consider in the present work only the heterogeneous photolysis as conducive for generation of a device quality layer. We present a Monte Carlo simulation for the deposition of amorphous SiN : H from a mixture of silane and ammonia by UV induced CVD technique. We relate the deposition conditions to bulk and surface characteristic of the film. Our physically based framework for UV CVD makes use of a previous phenomenological model, of photonucleation and early photodeposition, reported in [4]. The modeling techniques we

---

* Corresponding author. Tel.: 33 1 42 31 7285; fax: 33 142 53 49 30; e-mail: jean.flicstein@cnet.francetelecom.fr.

use are capable of simulating the deposition of thin films exceeding hundreds of monolayers. Understanding the surface dynamics for photonucleation and early photodeposition is important for controlling thin film microstructure. However, less attention has been paid to modeling of the photonucleation and UV induced early deposition. Two major reasons can be mentioned: (a) the complexity of heterogeneous photolysis; and (b) the difficulty to take into account the parasitic effect of homogeneous photolysis and photonucleation in heterogeneous photonucleation and photodeposition. Thus, in order to understand the processes by which observed structures evolve, experimental results should be completed by deposition models such as Monte Carlo simulations. While Monte Carlo simulations have been used for over 18 years to model liquid-phase epitaxy [6] they have not till now been applied to UV CVD. In the present work, the Monte Carlo simulation technique is inspired from the one used by Kotecki and Herman [7]. The purpose of this work is to gain insight into processing during UV CVD. We simulate both the layer structure and surface reconstruction and present the three-dimensional (3D) results for deposition simulation. In addition to conventional nucleation centers, the model takes into account the active charged centers (ACC) creation by UV-dependent stage of the photolysis. Concomitant precursor heterogeneous photolysis and surface diffusion processes are considered as well. As a test case is simulated here the deposition of silicon nitride by photolysis onto InP.

2. Experimental observations

An incoherent VUV light source produces a large-area irradiated region at the substrate surface [8]. During VUV chemical vapor deposition a substrate surface is in steady state with a precursor flow. Precursor molecules incident onto this reaction zone are subjected to heterogeneous photolysis into the deposit species plus volatile photoproducts. So after irradiation, the adsorbed phase is transformed to the photoproduct species to be deposited. In contrast to other conventional large-area deposition processes, in

UV CVD deposit species are laterally confined to the beam irradiated region. Because there are many short-lived intermediates and heterogeneous photoproducts, laboratory experiments are notoriously difficult to control, planify and predict. However, the qualitative features of photonucleation are remarkably robust. In particular it is observed that photonucleation occurs via substrate surface. The species resulted from photolysis are migrating toward their centers of nucleation. In general the grains are hemiellipsoidal. Small grains shrink and bigger grains grow. Any valid photodeposition model must reproduce these experimental features of the photodeposition process.

3. The model

A flow of precursors which operates under a constant pressure establishes a steady state concentration of *adspecies* (*adsorbed species*) per unit area on the substrate. In the following adspecies are considered as mono- or polyatomic and chemisorbed to the surface. Physisorbed adspecies have a vanishing contribution to the deposit [9].

3.1. The physico-chemical model for photodeposition

Low temperature photonucleation and photodeposition occurs by nine juxtaposed sequences which can describe the following physico-chemical model [4]: (1) arrival of the precursors flux into the photoreaction volume, this is composed of impinging of gas-phase species; (2) adsorption of the precursor(s) onto the substrate; (3) photon activation of the precursor decomposition at the substrate surface (heterogeneous photolysis) to produce adsorbed photoproducts; (4) relaxation by migration and incorporation or desorption of deposit species and volatile photoproducts; (5) transport of volatile photoproducts out of the photoreaction volume; (6) formation of clusters of photo-products with higher probability of survival onto photo-induced sites for nucleation; (7) coalescence clusters onto the substrate; (8) competition of clusters to produce deposit and; (9) early continuous photodeposition. Three principal mechanisms of early thin film deposition have been identified in the literature [10]: The island type deposition (or Volmer–Weber), when the deposit is more strongly bound to itself than to the substrate. The Frank–Van der Merwe mechanism is characterized by a layer-by-layer growth. The deposit is more strongly bound to the substrate than to itself. The intermediate deposition mechanism (or Stranski–Krastanov) where layer-by-layer deposition is followed by island deposition. Conform to the atomistic theory of nucleation the interrelation mechanism-deposit morphology appears in microscopic form as the binding energy between the nearest neighbors [11,12].

4. Computational method

4.1. Monte Carlo–Metropolis method

A real time Monte Carlo simulation is employed herewith the developed model, similar to that described initially by Abraham and White [13] for random determination of the events on the substrate surface.

The creation of active charged centers for nucleation under UV light exposure: The UV-light dependent stage of heterogeneous photolysis occurs on the bare semiconductor, e.g. InP, substrate and within a thin layer, like an oignon skin, that are part of the substrate (see later on in this section the definition of the surface lattice points at any site). Under UV light exposure, local charges are produced across this very thin semiconductor layer. We may consider the same forming mechanism as for charged semiconductor surfaces which constitutes an active charged center (ACC) [4]. The energy formation of a nucleus on a charged center is lower than on a neutral one [14]. Therefore probability of atoms sticking onto a charged center in the identical conditions increases. So, we suppose the nuclei to be generated with highest probability at ACCs. Hence in the photonucleation case a charged center became the preferential site for photonucleation. This working model includes the UV induced effect on ACCs surface density creation. ACCs are UV continuously created prior to and during the nucleation on

a III–V substrate. *Local supersaturation* of adsorbed species is defined at an atomic site and corresponds to the driving force $(\partial \mu)$ of the photonucleation process to create critical nuclei, where μ is the chemical potential [15]. Thus a defect free semiconductor flat surface has a low probability of atoms sticking on, which imposes an increase in the critical value of the *local supersaturation* in order to promote nucleation at the *nucleation threshold*. Inversely, this critical value of *local supersaturation* decreases at a center for nucleation. At higher luminance than the threshold, because increased surface density of attachment centers for adatoms are now available, the onset of nucleation starts when the adatom surface density attains a vanishing *critical value of local supersaturation*. In contrast to what happens in the dark, there is neither a *critical threshold* nor a critical flux density for clustering to start during UV impingement around an ACC. Each nucleus builds up a circular zone for adspecies migration. All other charged or conventional centers within this radius will be excluded from further activity, as inactive charged centers (ICC). Consequently the ICC are excluded from the photon activated nucleation (PAN).

4.2. Three-dimensional reconstruction of structures

For the sake of simplicity, a three-dimensional fcc lattice without surface steps is considered here, for the reconstruction of the film. Each lattice point may be occupied by a species or a void. The lattice point signifies the position of a particular species in the film below a given site. The lattice site signifies the (x, y, z) coordinates of a position on the film. Any long-range order is lacking in the occupation probability of a lattice point by a particular species or a void. The coordinated atoms in the species are selected among the highest probable photolysis end products. The amorphous character of the film results from the random distribution of the coordination atoms for each site in the lattice and from the random probability that a lattice site is occupied by either a species or a void. The structure of photodeposited dielectrics has been studied experimentally in detail in our laboratory [16,17]. Recent AFM (atomic force microscopy) results provide evidence that the surface roughness observed is due to both the bond strength to the substrate, of the adspecies, and the deposition temperature [18]. To check, the range values of the energy bond strength are taken from 1 to 8 eV. Concerning the adspecies, a distinction between adsorbates with one-, two- or three-atoms is addressed in this paper. Based on kinetical studies [18] and energy minimization calculations after [19,20], we predict the deposit constructed by three atoms adspecies for silicon nitride (SiN : H).

4.3. Precusor flow

The flow of neutral species, incident to the substrate, consists of an arbitrary choice of one selected species monoatomic or multiatomic. The results are limited to $SiNH_n$ $(1 \geq n \geq 0)$. We do not presently consider a mixture of different species. The flow of precursor is assumed to have the same chemical nature as the main product of heterogeneous photolysis incorporated in the film. The time increment between incident of precursor species onto the substrate surface is $\Delta t_F = (\phi L^2)^{-1}$, where ϕ is the total flow and L is the film square length, in a number of atomic sites of the lattice. It is assumed as an average lattice constant of 6.4 Å. Based on selection of a random number ran, $0 < ran \leq 1$, a species is directed isotropically at vertical incidence towards the substrate surface lattice for every Δt_F. Peculiar morphology comportment of the selected site is included and will be analyzed elsewhere [9].

4.4. Description of heterogeneous photolysis reactions

At low pressure (< 10 Torr), the heterogeneous photolysis reactions are effective on substrate surface with an *overall quantum yield* which corresponds to *the overall photolysis* reaction. The precursors photolyze into deposit species. This defines for a deposit species the *surface reaction efficiency* which includes the *overall quantum* yield. Following their physicochemical activity, the model classifies the substrate surface lattice points at any site in four distinct kinds: (a) site fully coordinated to other identical species, (b) site fully coordinated to other species but which are not necessary identical, (c) site exhibiting UV induced activity (ACC), and (d) defective site e.g. exhibiting one or more dangling bonds. A species incident on an (a) kind inert site is assumed to adsorb onto the surface. All species incident on a (c) or (d) kind activated sites may directly incorporate into the site. If the luminance is low, the corresponding fractional surface density of ACC is small. This condition is important in determining the surface morphology. The model ignores the UV contribution to cold annealing and the densification of the film [21].

4.5. Surface diffusion

The following assumptions are made concerning surface diffusion: (a) single species are allowed to diffuse, but only by surface migration, (b) species diffuse a length which corresponds to the column spacing in a single event, and (c) surface diffusion rates are isotropic. During the simulation adspecies, in the present case SiN : H, relax from surface site to surface site. Adspecies are moved from their current surface site with each increment of time. The direction of move, or *relaxation*, is determined by choice of a random number. An adspecies is allowed to a specified number of relaxations before being desorbed from the substrate surface. Then the number of relaxations is an input parameter to give a reasonable average for the sticking coefficient.

5. Simulation results and discussion

In this section we will present and discuss simulation results obtained from our model for UV deposition films from silane and ammonia precursors, when the precursor flow is specified as a parameter. We first begin the parametric discussion for typical nucleation phase as shown in Fig. 1. The *number of cycles* of species relaxation, for surface dynamics in this case, was subjected to the simulation. The simulated nucleation conditions, taken identical with the

<u>Results and discussion</u> :

TWO-DIMENSIONAL SIMULATION OF SILICON NITRIDE PHOTONUCLEATION

THE MAIN STAGES : island formation, coalescence, nuclei survival and competition

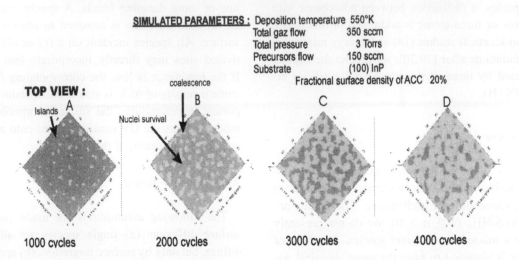

SIMULATED PARAMETERS : Deposition temperature 550°K
Total gaz flow   350 sccm
Total pressure   3 Torrs
Precursors flow   150 sccm
Substrate   (100) InP
Fractional surface density of ACC 20%

TOP VIEW :

Islands A   coalescence B   C   D

Nuclei survival

1000 cycles   2000 cycles   3000 cycles   4000 cycles

Fig. 1. Top view of 2D Monte Carlo simulation of the photonucleation main stages. The arrows are pointed on small nuclei, survival and coalescence onto the substrate $(50 \times 50$ atomic site^2).

experimental conditions, are substrate temperature 550 K, total gas pressure 3 Torr, and UV luminance 26 mW cm^{-2} sr. The total species flow incident to the substrate is 350 sccm in the mixture SiH$_4$/NH$_3$/N$_2$. The normalized frequencies for examination of the nearest, the next nearest neighbors and the next to the next nearest neighbors are: $\upsilon_n = 81.2\%$, υ_{nn}=17.8%, and $\upsilon_{nnn} = 1.0\%$, respectively. The calculated 2D-nucleation rate from the model using this parameters is 0.1 Å s^{-1}. The simulated 2D nucleated fraction of the surface is (a) 0.31, (b) 0.48, (c) 0.60 and (d) 0.70. The appearance of the 2D image agrees well with recent atomic force microscopy measurements (AFM) of UV deposited films. The phenomena of island formation, coalescence of nuclei and competition are evidenced by arrows. Our predicted morphology concerning photonucleated grains, however, is not always a cross section of an hemiellipsoid. We interpret this discrepancy, as concerning the exhibited morphology, because we did not include this constraint in the simulator.

5.1. Results of surface roughness

The surface roughness of SiN : H films is important, with respect to films used in optoelectronic devices and in multilayer structures, where the lattice period may be comparable to the surface roughness, which results in poor dielectric properties. UV CVD very high luminance conditions, like lasers, which have a predominance of producing species which have high sticking coefficients, e.g. species with many dangling bonds, generally result in rougher films. The low surface mobility for these species results in random surface features and magnify the surface roughness. Figs. 2(I)–(III) depicts a typical series of 3D simulated surface structures of early deposition mode for low, high and medium adhesion, respectively. The parameter is the dissociation energy, E_{aa} and E_{as}, respectively, for adspecies–adspecies and adspecies–substrate interactions [22]. We find that the surface roughness generally follows the three early deposition modes previously discussed (see Fig. 3). In summary,

Results and discussion : THREE-DIMENSIONAL SIMULATION OF EARLY PHOTODEPOSITION MODES

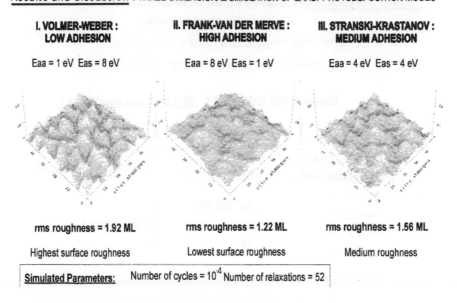

I. VOLMER-WEBER : LOW ADHESION	II. FRANK-VAN DER MERVE : HIGH ADHESION	III. STRANSKI-KRASTANOV : MEDIUM ADHESION
Eaa = 1 eV Eas = 8 eV	Eaa = 8 eV Eas = 1 eV	Eaa = 4 eV Eas = 4 eV

rms roughness = 1.92 ML	rms roughness = 1.22 ML	rms roughness = 1.56 ML
Highest surface roughness	Lowest surface roughness	Medium roughness

Simulated Parameters: Number of cycles = 10^{4} Number of relaxations = 52

Fig. 2. Three-dimensional simulation of early deposition modes exhibiting the surface roughness.

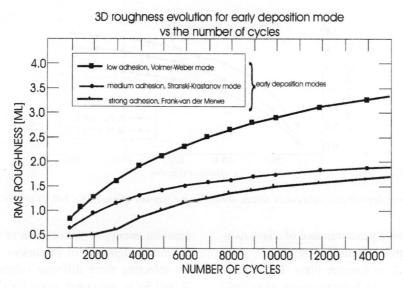

Fig. 3. Three-dimensional roughness evolution vs. the number of cycles.

we use the physico-chemical photodeposition model with Metropolis probability function on an 50×50 atomic sites fcc lattice. These results are shown in Fig. 2(I). To investigate the effect of adsorbant energy of dissociation we parametrized our model while varying the number of cycles of relaxations (see

Fig. 3). For medium and strong adhesion, we find that the surface roughness becomes quite constant after 8000 cycles, below which surface roughness increases markedly. These results imply that surface adsorbant energy of dissociation is most influential in determining the surface roughness. UV CVD

Results and discussion : roughness (rms) three-dimensional validation for silicon nitride photodeposition

Surface morphology microstructure and rms roughness

characterised by atomic force microscopy (1μm x 1μm)

	Experimental conditions :	3D-simulated conditions :
Substrate (100) InP	↓	↓
Substrate temperature, °K	550	550
Total pressure, Torrs	3	3
Total flow, sccm	350	350
Precursors flow	150 (NH₃/SiH₄)	150
Thickness, Å	500	140 (50000 cycles)
Roughness, Å	14	16

Fig. 4. Roughness (rms deviation) for silicon nitride photodeposition: experimental, simulated conditions and results.

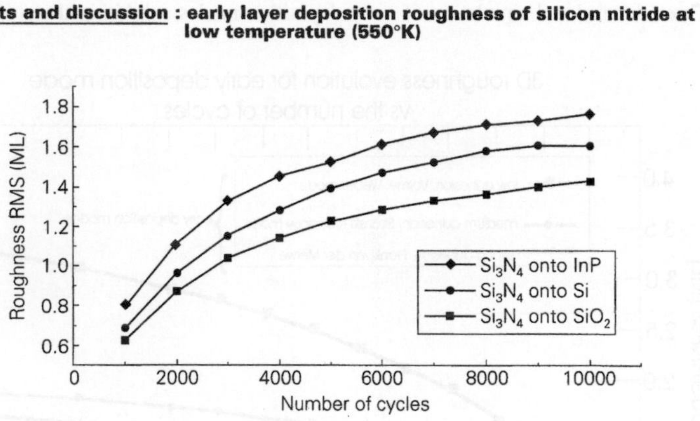

Results and discussion : early layer deposition roughness of silicon nitride at low temperature (550°K)

Fig. 5. Early layer deposition roughness of silicon nitride vs. the number of cycles for InP, Si and SiO₂ substrates.

conditions which have a predominance of silane radicals, which have high interaction energy ($E_{as} \sim 8$ eV [23]), generally result in rougher films. The deposition of silicon nitride by heterogeneous photolysis of a mixture $SiH_4 + NH_3$ is used as a test case for modeling and simulation. For the experimental and simulated conditions of silicon nitride photodeposition, see Fig. 4. Surface morphology microstructure and root-mean-square (rms) deviation were validated as described in the following. As the surface roughness is a function of number of cycles and substrate temperature, we performed our calculations for both constant energy of dissociation of the adsorbant and constant magnitude of precursors flow (see Fig. 5). By selecting three different substrates, namely InP, Si and SiO₂, we check again for the three modes of photodeposition, now in the case of silicon nitride ($E_{aa} = 4.9$ eV) onto InP ($E_{as} = 3.18$ eV). The surface fraction of ACCs of the simulated film is 0.2. Surface roughnes is a function of substrate temperature. In Fig. 6 we found a steep simulated evolution, by using incident precursors flow of 134 sccm at total pressure of 3 Torr, whereas this dependence on temperature has to be translated (for about 2 Å, for

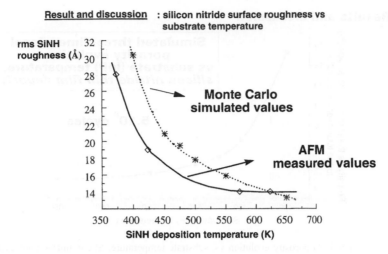

Fig. 6. Silicon nitride surface roughness vs. surface temperature for 3D simulated and measured values.

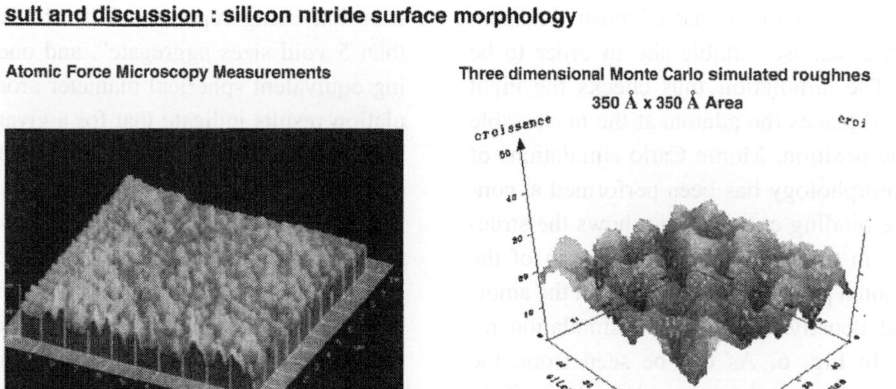

sult and discussion : silicon nitride surface morphology

Fig. 7. Silicon nitride 3D surface morphology for measured and simulated values.

the rms deviation) for similarity with atomic force microscopy measurements. For RT (rapid thermal) CVD [17] found that for conditions dominated by fast deposition, surface roughness increased with increasing film thickness; while the rms deviation of roughness in our case has only a weak dependence on layer thickness, for thicknesses greater than 40 monolayers, about 260 Å (see Drevillon in [17]). Our results are shown in Figs. 5–7. For UV CVD conditions they agree with the AFM measurements and

tend to confirm the results previously reported in the literature.

5.2. Results of simulation of structural evolution of silicon nitride density

Incorporation can occur only when the species have satisfied the stability criterion [19,20]. Because of this restriction on acceptable surface sites for deposition, a fraction of attempted incorporation events does not

Results and discussion

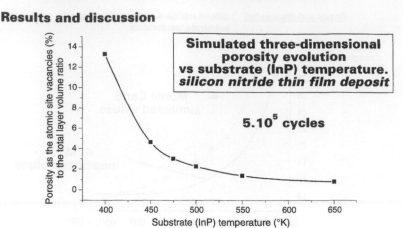

Fig. 8. Simulated 3D porosity evolution vs. substrate temperature. Slicon nitride photodeposit.

succeed. So we have assumed that a deposited adatom can relax to the nearest suitable site in order to be incorporated. The simulation thus checks the eight nearest sites and places the adatom at the most stable first acceptable position. Monte Carlo simulations of the thin film morphology has been performed at constant adsorbate binding energy. This shows the structural evolution in time during UV irradiation of the surface. Additional picture is obtained about the amorphous material density. Monte Carlo simulation result is shown in Fig. 6. As can be seen from, the bulk atomic arrangement changes with the progress in SiN : H deposition. It is well demonstrated that the adsorbed species gradually form the amorphous stochastic structure with repeating cycles of relaxation. The final structure fulfills the bulk and shows the porosity development of the SiN : H system.

Silicon nitride density evolution with temperature: In the present work we are careful to define the porosity. The measure of porosity is the fraction of free lattice sites, or defective porous sites, below the deposit surface. In this definition the surface roughness is ignored. So the porosity is the defect density below the deposit surface.

Void formation: Experimentally void fraction tends to increase with increasing deposition rate and decrease with the increasing substrate temperature [24]. The range number of the void sizes obtained in our simulations is approximately 5–60. 90% of this dis-

tribution being represented by the fraction of "less than 5 void sizes aggregate", and one void size having equivalent spherical diameter around 6.5 Å. Simulation results indicate that for a given flow ratio and adsorbate–substrate binding energy, for silicon nitride onto $(1\,0\,0)$ InP there is a steep porosity decrease with increasing temperature, up to 500 K (see Fig. 8) playing a direct influence onto the overall sample density. For temperatures lower than a *threshold*, low density silicon nitride is deposited while at a higher deposition temperature, high density silicon nitride results at constant luminance. Using the same silicon nitride samples but now voluntarily oxidized, the characterization by InfraRed Fourier Transform (FTIR) confirms that the porosity of the deposit is in the same proportion as is the simulated porosity. These results imply that the oxidation is effective in bulk only at the defective porous sites below the surface.

Evolution in time of silicon nitride thickness: To simulate silicon nitride films using realistic thickness we computed species flow (the solid line) fitting with the UV CVD model described in [18] (see the insert in Fig. 9). The deposition conditions were total pressure 3 Torr and a substrate temperature of 423 K with a $SiH_4/NH_3/N_2$ gas mixture 334 sccm. The luminance was kept constant as reported. Simulated film proporties for UV CVD of silicon nitride with a constant magnitude of precursor flow are shown in Fig. 9. The agreement with the fitted experiment by the model was

Results and discussion : silicon nitride photodeposited thickness onto InP vs time. Validation.

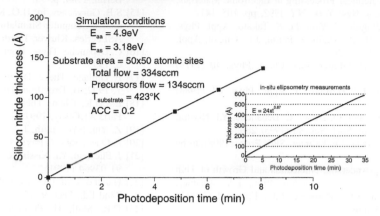

Fig. 9. Simulated silicon nitride thickness vs photodeposition time. Insert: experimental in situ ellipsometry measurements vs. time.

obtained. This result implies that the simulator proposed was validated for silicon nitride (SiN : H).

6. Conclusion

A model for photon induced nucleation during low pressure UV CVD has been applied to early layer deposition of amorphous silicon nitride (SiN : H) from silane and ammonia onto strongly adhering substrates such as SiO_2, onto weakly adhering substrates such as InP, and onto medium adhering substrates such as silicon. The model includes the UV induced concomitant effects of heterogeneous photolysis of precursors and creation of active charged centers (ACC). The Monte Carlo–Metropolis technique takes place, respectively, over an area of 50×50 atomic sites and a 3D simulation $50 \times 50 \times 50$ atomic sites at constant luminance and temperature. Computations were made for the low temperature case of photodeposition. The UV creation of active charged centers (ACC) provided stable nucleation sites for the self-assembled 2- or 3D islands. For constant heating, in the UV CVD temperature range suitable for the III–V substrates, the photonucleation stage of layer deposition is shown to strongly influence the early 3D deposition morphology and roughness. Another purpose of this work has been to identify a specific source that leads to pore formation. We have found a general trend. For low

temperatures, low density silicon nitride is photodeposited while at higher substrate temperature than a threshold, high density silicon nitride results at constant luminance. All predictions show considerable accuracy, compared with the experience, although the model did not regard the contributions from homogeneous photolysis and homogeneous nucleation in simulations. The agreement obtained between the simulated roughness and the surface roughness measured by Atomic Force Microscopy (AFM) from the photodeposited silicon nitride gives strong support to the validity of the model and the simulation technique. This is also the case for temperature influence on high to low density layer evolution.

Acknowledgements

The authors would like to thank Dr. M. Sigelle and Mr. J. Marquez (from ENST Paris) for helpful discussions, and Mr. C. Daguet for AFM measurements.

References

[1] S. Fujita, F.Y. Takeuchi, Sg. Fujita, Jpn. J. Appl. Phys. 27 (1988) L2019.
[2] M.E. Coltrin, R.J. Kee, J.A. Miller, J. Electrochem. Soc. 133 (1986) 1206.
[3] M. Petitjean, Thèse, Docteur en Sciences, Université Paris XI Orsay, 1991.

[4] J. Flicstein, J.E. Bourée, in: I.W. Boyd, R.B. Jackman (Eds.), Photochemical Processing of Electronic Materials, Academic Press, New York, NY, 1992, pp. 105–141.

[5] A. Tate, K. Jinguji, T. Yamada, N. Takato, Appl. Phys. A38 (1985) 221; B. Allain, J. Perrin, J.L. Guizot, Appl. Surf. Sci. 36 (1989) 205.

[6] J.D. Weeks, G.H. Glimer, Adv. Chem. Phys. 40 (1979) 157.

[7] D.E. Kotecki, I.P. Herman, J. Appl. Phys. 64 (1988) 4920.

[8] J. Flicstein, Y. Vitel, O. Dulac, C. Debauche, Y.I. Nissim, C. Licoppe, Appl. Surf. Sci. 86 (1995) 286.

[9] J. Flicstein, S. Pata, J.F. Palmier, SPIE Proc. (1998), to be published.

[10] B. Lewis, J.C. Anderson, Nucleation and Growth of Thin Films, Academic Press, London, 1978; J.A. Venables, G.L. Price, in: J.W. Mathews (Ed.), Epitaxial Growth, 1975, pp. 382–436.

[11] D. Walton, J. Chem. Phys. 37 (1962) 2182.

[12] S. Stoyanov, Thin Solid Films 18 (1973) 91.

[13] F.F. Abraham, G.M. White, J. Appl. Phys. 41 (1970) 1841.

[14] A.A. Chernov, Modern Crystallography III, Springer, Berlin, 1984, p. 79.

[15] S.B. Goryachev, in: H.O. Kirchner, L.P. Kubin, V. Pontikis (Eds.), Computer Simulation in Materials Science, Kluwer ASI series, Kluwer, Dordrecht, 1996, p. 17.

[16] C. Debauche, Thèse, Docteur en Sciences, Université Paris 6, 1993.

[17] F. Leblanc, Thèse, Docteur en Sciences, Université Paris 7, 1992; B. Drevillon, Thin Solid Films 130 (1985) 165.

[18] L.S.H.K. Chun, Thèse, Docteur en Sciences, Université Paris XI Orsay, 1997.

[19] Z. Yin, F.W. Smith, J. Vac. Sci. Technol. A 9 (1991) 972.

[20] P. Quémérais, J. Phys. 4 (1994) 1669.

[21] J. Flicstein, Y.I. Nissim, C. Licoppe, Y. Vitel, French Patent 91 03964 (1991).

[22] Handbook of Physics and Chemistry, Rubber Company, 74nd Ed., Sect. 9, 1992.

[23] D.R. Stull, H. Prophet (Eds.), Janaf Thermochemical Tables, 2nd. Ed., US GPO, Washington, DC, 1971.

[24] O. Kuboi, M. Hashimoto, Y. Yatsurugi, H. Nagai, M. Aratani, M. Yanokura, S. Hayashi, I. Kohno, T. Nozaki, Appl. Phys. Lett. 45 (1984) 543.

Computational Materials Science 10 (1998) 127–133

COMPUTATIONAL
MATERIALS
SCIENCE

Heat and mass transfer during crystal growth

Koichi Kakimoto *, Hiroyuki Ozoe

Institute of Advanced Material Study, Kyushu University, 6-1 Kasyga-Koen, Kasuga, 816 Japan

Abstract

Quality of semiconductor and oxide crystals which are grown from the melts plays an important role for electronic and/or optical devices. The crystal quality is significantly affected by the heat and mass transfer in the melts during crystal growth in a growth furnace such as Czochralski or horizontal Bridgman methods. This paper reviews the present understanding of phenomena of the heat and mass transfer of the melts, especially instability of melt convection from the detailed numerical calculation, which helps to understand the melt convection visualized using X-ray radiography. Large scale simulation of melt convection during crystal growth is also reviewed.

Characteristics of flow instabilities of melt convection with a low Prandtl number (ratio between momentum and thermal diffusivities) are also reviewed by focusing on the instabilities of baroclinic, the Rayleigh–Benard and the Marangoni–Benard, from the points of view of temperature, rotating and/or magnetic field effects during crystal growth. Oxygen concentration in grown crystals is also discussed how melt convection affects. Copyright © 1998 Elsevier Science B.V.

1. Introduction

The Czochralski (CZ) [1] and/or horizontal Bridgman (HB) [2] crystal growth technique are widely accepted for fabricating high-quality substrates for silicon (Si) VLSIs, gallium arsenide (GaAs) monolithic and integrated circuit devices, and optical devices. It is well known that the breakdown voltage of oxide layer grown on Si substrates strongly depends on the growth conditions under which the Si crystals were grown, such as the crystal pulling speed [3] or the temperature distribution in the growth furnace [4]. The melt convection should be controlled from the point of view mentioned above. However, an actual flow has been hard to be monitored for semiconductor or oxide melts on account of opaque melts.

So far, the distribution of impurities and point defects in grown crystals, which affects the degradation of the breakdown voltage for silicon devices, is dependent on the amplitude of temperature fluctuation at solid–liquid interface [3]. The fluctuation is mainly caused by the flow instability which contains laminar or turbulent flows. Therefore, the origin of the flow instability should be clarified, so that it can be controlled to obtain high-quality crystals. Several possibilities of the instabilities have been reported in the CZ crucible, for example Rayleigh–Benard and baroclinic instabilities by using model fluids [5]. Visualization using numerical simulation and experimental setup would be essential for correct understanding of the flow of actual molten silicon itself to confirm the origin of the instabilities.

The purpose of the present paper is to introduce hierarchy of possible flow instabilities in an actual CZ system which were clarified by a large scale computation

* Corresponding author. Tel.: +81 92 583 7836; fax: +81 92 583 7838; e-mail: kakimoto@cm.kyushu-u.ac.jp.

of fluid flow and experimental observation using X-ray radiography. Additionally, magnetic field effects on melt flow and oxygen distribution in the melt are also reviewed.

2. Numerical techniques and models

The control volume method was used for discretizing the governing equations such as continuity, Navier–Stokes, energy, and impurity transfer equations in the present calculation. Time-dependent calculation with three-dimensional geometry was performed by using alternating directional implicit (ADI) method [6] as a matrix solver. Governing equations are expressed in Eqs. (1)–(3). When the effect of axial magnetic fields was taken into account, Lorentz force (f_L) expressed by Eq. (4) was included in Eq. (3) as an external force,

$$\frac{\partial \rho}{\partial t} + \nabla \rho u = 0, \tag{1}$$

$$\rho \frac{\partial \Phi}{\partial t} + u(\rho \nabla \Phi)$$
$$= \nabla(\Gamma \nabla \Phi) + S_\Phi \quad (\Phi = u, T, c), \tag{2}$$

$$S_\Phi = -\nabla_p + F(= \rho g) + f, \tag{3}$$

$$f_L = J \times B, \tag{4}$$

$$J = \sigma(-\nabla \Psi + u \times B), \tag{5}$$

where ρ, T, Γ are density, temperature and diffusivity for the variables such as velocity, temperature and impurity concentration. p, f and g are pressure, an external force and gravitational acceleration, respectively. u, T and c are velocity, temperature and impurity concentration, respectively. S is a source term of each valuable of velocity, temperature and impurity concentration. J, Ψ, σ and B are electric current, scalar potential, electric conductivity and magnetic field, respectively.

When oxygen transfer in silicon melt was taken into account in the calculation, the following assumption of equilibrium concentration at an interface between the melt and a crucible, and flux at the boundary melt-

Table 1
Thermophysical properties for numerical simulation [12]

Density (Kg/m^3)	2520
Heat capacity (J/m^3 K)	2.39×10^6
Dynamic viscosity (Kg/ms)	7×10^{-4}
Thermal expansion coefficient (K^{-1})	1.4×10^{-4}
Latent heat of fusion (J/m^3)	4.14×10^9
Melting temperature (K)	1685
Thermal conductivity (W/mK)	45
Emissivity	0.3
Electrical conductivity (S/m)	12.3×10^5

gas was adopted as expressed by Eqs. (6) [7] and (7) [8],

$$o = 3.99 \times 10^{23} \exp(-2.0 \times 10^4/T) \text{ atoms/cm}^3, \tag{6}$$

$$q = h(O(\text{melt}) - O(\text{gas})), \tag{7}$$

where h and O are mass transfer coefficient at an interface between the melt and ambient gas, and oxygen concentration in the melt, respectively. h and $O(\text{gas})$ are imposed to 2 and 0 atoms/cm^3, respectively [8]. Thermophysical properties used in the calculation are listed in Table 1.

Numerical simulation with a three-dimensional configuration with a grid of $60 \times 40 \times 40$ in r, θ, z directions was carried out. The geometry of the present calculation was set identical to the experimental one with a 3 in. diameter crucible and 1.5 in. crystals. The temperature boundary conditions of the melt, which strongly affect the flow mode [9], were set to be axisymmetric to identify the origin of non-axisymmetric profile.

One of the authors has succeeded in the direct observation of molten silicon convection by using X-ray radiography method [10] including an effect of magnetic fields [11]. Two sets of X-ray sources and X-ray cameras are set to a furnace to visualize melt flow using a special tracer [10]. Flow velocity obtained from the visualization can be compared with numerical result which can check a validity of both experimental and numerical results.

3. Temperature and rotation effects

Flow visualization was carried out using the following condition to clarify temperature effects. Crystal and crucible rotated with angular velocities of +1 and −1 rpm, respectively, to study an effect of natural convection. Crystal growth was stopped to remove a modification of temperature distribution along boundaries.

Figs. 1(a) and (b) show plane views of calculated velocity vectors and temperature contours on 5 mm below the surface plane. If the flow mode is exactly axisymmetric azimuthal, velocity component should not be observed. However, alternative modulation of azimuthal velocity component which is almost one

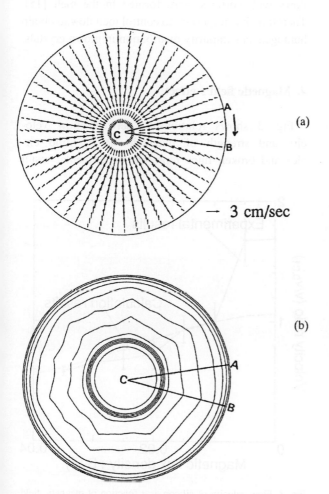

Fig. 1. Plane views of calculated velocity vectors (a) and temperature contours (b) on the plane of 5 mm below the surface.

order of magnitude smaller than radial velocity existed. The lines indicated by A and B in Fig. 1(a) are corresponding to the lines indicated by A and B in Fig. 1(b) [12]. It can be easily recognized that fluid is ascending from inside of the melt and spreading out in the azimuthal direction along line A at the top of the melt, although it is descending into inside of the melt along line B. Lines A–C and B–C correspond to positions in which flow is ascending and descending along the lines, respectively. Additionally, temperature in the ascending part is higher than that in the descending part. These results lead us to understand that cell structure is originated by the Rayleigh and/or thermocapillary Benard instabilities [12]. Almost the same velocity profile can be observed by experiments using X-ray radiography [10].

From the above results, we are able to conclude that the spoke pattern is formed in silicon melt in a CZ crucible. Additionally, we can conclude that the pattern penetrates into the bottom of the silicon melt although it has been thought to be terminated just beneath the surface for the case of oxide melts [13]. The reason for the increase of penetration depth may be attributed to the large flow velocity which transfer momentum of the melt readily to the bottom. Since silicon melt has small viscosity, the Rayleigh number ($Ra = g\beta\Delta T L^3/\alpha\nu$ which contains viscosity becomes large, where β, ΔT, L, α and ν are volume expansion coefficient, temperature difference between crystal and crucible wall, crucible radius, thermal diffusivity and kinematic viscosity, respectively. Consequently, the flow becomes inertial flow dominant. This means that the large momentum of fluid transfers from surface to the bottom readily. Therefore, the pattern reaches from the surface to the bottom. The origin of the pattern is thought to be both the Benard [12] and/or thermocapillary [12] instabilities, although it is difficult to identify which instability is dominant because reported values of temperature gradient of surface tension which originates surface-tension-driven flow have some sort of ambiguity. Therefore, it is important to obtain reliable value of the temperature gradient of surface tension which produces thermocapillary flow.

We previously reported that the flow becomes axisymmetric at a ΔT (= 55 K) for the small crucible case: $H = 3.84$ cm, $\omega_c = -1$ rpm, $\omega_s = 1$ rpm, $R_c = 3.75$ m, and $R_s = 1.75$ cm [14], where H, ω_c, ω_s, R_c, and R_s are melt height, crucible and crystal rotation rates, crucible radius and crystal radius, respectively. In the second calculation, both crucible and crystal rotation rates were set to $\omega_c = -4$ rpm, $\omega_s = -4$ rpm to study rotational effect. The calculated velocity vectors from rotational viewpoint with the same angular velocity as crucible rotation rate and from stationary point of view are shown in Figs. 2(a) and (b) [14],

respectively. Wavy structure with wave number of 2 in azimuthal direction can be recognized which can also be observed from experiments of X-ray radiography. It is also clarified that the vortices are penetrated into the bottom.

Benard convention was found in the melt which is attributed to temperature distribution in the melt when crucible and crystal rotation rates were small. Vortice's structure which was observed from a rotating view was observed when crucible rotation rate was large.

When crystals were grown under axisymmetric flow, small inhomogeneity of oxygen was found in grown crystals. However, large fluctuation of oxygen was observed in grown crystals when asymmetric flows with vortices were formed in the melt [15]. Therefore, it is important to control melt flow to obtain homogeneous impurity distribution in grown crystals.

4. Magnetic field's effect

Fig. 3 shows relationship between flow velocity and strength of magnetic fields. Solid circles and broken lines represent experimental result

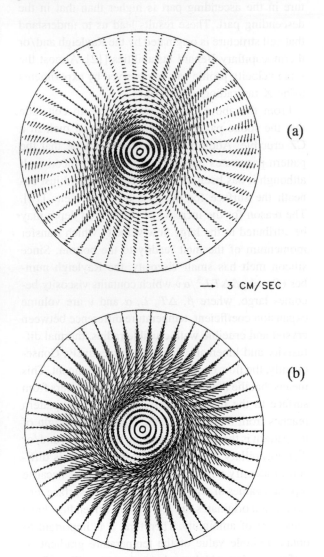

(a)

(b)

⟶ 3 CM/SEC

Fig. 2. Calculated velocity vectors from rotational viewpoint with the same angular velocity as crucible rotation rate (a) and vectors from stationary viewpoint (b).

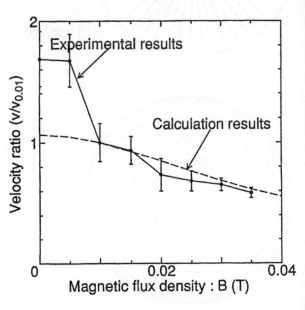

Fig. 3. Flow velocity of silicon as a function of magnetic field strength. Solid and broken lines are calculated and experimental results, respectively.

and numerically calculated results by using three-dimensional calculation [16]. The calculated results agree well with experimental one except at low magnetic fields. The discrepancy in low magnetic fields may be due to the following reason. We obtained a steady solution from numerical calculation even if we carried out transient calculation, however, the actual flow was time-dependent. This means the present calculation offers too stabilized solution. Therefore, it is necessary to use an algorithm with higher order accuracy for non-linear term in Navier–Stokes equation.

Trend of the reduction of flow velocity sometimes has been explained by a concept of "magnetic viscosity" [17], however, the concept does not directly express fluid motion. Fluid motion of molten silicon is able to express by "magnetic number" [17] contrary to the magnetic viscosity [17]. The concept is based on the inertial flow regime, therefore, thickness of velocity boundary layer near a crucible wall is almost constant with and without magnetic fields.

Temperature at the bottom of the crucible was set at two different values, 1412°C (melting point) which is identical to types A in Fig. 4(a), and 1430°C, which is identical to type B in Fig. 4(b) to clarify how Benard

convection occurs under the two different temperature boundary conditions of types A and B [16]. These two types of temperature distribution could be obtained by modification of heater system shown in Fig. 4.

An axisymmetric flow pattern was obtained from the numerical simulation with type A heating system within the magnetic field from 0 to 0.3 T. However, a non-axisymmetric flow pattern was observed in the simulation with type B heating system as shown in Figs. 5(a) and (b), which indicate profiles of the velocity and temperature distribution at the top of the melt under a magnetic field of 0.1 T. This pattern can be found in the range of magnetic fields above 0.1 T.

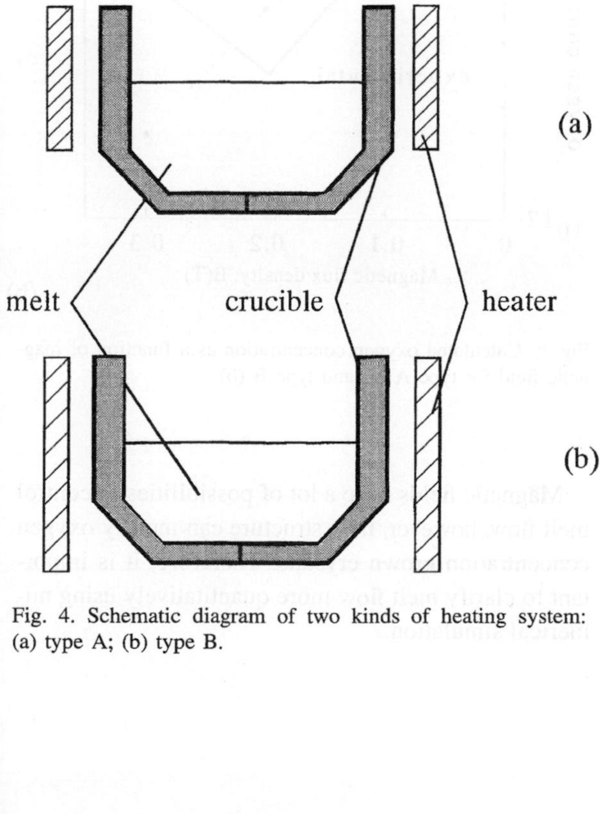

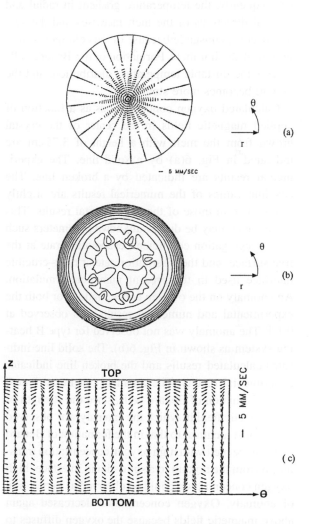

Fig. 5. Calculated velocity vectors (a), temperature distribution (b) and velocity vectors in a z–θ plane (c).

Fig. 4. Schematic diagram of two kinds of heating system: (a) type A; (b) type B.

When magnetic field becomes larger than 0.1 T, flow velocity becomes small, therefore, temperature profile becomes almost axisymmetric. To understand the non-axisymmetric structure in more detail, the velocity profile in z–θ plane is shown in Fig. 5(c) at a radius of 1 cm under a magnetic field of 0.1 T. A cell structure similar to Benard cells can be recognized [16]. This cell structure can be observed only with the type B heating system, and it was not observed with the type A system. This suggests that the origin of the cell structure was Benard instability. Since vertical magnetic fields were applied to the melt, radial flow was suppressed due to the Lorentz force [18]. Consequently, the temperature gradient in radial and vertical directions of the melt increases and the system becomes unstable because of the cold melt sitting on top of the hot melt. Formation of the Benard cells relaxes the unstable temperature distribution, and the system becomes more stable.

Calculated oxygen concentrations as a function of applied magnetic fields in the center of the crystal grown from the melt with a height of 3.75 cm are indicated in Fig. 6(a) by a solid line. The experimental results are indicated by a broken line. The absolute values of the numerical results are slightly different from those of the experimental results. This discrepancy may be due to unknown parameters such as the segregation coefficient, evaporation rate at the free surface, and the dissolution rate at melt–crucible interface used in the present numerical simulation. An anomaly on the oxygen concentration for both the experimental and numerical results was observed at 0.1 T. The anomaly was not observed for type B heating system as shown in Fig. 6(b). The solid line indicates calculated results and the broken line indicates experimental results. The anomaly can be attributed to the formation of Benard cells, since the strength of the magnetic field in which the anomaly was observed was almost identical to that in the formation of Benard cells. Since melt with high oxygen concentration transferred from the bottom to a solid–liquid interface, high oxygen concentration could be observed at a condition of anomaly. Oxygen concentration decreased again above magnetic fields because the oxygen diffuses to the top of the melt and evaporates into the gas phase.

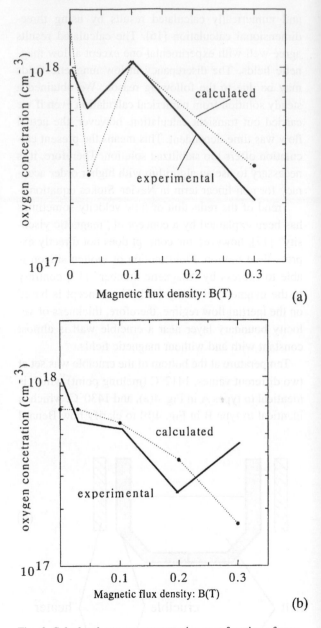

Fig. 6. Calculated oxygen concentration as a function of magnetic field for type A (a) and type B (b).

Magnetic fields have a lot of possibilities to control melt flow, however, flow structure can modify oxygen concentration grown crystals. Therefore, it is important to clarify melt flow more quantitatively using numerical simulation.

5. Conclusion

Large scale computation with time-dependent and three-dimensional including global model [19–21] is going to become a powerful tool to investigate fluid flow during crystal growth. The computation can offer more accurate result if it can use more appropriate algorithm for an actual flow of the melts of semi-turbulent flow. Therefore, new algorithms are expected which can express flow structure between laminar and turbulent flow.

Acknowledgements

The authors would like to acknowledge Mr. Minoru Eguchi and Dr. Yi Kung-Woo of Fundamental Research Laboratories of NEC Corporation for their skillful experiment and fruitful discussion. A part of this work was conducted as JSPS Research for the Future Program in the Area of Atomic-Scale Surface and Interface Dynamics.

References

[1] J. Czochralski, Z. Phys. Chem. 92 (1917) 219.
[2] W. Dietze, W. Keller and A. Muhlbauer, Crystals – Growth, Properties and Applications, ed. J. Grabmaier, Vol. 5 (Springer, Berlin, 1982).
[3] H. Oya, Y. Horioka, Y. Furukawa and T. Shingyoji, in: Extended Abstracts of the 37th Spring Meeting of The Japan Society of Applied Physics and Related Society (1990).
[4] E. Dornberger and W.V. Ammon, J. Electrochem. Soc. 143 (1996) 1636.
[5] J.R. Ristorcelli and J.L. Lumley, J. Crystal Growth 116 (1992) 447.
[6] D.W. Peaceman and H.H. Rachford, J. Soc. Indust. Appl. Math. 3 (1955) 28.
[7] H. Hirata and K. Hoshikawa, J. Crystal Growth 96 (1989) 747.
[8] K. Kakimoto, Y.-W. Yi and M. Eguchi, J. Crystal Growth 163 (1995) 238.
[9] M. Mihelcic, C. Wingerath and Chr. Pirpon, J. Crystal Growth 69 (1984) 473.
[10] K. Kakimoto, M. Eguchi, H. Watanabe and T. Hibiya, J. Crystal Growth 88 (1988) 365.
[11] K. Kakimoto and K.-W. Yi, Physica B 216 (1996) 406.
[12] K.-W. Yi, K. Kakimoto, M. Eguchi, M. Watanabe, T. Shyo and T. Hibiya, J. Crystal Growth 144 (1994) 20.
[13] A.D.W. Jones, J. Crystal Growth 65 (1983) 124.
[14] K.-W. Yi, V.B. Booker, M. Eguchi, T. Shyo and K. Kakimoto, J. Crystal Growth 156 (1995) 383.
[15] M. Watanabe, M. Eguchi, K. Kakimoto, H. Ono, S. Kimura and T. Hibiya, J. Crystal Growth 151 (1995) 285.
[16] K. Kakimoto, K.-W. Yi and M. Watanabe, J. Crystal Growth 163 (1996) 238.
[17] K.-W. Yi, M. Watanabe, M. Eguchi, K. Kakimoto and T. Hibiya, Jpn. J. Appl. Phys. 33 (1994) L487.
[18] H.P. Utech and M.C. Fleming, J. Appl. Phys. 37 (1966) 2021.
[19] R.A. Brown, T. Kinney, P. Sackinger and D. Bornside, J. Crystal Growth 97 (1989) 99.
[20] F. Dupret, P. Nicodeme, Y. Ryckmans, P. Wouters and M.J. Crochet, Int. J. Heat and Mass Transfer 33 (1990) 1849.
[21] A. Virzi and M. Porrini, Mat. Sci. Eng. B 17 (1993) 196.

Computational Materials Science 10 (1998) 134–138

COMPUTATIONAL
MATERIALS
SCIENCE

ELSEVIER

The composition changes induced by surface roughening and mixing during the ion profiling of multilayers

A. Galdikas [a,c,*], L. Pranevičius [a], C. Templier [b]

[a] Department of Physics, Vytautas Magnus University, 28 Daukanto st., LT-3000 Kaunas, Lithuania
[b] Université de Poitiers, 86960 Futuroscope Cedex, France
[c] Department of Physics, Kaunas University of Technology, 73 Donelaičio st., LT-3006 Kaunas, Lithuania

Abstract

The kinetics of the surface composition during the depth profiling of multilayered structures is considered by the proposed phenomenological models. In order to emphasize the composition changes on the surface produced by surface roughness development and ion mixing separately, the calculated results from the two different models are compared. The first one includes the processes of sputtering and surface migration of atoms, and second one includes the processes of ion mixing between layers and sequential removal of surface monolayers (layer by layer). The main conclusion made from the qualitative analysis of the results is that the effects induced by these two different processes are quite similar and the interpretation of such experimental results is not so obvious in many cases. Copyright © 1998 Elsevier Science B.V.

1. Introduction

One of the limitations of the depth profiling techniques, i.e., the sequential ion etching and analysis of the surface by means of a surface sensitive method like secondary ion mass spectroscopy (SIMS) or Auger electron spectroscopy (AES), is that the ion beam, besides the removal of the atoms, modifies the studied specimen. Therefore, in the course of the depth profiling study, the specimen gets altered [1,2]. Several factors and physical phenomena influence the sputter depth profiling process: atomic cascade mixing, projectile implantation, preferential sputtering, preferential recoil implantation, surface roughness, radiation

enhanced diffusion etc. [3–9]. A direct consequence of these processes induced by sputter profiling is the broadening of an originally sharp interface.

The sputtering process is fundamentally a stochastic process and is random in space and time [10,11], and leads to the development of surface roughness. The surface roughening is a strong effect having significant influence on layer broadening. In most experiments this effect is reduced (but not excluded) by rotating the sample and by grazing ion incidence angle [9]. During sputtering of the multilayered structures, since the removal of atoms takes place from the various uncovered (outermost) monolayers having complex topography, the interface between layers is crossed by moving surface boundary at different places at different moments of time (Fig. 1). It follows that the time needed to clean the substrate of all atoms of a thin layer is much longer than in the case that the removal of atoms takes

* Corresponding author. Address: Department of Physics, Vytautas Magnus University, 28 Daukanto St. LT-3000, Kaunas, Lithuania. Tel.: +370-7-796958; fax: +370-7-203858; e-mail: arvaidas.galdikas@vdu.it.

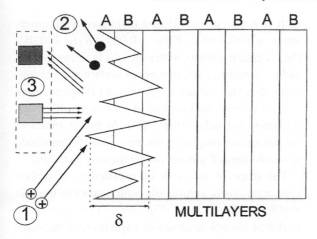

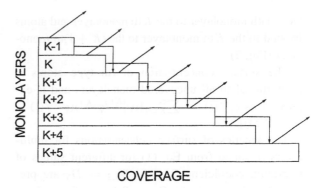

Fig. 2. The schematic presentation of the processes of sputtering and surface atom migration, as the population of surface recoils, which during ion bombardment are transported across the surface.

Fig. 1. The schematic presentation of the process of depth profiling of multilayered structures taking into account the development of the surface roughness during ion bombardment: 1 – bombarding ions, 2 – sputtered particles, 3 – surface composition analysis technique. The surface roughness δ is defined.

place homogeneously, monolayer by monolayer, with constant velocity. In drawing conclusions about how the geometrical distribution of atoms at the interface affects the elemental composition on the surface measured by means of the ion sputtering technique, one can make mistakes by attributing geometrical effects of the surface topography development to the distribution effects of atoms in the interface region.

In order to emphasize the composition changes on the surface during depth profiling of multilayers produced by surface topography development and ion mixing, two different models are proposed. The first includes the processes of sputtering and surface migration of atoms, and the second includes the processes of ion mixing of multilayers and sequential removal of surface monolayers (layer by layer).

2. Surface atom migration

An ideal solid without any defects, with the sites occupied by atoms is considered. At time $t = 0$ the coverage $\varphi^{(K)}$ of all monolayers K (where K is the number of monolayer), including the top monolayer $K = 1$ is $\varphi^{(K)}(0) = 1$. For multicomponent targets the partial coverage, i.e. the coverage by ith type atoms

is defined as $\varphi_i^{(K)} = n_i^{(K)} \varphi^{(K)}$ ($\varphi^{(K)} = \sum_{i=1}^{m} \varphi_i^{(K)}$), where $n_i^{(K)}$ is the relative concentration of ith type atoms in Kth monolayer. The sputtering process during the depth profiling removes atoms from the surface with frequency probability $w_i = Y_i i_0$ (s^{-1}) where Y_i is the sputtering yield of ith type atoms and $i_0 = I_0/C$ is the relative ion flux (s^{-1}), where I_0 is the ion flux (cm^{-2} s^{-1}) and C is the surface atom concentration (cm^{-2}). During the ion bombardment the process of surface atom migration (as the interlayer diffusion) which smoothes the surface roughness takes place. In the model this process is included as the jump of atoms from the one uncovered surface monolayer to another (Fig. 2) and is defined by the surface migration coefficient $D_{S,i}$ (cm^2 s^{-1}). These processes give the following system of equations for a kinetics of the Kth monolayer coverage by the ith type atoms:

$$
\begin{aligned}
\frac{d\varphi_i^{(K)}}{dt} = &-w_i(\varphi_i^{(K)} - \varphi_i^{(K-1)}) \\
&+ \frac{D_{S,i}}{h^2}[\varphi_i^{(K-1)}(\varphi^{(K+1)} - \varphi^{(K)}) \\
&- \varphi_i^{(K)}(\varphi^{(K+2)} - \varphi^{(K+1)})],
\end{aligned}
\tag{1}
$$

where h is the thickness of one monolayer.

The first term of Eq. (1) describes the changes of Kth monolayer coverage as a result of sputtering. The second one describes the process of surface atom migration, i.e. the transport of atoms located in the

$(K-1)$th monolayer to the Kth monolayer and atoms located in the Kth monolayer to the $(K+1)$th monolayer (Fig. 2).

The surface concentration of ith type atoms is proportional to all uncovered monolayers and is expressed as $c_i^{(S)}(t) = \sum_{K=1}^{\infty}[(\varphi_i^{(K)}/\varphi^{(K)})(\varphi^{(K)}(t) - \varphi^{(K-1)}(t))]$.

The kinetics of surface concentrations for multilayers obtained from Eq. (1) for different values of migration coefficient $D_{S,1} = D_{S,2} = D_S$ are presented in Fig. 3(a). At $D_S = 0$ (no surface migration) the amplitude of oscillations of surface concentrations monotonously decreases. This is because the surface roughness $\delta(t)$ (which is defined as a distance expressed in monolayers between monolayer

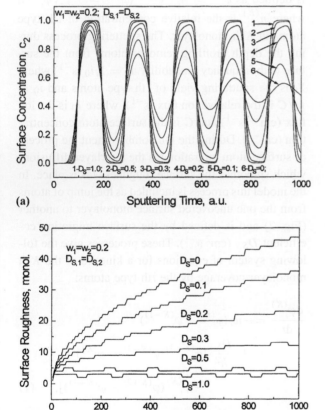

Fig. 3. (a) The kinetics of surface concentration $c_2(t)$ obtained from Eq. (1) for different values of parameter D_S. (b) The kinetics of surface roughness $\delta(t)$ obtained from Eq. (1) for different values of parameter D_S.

with coverage $\varphi = 0.95$ and monolayer with coverage $\varphi = 0.05$: $\delta(t) = K_2(\varphi = 0.95) - K_1(\varphi = 0.05)$, Fig. 1) goes to infinity with time as a function $\sqrt{t}$ [12] and is unrealistic. The process of surface atom migration reduces the surface roughness and at higher values of D_S the steady state regime of the kinetics of surface roughness is reached (Fig. 3(b)). At higher values of D_S the surface roughness decreases and the amplitude of oscillations of surface concentrations increases.

The shape of the peak of oscillations of surface concentrations is asymmetric with respect to the axis parallel to the ordinate (Fig. 3(a)). This asymmetry is almost always observed in experimental measurements of surface concentrations of constituents for multilayers during ion bombardment and, generally, is attributed to the process of ion mixing taking place in deeper layers. From this, it is interesting to result that such kinds of peak asymmetry occur as a result of surface atom migration, and is obtained from the model where ion mixing in deeper layers is not included. Generally, ion mixing always takes place and gives some corrections to the amplitude and shape of the peak.

3. Ion mixing

Let us consider the processes of atom removal and ion mixing. In this case the processes of surface roughening will not be taken into account, i.e. the atoms are removed from the surface monolayer by monolayer. The model works according to the following scheme (Fig. 4): (1) ion mixing at the given distance from the moving surface takes place; (2) subsequent removal of monolayers (monolayer by monolayer) occurs. The first process gives the changes of elemental composition in the near-surface layers and the second one gives the kinetics of surface composition, which is a final result of depth profiling.

The process of ion mixing is defined by the mixing coefficient D_m (cm^2 s^{-1}) [5,7]. Taking into account the range of ion penetration depth, the mixing coefficient is the function of distance from the surface $D_m(K)$. At the first approach this function may be assumed having the Gaussian distribution form, with

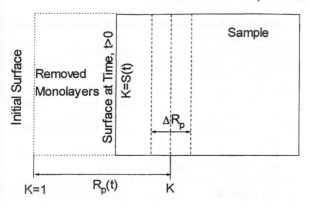

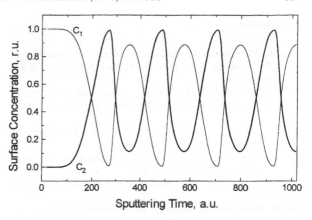

Fig. 4. Schematic presentation of the model including sputtering and ion mixing. The time-dependent ion penetration depth $R_p(t)$ is defined.

Fig. 5. The kinetics of surface composition obtained from Eq. (2). The calculation parameters: $w_1 = 0.05$, $w_2 = 0.5$, $N_1 = 20$, $N_2 = 40$, $D_{m,0} = 0.5$, $R_{p,0} = 15$.

time-dependent range of ion penetration depth $R_p(t)$ (with respect to initial surface $K = 1$, Fig. 4), which may be obtained from the sputtering rate V_s with initial conditions $R_p(t = 0) = R_0$: $dR_p/dt = V_s = w_1 c_1^{(S)} + w_2 c_2^{(S)}$, where $c_i^{(S)}$ are the concentrations of constituents in monolayer $K = S(t)$ which represent the surface monolayer at fixed time t. The monolayer $S(t)$ is obtained from the following expression: $S(t) = R_P(t) - R_0 + 1$

The balance equation describing the kinetics of constituent concentrations in monolayers due to the process of ion mixing has the next form ($D_m^{(K)} = 0$ if $K < S$)

$$\frac{dc_i^{(K)}}{dt} = \frac{D_m^{(K-1)}}{h^2}(c_i^{(K-1)} - c_i^{(K)})$$
$$- \frac{D_m^{(K)}}{h^2}(c_i^{(K)} - c_i^{(K+1)}), \quad K > S. \quad (2)$$

Eq. (2) gives the changes of elemental composition in the bulk of sample as a result of ion mixing. The kinetics of surface composition is obtained by the construction of the timescale for monolayers $K \le S$ (i.e. for removed monolayers). The compositions of these monolayers are known from Eq. (2) and the time to remove Kth monolayer is equal to $t = 1/(w_1 c_1^{(K)} + w_2 c_2^{(K)})$, $K \le S$.

The typical results obtained from Eq. (2) are presented in Fig. 5. It is important to note, that, despite, the mixing coefficients of both components being the same $D_{m1} = D_{m2}$, the asymmetric peaks of concentra-

tion oscillations with respect to ordinate are observed. This asymmetry depends on time of mixing, i.e. on the rate of surface moving and is less pronounced at the higher values of parameter R_p. The change of R_p is related with the change in ion energy or ion incident angle. The calculations from the above model have shown that the amplitude of the oscillations of surface concentrations decreases with increasing of R_p and D_m, and increases with increasing w_i. All these parameters depend on ion energy.

4. Conclusions

The surface atom migration and ion mixing take place simultaneously during ion bombardment, and the real distribution of components obtained from the depth profile measurements based on ion sputtering technique is not so obvious. The increase in ion energy increases the parameters D_s and D_m, however, the first one gives the increase and the second one gives the decrease of the amplitude of the surface concentration oscillations. This conclusion can be applied for the profiling results obtained at different temperatures. At elevated temperatures the surface roughness is reduced as a result of surface atom migration and it follows that the amplitude of surface concentration oscillations increases.

A. Galdikas et al. / Computational Materials Science 10 (1998) 134–138

References

[1] M. Menyhard, A. Barna, A. Sulyok, K. Järrendahl, J.-E. Sundgren and J.P. Biersack, Nucl. Instr. Meth. B 85 (1994) 383.

[2] M. Menyhard, A. Konkol, G. Gergely and A. Barna, J. Electron Spectroscopy and Related Phenomena 68 (1994) 653.

[3] N. Tamovic, L. Tamovic and J. Fine, Nucl. Instr. Meth. B 67 (1992) 491.

[4] M. Menyhard, A. Barna and J.P. Biersack, J. Vac. Sci. Technol. A 12 (4) (1994) 2368.

[5] M. Nastasi and J.W. Mayer, Mat. Sci. Eng. R 12 (1) (1994) 1.

[6] R. Kelly and A. Miotelo, in: Materials and Processes for Surface and Interface Engineering, ed. Y. Pauleau, NATO ASI Series, Vol. 290 (Kluwer Academic Publishers, Dordrecht, 1994) p. 67.

[7] W. Bolse, Mat. Sci. Eng. R 12 (2) (1994) 53.

[8] D. Marton, J. Fine and G.P. Chambers, Phys. Rev. Lett. 61 (23) (1988) 2697.

[9] K. Wittmark, in: Sputtering by Particle Bombardment III, eds. R. Behrisch and K. Wittmark, Topics in Applied Physics, Vol. 64 (Springer, Berlin, 1991) p. 161.

[10] G. Carter and M.J. Nobes, Nucl. Instr. Meth. B 90 (1994) 456.

[11] L. Tanovic, N. Tanovic, G. Carter and M.J. Nobes, Nucl. Instr. Meth. B 90 (1994) 462.

[12] A. Galdikas, L. Pranevičius, C. Templier, Appl. Surf. Sci., 103 (1996) 471.

Computational Materials Science 10 (1998) 139–143

Simulation of the deposition and aging of thin island films

P. Bruschi *, A. Nannini

*Dipartimento di Ingegneria dell'Informazione: Elettronica, Informatica, Telecomunicazioni,
Università degli Studi di Pisa, via Diotisalvi 2, I-56126 Pisa, Italy*

Abstract

Two-dimensional atomistic simulations of the growth and post-deposition behaviour of island metal films are described. The program includes a module for the calculation of the film resistance on the basis of a charge limited tunnelling model. The dependence of the resistance on various deposition parameters is investigated. Examples of simulated post-deposition resistance drift are shown. Copyright © 1998 Elsevier Science B.V.

1. Introduction

Island metal films have been studied for their intriguing electrical and optical properties[1]. Recently these films have been the object of extensive studies since they represent the first stage of growth of many continuous films of practical interest [2,3]. Discontinuous metals have also raised interest for practical applications in the field of strain sensors for their high gage factor [4]; so far, long resistance drift due to aging effects, high sensitivity to temperature and lack of reproducibility has prevented the fabrication of reliable devices based on these materials.

In this paper we present a series of two-dimensional Monte Carlo simulation of the growth and post-deposition aging of island films. The program can be divided into two main modules: the first is aimed to simulate the microscopical film evolution, the second is used to calculate the resistance of the film. The program permits to relate important parameters such

as the deposition rate and temperature not only to the morphology of the clusters but also to the overall resistance. The structure of the first module is described in details in [5] where it is used for predicting the effects of some deposition parameters on the cluster size distribution and morphology.

In our program, electrical conduction of the films is ascribed, as generally accepted, to a process of charge limited tunnelling of electrons between the islands [6]. The problem of calculating the resistance of a film portion is solved by replacing the film itself with an electrical network where the charge exchange between the islands is modelled by equivalent conductances which depend on the size and spacing of the grains. The novelty of this work with respect to previous studies based on a similar approach [7,8] stands in the combination of a very efficient program for morphological simulations with a resistance extractor which operates directly on the simulated films.

The simulations have been devoted to show the dependence of the resistance on the fractional coverage of films deposited at different substrate temperatures. In order to reduce statistical fluctuations the

---

* Corresponding author. Tel.: +39 50 568538; fax: +39 50 568522; e-mail: bruschi@iet.unipi.it.

resistances has been averaged over a set of values obtained by changing only the seed of the random number generator. The standard deviation of each set of values has been reported to provide an indication of the reproducibility.

Preliminary data about the application of the program to the prediction of the post-deposition behaviour of the resistance have also been shown.

2. Model of film growth and electrical conduction

The substrate is modelled as an $N \times N$ array of sites mapped on a rectangular portion of a triangular lattice.

At each simulation step, three type of transitions are allowed: (i) arrival of an atom at the substrate, (ii) diffusion of an adatom to an empty nearest neighbour and (iii) re-evaporation of an adatom from the substrate. The rate r_a of atom arrivals is fixed and given by $r_a = R \times N^2$, where R is the number of impinging atoms per site and unit time. The rate $r^{(d)}$ for a diffusion transition is given by the equation

$$r^{(d)} = r_0^{(d)} \exp\left(\frac{-E^{(d)}}{k_B T}\right), \tag{1}$$

where $r_0^{(d)}$ is an attempt frequency which is considered to be independent of the occupancy of the neighbours [9], k_B is the Boltzmann constant and T the absolute temperature. Indicating with c_1–c_5 the occupancies of the five nearest neighbours of hopping particles, as shown in Fig. 1(a), the diffusion activation energy is calculated by the following formula:

$$E^{(d)} = E_0^{(d)} + (c_3 + \overline{c_5}c_4 + \overline{c_1}c_2)E_B$$
$$+ (c_1 + c_5)E_L, \tag{2}$$

where $E_0^{(d)}$ is a fixed contribution due to the interaction with the substrate, E_B the binding energy between two adatoms and E_L is the activation energy for diffusion of adatoms along cluster borders. The occupancy c_i is defined as 0 if site i is empty and 1 if it is occupied and $\overline{c_i} = 1 - c_i$. The distinction between the two different contributions E_B and E_L of the neighbours to the activation energy is also present in the models

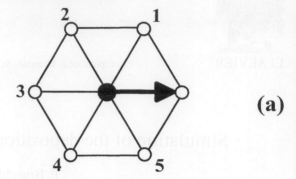

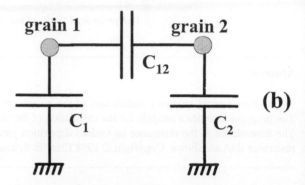

Fig. 1. (a) Symbolic representation of a particle hop in a triangular lattice showing the five neighbours affecting the activation energy. (b) Equivalent electrical circuit used to calculate the energy of a pair of charged grains.

of Amar an Family [2] and Breeman et al. [3] dealing with square lattices.

The algorithm used to determine the rate for re-evaporation transitions is described in [5] where it is shown that introducing re-evaporation leads to cluster size distributions which are not in agreement with experimental data. For this reason re-evaporation rate is set to zero in this work.

A transition is randomly selected according to a probability proportional to the transition rate. At each step the simulation time is incremented by an amount given by the inverse of the sum of all the transition rates.

The equivalent electrical network utilized for calculating the resistance is obtained by associating a node to each cluster and introducing conductances between the nodes to model the electron exchange occurring by tunnel effect. The conductance between a pair of clusters is calculated by summing up the rates of all the possible charge transition involving hopping of one electron from a grain to the other. As shown in [10], transitions can be grouped in "pair of antagonist transitions"; each pair gives the following contribution G_i to the total low field conductance:

$$G_i = G_0 \exp(-2\chi s)$$
$$\times P_i \frac{(\Delta E/kT - 1)\exp(\Delta E/kT) + 1}{(\exp(\Delta E/kT) - 1)^2} \qquad (3)$$

where G_0 is a constant, χ the decay rate of the electron wave function in the insulator, s the minimum distance between the two clusters, P_i the occurrence probability of the initial state of charge and ΔE is the electrostatic energy variation caused by the transition. The distance s is the length of the shortest line segment connecting the two grains without intersecting other clusters. Conductances are included only if the distance s is smaller than a cut-off distance s_{off}.

The electrostatic energy of a given state of charge is calculated modelling each pair of clusters with the circuit shown in Fig. 1(b): the capacitor C_{12} represents the capacitance of cluster 1 to cluster 2 while capacitors C_1 and C_2 represent the self-capacitance of the clusters which are estimated by summing up all the capacitance to the surrounding clusters. The capacitance between two clusters is calculated by replacing them by two spheres of equivalent volume spaced by s. The volume of the clusters is estimated by assuming that each adatom gives a contribution equal to a^3, where a is the lattice constant. An exact solution for the capacitance between two spheres is adopted [8]. Capacitance between clusters are included only if their distance s is smaller than the cut-off distance s_{off}.

A given state of charge is identified by the total charges Q_1 and Q_2 in node 1 and 2, respectively; the set of possible states is restricted by allowing Q_1 and Q_2 to assume only the values $0, +e, -e$, where e is the electron charge. The occurrence probability P_i of a given state of energy E_i is calculated by the Boltzmann statistics:

$$P_i = \frac{\exp(E_i/kT)}{\sum_j \exp(E_j/kT)}, \qquad (4)$$

where the sum includes all the possible states of charge.

Two extra nodes are included in order to represent two parallel contacts placed at two opposite sides of the sample: each contact is connected by a conductance to all the clusters which are less than s_{off} away from the corresponding side. The electrostatic energies required to determine these conductances are calculated by modelling the contacts as infinite conductive planes. The network is then solved using the Fogelholm method [11] to obtain the equivalent resistance between the contacts.

3. Results

The deposition simulations described in this paper are completely characterized by a set of five dimensionless parameters, namely the dimensionless temperature T/T_0 (where $T_0 = E_0^{(d)}/k_B$), the dimensionless deposition time $t_d r_0^{(d)}$ (where t_d is the deposition time) and the ratios $E_B/E_0^{(d)}$, $E_L/E_0^{(d)}$, and $R/r_0^{(d)}$. All the simulations have been performed on 400×400 samples.

The ratios $E_L/E_0^{(d)}$ and $E_B/E_0^{(d)}$ are fixed to 0.2 and 0.5, respectively, in conformity with the assumptions made in [5].

As far as the conduction parameters in Eq. (3) are concerned, χ is set to 0.3×10^{10} m^{-1} and the lattice constant to 0.5 nm. All the resistances are calculated at 300 K.

A first series of simulations was devoted to determine the effect of the deposition parameters on the resistance of the films. In Fig. 2 the calculated resistance, normalized to $R_0 = G_0^{-1}$, is plotted as a function of the fractional coverage x, for four different settings of the deposition parameters, indicated in the figure. Since each curve refers to a fixed deposition time, the coverage has been varied by changing the deposition rate R. Each resistance value is averaged over a set of nine samples. For each curve the maximum coverage shown approaches the

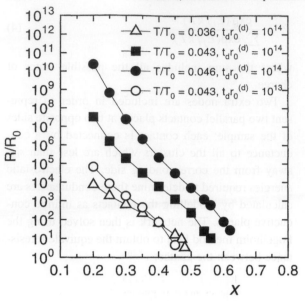

Fig. 2. Plot of the resistance, normalized to $R_0 = G_0^{-1}$, as a function of the fractional coverage x for various dimensionless temperatures and deposition times.

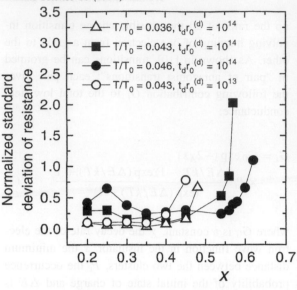

Fig. 3. Standard deviation of the resistance normalized to the average resistance, calculated over a set of nine samples. The deposition parameters are indicated in the legend.

percolation threshold for the corresponding deposition parameters.

Comparison with the experimental dependence of the sheet resistance of island film on the equivalent thickness [1,4] reveals a qualitative agreement. It can be observed that, for a given coverage, the higher the deposition temperature, the higher the resistance. This is clearly the result of temperature activated island agglomeration [12] occurring during film growth and causing the inter-island distance to increase. Reducing the deposition time produces similar effects to reducing the deposition temperature since less time is available for agglomeration.

The reproducibility of the deposition experiments is represented in Fig. 3, where the normalized standard deviation of the resistance, calculated over sets of nine samples is shown for the same deposition conditions as in Fig. 2. The high uncertainty is due to the small size of the samples compared with macroscopical films. The curves exhibit an abrupt increase at the percolation threshold: this can be a limiting factor for exploiting the small sensitivity to temperature which can be theoretically predicted for films near the transition to a metallic regime of conduction [1]. For

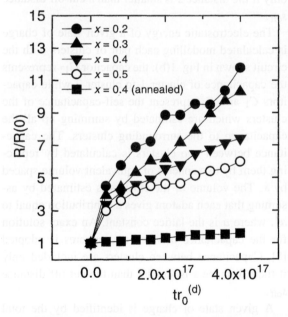

Fig. 4. Simulated post deposition resistance variations, normalized to the initial value, as a function of the dimensionless time $t \times r_0^{(d)}$ for samples of different fractional coverage x. The deposition and aging temperature is $T/T_0 = 0.043$ for all the curves. In one case, indicated in the legend, an annealing at $T/T_0 = 0.052$ for a dimensionless time $t \times r_0^{(d)} = 10^{15}$ was performed before the aging simulation.

coverages far from the percolation threshold it can be roughly observed that the lower the temperature, the lower the uncertainty. This was experimentally observed for gold films [13] and suggested as a viable method to improve reproducibility.

Simulation of the post-deposition aging are performed by the same program, setting the deposition rate R to zero. Fig. 4 shows the resistance variations of samples of different coverage maintained at $T/T_0 = 0.043$ for a dimensionless time $t \times r_0^{(d)} = 4 \times 10^{17}$. All the samples were deposited at $T/T_0 = 0.043$ for a deposition time $t_d \times r_0^{(d)} = 10^{14}$. One aging simulation, indicated in the figure, was performed after annealing the samples at $T/T_0 = 0.052$ for a dimensionless time of 10^{15}: the annealing produced a large resistance increase (about 28 times the initial value) but improved the stability. All the curves have been averaged over a set of four independent experiments.

These preliminary results provide an atomistic support to the hypothesis that the continuous resistance increase observed in ultra-thin metal films [14] is due to cluster aggregation caused by surface diffusion of adatoms and clusters.

References

[1] Z.H. Meiksin, in: Phys. of Thin Films, eds. G. Hass, M. Francombe and R.W. Hoffman, Vol. 8 (Academic Press, New York, 1975) p. 9.

[2] J.G. Amar and F. Family, Thin Solid Films 272 (1996) 208.

[3] M. Breeman, G.T. Barkema, M.H. Langelaar and D.O. Boerma, Thin Solid Films 272 (1996) 195.

[4] G.R. Witt, Thin Solid Films 22 (1974) 133.

[5] P. Bruschi, P. Cagnoni and A. Nannini, Phys. Rev. B 55 (1997) 7955.

[6] C.J. Adkins, J. Phys. C 15 (1982) 7143.

[7] P. Sheng, Philosofical Magazine B 65 (1992) 357.

[8] P. Šmilauer, Thin Solid Films 203 (1991) 1.

[9] A.M. Bowler and E.S. Hood, J. Chem. Phys. 94 (1991) 5162.

[10] P. Bruschi and A. Nannini, Thin Solid Films 201 (1991) 29.

[11] R. Fogelholm, J. Phys. C 13 (1980) L571.

[12] J.G. Skofronick and W.B. Phillips, J. Appl. Phys. 38 (1967) 4791.

[13] Z.H. Meiksin, E.J. Stolinski, H.B. Kuo, R.A. Mirchandani and K.J. Shah, Thin Solid Films 12 (1972) 85.

[14] M. Pattabi, M.S. Murali Sastry and V. Sivaramakrishnan, J. Appl. Phys. 63 (1988) 983.

ELSEVIER

Computational Materials Science 10 (1998) 144–147

COMPUTATIONAL
MATERIALS
SCIENCE

Monte Carlo simulation of pulsed laser ablation into an ambient gas

T.E. Itina *, W. Marine, M. Autric

Laboratoire Interdisciplinaire Ablation Laser et Applications, IRPHE-LP3 UMR CNRS 6594, and GPEC UMR CNRS 6631, Case 918, Parc Scientifique et Technologique de Luminy, 13288, Marseille, Cedex 9, France

Abstract

Laser ablation from one- and two-component targets into a diluted gas background is simulated by combined direct simulation-test particle Monte Carlo method. The spatial and velocity distributions of particles deposited at the plane substrate are calculated. The approach developed has allowed us to consider the influence of the collisions both among the ablated particles and between the ablated and ambient gas particles on the uniformity of film stoichiometry. It is found that the increase of the background gas pressure results in the more uniform distribution of the stoichiometrical ratio of the deposited particles with different masses. Copyright © 1998 Elsevier Science B.V.

1. Introduction

The process of laser ablation has a specific application for growing thin films of a wide variety of materials. A careful choice of the experimental parameters (laser fluence, spot size, ambient gas pressure, etc.) is known to be important for developing high-quality thin films by pulsed laser deposition (PLD) technique. Numerous studies have shown that the gas-phase collisions give rise to the formation of the forward peaked angular distribution of the ablated particles, that can be directly related with the shape of the deposited profile, if the deposition is performed in vacuum. In the presence of a background gas, additional interactions of the ablated particles with the particles of the filling gas can affect the distribution of the deposited material,

as well as the composition of the films [1,2]. It should be noted that adiabatic models used by a number of researchers to describe the dynamics of the laser ablated flow in vacuum are not valid due to heating of the filling gas by ablated particles. The problem presents a particular interest, since high-quality thin films are usually grown in the presence of inert or reactive atmosphere.

The aim of this work is to analyze the influence of the ambient gas on the characteristics of the laser ablated flow of particles. In particular, the influence of the background gas on the film thickness distribution, velocity distributions of the particles deposited at the substrate, film uniformity and composition are investigated. To consider both the collisions among the ablated particles and the interactions of the ablated particles with the particles of the ambient gas a combined direct simulation-test particle simulation Monte Carlo method (DSTP) is proposed. The results of the modeling can support better understanding of the laser ablation from multi-component targets and the role of

---

* Corresponding author. On leave from The Moscow Institute of Physics and Technology. Permanent address: 126, Ozernaya 10, Sergiev Posad, Moscow reg., 141307, Russia. Tel.: +7 096 54 5 63 13; fax: +7 095 135 03 76; e-mail: sr@itin.msk.ru.

the background gas parameters in the developing stoichiometrical thin films.

2. Simulation

We assume that the process of laser ablation in the diluted ambient gas consists of two parts: (i) desorption and initial expansion when collisions of the ablated particles among themselves are significant and the influence of the ambient gas can be disregarded; (ii) the expansion of the gas cloud in the ambient gas when collisions of the ablated particles among themselves can be disregarded. Based on these assumptions, the Monte Carlo simulation was performed using two different techniques. The desorption and initial expansion were modeled by direct simulation Monte Carlo (DSMC) method based on Bird's algorithm [3] in the finite volume V_0 that was subdivided into a network of annular cells. The calculations started when the volume was empty and the particles desorbed from the surface (laser spot area πr_0^2) during the pulse time τ. We supposed that the target material is composed of two species with the masses m_1 and $m_2(m_1 < m_2)$. The initial velocity distribution is Maxwell–Boltzmann with the same temperature T_0 for both species. The collisions between the ablated particles (a–a collisions) were simulated using a hard-sphere collision dynamics with the identical collision cross sections Σ_{aa} for both species. The number of monolayers [4] desorbed during one laser pulse is $\Theta = \Phi \cdot \Sigma_s \cdot \tau$, where $\Sigma_s = \Sigma_{aa}/4$ is the area the atom occupies at the surface, $\Phi = (n_1 V_1 + n_2 V_2)/2$ the total desorption flux, V_i the thermal velocity of the component i, and n_i is the density of the desorbed gas immediately in front of the surface (we assume $n_1 = n_2$). To determine the radius of the evaporated area we used the parameter $b = r_0/V_1\tau$, where $V_1\tau$ is the length of the light particle cloud at the end of laser pulse. The backscattered particles were assumed to recondense on the irradiated surface. The motion of the particles outside V_0 was simulated with test particle Monte Carlo method (TPMC) [5]. The calculations were performed for each of the ablated particles leaving V_0 in the outward

direction and to be followed in the background gas (pressure P_{amb}). We suppose that the ambient gas has the equilibrium velocity distribution with temperature T_{amb}. Collisions with ambient gas particles (a–b collisions) were calculated using a hard-sphere collision cross section Σ_{ab}. The particles were followed until they reached the substrate or they were redeposited on the irradiated surface. In the present modeling, we neglected the difference in the sticking probability for species with different masses.

We note that the approach described above is justified if (i) the mean free path in the background gas is longer than a cloud dimension at the end of the first expansion process; (ii) beyond the volume V_0 collisions of the ablated particles among themselves can be disregarded. It can be shown [6] that these assumptions are valid if the background pressure is less than about 100 Pa = 760 mTorr.

The results presented were obtained with following parameters: $\Theta = 3, b = 2.5, \tau = 3 \cdot 10^{-8}$ s, $T_0 = 2901.25$ K and 5802.5 K; $m_2 : m_1 = 2 : 1$, where $m_1 = 63.54$ a.m.u. (Cu atoms) $m_3 : m_1 = 0.63$ (Ar atoms), $T_{amb} = 273$ K, $\Sigma_{ab} = 25.1$ Å^2, and the target-substrate distance $L = 5$ cm. As a rule we used the total number of 10^6 particles and performed five

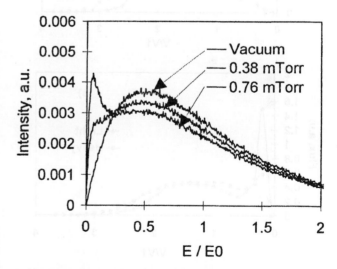

Fig. 1. Kinetic energy distribution of particles arriving at the target in the case of laser ablation of the one-component (Cu) target in Ar atmosphere with different pressure. Here we assumed that surface temperature $T_0 = 5802.5$ K, $E_0 = 2kT_0$.

repetitions to gain enough statistics. For all the simulation results, the 95% confidence error bars are smaller than the size of the symbols.

3. Results and discussion

Fig. 1 shows the distributions of the kinetic energy E of particles deposited at the plane substrate for the ablation of one-component target. The decrease of the kinetic energy of the ablated particles can be explained by the energy lost in collisions with the background gas. Due to these collisions, the part of particles reaching the substrate is thermalized. The thermalization phenomenon for a binary flow can be observed in Fig. 2. When the pressure increases, the fraction of the

thermalized particles increases until all the particles are in equilibrium with the background temperature.

Fig. 3 displays the deposition profiles $H(x)$ of particles arriving at the plane substrate. It can be observed that the number of both species reaching the substrate decreases with the increase of the pressure. Moreover, the distributions of both species become less peaked as the pressure increases. The broadening is more pronounced for light particles since they are more probably scattered at large angles in a–b collisions. Thus, a smaller background pressure is required for thermalization of light component. Since a substantial fraction of particles is scattered back and is

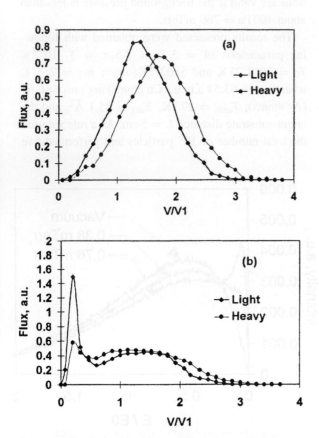

Fig. 2. Velocity distributions of particles arriving at the center of the target (flying close to the surface normal) in the case of binary target; (a) $P_{amb} = 0.0001$ m Torr (vacuum), (b) $P_{amb} = 10$ m Torr.

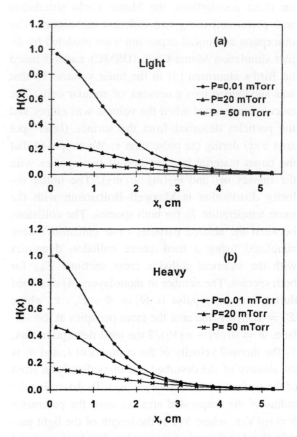

Fig. 3. Spatial distributions $H(x)$ of the number of particles deposited per unit area at the plane substrate, where x is a radial distance from the center of the substrate. (a) the distribution of light particles for different values of the ambient gas pressure; (b) the same for heavy particles. All distributions were normalized by $H_0 = H(x = 0)$ for heavy particles at $P_{amb} = 0.01$ m Torr.

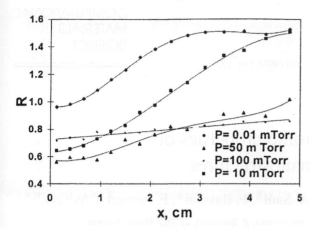

Fig. 4. The spatial distribution of the ratio $R(x)$ of the number of light particles deposited at the plane substrate to this one of heavy particles. Here x is a radial distance from the center of the substrate.

recondensed on the irradiated surface, the deposition profiles become thinner.

The distributions of the stoichiometrical ratio $R(x)$ of the particles deposited at a plane substrate are shown in Fig. 4. One can see that at low ambient gas pressure the ratio $R(x)$ is larger than the one at higher P. The lack of light particles at the distances close to the substrate center can be also observed at low background gas pressures. With the increase of the pressure the ratio R diminishes and becomes more uniformly distributed along the substrate. The lack of light species at the distances close to the substrate center at low gas pressures can be explained by the fact that the angular distribution of heavy species formed after the initial expansion process is more focused toward the surface normal [4]. The stoichiometrical ratio $R(x)$ decreases as a result of the more intense backscattering of light particles.

The results obtained are in agreement with several experimental findings. For example, Lichtenwalner et. al. [7] found that at low laser power, the deposition rate decreases and the plume broadens as the gas pressure is increased. The increase of the film uniformity with increasing the background pressure was experimentally observed by a number of researchers (see, for example, [8]).

4. Conclusions

The process of laser ablation into a diluted background gas has been investigated by Monte Carlo simulation. The simulation procedure developed in the paper has allowed us to calculate the collisions of the particles inside the ablated plume as well as the scattering of the ablated particles in the background gas with Maxwell velocity distribution. It has been shown that the velocities of the ablated particles decrease and the film thickness distributions are spread in the presence of the gas. The low-energy part of the ablated species was shown to appear as a result of the interactions with the background particles. When the pressure is sufficiently high, the thermalization of the ablated particles and further diffusion toward the substrate were observed. In addition, the number of particles scattered back was found to increase with increasing ambient gas pressure. These effects are more pronounced for light component of the ablated plume than for heavy one, resulting in the change of the composition of the films. The random scattering processes in the ambient gas are shown to increase the uniformity of the stoichiometrical ratio distribution along the substrate. These results present particular interest for the PLD of the complex materials with critical stoichiometrical requirements.

References

[1] J. Gonzalo, C.N. Afonso, F. Vega, D. Matinez Garcia and J. Perrière, Appl. Surf. Sci. 86 (1995) 40.
[2] J. Gonzalo, F. Vega and C.N. Afonso, J. Appl. Phys. 77 (1995).
[3] G.A. Bird, Molecular Gas Dynamics (Clarendon Press, Oxford, 1976).
[4] H.M. Urbassek and D. Sibold, Phys. Rev. Lett. 70 (1993) 1886.
[5] K. Koura, Phys. Fluids 6 (10) (1994) 3473.
[6] J.C.K. Kools, J. Appl. Phys. 74 (1993) 6401.
[7] D.J. Lichtenwalner, O. Auciello, R. Dat and A.I. Kingon, J. Appl. Phys. 74 (1993) 7497.
[8] C.M. Rouleau, D.H. Lowndes, J.W. McCamy, J.D. Budai, D.B. Poker, D.B. Geohegan, A.A. Puretzky and S. Zhy, Appl. Phys. Lett. 67 (17) (1995) 2545.

ELSEVIER

Computational Materials Science 10 (1998) 148–153

COMPUTATIONAL
MATERIALS
SCIENCE

Grain effect in electronic properties of silicon epitaxial nanostructures

A.B. Filonov [a], A.N. Kholod [a], V.E. Borisenko [a], A. Saúl [b], F. Bassani [b], F. Arnaud d'Avitaya [b],*

[a] *Belarusian State University of Informatics and Radioelectronics, P. Browka 6, 220027 Minsk, Belarus*
[b] *Centre de Recherche sur les Mecanismes de la Croissance Cristalline, Campus de Luminy – case 913, 13288 Marseille cedex 9, France*

Abstract

The electronic properties of grained nanocrystalline silicon (111) films were theoretically studied within the self-consistent semiempirical LCAO method. Grains in the films were found to provide a direct band gap and their interaction results in the gap reduction with respect to the one in the isolated grain. The gap value varied from 1.55 to 3.04 eV depending on the film thickness as well as on the lateral size of the grains. Grained layer stacking inside the film induces a considerable increase of the gap as compared to the unstacked film of the same effective thickness. New simple approach based on the effective mass theory (EMT) has been developed and successfully applied to simulate the electronic properties of nanocrystalline films accounting for confinement effects and interaction between the grains. It consistently reproduces main features of the properties of the grained films derived from LCAO calculations. Copyright © 1998 Elsevier Science B.V.

Keywords: Silicon nanostructure; Electronic properties

Since the discovery of the luminescence of porous silicon [1], a lot of silicon based nanosize structures have demonstrated efficient light emission in the visible range at room temperature (silicon clusters in CaF$_2$ and in SiO$_2$, silicon nanopillars, etc.) [2]. Although those systems are different, they all present two common features: the small (nanometric size) silicon inclusions, and the need of an efficient passivation of the surfaces (oxygen and/or hydrogen). Recently, nanocrystalline Si/CaF$_2$ multilayers were fabricated and exhibited a tunable luminescence in the visible at room temperature [3] and also different optical

absorption, as a function of the silicon film thickness in the range 5–25 Å. First-principles calculations for monocrystalline silicon slabs embedded in CaF$_2$ [4] have shown that they have properties of an indirect or quasidirect band gap semiconductor. Moreover, the calculations predict band gap values different from those experimentally measured. This difference can be explained by the fact that these films consist of small silicon grains. Their size does not exceed a lateral dimension of 15 Å and the height is close to the layer thickness, as it was estimated by EXAFS measurements [3]. It was also shown that the energy gap, as deduced from optical absorption and PL measurements as well as their change with the silicon film thickness, could not be explained by simple emission from the isolated confined grains.

* Corresponding author. Tel.: +33 91 17 28 61; fax: +33 91 41 89 16; e-mail: davitaya@crmc2.univ-mrs.fr.

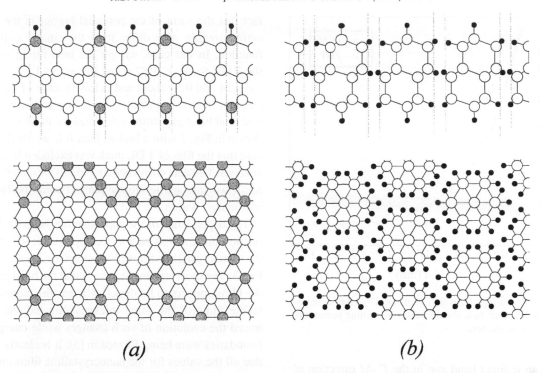

Fig. 1. Cross section and plane view of mono- (a) and nanocrystalline (b) 3 DL silicon film. Open circles and dots are silicon and hydrogen atoms, respectively. Some silicon atoms (grey coloured) within the region indicated by vertical lines are removed and dangling bonds are saturated with hydrogen atoms.

In order to understand the mechanisms responsible for this effect, we have investigated by different calculation methods [5] the effect of the grain interaction within the layer. We have found that by taking into account this interaction, the deduced energy gap was drastically reduced as compared to that of isolated grains of the same size. Now we extended the analysis in order to establish the dependence of the gap value on the confinement parameters as well as on the grain interaction in the epitaxial silicon nanocrystalline films, which is related to the variation of film thickness, lateral size of the grains and layer stacking inside the films. Self-consistent semiempirical LCAO calculations have been performed for that. Details of the procedure and its efficiency for electronic properties simulation of bulk silicon, silicon nanosize films, wires, and clusters can be found elsewhere [6]. However, these calculations are time consuming and fastidious, limiting the modelling of large structures. Thus, we have also developed and demonstrate here that a quite simple analytical approach based on the effective mass theory (EMT) can be effectively applied to the property simulation of complicated silicon nanostructures.

Electronic band structure calculations were performed for free standing silicon (1 1 1) films. The films are constituted of (1 1 1) "double-layers" (DL) of silicon, which are stacked as in the bulk one. Dangling bonds of silicon at the surface are passivated by hydrogen. The way we have considered the intergrain boundaries is illustrated by Fig. 1. In order to simulate the nanocrystalline structure we removed some silicon atoms from their inlayer positions and replaced them with four hydrogen atoms to saturate each dangling bond.

For the monocrystalline silicon (1 1 1) films the band gap dependence on the effective film thickness, as deduced from our LCAO calculations, is shown in Fig. 2 by empty dots. The film thickness was varied from 2 to 10 DL. All the structures are characterised

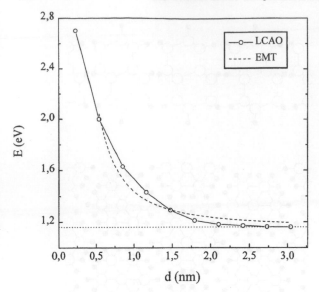

Fig. 2. Total band gap versus film thickness. Dotted lines reproduce the appropriate bulk silicon values. The fitting parameters are discussed in the text.

to have an indirect band gap in the Γ–M direction of the two-dimensional (2D) hexagonal Brillouin zone, which corresponds to the Γ–X direction in the bulk silicon. The gap value rapidly decreases when the film thickness is increased. It is found to reproduce its bulk value starting from the 10 DL thick of the films.

This behaviour explained within the EMT. If one assumes that the free carrier motion is confined in the (1 1 1) direction and the confinement affects both electrons and holes, the following formula may be applied for the gap (E_g) evaluation of the system

$$E_g = E_{go} + E_{g1} = E_{go} + \left(\frac{1}{m_e^*} + \frac{1}{m_h^*}\right) \cdot \frac{\pi^2 \hbar^2}{2d^n},$$

(1)

where E_{go} is the band gap of the bulk material, d is the film thickness, m_e^* and m_h^* are the effective masses in the confinement direction of electrons and holes, respectively. The exponential power $n = 2$ in the case of ideal infinite rectangular well. However, some deviation from the $E_g \sim 1/d^2$ law was observed for various silicon nanostructures [7]. Usually the value of n is smaller than 2 for real structures accounting for the

fact that the form of the potential barrier in the near surface region may differ from the simple step-like function. In our case, we got the best fitting within the EMT for $n = 1.87$ when the m_e^* and m_h^* values are 0.24 and 0.54 (here and below in units of the free electron mass), respectively as in the bulk silicon. [1] The total band gap fitting calculated for these values is shown in Fig. 2 with a broken line. It is worthwhile to note that the film of 1 DL thick was excluded from the analysis. The data corresponding to this extremely thin film (two interacting atomic planes) could be hardly reproduced within the EMT.

The dependence of the gap value for the grain films on their thickness as well as the appropriate values for the monocrystalline films for comparison are shown in Fig. 3. The most striking feature is that the band gap of the films appears to be direct, whereas the monocrystalline films of the same thickness are indirect. We traced the evolution of such changes while intergrain boundaries were being formed in [5]. It is clearly seen that all the values for the nanocrystalline films are located somewhat higher on the energy scale. The curves corresponding to the films with larger grains approximately follow the behaviour of E_g versus d for the monocrystalline film with the constant unique upshift in each case. We ascribe this new upshift to additional confinement conditions within the plane of the film. In these cases we have a 2D set of interacting quantum boxes, which are associated with the silicon grains. The intergrain regions may be treated as potential barriers of some height. Therefore, within the EMT the gap increasing of such films can be expressed as an additional term (E_{g2}) which shall be added to (1), so that

$$E_{g2} = \left(\frac{1}{m_e^*} + \frac{1}{m_h^*}\right) \cdot \frac{\hbar^2 \mathbf{K}^2}{2},$$

(2)

---

[1] We have assumed that the down shift of the valence band is mainly defined by heavy holes because the states corresponding to the light ones are situated somewhat lower in energy. The electron effective mass values in the direction of the confinement we have estimated from calculations, as far as our calculated values of the longitudinal and transverse effective mass of electrons (0.86 and 0.18) are close to the experimental ones.

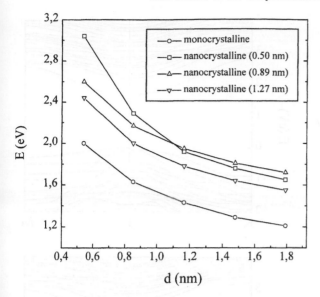

Fig. 3. LCAO results of the total band gap versus film thickness for different silicon (111) films considered.

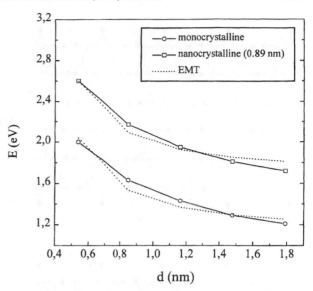

Fig. 4. LCAO results in comparison to EMT fitting of total band gap dependence on silicon (1 1 1) film thickness. The fitting parameters are discussed in the text.

where m_e^* and m_h^* are the electron and hole effective masses in the plane of the film, respectively. The value of $\mathbf{K}^2 = K_x^2 + K_y^2$ can be obtained by solving the 2D Schrödinger equation with appropriate Bloch's boundary conditions. For the general case, an analytical solution is not possible and one needs to solve the equation numerically. But, we found that, for our purposes, one can get qualitatively the same results if the grain films considered, which have the hexagonal symmetry, are represented by an effective 2D set of interacting square wells with rectangular barriers. Thus, to estimate the values of K_x^2 and K_y^2, for the square lattice (see Appendix A) inside the unit cell one has two independent Kronig–Penney equations for both electron and hole subsystems.

Applying the above mathematics to the film with the grain size $a = 0.89$ nm and assuming $b = 0.26$ nm, it appeared to be possible to fit the LCAO data for the grained films (Fig. 4). The barrier height V_0 was used as a fitting parameter. It was estimated to be 0.68 eV. The procedure was also successfully applied to fit the data for the films with larger grains characterising by $a = 1.27$ nm. It is important to note that we got almost the same value of the barrier height for these larger grain structures ($V_0 = 0.67$ eV). Moreover, in both

cases the results are very weakly affected by the choice of the appropriate carrier effective masses.

It was mentioned above that there were some features in the band gap dependence on the thickness for the films with the smallest effective grain size. In our opinion, within this specific case more complicated behaviour is due to the fact that for the region of the film thickness considered there is a transition from interacting dot-like to wire-type structures. Normally, the latter are characterised by lower values of the gap in comparison to the dots [7]. To check up this point, we calculated the band structure when 6 DL film with the grain size of 0.50 nm consisted of two appropriate 3 DL films (Fig. 5). In this case the gap value considerably increases from 1.65 up to 2.11 eV. Thus, we found that possible stacking effect results in a significant increase of the corresponding gap value. It is considered as an additional confinement condition inside the film.

In conclusion, nanocrystalline silicon (1 1 1) films with intergrain boundaries passivated by hydrogen have been found to have a direct band gap structure, while a monocrystalline film with the same thickness appears to be characterised by an indirect gap typical for the bulk silicon. Details of quantum

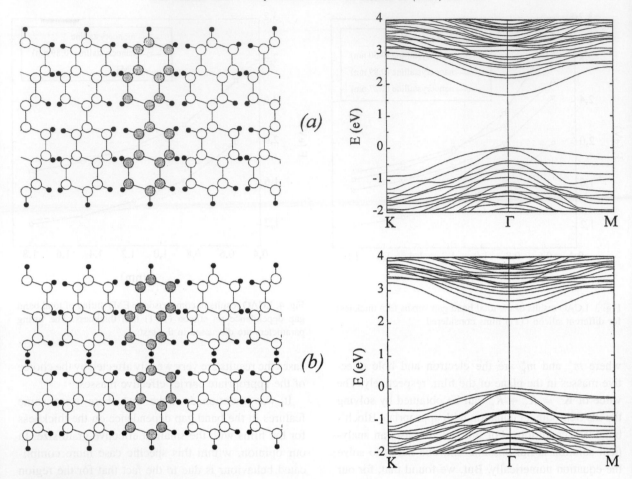

Fig. 5. Evolution of the atomic configuration and related electronic band structure in 6 DL silicon film demonstrating stacking effect. 1 × 6 DL (a), 2 × 3 DL (b) films.

confinement and related changes in electronic properties of the nanocrystalline films were investigated by using LCAO calculations. Their gap dependence on grain size and film thickness was traced. Moreover, a new and simple approach based on the EMT has been developed. Such an approach allows for a substantial reduction of the calculation parameters, and it gives them a physical sense. In this way, it is easier to predict the energy gap variation for more sophisticated structures. It was shown that the main features of the gap behaviour obtained within the LCAO method can be well reproduced and simply explained in terms of confinement effect for film-type structures and cluster interaction.

Appendix A

The Schrödinger equation for a 2D set of quantum wells is

$$\Delta\Psi + 2m/\hbar^2(E - U(\mathbf{r}))\Psi = 0,$$

where $U(\mathbf{r}) = \sum_{\mathbf{R}} V(\mathbf{r} + \mathbf{R})$, $\mathbf{R}$ is the translation superlattice vector, $V(\mathbf{r})$ the potential of a single well, m the carrier effective mass, and E is the energy. Assuming $V(\mathbf{r}) = 0$ inside the well, within the unit cell one has

$$\Delta\Psi + \mathbf{K}^2\Psi = 0 \quad \text{inside the well,}$$
$$\Delta\Psi - \chi^2\Psi = 0 \quad \text{outside the well,}$$

where the momentum $\mathbf{K}^2 = 2mE/\hbar^2$ and $\chi^2 = 2m(V(\mathbf{r}) - E)/\hbar^2$. Appropriate Bloch's boundary conditions are

$$\Psi(\mathbf{r} + \mathbf{R}) = e^{ikR}\Psi(\mathbf{r}),$$

where $\mathbf{k}$ is the quasimomentum vector.

For the square lattice, taking into account that a variable separation is possible for a potential of the form $V(\mathbf{r}) = V_0(\Theta(x) + \Theta(y))$, where Θ is the step-like function which is equal to 0 and 1 inside and outside the well, respectively, one has two independent 1D set of equations [8]. Moreover, assuming that $\Psi(\mathbf{r}) = \Psi(x)\Psi(y)$, Bloch's boundary conditions can be split into two ones for $\Psi(x)$ and $\Psi(y)$ as well. Finally, one has two independent 1D Kronig–Penney problems for K_x and K_y. For example for x-direction:

$$\cosh(\chi_x b)\cos(K_x a) + \sinh(\chi_x b)\sin(K_x a)(\chi_x^2 - K_x^2)/(2\chi_x K_x) = \cos(k_x c),$$

where $\chi_x^2 = 2mV_0/\hbar^2 - K_x^2$. The parameters a and b are the lateral size of the well and the barrier, respectively and $c = a + b$. The quasimomentum projection k_x varies in between $\pm\pi/c$. So, it is easy to obtain K_x^2 at the appropriate band edge. Therefore, for the square lattice one has two identical solutions, K_x and K_y, which are the same. This corresponds to a lattice symmetry with respect to $\pi/2$ rotations.

References

[1] L.T. Canham, Appl. Phys. Lett. 57 (1990) 1046.

[2] F. Arnaud d'Avitaya, L. Vervoort, F. Bassani, S. Ossicini, A. Fasolino and F. Bernardini, Europhys. Lett. 31 (1995) 25; D.J. Lockwood, Z.H. Lu and J.-M. Baribeau, Phys. Rev. Lett. 76 (1996) 539; M. Watanabe, F. Iizuka and M. Asada, IEICE Trans. Electron. E79-C (1996) 1562; A.G. Nassiopoulos, S. Grigoropoulos and D. Papadimitriou Appl. Phys. Lett. 69 (1996) 2267.

[3] L. Vervoort, F. Bassani, I. Mihalcescu, J.C. Vial and F. Arnaud d'Avitaya, Phys. Stat. Sol. B 190 (1995) 123; F. Bassani, L. Vervoort, I. Mihalcescu, J.C. Vial and F. Arnaud d'Avitaya, J. Appl. Phys. 79 (1996) 4066.

[4] S. Ossicini, A. Fasolino and F. Bernardini, Phys. Rev. Lett. 72 (1994) 1044.

[5] A.B. Filonov, A.N. Kholod, V.A. Novikov, V.E. Borisenko, L. Vervoort, F. Bassani, A. Saúl and F. Arnaud d'Avitaya, Appl. Phys. Lett. 70 (1997) 744; L. Vervoort, F. Bassani, A. Saúl and F. Arnaud d'Avitaya, Thin Solid Films, to be published.

[6] A.B. Filonov, I.E. Tralle, G.V. Petrov and V.E. Borisenko, Model. Simul. Mat. Sci. Eng. 3 (1995) 45; A.B. Filonov, G.V. Petrov, V.A. Novikov and V.E. Borisenko, Appl. Phys. Lett. 67 (1995) 1090; A.B. Filonov, in: Physics, Chemistry, and Application of Nanostructures, eds. V.E. Borisenko et al. (Minsk, 1995) pp. 50–55.

[7] J.P. Proot, C. Delerue and G. Allan, Appl. Phys. Lett. 61 (1992) 1948; F. Buda, J. Kohanoff and M. Parrinello, Phys. Rev. Lett. 69 (1992) 1272; B. Delley and E.F. Steigmeier, Appl. Phys. Lett. 67 (1995) 2370.

[8] P.M. Morse and H. Feshbach, Methods of Theoretical Physics, Pt I (McGraw-Hill Kogakusha, Tokyo, 1953) p. 498.

Computational Materials Science 10 (1998) 154–158

Interaction of Cu(1 1 1) surface states with different extended defects

Andrea Barral [a,*], Ana María Llois [a,b]

[a] *Departamento de Física "Juan José Giambiagi", Facultad de Ciencias Exactas y Naturales, Universidad de Buenos Aires, Ciudad Universitaria, Pabellón I, 1428 Buenos Aires, Argentina*
[b] *Departamento de Física, Comisión Nacional de Energía Atómica, Avda del Libertador 8250, 1429 Buenos Aires, Argentina*

Abstract

We study numerically the interaction of the Shockley surface state at the Γ-point of Cu(1 1 1) with periodic arrangements of adsorbed/absorbed Cu and Fe atoms. We use an spd tight-binding hamiltonian with the local basis enlarged to account for electron spillover. Copyright © 1998 Elsevier Science B.V.

Keywords: Surface states; Defects; Quantum corrals

1. Introduction

The study of the electronic properties of low dimensional systems has attracted much attention from both the theoretical and experimental side in the last decade. In particular surface states have been studied for a long time using, among other techniques, photoemission and inverse photoemission spectroscopy. Using scanning tunnelling microscopy (STM) it has been observed that the well-known Shockley states, that appear on the (1 1 1) surface of noble metals, can form a 2D gas which reveals surface defects giving rise to interference diagrams [1,2]. For instance, Fe atoms have recently been arranged on a Cu(1 1 1) surface into circular "quantum corrals" and standing surface waves have been observed inside the Fe corrals [3].

Shockley states arise from special boundary conditions introduced by vacuum/metal interface and occur in the sp-gap of the projected bulk band structure of

noble metals [4]. They are recognized because of their almost parabolic dispersion relations.

In this contribution we study the evolution of the sp-Shockley state at the Γ-point of Cu(1 1 1) in the presence of periodically deposited Cu rows, Cu steps, small rombohedral corrals and Fe rows ad- and absorbed on the surface. It is seen that this state disappears for a high density of adsorbed defects and that it begins to reappear when defects form a reasonable large corral, while it is not affected by the presence of absorbed defects. This effect seems to be independent of the type of 3d late transition metal atoms used to build the corrals.

2. Method of calculation

We study the interaction of the Cu(1 1 1) Shockley state at Γ with different kinds of defects doing slab calculations. As we are handling with large unit cells, we use a Hubbard tight-binding formalism in

---

* Corresponding author. Tel.: 54 1 754 7092/93/94; fax: 54 1 754 7121; e-mail: barrral@dfuba.df.uba.ar.

the unrestricted Hartree–Fock approximation including spd-orbitals, easily parametrized to reproduce bulk ab initio results [5,6]. For Fe the exchange integral is taken different from zero only for d orbitals and it is fitted to give the experimental bulk magnetization value [7].

In order to model the electron spillover at the surface we introduce empty s-like orbitals, from now on called s'-orbitals, outside the surface to which the delocalized sp-electrons of the surface can be transferred [7,8].

The extra layer of s-type orbitals follows the original lattice geometry. We also introduce s'-orbitals if a superficial vacancy is present. The onsite energy of the s'-orbitals, $\varepsilon_{s'}^0$, is chosen in such a way that the Shockley state at the Γ-point of the perfect Cu(1 1 1) surface lies close to the experimental value. Instead of using a Madelung term to account for charge transfers we align the Fermi level of the outer atomic layers with the Fermi energy of Cu bulk as in [9].

In the absence of s'-orbitals ($\varepsilon_{s'}^0 = \infty$) the Shockley states at the clean (1 1 1) transition metal surfaces appear too high in energy, located in between the edges of the sp-gap at Γ. With the introduction and proper parametrization of the s'-orbitals the energy of these states can be brought down to the experimental value, which in the case of Cu(1 1 1) is (0.4 ± 0.1) eV below the Fermi level and around 0.5 eV above the top of the bulk bands. By setting $\varepsilon_{s'}^0 \sim 13$ eV, measured with respect to the Fermi level of Cu bulk we obtain the Shockley state at the desired energy position, this implies allowing for a d-band filling of $+0.2e^-$ of the Cu(1 1 1) monolayer with respect to the bulk value [10].

In our slab calculations we always use 18 layers including the s'-ones. This guarantees the identification of the slowly decaying surface sp-states. Due to the slab geometry two surface states are obtained in our calculations. We take into account 2D-periodicity and work in reciprocal space.

3. Results and discussion

In Fig. 1 we show the dispersion of the Shockley state around Γ and compare it with the experiment

of [11]. In the perfect surface the Shockley state is of sp_z composition and decays 76% within the first four outer Cu layers (two Cu's + one s' layers on each side of the slab). To follow the evolution of this state in the presence of defects we analyze therefore the orbital composition of the eigenstates looking for those of mainly sp_z composition which are localized on the surface layers. The z-direction is given by the perpendicular to the (1 1 1)-surface.

In Fig. 2 we show schematically the geometric arrangement of the different kinds of surface defects studied. We consider the following systems: (a) rows of Cu atoms adsorbed on the perfect Cu surface separated by N rows of s'-orbitals, N ranging from 1 to 7; (b) one missing Cu row every four Cu rows; (c) same as system (a) but with $N = 1$ and taking out 50% of the Cu's in each row in such a way that each surface Cu has no Cu nearest neighbors in the plane; (d) rows of Fe-like atoms adsorbed on perfect Cu separated by N rows of s'-orbitals; (e) rows of Fe-like atoms absorbed in the surface layer separated by N Cu rows. In all cases there is a complete layer of s'-orbitals on each side of the considered slabs, which is not shown in Fig. 2.

When the size of the unit cell increases laterally, it becomes more and more cumbersome to distinguish among states. As we are sorting out just those states which are localized on the surface layers and that are mainly of sp_z composition, we compare the energy spectrum of the clean surface with the spectra of the surfaces with defects. The spectra are obtained by using the same supercell configuration for the clean as for the modified surfaces in each case. In Table 1 we show the characteristics of the sorted out states for some of the different situations under study. We give the shift of their energies with respect to the energy of the Shockley state of the perfect surface, ΔE, the percentage of sp_z composition and the weight of the eigenstate on the three first surface layers of the slab. The weight of the spillover is assigned to the first layers.

In most of the examples the localized sp_z states lie beyond the Fermi level and from the data shown in Table 1 (a)–(c) we infer that the upward energy shift of this state with respect to the energy position of the Shockley one of the clean surface depends on the

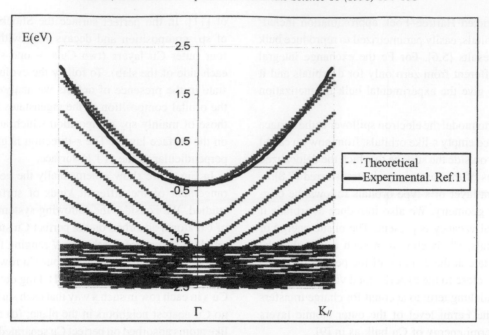

Fig. 1. Energy bands around the Γ-point of Cu(1 1 1) along the Γ-K direction. The theoretical curves from the present calculations are compared with the fitted experimental results for the Shockley state taken from [11]. Energies are given with respect to the Fermi level. The calculated energy dispersion of the Shockley state agrees very well with the experimental curve, both curves fall at −0.4 eV at the Γ-point. The other calculated bands, lying lower in energy, are projected bulk bands.

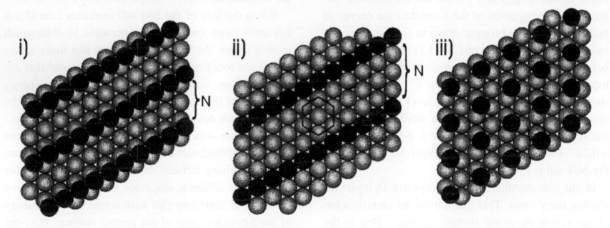

Fig. 2. (i) Geometric arrangement for cases (a) and (d). The dark spheres indicate Cu or Fe separated by N missing Cu rows. The light spheres indicate the Cu atoms lying underneath the Cu or Fe rows. (ii) Geometric arrangement for cases (b) and (e). In case (b) the dark spheres indicate missing Cu rows separated by a step of N Cu rows (light spheres). In case (e) dark spheres stand for Fe atoms. (iii) Illustrates case (c). Dark spheres are isolated Cu atoms on the first layer and light spheres Cu atoms of the second layer.

local coordination number of the respective defect. In cases (a)–(d) the decaying states of sp_z composition are mainly localized on the defect sites, so that they could be considered as defect states.

We expect that for N sufficiently large, the sp_z-Shockley state of the clean surface should develop again. In case (a) and $N = 7$ we effectively observe several states of the due orbital composition lying

Table 1
Data of localized sp_z states for the clean Cu(1 1 1) surface and surfaces with different kinds of defects shown in Fig. 2

System	N	ΔE (eV)	sp_z (%)	Layer weight		
				1st	2nd	3rd
Cu(1 1 1)		0.0	94	0.56	0.20	0.10
(a)	1	1.5	90	0.42	0.36	0.14
	2	1.5	91	0.52	0.28	0.12
	3	1.5	87	0.40	0.28	0.12
	4	1.5	85	0.24	0.2	0.12
	7	1.8	85	0.42	0.28	0.14
	7*	0.3	64	0.09	0.20	0.12
(b)	4	0.3	67	0.31	0.14	0.11
(c)		$\succ$ 2.0	92	0.53	0.28	0.11
(d)	1	1.5	88	0.54	0.28	0.10
	2	1.8	84	0.46	0.20	0.10
(e)	1	0.0	92	0.58	0.20	0.10
	2	0.0	88	0.55	0.19	0.10
	3	0.0	92	0.57	0.10	0.06
	4	0.0	90	0.56	0.20	0.10

ΔE(eV): energy shift of the sp_z state with respect to the Shockley state of the perfect surface, sp_z (%): percentage of sp_z orbital composition. The weight of the wave function on the first three layers is given.

below the Fermi level but still not highly localized. These states are not present in the folded energy spectrum of the perfect surface and their largest weight is on the first complete Cu layers on both sides of the slab. In Table 1, we give the *defect state*, lying high above the Fermi energy, and we give also the data for one of the mentioned states denoted with 7*.

In case (b), for which we show the results for a step-like defect with $N = 4$, the state that appears at nearly the Fermi energy has the expected composition and is going to transform into the Shockley state for larger N. The surface in this example is less perturbed than in the cases (a) and because of this the surface sp_z state has not moved as much.

In case (d), that is Fe rows adsorbed on the Cu surface, the results obtained for spin-up and spin-down sp_z states are similar to those of case (a). We give just the spin-up results. In case (e), that is Fe rows absorbed in Cu the results for the energy shift and the decaying behavior of this sp_z state is not much different from what has been obtained for the clean surface. This is clearly observed in Table 1 when comparing with the Shockley state of the clean surface. The wave functions of the spin-up states are more confined on the Cu atoms than on the Fe atoms. If one consid-

ers the weight of the sp_z state just on the first layer, the average weight on the Cu atoms is 1.1–1.5 the weight on the Fe atoms. The spin-down states always show a larger amount of d-orbital composition than the spin-up ones due to hybridization, this counterbalances the slight Cu confinement present in the spin-up states.

4. Conclusions

In this contribution we show that the Shockley state on Cu(1 1 1) already appears for a rather high density of transition metal defects. Corrals with diameter larger than seven times the nearest neighbor distance should be large enough for the surface state to develop. This situation is largely given in the experiments reported in the literature.

The upward energy shift of the sp_z surface states is a function of the local coordination of defects, this implies that in the case of Cu steps the Shockley state should develop for step diameters smaller than those needed for corrals.

The Shockley state sees nearly no difference between Fe and Cu as it couples mostly with the sp_z-orbitals, which are similar in both kinds of atoms. We expect the same behavior for Co and Ni as for Fe because of the same reason.

References

[1] M.F. Crommie, C.P. Lutz and D.M. Eigler, Science 262 (1993) 218.

[2] Y. Hasegawa and P.H. Avouris, Phys. Rev. Lett. 71 (1993) 1071.

[3] M.F. Crommie, C.P. Lutz and D.M. Eigler, Nature 363 (1993) 524.

[4] P.P. Gartland and B.J. Slagsvold, Phys. Rev. B 12 (1975) 4047.

[5] G. Fabricius, A.M. Llois and M. Weissmann, Phys. Rev. B 44 (1991) 6870.

[6] O.K. Andersen and O. Jepsen, Phys. Rev. Lett. 53 (1984) 2471.

[7] G. Fabricius, A.M. Llois, M. Weissmann and A. Khan, Phys. Rev. B 49 (1994) 2121.

[8] R. Gomez Abal, A.M. Llois and M. Weissmann, Phys. Rev. B 53 (1996) R8844.

[9] F. Fabricius, A.M. Llois, M. Weissmann, M.A. Khan and H. Dreyssé, Surf. Sci. 331–333 (1995) 1377.

[10] A. Barral, Master Thesis (1996).

[11] S.D. Kevan, Phys. Rev. Lett. 50 (1983) 526.

COMPUTATIONAL
MATERIALS
SCIENCE

ELSEVIER   Computational Materials Science 10 (1998) 159–162

Treatment of electrostatic interactions at the Si(1 0 0)–SiO$_2$ interface

Serguei A. Prosandeyev [a,*], Gérard Boureau [b], Stéphane Carniato [b]

[a] *Deparment of Physics, Rostov State University, 5 Zorge St., 344090 Rostov on Don, Russian Federation*
[b] *Laboratoire de Chimie Physique, Matière et Rayonnement, Université Pierre et Marie Curie,*
11, rue Pierre et Marie Curie, 75231 Paris, Cedex 05, France

Abstract

Silica in equilibrium with silicon is known to be partly crystalline. Because of crystallographic constraints, models proposed so far (cristobalite as well as tridymite) deal with polar surfaces of silica, which are expected to be highly unstable. In this paper, using both exact treatments of simplified models and Monte-Carlo simulations of the real system with effective potentials, we show how the system compromises to decrease the electrostatic energy. We also investigate the role of electrostatic interactions on the formation of oxygen vacancies at the Si–SiO$_2$ interface. Using a tight binding approach, implications of these interactions on mid-gap surface states in silica are also discussed. Copyright © 1998 Elsevier Science B.V.

1. Introduction

The silicon–oxygen bond has been known for a long time to have a partly ionic character [1]. Therefore electrostatic effects have to be taken into account for the study of surfaces and interfaces. The Si(1 0 0)–SiO$_2$ interface is known to be a good electrical interface [2]. Nevertheless the nature of the silica phase in the vicinity of silicon is far to be known. As it may be seen in Fig. 1, the two epitaxial oxides considered so far (cristobalite and tridymite) are highly polar. They belong to the type 3 of interfaces of the classification of Tasker [3], which implies that they are highly unstable. Physical systems have a number of ways to accomodate such constraints [4]:
– rumpling
– point defects
– charge transfer

– adjunction of a surface dipole
– amorphisation
– reconstruction

The goal of the present study is to evaluate in a qualitative way the importance of these effects and make some predictions on its influence on the appearance of mid-gap states.

2. Purely electrostatic calculations

This treatment follows earlier studies of analogous cases [5,6]. In the first step, we shall disregard the lattice structure and shall consider that silica (cristobalite or trydimite) is made of uniformly charged layers. The electrical potential is a function only of the coordinate z perpendicular to the planes:

$$\Phi_0(z) = 2\pi \sum_\alpha \sigma_\alpha \mid z - z_{0\alpha} \mid \qquad (1)$$

---

* Corresponding author.

cristobalite   trydimite

	O O -4	
	Si +4	
	O O -4	
	Si +4	
	O O -4	
	Si +4	
	O O -4	
	Si +4	

	O -2
	SiO +2
	O -2
	SiO +2
	O -2
	SiO +2
	O -2
	SiO +2

Fig. 1. Polar character of cristobalite rotated around the [1 0 0] axis by 45° with respect to the [1 0 0] axis of silicon and of trydimite with the [0 0 1] lattice vector parallel to the interface.

where σ_α is the surface charge density in the αth layer. This value due to the macroscopic field remains unchanged if a surface appears in z. In this case, the additional potential due to the surface is zero.

If now we consider the lattice, the difference, due to the microscopic field, is no more zero. This difference gives rise to an additional (surface) potential at the surface, which can be calculated from the Ewald method as now all the planes are neutral. The difference is purely local:

$$\Delta\Phi^{\text{loc}}(\mathbf{r}, z) = -\sum_{\alpha \notin V} q_\alpha(\mathbf{r}, z)\psi^{\text{loc}}(\mathbf{r} - \mathbf{r}_{0\alpha}, z - z_{0\alpha})$$

(2)

with

$$\psi^{\text{loc}}(\mathbf{r}, z) = \psi(\mathbf{r}, z) + V(z)$$

(3)

and

$$\psi(\mathbf{r}, z) = \sum_{\mathbf{l}}[|\mathbf{r} - \mathbf{l}|^2 + z^2]^{-1/2}.$$

(4)

$V(z)$ is the electrostatic potential (macroscopic potential) due to a uniformly distributed compensating charge adds on each monolayer and which make it electroneutral. From the electrostatic potential expression established by Prosandeyev [6], in the case of two-dimensional lattice systems, the final expression for calculating the two-dimensional lattice sum may be written ($z > 0$)

$$\psi(\mathbf{r}, z) = \frac{2\pi}{S}\sum_{g \neq 0}e^{i\mathbf{g}\mathbf{r}}\frac{1}{g}e^{-zg}$$

(5)

where $\mathbf{g}$ is a translation vector in the reciprocal space.

3. Applications

3.1. Cristobalite surface

We consider the crystal structure of the β-cristobalite with the straightened Si–O–Si angles (Fig. 2). We use the lattice parameter 5.06 Å. The charges used coincide with the formal charges of the ions O^{2-} and Si^{4+}.

There are two possible terminating layers in the case of the (0 0 1) surface. The former consists of the Si ions while the latter does consists of the O ions. Table 1 lists the results obtained for both the surfaces. In the former case, when the Si ions cover the surface, the surface potential on the Si site is rather strong and is of negative sign. It implies that the electron states of the surface Si ions will have lower energy than in the bulk and prove to be in the forbidden gap.

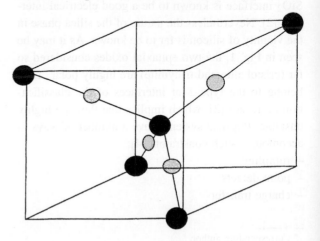

Fig. 2. The cristobalite cell.

Table 1
Additional electron potential on the (0 0 1) surface of cristobalite

Terminating layer	Atomic position	Layer's number	Electron potential (eV)
Si	Si	1	−6.12
	O	2	−0.38
	Si	3	+0.41
	O	4	+0.18
	Si	5	+0.08
	O	6	+0.01
O	O	1	+5.03
	Si	2	−0.71
	O	3	−0.62
	Si	4	−0.39
	O	5	−0.08
	Si	6	−0.01

Table 2
Additional electron potential on the (0 0 1) surface of compressed tridymite

Terminating layer	Atomic position	Layer's number	Electron potential (eV)
O1 [a]	O1	1	+7.2(4.4) [b]
—	Si	2	−0.3
—	O2	2	−1.2
—	O1	3	+0.2
Potential on the	Si [c]	—	+5.2
Interfacial ions	Si [d]	—	−0.03

[a] The O1 ions are on the top of the trydimite semi-infinite crystal.
[b] In the parentheses there is a value obtained with taking into account charging the interfacial Si ions.
[c] The Si lies between the silicon semi-infinite crystal and trydimite semi-infinite crystal.
[d] The Si ion lies on the top of pure silicon.

From the data listed in Table 1, it is seen that the absolute value of the additional potential sharply decreases when going inside the crystal. Only a few top layers are disturbed by this potential. It implies that the surface states will be localised in a nearest vicinity of the crystal's top.

Contrary to foregoing, in the latter case, when the O ions cover the surface, the additional surface potential is of positive sign. As a consequence, the oxygen energy levels will be raised above the top of the valence band. These levels could be traps for holes. As seen in the previous case, the potential sharply decreases when going inside the bulk.

3.2. Mid-gap states

As shown elsewhere [7], by using a simplified tight-binding model of the electronic structure of silica (cristobalite) similar to the Weaire and Thorpe model [8] devoted to silicon the breaking of Si–O bonds does not lead to the appearance of mid-gap states. This result differs from results concerning silicon. The mid-gap states found by Ciraci and Ellialtioglu [9] can have their origin in an increase of the Si–Si or O–O interactions which are likely connected to electrostatic interactions.

3.3. (1 0 0) Si-tridymite interface

In this section we deal with a model of the Si/SiO₂ interface proposed by Ourmazd et al. [10]. In this model, the top of the tridymite crystal at the interface is made of the O ions (O1). Between the O1 ions and pure silicon, there is a layer of interfacial Si atoms.

Table 2 shows the results of calculating the additional electron potential at the tridymite's top. We have used the same geometry of the compressed tridymite as in [11,12]. Let us first suppose that the interfacial silicon ions are neutral. It is seen that the electron potential on the O1 site is very large and is of positive sign. It means that the oxygen energy levels are raised with respect to the bulk's levels. As a consequence, the O1 ions could trap holes and become single-charged or even neutral. One can take into account that, therewith, the atomic size of the oxygen ions in the interface is to be strongly reduced. It implies that the creation of oxygen vacancies is alleviated in this plane.

We have also computed the potential produced by the tridymite semi-infinite crystal on the interfacial Si ions. This potential was found to be of positive sign. This will result in taking up the interfacial Si ions levels that will lead to charging the Si ions with respect to the neutral Si ions in the bulk of Si. However, the potential is not very large as in the bulk of SiO₂ that is in qualitative agreement with the assumption that the interfacial Si ions are single-charged.

Finally, the potential on the Si ions lying on the top of silicon semi-infinite crystal turns out to be very small as well as the additional potentials on the second and third layers of the tridymite. It means that only a thin region near the interface is disturbed by the long-range potential. Nevertheless, we believe this potential to play a very important role in the formation of the interface electron states.

3.4. Comparison with Monte-Carlo simulation

In two recent studies, Carniato et al. [11,12] have performed Monte-Carlo simulation of the (100) Si-tridymite interface. They have used semi-empirical potentials with explicit charges (four times smaller than the formal charges considered in the previous section and dependent on the environment) and have shown that it was easier to create an oxygen vacancy at the interface than at its vicinity. They have found a difference of 1.4 eV. This result is in good agreement with the conclusions of the previous section. In principle, in such a simulation, the various ways to accomodate constraints mentioned in the introduction have been taken into account. The silicon atoms at the interfaces have been considered to have a formal oxidation number equal to 1. Some relaxation and reconstruction occur and oxygen vacancies are the way to have surface dipoles which decrease the electrostatic interactions.

4. Conclusion and summary

It may be concluded that electrostatic effects give rise to an additional surface potential at the surface.

This additional potential is caused by the difference between the local and average electrostatic potential. The long-range surface potential in silica crystal causes the appearance of mid-gap states by a lowering of the Si levels if the Si ions cover the surface and rising the oxygen levels if the oxygen ions terminate the crystal.

Our calculations of the long-range potential near the Si–SiO_2 interface have shown that the interfacial Si- and O- have different oxidation states with respect to the ions in the bulk. This result is in good agreement with the well-known experiments of X-ray photoelectrons spectoscopy (XPS) [13].

References

[1] L. Pauling, Am. Mineral 65 (1980) 321.
[2] E.H. Pointdexter, Semicond. Sci. Technol. 4 (1989) 961.
[3] P.W. Tasker, J. Phys. C 12 (1979) 4977.
[4] C. Noguera, Physics and Chemistry of Oxide Surfaces (Cambridge University Press, Cambridge, 1996).
[5] S.A. Prosandeyev and I.M. Tennenboum, Phys. Rev. B 52 (1995) 4545.
[6] S.A. Prosandeyev, Surf. Sci. Lett. 340 (1995) 978.
[7] S.A. Prosandeyev, G. Boureau and S. Carniato, to be published.
[8] D. Weaire and M.F. Thorpe, Phys. Rev. B 4 (1971) 2508.
[9] S. Ciraci and S. Ellialtioglu, Solid State Comm. 40 (1981) 10 915.
[10] A. Ourmazd, D.W. Taylor, J.A. Rentschler and J. Bevk, Phys. Rev. Lett. 59 (1987) 213.
[11] S. Carniato, G. Boureau and J. Harding, Radiation Effects and Defects in Solids 134 (1995) 179.
[12] S. Carniato, G. Boureau and J. Harding, Philosophical Magazine A 75 (1997) 1435.
[13] F.J. Himpsel, F.R. McFeely, A. Taleb-Ibrahimi, J.A. Yarmoff and G. Hollinger, Phys. Rev. B 38 (1988) 6084.

Computational Materials Science 10 (1998) 163–167

COMPUTATIONAL
MATERIALS
SCIENCE

Use of semi-empirical potentials to simulate the Si(1 0 0)–SiO$_2$ interface

N. Capron *, A. Lagraa, S. Carniato, G. Boureau

*Laboratoire de Chimie Physique, Matière et Rayonnement, Université Pierre et Marie Curie,
11, rue Pierre et Marie Curie, 75231 Paris, Cedex 05, France*

Abstract

The rapid development of ab initio methods makes possible the test of semi-empirical potentials which retain some advantages in the simulation of large systems. Even systems involving oxygen vacancies can be studied with semi-empirical methods. Modelling of the Si(1 0 0)–SiO$_2$ interface raises some specific problems due to charge transfers and severe simplifications are still necessary. Copyright © 1998 Elsevier Science B.V.

Keywords: Si–SiO$_2$ interface; Potentials; Oxygen vacancy; Charge transfer

1. Goal of the present paper

Due to its importance in MOS technology [1], the Si(1 0 0)–SiO$_2$ interface is one of the most studied interfaces. Because of the continuous miniaturisation, a better control of interface is necessary. Experimental investigation is made difficult by the existence of amorphous silica at some distance of the interface [2]. Therefore computer simulation is a useful investigation tool. There have been a number of simulation studies (Monte-Carlo and molecular dynamics) dealing with the Si–SiO$_2$ interface: ab initio [3] as well as semi-empirical approaches [4–6] have been used. Both approaches have their merit. In order to simulate a real system, it would be of interest to deal with a large system with temperature, pressure and oxygen chemical potential constraints. Even with semi-empirical potentials this is a very difficult task. The present paper is a step in that direction.

While oxygen vacancies are in very small amount in the bulk, they are easy to form near the interface [6,7]. Moreover oxygen vacancies play an important role in diffusion processes. The two epitaxial oxides considered so far (cristobalite and tridymite) are highly polar. They belong to the type 3 of interfaces of the classification of Tasker [8], which implies that they are highly unstable. Charge transfer and adjunction of a surface dipole are among the usual ways to make this type of interface acceptable [9]. Therefore, if we want to work with a variable number of oxygen atoms, we shall have to use semi-empirical potentials in unusual conditions. As it is well known that empirical potentials have only a limited range of validity [10], the prerequisite to the simulation of the interface in the grand canonical ensemble is to test the ability of these potentials to handle the creation or the annihilation of oxygen vacancies and charge transfers. We shall now examine these two points in the following two sections.

---

* Corresponding author.

PII S0927-0256(97)00141-9

2. Bulk silica

2.1. Perfect silica

Classical potential models have often been used for silica. It has usually been found necessary to include three-body terms. Examples of their success include the simulation of the vitreous state (static structure factor, first sharp diffraction peak) [11] and studies of molten silica at high pressure [12].

Tsuneyuki et al. [13] have designed a *pair* potential with a simple analytical expression. This potential has been extracted from ab initio cluster calculations. This calculated potential is a Buckingham potential with non-integral charges on the ions.

This potential has been extremely successful in a number of applications: such as the α to β structural phase transition of quartz [14], the lattice dynamical properties of α-quartz [15], the structural properties of SiO_2 in the stishovite structure [16], the liquid–vapour coexistence curve [17], the high-pressure densification of silica glass [18], the behaviour of α-quartz under hydrostatic and non-hydrostatic high-pressure conditions [19].

2.2. Vacancy in bulk silica

2.2.1. Used criteria

In order to study the relevance of semi-empirical potentials to deal with vacancies, we have to add some new tests to the numerous ones mentioned in the previous section. In the present study, we shall restrict ourselves to the requirement of getting an acceptable value of the formation energy of an oxygen vacancy. By acceptable, we mean in reasonable agreement with the best available ab initio calculations [20–22] and with thermodynamic constraints: Boureau and Carniato [23] have shown that this formation energy is at least 7.5 eV. Allan et al. [20] and Allan and Teter [21] carried out total energy calculations and defects studies on different silica polymorphs by DFT-LDA method and got 7.85 and 8.41 eV for the energy of formation of this oxygen vacancy for α-quartz and α-cristobalite respectively. Pacchioni et al. [22] found 8.5 eV using

a Hartree–Fock method within a cluster model approach.

2.2.2. Covalent potentials

We have to make a distinction between two families of semi-empirical potentials. The first family deals only with neutral species without any charge attributed to individual ions. The Lee potential [4] and the first version of Stixrude [24] potential belong to this family. From the previous section, it may be concluded that the formation energy of an oxygen vacancy is not too different from half the cohesion energy (7.3 eV). Therefore, as the cohesion energy of oxides is one of the physical quantity fitted to build these potentials, we may expect to get reasonable values of the energies of formation of oxygen vacancies. It is indeed the case. A mere calculation at 0 K without relaxation provides 8.6 eV for the Stixrude potential and 8.1 eV for the Lee potential. These values are still decreased if high temperature relaxation is allowed. For instance, we have performed Monte-Carlo simulation at 1000 K with a cell containing 192 atoms for the Lee potential. The new value is 6.0 eV.

2.2.3. Ionic potentials

The silicon–oxygen bond has been known for a long time to have a partly ionic character [25]. Therefore the use of partly ionic interatomic potentials may look attractive. Unfortunately, the straightforward Mott and Littleton [26] strategy generally used to deal with defects makes no sense if non-integral charges are used on the ions, as is often the case in potentials used for silica. Moreover a number of problems connected with the affinity of oxygen which is an ill-defined quantity remain [27].

Semi-empirical potentials are unable to make predictions concerning charge transfers. Therefore, we have to make guesses. The simplest assumption is to consider that the electronic distribution is essentially unchanged. We have a so-called neutral vacancy (Fig. 1 case (b)). The Stixrude potential provides a value equal to 7.5 eV, for the idealised β-cristobalite, without relaxation, at 0 K. This is a quite reasonable value, the cause being that in this potential, the charges

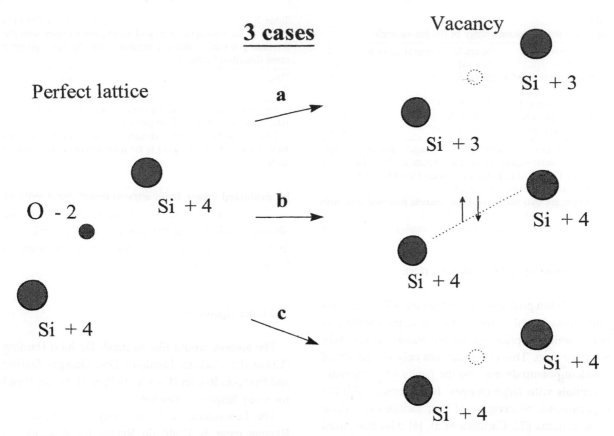

Fig. 1. Scheme of the three limiting cases considered in the paper. Model a of oxygen vacancy has been considered at interface. Model (b) (without charge transfer) has been used in bulk. Model (c) considers charges as pure calculation intermediates and does not respect electrical neutrality.

attributed to atoms (+1 for silicon atoms and −0.5 for oxygen atoms) are small in agreement with the work of Pauling. Therefore, short range contribution play an important role, which allows to get large enough values of the energy of formation. This is not the case if other interatomic potentials using large charges roughly equal to Mulliken charges [28,29] are used. In fact these formal charges are essentially a cheap way to introduce into these semi-empirical potentials repulsion between oxygen atoms, which forces the respect of the tetrahedron geometry without the burden of three-body terms.

Another way to use these potentials is to consider that their analytical form is purely empirical and that the physical picture underlying the Buckingham potential is lost. This is not surprising if one remembers

that the analytical expression has a term of the form $-c_i c_j / r^6$ which is usually considered as a dispersion interaction term whereas the HF-SCF method used to build this potential does not include such a feature. Therefore, it is legitimate to consider that the energy of formation may be evaluated by removing the term related to the oxygen without trying to make electrostatic corrections and without any attempt to respect the formal electrical neutrality. In this way, reasonable values for the energy have been obtained as it may be seen in Table 1. The crystalline structure used for this calculation is the idealised β-cristobalite. Calculations were carried out at 0 K, without relaxation, using periodic conditions, comparing ideal lattice and defective one, after removing an oxygen atom from the bulk and placing it at infinity (see Table 2).

Table 1
Values of the formation energy of oxygen vacancies

Covalent potentials	Stixrude[a] – neutral version –	8.6
	Lee [4]	8.1
Without charge tranfer (Fig. 1 case (c))	Stixrude [24]	7.5
With mere removal of the oxygen term (Fig. 1 case (b))	Vashista [11]	9.75
	Tsuneyuki [28]	18.8
	Stixrude [24]	11.2

No relaxation has been allowed. Charges attributed to silicon atoms surrounding the oxygen vacancies have been kept constant. The structural phase considered for silica is the ideal β-cristobalite. Energies are in eV.

[a] There are two versions of the Stixrude potential: one with charges of $+1$ and -0.5 on Si and O atoms, respectively and one without any charge (values of constants employed are different).

3. Problems specific to the interface

Simulation problems at interface are still more complicate. As stated in the introduction, the interface is polar, which implies that charge transfer is certainly no more zero. This eliminates not only the family of covalent potentials but also the family of interatomic potentials with large charges. In agreement with the experimental observation of four varieties of silicon environments [2], Carniato et al. [6] who have used the second version of the Stixrude potential [24] have considered that the effective charge of surrounding silicon atoms was decreased by half the charge of missing oxygen. In such a way, using a Monte-Carlo procedure, they found 12.8 eV for unrelaxed defects and 11.4 eV for relaxed defects at 1000 K. If such a charge transfer was used for bulk silica with the same potential, the figures would be 16.9 eV for unrelaxed defects and 15.8 eV for relaxed defect.

It should be noted that this approach is unable to predict the amount of charge transfer which has to be fixed.

4. Conclusion

While semi-empirical potentials have been extremely successful in the modelling of a number of properties of bulk silica, most of them are of no help for the study of interfaces. The charge transfer has to

Table 2
Values of the formation energy of an oxygen vacancy with the Stixrude potential – charged version – for the three possible cases described in Fig. 1

Case	(a)	(b)	(c)
E_f	16.9	7.5	11.2

Crystalline structure studied is the idealised β-cristobalite, at 0 K. without relaxation. Energies are in eV.

Case (a) is the positively charged vacancy; (b) the so-called neutral vacancy; and case (c) is the mere removal of an oxygen term.

be evaluated either from experimental arguments or from ab initio calculations. In this way, the picture obtained looks reasonable and large scale calculations in the frame of a grand canonical ensemble seem to be at hand.

Acknowledgements

The authors would like to thank Dr John Harding (University College, London), Drs. Georges Dufour and François Rochet (University P. et M. Curie, Paris) for many helpful discussions.

The Laboratoire de Chimie Physique, Matière et Rayonnement is Unité de Recherche Associée au CNRS No. 176.

One of us (N C) would like to thank Ministère de l'Enseignement Supérieur et de la Recherche for providing a doctoral grant.

References

[1] E.H. Pointdexter, Semicon. Sci. Technol. 4 (1989) 961.
[2] F.J. Himpsel, D.A. Lapiano-Smith, J.F. Morar, J. Bevk, in: The Physics and Chemistry of SiO$_2$ and the Si–SiO$_2$ Interface, eds. C.R. Helms and B.E. Deal, Vol. 2 (Plenum Press, New York, 1993) p. 237.
[3] A. Pasquarello, M.S. Hybertsen and R. Car, Appl. Phys. Lett. 68 (1996) 625.
[4] C. Lee, Phys. Rev. B 36 (1987) 2793.
[5] S. Carniato, G. Boureau and J. Harding, Radiation Effects and Defects in Solids 134 (1995) 179.
[6] S. Carniato, G. Boureau and J. Harding, Phil. Mag. A 75 (1997) 1435.
[7] R.A.B. Devine, Appl. Phys. Lett. 68 (1996) 3108.
[8] P.W. Tasker, J. Phys. C 12 (1979) 4977.
[9] C. Noguera, Physics and Chemistry of Oxide Surfaces (Cambridge University Press, Cambridge, 1996).

[10] J. Harding, in: Computer Simulations in Materials Science, eds. M. Meyer and V. Pontikis, NATO ASIE, Vol. 205 (1991) p. 159.

[11] P. Vashishta, R.K. Kalia, J.P. Rino and I. Ebbsjo, Phys. Rev. B 42 (1990) 12 197.

[12] B. Vessal, M. Amini and H. Akbarzadeh, J. Chem. Phys. 101 (1994) 7823.

[13] S. Tsuneyuki, M. Tsukada, H. Aoki and Y. Matsui, Phys. Rev. Lett. 61 (1988) 869.

[14] S. Tsuneyuki, H. Aoki, M. Tsukada and Y. Matsui, Phys. Rev. Lett. 64 (1990) 776.

[15] E.R. Cowley and J. Gross, J. Chem. Phys. 95 (1991) 8357.

[16] N.R. Keskar, N. Troullier, J.L. Martins and J.R. Chelikowsky, Phys. Rev. B 44 (1991) 40 811.

[17] Y. Guissani and B. Guillot, J. Chem. Phys. 104 (1996) 7633.

[18] R.G. Della Valle and E. Venuti, Phys. Rev. B 54 (1996) 3809.

[19] J. Badro and J.L. Barrat and P. Gillet, Phys. Rev. Lett. 76 (1996) 772.

[20] D.C. Allan, M.P. Teter, J.D. Joannopoulos, Y. Bar Yam and S.T. Pantelides, Mater. Res. Soc. Symp. Proc. 141 (1989) 255.

[21] D.C. Allan and M.P. Teter, J. Am. Ceram. Soc. 73 (1990) 3247.

[22] G. Pacchioni, A.M. Ferrari, G. Ieranò, Faraday Discuss. 106 (1997) in press.

[23] G. Boureau and S. Carniato, Solid State Comm. 98 (1996) 485, 1996 and erratum 99, i–iv.

[24] L. Stixrude and M.S.T. Bukowinski, Phys. Chem. Minerals 16 (1988) 513.

[25] L. Pauling, Am. Mineral 65 (1980) 321.

[26] N.F. Mott and M.J. Littleton, Trans. Faraday Soc. 34 (1938) 485.

[27] J. H. Harding and N.C. Pyper, Phil. Mag. Lett. 71 (1995) 113.

[28] S. Tsuneyuki, M. Tsukada, H. Aoki and Y. Matsui, Phys. Rev. Lett. 61 (1988) 869.

[29] B.W.H. van Beest, G.J. Kramer and R.A. van Santen, Phys. Rev. Lett. 64 (1990) 1955.

ELSEVIER

Computational Materials Science 10 (1998) 168–174

COMPUTATIONAL
MATERIALS
SCIENCE

Numerical studies of polymer networks and gels

Kurt Kremer[1]

Max-Planck-Institut für Polymerforschung, 55120 Mainz, Germany

Abstract

The elastic and relaxational properties of polymer networks are strongly influenced by quenched disorder, introduced by the crosslinks in the systems. By extensive computer simulations of appropriately tailored model systems these effects are systematically investigated. Copyright © 1998 Elsevier Science B.V.

Keywords: Networks; Polymers; Disordered systems; Elasticity

1. Introduction

Macromolecular materials now play a very prevalent part in our daily lives. They show up in many everyday products ranging from plastic shopping bags to modern high tech organic composite materials used to replace steel in automobiles and polymer glasses like the polycarbonates used to make compact discs. They are also used as food constituents and additives. Many biological materials such as DNA and proteins are other examples of complex macromolecules. Because of their importance, there has been considerable effort in recent years to obtain a deeper understanding of the physical mechanisms, which control the very versatile properties of these systems. However, the wealth of different systems, which is considerably larger than in any other classical physics field, causes enormous technical and conceptual problems in determining some overall principles.

It is the purpose of this short review to discuss just one aspect, namely, the effect of the noncrossability of chains and the frozen disorder in networks on the elasticity of polymer networks. For many questions of interest detailed knowledge of the chemical structure is not needed. Many properties of polymers are *universal*. This means that chemical details do not play a role for general qualitative properties, as long as we are investigating systems on time and length scales much larger than the monomeric (microscopic) dimensions. While such ideas were frequently employed in the pioneering works of scientists like Flory and Edwards, this universality was formally shown by de Gennes, 1991 Nobel Laureate, in 1972 [1]. It can easily be illustrated by two examples. First let us consider the spatial extension of the chains as given by the mean square end-to-end distance $\langle R^2(N) \rangle$, where N is the number of monomers. For $N \gg 1$ one finds

$$\langle R^2(N) \rangle = C N^{2\nu}.$$

The exponent ν is universal and depends only on solvent quality. For good solvent ν is rather well given by the Flory formula $\nu = 3/(2 + d)$, giving $\nu = 3/5$

[1] E-mail: kremer@mpip-mainz.mpg.de; fax: 49 6131 379340.

for $d = 3$ while for polymer melts and Θ solutions $\nu = 1/2$. The local flexibility as well as all other local chemistry dependent properties are hidden in the prefactor C. C can vary significantly for different chemical systems or as a function of temperature. In a very similar way dynamical properties are universal. Let us consider a melt of chains well above the glass transition temperature. For the chain diffusion constant $D(N)$ one finds

$$D(N) = \begin{cases} D_0/N, & N < N_e, \\ D_1/N^2, & N \gg N_e. \end{cases}$$

Again D_0 and D_1 and N_e, the so-called entanglement molecular weight, are dependent on the local chemical structure, while the general N-dependency of the chain diffusion constant is universal.

This leads to two important consequences for simulations [2]. First, for most simulations it is sufficient to employ very simple model systems. Second, the natural dynamics of polymeric systems displays a drastic slowing down with increasing chain length. While for many physics questions these universal properties are of foremost interest, one should keep in mind that the broad application strength of macromolecular materials is a consequence of the interplay of universal (entropy dominated) and material specific (energy dominated) properties.

Microscopic simulations containing all chemical details of the chains can only study properties and dynamics on very small time and length scales, which only in an averaged manner are related to the overall chain relaxation. Consequently, the questions to ask are not considered with the global dynamics of the chains, but more with local aspects, like local kink jump dynamics or the diffusion of small molecules through a polymeric matrix [3].

If we are interested in the dynamics of the chain as a whole, e.g. the relaxation of the end-to-end vector, the diffusion of chain monomers or stress relaxation of a network, one has to "coarsen" the view. This means that we have to go away from a chemically detailed model to more idealized model systems. Universality tells us that this is allowed.

2. Simulation model

At high density or for long range interactions the *molecular dynamics* (MD) method is an interesting alternative to dynamic MC algorithms. The simulations using the MD method, which are discussed here [4], employ a simple bead-spring model, where each monomer is weakly coupled to a heat bath. Each polymer chain consists of N monomers of mass m, connected by an anharmonic spring as bond potential:

$$U^{\text{bond}}(r) = \begin{cases} -0.5kR_0^2 \ln\left[1 - \left(\dfrac{r}{R_0}\right)^2\right], & r \le R_0, \\ \infty, & r > R_0. \end{cases}$$

The excluded volume interaction is taken into account between all monomers by introducing a simple purely repulsive Lennard–Jones potential

$$U_{\text{LJ}}(r) = \begin{cases} 4\epsilon\left[\left(\dfrac{\sigma}{r}\right)^{12} - \left(\dfrac{\sigma}{r}\right)^6 \right. \\ \left. \quad - \left(\dfrac{\sigma}{r_c}\right)^{12} + \left(\dfrac{\sigma}{r_c}\right)^6\right], & r \le r_c, \\ 0, & r > r_c, \end{cases}$$

where r_c is the interaction cutoff ($r_c = 2^{1/6}\sigma$). For melts the parameters k and R_0 are chosen to prevent any crossing of the chains ($k = 30\epsilon/\sigma^2$ and $R_0 = 1.5\sigma$). Denoting the total potential of monomer i by V_i, the equation of motion for monomer i is given by

$$m\frac{d^2\mathbf{r}_i}{dt^2} = -\nabla V_i - \Gamma\frac{d\mathbf{r}_i}{dt} + \mathbf{W}_i(t).$$

Here Γ is the bead friction which acts to couple the monomers to the heat bath. $\mathbf{W}_i(t)$ describes the random force acting on each bead. Standard MD would not have a friction and heat bath term. This approach is necessary in order to stabilize the runs. In addition, since $\mathbf{W}$ and Γ are coupled via the fluctuation dissipation theorem, we simulate a canonical thermodynamic ensemble. MD runs for polymers typically exceed the stability limits of a microcanonical simulation. In order not to manipulate the dynamics of the chains, the background friction Γ has to be much smaller than the monomer friction ζ, which results from the particle–particle interaction. The equations of motion can be solved using a variety of methods, though we use the velocity Verlet algorithm. Most of the programs for runs with chemical details use some sort of thermostat to stabilize the runs, however for very long runs, as for the above cases, the prescribed procedure is more useful. Both methods do not conserve momentum, which makes them inappropriate for dynamic properties of dilute or semi-dilute solutions where hydrodynamics plays an important role. In this case one has to use the conventional microcanonical method.

These programs can be vectorized and parallelized fairly well, yielding very efficient codes. Our fastest Cray T3E code runs at about $130\,000/(MN)$ time steps per second and processor where M is the number of chains of N monomers in the melt or network. Since we are using short range interaction potentials, there is no significant overhead from parallelization [5].

3. Polymer melts

The knowledge of melt properties is needed in order to proceed to networks. The dynamics of polymer melts is observed experimentally to change from an apparent Rouse-like behaviour to a dramatically slower dynamics for chains exceeding a characteristic length N_e. There are several theoretical models which try to explain this behaviour. However, only the reptation concept of Edwards and de Gennes and variants of this approach take the noncrossing of the chains explicitly into account [6,7]. It is the only one, which at least qualitatively, can account for a wide variety of different experimental results such as neutron spin echo scattering, diffusion and viscosity. While it cannot explain *all* experimental data it does remarkably well, particularly considering its conceptual simplicity. (A good experimental overview is given in [8].) For short chains the topological constraints do not play an important role. The dynamics is reasonably well described by the Rouse model. This means the melt dynamics can be understood in terms of the Langevin dynamics of individual random walks! All the complicated inter-chain effects average into a structureless background friction. This experimental and simulational observation is surprising and still not understood. For longer chains, exceeding a character-

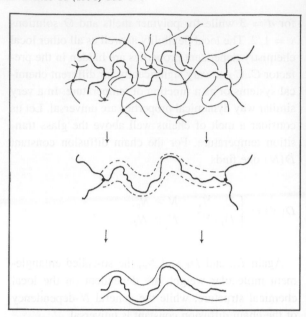

Fig. 1. History (from top to bottom) of the topological constraint idea and the reptation concept. Originating from the topological constraints in networks, Edwards developed the idea of the confinement in a tube. de Gennes later realized that for intermediate time scales and very long chains, the confinement should be the same for polymer melts, since the density if free ends becomes very small (from [9]).

istic entanglement length N_e, one observes a dramatic slowing down from $D \propto N^{-1}$ to $D \propto N^{-2}$ and a viscosity of $\eta \propto N^{3.4}$. The idea of the reptation model is that the topological constraints of each chain, as imposed by the surrounding eventually, cause a motion along the polymers own coarse-grained contour. Historically, the idea originates from the discussion of the influence of topological constraints in polymer networks, as explained in Fig. 1. The diameter of the tube, which the chain is constrained to, is the diameter of a subchain of length N_e, namely $d_T \propto N_e^{1/2}$. The chains follow the Rouse relaxation up to the time $\tau_e \propto N_e^2$ (the Rouse relaxation time of a chain of length N_e). For longer times the constraints are supposed to become dominant and the chain moves along its own coarse grained contour. In order to leave the tube, the chain has to diffuse along the tube a distance of the order of its own contour length, $d_T N/N_e$. This gives $D \propto N^{-2}$ and $\eta \propto N^3$. The difference between the predicted and the measured exponent for $\eta \propto N^{3.4}$ is still not completely understood [10].

One way to characterize the transition from Rouse to reptation is to investigate the diffusion constant $D(N)$. In order to be able to make a significant contribution by simulation, the data must be capable of covering the crossover from $D(N) \sim N^{-1}$ for small N to $D(N) \sim N^{-2}$ for $N > N_e$. As noted above, $D(N)N$ should define a plateau for small N, giving the monomeric friction or mobility. To compare results from different simulations and also experiment, a plot of $D(N)/D_{\text{Rouse}}(N)$ versus N/N_e or M/M_e, respectively should give one universal curve since N_e is thought to be the only characteristic length of the crossover. Here $D_{\text{Rouse}} = k_B T / \zeta N$ and M_e is the experimental entanglement mass. This mapping is important for our understanding, since experiment and simulation use different methods to estimate M_e or N_e. Experiments use the plateau modulus of the viscoelastic response of the melt, meaning the long time limit, while simulations use the mean square displacements of the monomers, which are typically "short time" data. The scaling of the different data onto one curve is shown in Fig. 2 for different simulations as well as experiment. This shows that for both cases, experiment and simulation, the same characteristic length is used. The experimental data are from pulsed field gradient NMR measurements of the diffusion constant by Pearson et al. for polyethylene [12].

A direct comparison to experiment and the standard theoretical approaches can be made by investigating the dynamic scattering function $S(q, t)$ of the chains. This function can directly be measured by neutron spin echo measurements. For short chains, $S(q, t)/S(q, 0)$ is only a function of the Rouse variable $q^2 \sqrt{(t)}/6$ and decays exponentially. For the reptating case the relaxation is expected to follow the Rouse decay up to the time τ_e and then splits into q-dependent plateaus. The scatterer measures a smeared out density of the chain along the tube for intermediate times. Fig. 3 shows that this is also observed for simulations. The slowing down in the motion of the monomers on intermediate time scales, actually was verified first by simulation (in real space) and later by experiment [4].

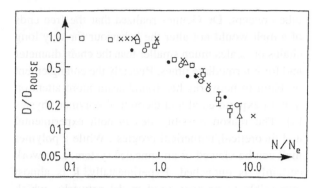

Fig. 2. The diffusion constant normalized to the Rouse diffusion constant for MD data (squares, $N_e = 35$) and MC data at two different densities (open circles, $N_e = 30$, full circles $N_e = 40$) and for PE melts (crosses, $N_e = 96$), as a function of N/N_e (from [11]).

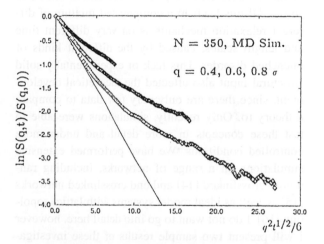

Fig. 3. Dynamic scattering function $S(q, t)/S(q, 0)$ of individual chains in a melt. Data from an MD simulation of chains of length $N = 10N_e = 350$. For Rouse chains the scattering function shows a single exponential decay as a function of the Rouse scaling variable $q^2 \sqrt{(t)}/6$. This decay, as found for short chains, is indicated by the straight line (from: [9]).

4. Polymer networks

While the test of the reptation concept played an important role in polymer physics over the last years, the original idea of the confinement of the chains to a tube like region in space was developed by Edwards [6] for polymer networks. His ansatz was to describe the relaxation of chain conformation in a network of very long chains. The constraint, that the chains cannot pass through each other led to the original tube concept. De Gennes realized that the free ends of a melt would not alter the behaviour for very long chains on scales much smaller than the chain diameter and for intermediate times. Recently the entanglement problem in networks has found again more attention in both experimental and theoretical polymer physics [3]. The reason possibly lies in both experimental and theoretical/numerical progress. While a polymer melt can be viewed as an annealed system, a network certainly is quenched. Experimentally, it is almost impossible to prepare good model networks, which would allow a detailed comparison to theoretical models. The strands between crosslinking points are highly polydisperse, one encounters network defects, in many cases clustering of vulcanization sites and so on. All this leads to a complicated mixture of different relaxation mechanisms on very different time and length scales, caused by the different kinds of quenched disorder. This lack of experimentally solid structural input also affected the theoretical development, since there are only very few data to compare a theory to. Only recently, simulations were able to test these concepts in more detail and under more controlled conditions. We have performed extensive simulations on a range of networks, including randomly crosslinked [14] and end crosslinked networks [15] as well as ideal model systems with lattice topology [16]. I do not want to go into detail here, however I will present two sample results of these investigations, namely the elastic modulus as a function of the strand length N_s between crosslinks and the role of random knots in model networks.

Traditionally, rubber elasticity is accounted for by the entropic elasticity of random walks. The number of possible conformations for walks is a function of the distance between the two endpoints (Gaussian distribution for random walks). The resulting change in entropy under deformation is the origin of the entropic elasticity, leading to an elastic modulus $G^\circ \propto k_B T / N_s$. For longer chains, $N_s > N_e$, this simple argument is not any longer expected to hold trivially. The confinement of the network strands into a tube of diameter $d_T \propto N_e^{1/2}$ means that the modulus should not decay with $1/N_s$, but converge to a limiting value around $k_B T / N_e$. A direct experimental

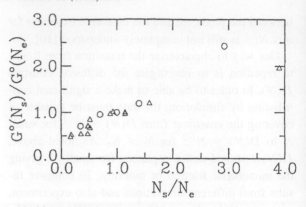

Fig. 4. Normalized moduli from simulation of endlinked polymer melts (o) and experiments on PDMS end linked model networks (△) (from [15]).

test is rather difficult, since the argument assumes that the crosslinking procedure does not alter the melt structure. Simulations are currently the only method, where both static and relaxational properties can be checked in detail. To look at this problem, we investigated end cross linked polymer melts. The crosslinking procedure was continued, until almost no free ends remained. To our knowledge, there is only one experiment, which attempts to control the crosslinking amount carefully enough [17]. Fig. 4 shows a direct comparison of the simulation results to the experiments of Patel et al. This again is possible due to the normalization of the data for the PDMS networks and the simulation to a unique scale, the entanglement chain length N_e or M_e, respectively. The agreement is very good. A comparison to analytic theories now allows for quantitative predictions.

For end linked systems one can reach the situation of almost defect free networks, meaning that there are no dangling chain ends. However, there are still several sources for quenched disorder. Thinking of shortest paths through a system, which might be overstretched first under strong elongation, one gets two kinds of paths. First chemical short paths, which percolate through the system and second "topological short paths", as a result of random links between network loops in the system. The latter were the aim of a recent study on even more idealized model networks [16]. In order to isolate the effect of random links, interpenetrating networks with connectivity of

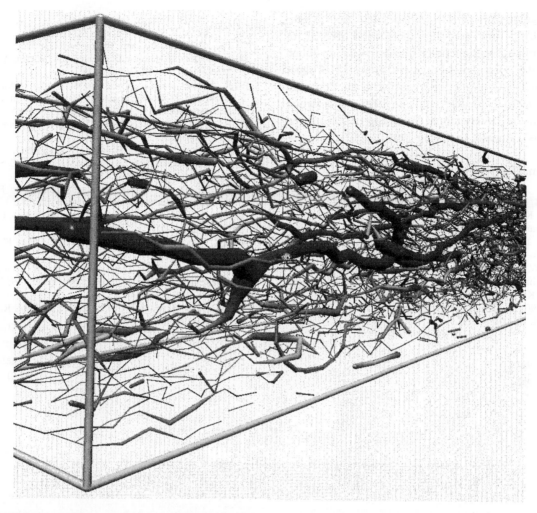

Fig. 5. Illustration of the contribution of topologically short paths to the restoring force for diamond lattice model networks. The colour code indicates the variation from weak stress (thin) to high stress (thick) (from [16]).

a diamond lattice were studied. Thus all short chemical paths are of the same length. The only source of disorder are random links originating from random interpenetration of the subsystems. Under strong elongation linked loops, which compose short "topological paths", are overstretched first and cause a strong deviation from the standard rubber force law. Fig. 5 shows a direct visualization of such a situation.

While these results are the first direct tests of the influence of links and the tube concept for polymer networks, experimental systems under high stress will certainly be dominated by a combination of effects from both chemical and topological short paths. With increasing strand length however, we expect the latter to become more and more dominant.

Acknowledgements

During the course of my research it was a great pleasure to collaborate and discuss with many colleagues, especially G.S. Grest, B. Dünweg, E.R. Duering, R. Everaers, J. Eckert, M. Pütz, and H.P. Wittmann. This work was supported by a NATO travel grant. Part of this work resulted from a generous CPU time grant for the Cray YMP8/32 of the German Supercomputer Center, HLRZ, in Jülich within the "Disordered Polymers" project. This paper is an updated and modified version of the Foreword of the internal IFF-KFA-Bulletin 1995 and a lecture given at the "International Workshop on Disordered Systems", Andalo, in Spring 1997.

References

[1] P.G. deGennes, Phys. Lett. A 38 (1972) 339.

[2] K. Kremer, G.S. Grest, in: R.J. Roe (Ed.), Computer Simulations of Polymers, Prentice-Hall, Englewood Cliffs, NJ, 1991.

[3] A.A. Gusev, F. Müller-Plathe, W.F. van Gunsteren, U.W. Suter, in: Advances in Polymer Science, vol. 116, Springer, Berlin, 1994.

[4] K. Kremer, G.S. Grest, J. Chem. Phys. 92 (1990) 5057 (1990); K. Kremer, G.S. Grest, I. Carmesin, Phys. Rev. Lett. 61 (1988) 566.

[5] M. Pütz, A. Kolb, B. Dünweg (1997).

[6] S.F. Edwards, Proc. Phys. Soc. 92 (1967) 9.

[7] P.G. deGennes, J. Chem. Phys. 55 (1971) 572.

[8] L.J. Fetters, D.J. Lohse, D. Richter, T.A. Witten, A. Zirkel, Macromolecules 27 (1994) 4639.

[9] K. Kremer and G.S. Grest, in: K. Binder (Ed.), Monte Carlo and Molecular Dynamics Simulations in Polymer Science, Oxford University Press, Oxford, 1995.

[10] R. Colby, L.J. Fetters, W.W. Graessley, Macromolecules 20 (1987) 2226.

[11] W. Paul et al., J. Phys, Paris (II) 37 (1) (1991).

[12] D.S. Pearson, G. ver Strate, E. von Meerwall, F.C. Shilling, Macromolecules 20 (1987) 1133.

[13] J.E. Mark, B. Erman (Eds.), Elastomeric Polymer Networks, Prentice-Hall, Englewood Cliffs, NJ, 1992.

[14] E.R. Duering, K. Kremer, G.S. Grest, Phys. Rev. Lett. 67 (1991) 3531.

[15] E.R. Duering, K. Kremer, G.S. Grest, J. Chem. Phys. 101 (1994) 8169.

[16] R. Everaers, Ph.D. Thesis, University of Bonn, 1994; R. Everaers, K. Kremer, Macromolecules 28 (1995) 7291; Phys. Rev. E 53 (1996) R37.

[17] S.K. Patel, S. Malone, C. Cohen, J.R. Gillmor, R. Colby, Macromolecules 25 (1992) 5241.

ELSEVIER

COMPUTATIONAL
MATERIALS
SCIENCE

Computational Materials Science 10 (1998) 175–179

An extended BFM model for simulation of copolymers at an interface

E. James *, C.C. Matthai

Department of Physics and Astronomy, University of Wales, Cardiff, Cardiff CF2 3YB, UK

Abstract

We have extended the original bond fluctuation model (BFM) of Carmesin and Kremer [Macromolecules 12 (1988) 2819] so that it allows for a threefold an increase in the possible bonding configurations. This appears to be computationally more efficient in determining equilibrium configurations. We have used this modified BFM to simulate random copolymers at the interface of two good solvents. Copyright © 1998 Elsevier Science B.V.

1. Introduction

In recent years there has been an increase in computer simulation studies of polymer systems. Because of the different time scales in the motions of polymers, it is virtually impossible to simulate realistic long chain polymers even with simple interaction potentials. Much work has therefore concentrated on different aspects of the motion and coarse-grained models, where the details of the chemical structure on the monomer length scale are omitted, have formed the basis of these simulations. In these approaches, a monomer unit can be comprised of many molecules and bonds. Lattice models, in which the monomer units are associated with lattice sites, are computationally more efficient than continuous space models and even though they are slightly less realistic, the results obtained using these models are of interest. Also, the Monte Carlo method is well suited to performing simulations on lattice models. The bond fluctuating model (BFM) advanced by Carmesin and

Kremer [1] and Deutsch and Binder [2] has all the advantages of simple lattice models but in allowing for a greater choice of bond vectors is able to give a better description of polymer dynamics enabling it to be used in the study of a wide variety of polymer systems. The BFM has the advantage of using the efficiency of lattice models and adding the flexibility of the continuum models. In this paper, we report on how the BFM may be extended to give even more flexibility without losing any efficiency. The extension is described in the next section which is followed by the application of the model to study copolymers at an interface.

2. The extended BFM model

In the original BFM, polymers of length n are represented by beads (monomers) connected by $(n - 1)$ bonds. The relationship between the model and a real polymer is that the beads and bonds represent Kuhn segments which are a sequence of a few tens of monomers of the real chain. These segments should be

---

* Corresponding author. E-mail: E.James@cf.ac.uk.

allowed to have a range of lengths to allow for kinking and stretching of the real chain, and there should be very little restriction on the possible angle between two successive segments. The polymer is mapped on to a simple cubic lattice and each monomer or bead occupies a $(2 \times 2 \times 2)$ cube containing eight lattice sites. The cubes of the neighbouring monomers are not allowed to overlap which in effect models excluded volume constraints. Within this framework, it is possible to have many more bond lengths and bond angles than is possible in a simple single site lattice model. The initial polymer configuration is set up by restricting the bonds to an allowed set of vectors which are obtained by requiring that no two bonds to cross each other and that all bonds are reachable from each other by successive bead jumps. The Monte Carlo method is used to perform the simulations. Moves are made by selecting a monomer and moving it one lattice spacing in the direction of a lattice basis vector. A move is accepted if it does not touch or intersect another monomer and if any bonds associated with the monomer remain in the allowed bond set.

Although this model has proved to be very successful in simulating many polymer systems, it is of interest to see if this flexibility in bonding configurations and dynamics can be further increased without affecting the computational efficiency. The obvious way to do this is to increase the size of the unit cell occupied by the monomer unit. Of course, in the limiting case of the lattice spacing becoming infinitely small, we recover the continuum limit but there will be much reduced efficiency. We therefore propose an extension to the BFM in which the centre of each monomer is situated on a lattice site (unlike in the original model) and occupies a $(3 \times 3 \times 3)$ cube containing 27 lattice sites. Then, by following the rules of Deutsch and Binder [2] for determining the allowed set of vectors for the bonds and using their notation, we find that these are given by

$$\mathbf{B} = \mathbf{P}(3, 0, 0) \cup \mathbf{P}(3, 1, 0) \cup \mathbf{P}(3, 1, 1) \cup \mathbf{P}(3, 2, 0)$$
$$\cup \mathbf{P}(3, 2, 1) \cup \mathbf{P}(3, 2, 2) \cup \mathbf{P}(3, 3, 0) \cup \mathbf{P}(3, 3, 2)$$
$$\cup \mathbf{P}(4, 0, 0) \cup \mathbf{P}(4, 1, 0) \cup \mathbf{P}(4, 1, 1) \cup \mathbf{P}(4, 2, 0)$$
$$\cup \mathbf{P}(4, 2, 2) \cup \mathbf{P}(5, 0, 0) \cup \mathbf{P}(5, 1, 0)$$

where $\mathbf{P}(x, y, z)$ stands for the set of all permutations and sign combinations of $(\pm x, \pm y, \pm z)$. This results in 318 possible bonds, which is almost three times that in the original model. With this modification and leaving all the rules of moves the same, we tested the model for computer efficiency before performing the simulations of polymers at an interface.

3. Single chain test

As a test of the model for computer efficiency, we considered a single free polymer chain of 64 units and performed the MC simulations to equilibrate the chain using both the original and extended BFM. The chain was allowed to evolve for 200 000 moves and the bond angles between successive segments and the bond lengths were recorded every 10 moves. The resulting distributions are displayed in Figs. 1 and 2, respectively. It can be seen that with the new model, there are many more structural possibilities. The number of possible bond lengths increases from 5 to 13 and the angles from 100 to 600. This is only to be expected in view of the finer lattice mesh relative to the polymer chain. What was interesting, however, was that the acceptance ratio for moves in the old model was only 0.27 as compared to 0.37 with the modification. The consequence of this was that although marginally more time is spent searching through all the allowed bond vectors for each move, because the acceptance ratio is greater, the overall time taken to reach an equilibrium configuration is actually reduced. We expect this advantage to become more pronounced for denser systems.

4. Simulation of a random copolymer at an interface

Having demonstrated that this extended BFM was capable of performing simulations of polymer systems with increased efficiency we focused on a particular system of interest. For many technical applications it is important to have a knowledge of how copolymers

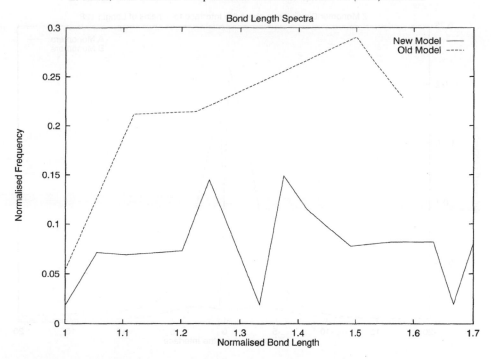

Fig. 1. Normalised bond length distributions of a single polymer chain using the original and modified BFM. The peaks correspond to the recorded bond lengths.

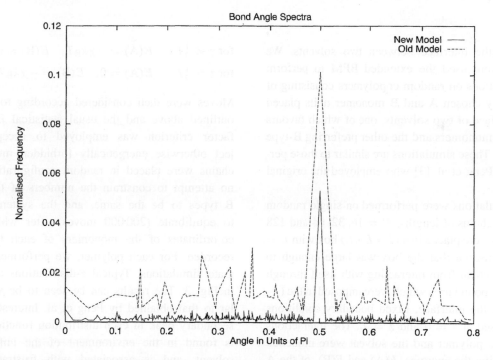

Fig. 2. Bond angle distributions of a single polymer chain using the original and modified BFM. The peaks correspond to the recorded angles.

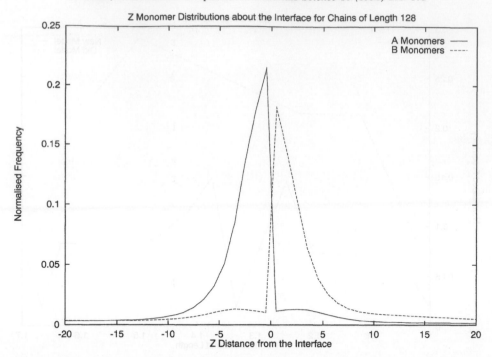

Fig. 3. The normalised distribution of A (full) and B (dashed) type monomers for a copolymer chain of length $N = 128\%$ at an interface. Note that the different peak heights are due to the different number of A and B type monomers in the polymer.

behave at the interface between two solvents. We have therefore used the extended BFM to perform MC simulations on random copolymers consisting of N randomly chosen A and B monomer units placed at the interface of two solvents, one of which favours the A-type monomers and the other preferring B-type monomers. These simulations are similar to those performed by Peng et al. [3] who employed the original BFM.

The simulations were performed on single random copolymer chains of lengths, $N = 16, 32, 64$ and 128 units which were placed in a $(L \times L \times L)$ box with $L = 6N$. This ensured that the box was large enough to prevent the chain from interacting with itself through the periodic boundaries which were imposed in the $(x–y)$ plane of the interface. Fixed boundary conditions were imposed at $z = 0$ and $z = L$. The interactions between the polymer and the solvent were modelled by prescribing the energies, $E(A)$ and $E(B)$, of the A and B monomers as follows:

$$\text{for } z < \tfrac{1}{2}L, \quad E(A) = -\chi k_B T, \quad E(B) = 0,$$
$$\text{for } z > \tfrac{1}{2}L, \quad E(A) = 0, \quad E(B) = -\chi k_B T.$$

Moves were then considered according to the rules outlined above and the usual statistical Boltzmann factor criterion was employed to accept or reject otherwise energetically forbidden moves. The chains were placed in random configurations, with no attempt to constrain the numbers of the A and B types to be the same, and the system allowed to equilibrate (200 000 moves) after which the z co-ordinates of the monomers of each type were recorded. For each polymer, we performed 24 separate simulations. Typical z-distributions are shown in Fig. 3. The results can be seen to be very similar to that obtained by Peng et al. Interestingly, the secondary peak in each distribution function, which is found in the environment of the unfavourable solvent, and is associated with frustration, persists even in our finer grain model suggesting that it is not an artefact of the lattice structure of the simulations.

5. Conclusion

By extending the BFM, we have demonstrated that there may be some optimum lattice mesh in coarse-grained model simulations which, with little or no increase in computation time, can give a more detailed structure of the polymer systems being simulated. We tested the 8 and 27 site models on a simple single chain system and found that computational efficiency was increased with the new 27 site model. We have used this model to simulate the random copolymer configuration at an interface of two solvents and found that the results are in good agreement with earlier work.

Although the BFM method may not be very efficient in generating static configurations, it is able to give dynamical information which is almost impossible with other Monte Carlo methods.

Acknowledgements

EJ acknowledges financial support from the University of Wales, Cardiff.

References

[1] I. Carmesin, K. Kremer, Macromolecules 12 (1988) 2819.
[2] H.P. Deutsch, K. Binder, J. Chem. Phys. 94 (1991) 2294.
[3] G. Peng, J.U. Sommer, A. Blumen, Phys. Rev. E 53 (1996) 5509.

ELSEVIER

Computational Materials Science 10 (1998) 180–183

COMPUTATIONAL
MATERIALS
SCIENCE

Ultra strong polymer fibers: Ab initio calculations on polyethylene

J.C.L. Hageman [a,*], R.A. de Groot [a], Robert J. Meier [b]

[a] ESM Group, Research Institute of Materials, University of Nijmegen, Toernooiveld 1, 6525ED Nijmegen, Netherlands
[b] DSM Research, PO Box 18, 6160MD Geleen, Netherlands

Abstract

The Car–Parrinello technique is used to study the electronic structure of orthorhombic polyethylene as well as the elastic modulus. The theoretical band structure and density of states are in very good agreement with experiments. The best experimentally realized elastic modulus is better than 86% of the presented theoretical value. Copyright © 1998 Elsevier Science B.V.

Keywords: Polyethylene; Carr–Parrinello; Mechanical properties; Electronic properties

1. Introduction

High performance polymer fibers form an important class of materials. The fibers are light and yet have excellent mechanical properties. For example, the elastic modulus of such fibers can exceed the modulus of steel. With special techniques the fibers are processed from ultra-high molecular weight polymers, in such a way that the long molecular chains have near crystalline orientation. An increase in the crystallinity and in the chain length improves the performance. Hence, the physical properties of the best polymer fibers will resemble the properties of the 100% crystalline polymer comprised of chains of infinite length. The study of this ideal material can provide a clue how the fibers can be improved and how much can be gained in the performance of the actual material.

The Car–Parrinello technique [1] is very useful in studying the properties of the ideal material: it allows the molecular chains to relax under forces which are calculated from the electronic structure, computed within the density functional theory. In this way, it can provide values for mechanical properties of the ideal material, like the elastic modulus, as well as information on the electronic structure.

The polymer under consideration here is polyethylene, as it is the simplest polymer and yet of technological relevance. A detailed report on the elastic modulus is reported elsewhere [2] and will be briefly summarized. The details on the electronic structure are presented here.

2. Computational details

The Car–Parrinello technique is based on the density functional theory and the local density approximation is used to describe the exchange–correlation energy. In the present study the fhi93cp code [3] was used. Only the valence electrons are treated explicitly; the ionic core is described by BHS-pseudopotentials

* Corresponding author. Tel.: +31 243652805; fax: +31 243652120; e-mail: joosth@sci.kun.nl.

[4] as given in [5]. The electronic wave functions are expanded in a plane wave basis set. The size of this set is controlled by a cut-off energy; in this case 54 Ry was sufficient to converge the elastic modulus. To sample the Brillouin-zone integration 12 k-points are used. More technical details can be found in [2].

The ideal material which corresponds to the high performance polyethylene fibers is the orthorhombic crystal with two monomers per unit cell. The a-axis and b-axis, perpendicular to the chain direction (c-axis), are fixed at the experimental values measured at a temperature of 4 K ($a = 7.12$ Å and $b = 4.85$ Å) [6], as no temperature effects are included. The c-axis is optimized ($c = 2.53$ Å). The orientation of the chains is such that one chain makes an angle $\phi = 42°$ with the ac-plane and the other an angle of $-\phi$. No symmetry constrains are imposed during simulation, only the form of the unit cell is kept fixed.

3. The elastic modulus

The axial elastic modulus (Y_0), also known as Young's modulus, is defined as

$$Y_0 = \frac{c_0}{A} \left. \frac{d^2 E}{dc^2} \right|_{c=c_0},$$

where A is the area of the unit cell perpendicular to the chains, c_0 the equilibrium unit cell length and E the elastic strain energy per unit cell, which is at zero temperature equivalent to the total energy.

The total energy curve is calculated by taking unit cells of different length in the c-direction and relaxing the atomic positions for each unit cell. From this energy curve the equilibrium unit cell length can be extracted as well as the second derivative of the energy at equilibrium length. The value of c_0 is found to be 2.53 Å, which is less than 1% smaller than the experimental value at 4 K ($c_0 = 2.548$ Å) [6]. Eq. (1) gives the elastic modulus $Y_0 = 334$ GPa.

To compare this value with other theoretical values, one should notice that almost all other theoretical values are calculated for a single chain and the value of A is chosen to be of room temperature ($A = 36.48$ Å^2 [7]). Using this value in Eq. (1) Young's modulus

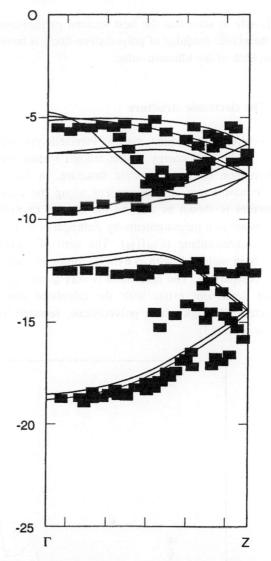

Fig. 1. The calculated valence band structure (solid lines) in the chain direction compared to the experimental band structure from ultraviolet photoelectron spectroscopy for the paraffin $C_{36}H_{74}$ (■) [12]. On the vertical axis the energy is given in eV.

becomes 316 GPa, this procedure is validated in [2]. This is 5% larger than the value of the most recently reported ab initio calculation, $Y_0 = 300$ GPa [8]. It is about 10% smaller than calibrated semi-empirical calculations, 349 GPa [9] and 343 GPa [10].

The best performance of the actual fibers at low temperature is 288 GPa [11]. This is 86% of our presented value. As temperature effects will lower the modulus, it is safe to state that the best realized performance in the elastic modulus of polyethylene fibers is better than 86% of the ultimate value.

4. The electronic structure

As already mentioned, the Car–Parrinello provides next to the total energy of the relaxed system also information on the electronic structure. In Fig. 1 the bandstructure of polyethylene along the chain direction is shown as well as the angular resolved photoemission measurements by Zubrägel et al. [12] on hexatriacontane ($C_{36}H_{74}$). The zero of energy has been adjusted with 5.8 eV in order to facilitate comparison. The agreement is very good. Also, good agreement exists with the calculated bandstructure of single chain polyethylene, reported by Miao et al. [13].

The fact that this work comprises a polyethylene lattice implies that there is an influence of the interchain interaction. The interaction is most notably in the Γ–Y-direction. The occupied state of mostly H character shows a dispersion of 1.24 eV, indicating a next nearest neighbor interaction of 0.62 eV. This interaction does not necessarily indicate bonding, however. As a consequence of the interchain interaction, the direct bandgap does not occur at Γ but along the Γ–Y axis, close to Y (6.0 eV). The position of this point in k-space is determined by a subtle interplay of nearest and next nearest chain interactions.

Fig. 2 shows the density of states, convoluted with a Gaussian of FWMH (full width at half maximum) of 0.5 eV in order to facilitate with the X-ray photoelectron data of Endo et al. [14]. Very good agreement exists if one considers that the peak at 15 eV is not visible because of the small cross section of H and the

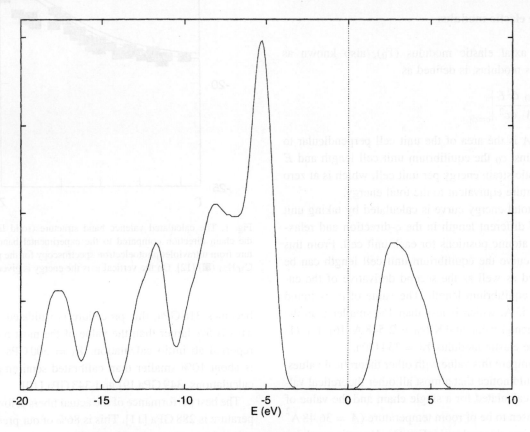

Fig. 2. The density-of-states for polyethylene, convoluted with a Gaussian of FWMH of 0.5 eV.

structure from 0 to 10 eV greatly reduces in intensity for similar reasons.

5. Conclusions

It is shown that the Car–Parrinello method is a very useful technique for polyethylene in studying the mechanical properties like the elastic modulus as well as in providing electron structure information. The properties of the crystalline polymer can be related to the experimental properties of actual materials. The bandstructure is in very good agreement with experiment. The best realized performance for the elastic modulus is really close to the ultimate performance.

References

[1] R. Car and M. Parrinello, Phys. Rev. Lett. 22 (1985) 2471.
[2] J.C.L. Hageman, R.J. Meier, M. Heinemann and R.A. de Groot, to be published.
[3] R. Stumpf and M. Scheffler, Comput. Phys. Commun. 79 (1994) 447.
[4] G.B. Bachelet, D.R. Hamann and M. Schlüter, Phys. Rev. B 26 (1982) 4199.
[5] X. Gonze, R. Stumpf and M. Scheffler, Phys. Rev. B 44 (1991) 8503.
[6] G. Avitable, R. Napolitano, B. Pirozzi, K.D. Rouse, H.W. Thomas and B.T.M. Wills, J. Polym. Sci., Polym. Lett. Ed. 13 (1973) 351.
[7] C.W. Bunn, Trans. Faraday Soc. 35 (1939) 482.
[8] B. Crist and P.G. Hereña, J. Polym. Sci. Part B 34 (1996) 449.
[9] R.J. Meier, Macromolecules 26 (1993) 4376.
[10] T. Horn, W.W. Adams, R. Pachter and P.D. Haaland, Polymer 34 (1993) 2481.
[11] P.J. Barham and A. Keller, J. Polym. Sci., Polym. Lett. Ed. 17 (1979) 591.
[12] Ch. Zubrägel, F. Schneider, M. Neumann, G. Hähner, Ch. Wöll and M. Grunze, Chem. Phys. Lett. 219 (1994) 127.
[13] M.S. Miao, P.E. van Camp, V.E. van Doren, J.J. Ladik and J.W. Mintmire, Phys. Rev. B 54 (1996) 10430.
[14] K. Endo, Y. Kaneda, M. Aida and D.P. Chong, J. Phys. Chem. Solids 56 (1995) 1131.

ELSEVIER

Computational Materials Science 10 (1998) 184–187

COMPUTATIONAL
MATERIALS
SCIENCE

Theoretical study of charge-induced defects at metal/polymer interface

Marta M.D. Ramos *, Judite P.P. Almeida

Departamento de Física, Universidade do Minho, Largo do Paço, 4709 Braga Codex, Portugal

Abstract

Conducting polymers have attracted much attention concerning the possibility of their use as active components of electronic and optoelectronic devices. We use a molecular dynamics method with semi-empirical quantum chemistry at CNDO (complete Neglect of differential overlap) level to study the chemical interactions between aluminium atoms and *trans*-polyacetylene during the interface formation. Our results suggest that aluminium dimer (Al$_2$) bound to a polymer chain is energetically more favourable than the adsorption of isolated aluminium atoms. In both cases, the compound formation is accompanied by charge transfer between metal and polymer. As a result charge rearrangement among the polyacetylene atoms is induced. We shall describe the charge-induced structural relaxation of the *trans*-polyacetylene backbone which is accompanied by a local change in the electronic structure of the polymer, commonly called defects. Copyright © 1998 Elsevier Science B.V.

Keywords: Atomistic modelling; Defects; Polyacetylene; Interfaces

1. Introduction

Conjugated polymers exhibit metal-type conductivities while retaining the physical and mechanical properties characteristic of polymer materials. Because of their low cost and easy processing, these conducting polymers constitute a field of increasing scientific and technological interest, offering the potential to be used as electro-active materials [1–3]. Several difficulties exist in the interpretation of the observed device characteristics, namely those which arise from the formation of interfaces through reaction of metal electrodes with the polymer. Since the interfacial properties are directly influenced by the way in which the interface is formed, atomistic modelling can play an impor-

tant role in understanding the nature of metal–polymer interface, as well as it may suggest new methods for controlling or optimising interfaces.

The softness of conjugated polymers leads to strong coupling between the polymer's electrons and lattice vibrations (phonons) [4,5]. Therefore, it is necessary to perform self-consistent calculations of electronic wave functions and atomic positions in order to study the charge-induced defects in conjugated polymers as a result of chemical bonding at metal–polymer interface. We report here a self-consistent quantum chemistry molecular dynamics calculation of the interaction of aluminium atoms with an individual *trans*-polyacetylene strand. The chemical interaction of aluminium atoms with doped *trans*-polyacetylene is also discussed. It seems likely that the principles discussed will be applicable to other conducting polymers, with appropriate modifications dictated

* Corresponding author. Tel.: ++351 53 604330; fax: ++351 53 678981; e-mail: marta@fisica.uminho.pt.

by differences in chemical composition, molecular structure and morphology. This study is also the stepping stone for further investigation of charge injection mechanisms in conjugated polymers.

2. Theoretical methods

Because of the very nature of the atomic processes, studies of both the chemical and the electronic structure of interfaces are essential. The approach we adopt, based on the Harwell CHEMOS code, consists in the self-consistent calculation of both electronic structure and molecular geometry of metal/polymer system, using a semi-empirical molecular orbital method [5–7]. The molecular orbital calculations were performed at CNDO (complete neglect of differential overlap) level using a linear combination of atomic orbitals (LCAO) and a cluster model framework. A molecular dynamical method was used in parallel to perform geometry optimisation. Forces on atoms are evaluated from electronic quantum-mechanical calculations. Our earlier work [8,9] gave results for static and dynamic behaviour of *trans*-polyacetylene in contact with metal electrodes modelled as a half-space with a well-defined Fermi level. We have extended our previous work by including chemical interactions between metal atoms and between those and the polymer.

3. Results and discussion

3.1. The isolated t-PA chain

In the course of modelling aluminium–polyacetylene interfaces, we will concentrate on the properties of an individual *trans*-polyacetylene (*t*-PA) strand. Before any chemical interaction is considered we have relaxed the geometry of a finite *t*-PA chain to equilibrium, starting off the planar configuration.

The charge transfer between different species is induced by their chemical potential difference. Since the calculated chemical potential of *t*-PA decreases as a function of chain length to a nearly constant value for

a chain of 20 carbon atoms [8], the ground state geometry of a free-defect *t*-PA chain was found examining a t-$C_{20}H_{22}$ molecule. Our results are in good agreement with experimental observations of the dimerisation amplitude [10].

We have addressed the interaction of aluminium atoms with doped *t*-PA by examining the behaviour of the t-$C_{19}H_{21}$ molecule as a model for a soliton in *t*-PA. The (positively and negatively) charged and uncharged solitons have been considered. These are the defects generated in individual polyacetylene molecules by n-type and p-type dopants [11].

3.2. Aluminium atoms on t-PA

In order to study the interactions of evaporated aluminium atoms with *t*-PA, we have simulated the bonding of two Al atoms to an individual *t*-PA molecule, in accordance with low coverage UPS experiments. The aluminisation of *t*-PA was modelled in the following way. Two aluminium atoms were brought close to the polymer, while the *t*-PA atoms were kept frozen. When the equilibrium was reached for metal atoms, the entire system was allowed to relax.

3.2.1. Preferential reaction sites

Aluminium atoms are found to interact strongly with *t*-PA chain either isolated or as dimers (Al_2), both bonding positions being stable. A binding energy difference of approximately 3 eV was obtained for the adsorption positions indicated above. The fundamental interaction unit predicted corresponds to the formation of Al_2–polyacetylene complex. Our results also suggest that the most stable position for the aluminium dimer is out of the polymer plane with both aluminium atoms adsorbed onto twofold site. These results hold for *t*-PA with and without a soliton-type defect.

Preliminary calculations suggest serious overestimates of the magnitudes of binding energies [12]. Whilst we should not regard binding energies as accurate, we expect the predicted trends for preferred reaction sites to be reliable.

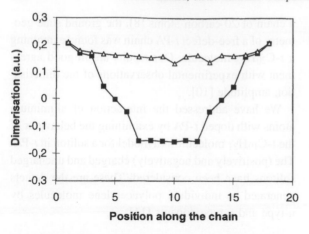

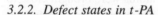

Fig. 1. Dimerisation patterns in a t-PA chain with 20 carbon atoms when: (a) two isolated aluminium atoms are bound to t-PA (square); (b) an aluminium dimer (Al$_2$) is bound to the polymer chain (triangle). The dimerisation is defined [5,6] by the difference in C–C bond lengths at each carbon (e.g. left-hand bond length minus right-hand bond length, so the dimerisation changes sign across a soliton). The marks indicate the data points that were calculated explicitly. The curves are simply a guide to the eye.

3.2.2. Defect states in t-PA

The chemical interaction between aluminium and t-PA leads to a distortion of the t-PA fragment both parallel and perpendicular to the molecular plane, which, in turn, leads to a local change in the electronic structure of the polymer. The new localised electronic states, commonly called defects, arise within the t-PA energy gap due to electron–lattice coupling.

The geometrical distortion in the bond length distribution of a free-defect t-PA chain corresponding to the aluminium adsorption indicated above is shown in Fig. 1. The resulting distortion patterns show three main features. First, the degree of dimerisation increases slightly towards the ends of the polymer chain. Second, the distortion due to an aluminium dimer bound to a single t-PA chain loosely resembles that of a polaron-type defect [4,5], with a more marked distortion on a few atoms at the centre of the defect. Third, the adsorption of two isolated Al atoms to t-PA gives two well-separated solitons. Both chemical interactions lead to a molecular distortion perpendicular to the polymer plane as well as in-plane distortion, the latter being the dominant term.

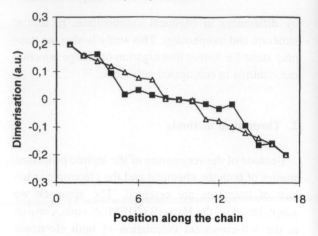

Fig. 2. Dimerisation patterns for a positive soliton in t-PA when: (a) two isolated aluminium atoms are bound to t-PA (square); (b) an aluminium dimer (Al$_2$) is bound to the polymer chain (triangle). The dimerisation is defined [5,6] by the difference in C–C bond lengths at each carbon (e.g. left-hand bond length minus right-hand bond length, so the dimerisation changes sign across a soliton). The marks indicate the data points that were calculated explicitly. The curves are simply a guide to the eye.

The effects of aluminium bonding on dimerisation pattern of positive soliton in t-PA are shown in Fig. 2. The calculated dimerisations loosely resemble that of a soliton in t-PA [4,5]. Moreover, the dimerisation is reduced in the centre of the positively charged t-C$_{19}$H$_{21}$ molecule when two isolated Al atoms are bound.

3.2.3. Aluminium-induced charge transfer

When a free-defect t-PA molecule is put in contact with aluminium atoms charge will flow until the chemical potentials of both materials are equal [13]. The electron affinity $A(N)$ and ionisation energy $I(N)$ can be combined to yield the chemical potential, and the arithmetic average of $I(N)$ and $A(N)$ [8]. Electron transfer from aluminium to t-PA is predicted for the adsorption of two isolated aluminium atoms while electron transfer in opposite direction is predicted for an aluminium dimer bound to a t-PA chain (containing or not a defect).

The charge transfer from or to a free-defect t-PA chain is found to be about 0.4 electrons. Our results also suggest that approximately 30% of the charge is stored on the hydrogen atoms, rather than the entire charge resting on the carbon atoms of chain. The

Table 1
Atomic (Mulliken) charges of aluminium, carbon and hydrogen atoms when an aluminium dimer (Al_2) is bound to a (positively and negatively) charged and uncharged t-PA molecule with a soliton-type defect (chain of 19 carbons)

Molecular charge	Atomic charge (electrons)		
	Aluminium atoms	Carbon atoms	Hydrogen atoms
Positive	−0.1392	0.7659	0.3729
Negative	−0.4466	−0.1384	−0.4150
Neutral	−0.3497	0.3414	0.0081

amount of charge transferred to a soliton in t-PA due to the adsorption of an aluminium dimer is shown in Table 1. For a positive or a negative soliton in t-PA we find a charge alternation on carbon atoms, which is consistent with the interpretation of XPS experiments [14]. The charge stored on hydrogen atoms do not display this alternation.

4. Conclusions

We have studied localised electronic states in individual t-PA molecules as a result of chemical bonding at aluminium–polyacetylene interface. The reaction of aluminium atoms with a free-defect t-PA chain could lead to the formation of either a polaron-type defect or two well-separated solitons. In the case of doped t-PA, changes in soliton deformation pattern apparently occur. Moreover, the metal induces electron transfer from or to t-PA depending on the relative position of the Fermi levels of both materials.

Self-consistent quantum chemistry molecular dynamics calculations, such as those reported here, are a useful tool for atomistic modelling of interfacial

bonding at metal–polymer interface as well as for providing data which is currently impossible to obtain experimentally.

Acknowledgements

This work was supported in part by FEDER and PRAXIS XXI under the project no. PRAXIS/2/2.1/ FIS/26/94 and by JNICT under the project no. PBIC/C/FIS/2151/95.

References

[1] D.S. Soane, Z. Martynenko, Polymers in Electronics: Fundamentals and Applications, Elsevier, Amesterdam, 1989.
[2] E. Ruiz-Hitzky, Adv. Mater. 5 (1993) 334.
[3] M.S. Weaver, D.G. Lidzey, T.A. Fisher, M.A. Pate, D. O'Brien, A. Bleyer, A. Tajbakhsh, D.D.C. Bradley, M.S. Skolnick, G. Hill, Thin Sol. Films 273 (1996) 39.
[4] A. Fisher, W. Hayes, D.S. Wallace, J. Phys.: Condens. Matter 1 (1989) 5567.
[5] D.S. Wallace, A.M. Stoneham, W. Hayes, A. Fisher, A. Testa, J. Phys.: Condens. Matter 3 (1991) 3905.
[6] D.S. Wallace, D. Phil. Thesis, University of Oxford, 1989.
[7] D.S. Wallace, A.M. Stoneham, W. Hayes, A. Fisher, A.H. Harker, J. Phys.: Condens. Matter 3 (1991) 3879.
[8] A.M. Stoneham, M.M.D. Ramos, J. Sol. State Chem. 106 (1993) 2.
[9] M.M.D. Ramos, A.M. Stoneham, A.P. Sutton, Synthetic Metals 67 (1994) 137.
[10] C.S. Yannoni, T.C. Clarke, Phys. Rev. Lett. 51 (1983) 1191.
[11] A.N. Chuvyrov, G.I. Yusupova, JETP Lett. 50 (1989) 420.
[12] J.A. Pople, D.L. Beveridge, Approximate Molecular Orbital Theory, McGraw-Hill, New York, 1970.
[13] W.J. Brennan, J. Lowell, G.W. Fellows, M.P.W. Wilson, J. Phys. D 28 (1995) 2349.
[14] M. Sasai, H. Fukotome, Solid State Commun. 58 (1986) 735.

ELSEVIER

COMPUTATIONAL
MATERIALS
SCIENCE

Computational Materials Science 10 (1998) 188–197

Oscillatory behavior of interface exchange coupling caused by finite caps of variable thickness

J. Kudrnovský [a,b,*], V. Drchal [a,b], P. Bruno [c], I. Turek [d], P. Weinberger [b]

[a] *Institute of Physics, Academy of Sciences of the Czech Republic, Na Slovance 2, CZ-180 40 Praha 8, Czech Republic*
[b] *Institute for Technical Electrochemistry, Technical University of Vienna, Getreidemarkt 9, A-1060 Vienna, Austria*
[c] *Institut d'Électronique Fondamentale, CNRS URA 22, Bât. 220, Université Paris-Sud, F-91405 Orsay, France*
[d] *Institute of Physics of Materials, Academy of Sciences of the Czech Republic, Žižkova 22, CZ-616 62 Brno, Czech Republic*

Abstract

The effect of non-magnetic cap-layers on the periods and the amplitudes of the oscillations of interlayer exchange coupling (IEC) is studied theoretically using an ab initio spin-polarized surface Green function technique within a tight-binding linear muffin-tin orbital method and the Lloyd formulation of the IEC. Applications are made to the free-electron like model as well as to Co/Cu/Co(0 0 1) trilayers with a cap interfacing vacuum through the dipole barrier. The results are analyzed in terms of a discrete two-dimensional Fourier transformation which confirms a pronounced oscillatory behavior of the IEC with respect to the thickness of the cap- and the spacer-layers. The results are in agreement with available experimental data as well as with predictions of the electron confinement model of the IEC. Copyright © 1998 Elsevier Science B.V.

1. Introduction

The oscillatory interlayer exchange coupling (IEC) between magnetic layers separated by a non-magnetic spacer has recently attracted considerable attention. The physical origin of such oscillations is attributed to quantum interferences due to spin-dependent confinement of electrons in the spacer [1–4]. An important conclusion, namely that the periods of the oscillations with respect to the spacer thickness are determined by the spacer Fermi surface, has been confirmed by numerous experiments.

Recently increasing interest was devoted to the study of the influence of the cap [5–7] on the IEC, and first model theories have appeared based on simple free-electron-like models [6–9]. These theories predict a novel oscillatory behavior of the IEC with respect to the thickness of the cap, which is attributed to electron confinement in the non-magnetic cap due to the vacuum barrier [9]. The periods of these oscillations are related to the Fermi-surface of the cap material. Experimental observations of the oscillations of the IEC with respect to the cap thickness in a

---
* Corresponding author. Address: Institute of Physics, Academy of Sciences of the Czech Republic, Na Slovance 2, CZ-180 40 Praha 8, Czech Republic. Tel.: +420-2-6605 2905; fax: +420-2-821227; e-mail: kudrnov@fzu.cz.

variety of materials [5–7] support this view of the IEC. In this paper we present a detailed ab initio study of such oscillations.

2. Formalism

The system considered consists of a stack of layers, namely from the left to the right: (i) a semi-infinite (non-magnetic) substrate, (ii) a left ferromagnetic slab of thickness M (in monolayers, MLs), (iii) a non-magnetic spacer of thickness N, (iv) a right ferromagnetic slab of thickness M', (v) a non-magnetic cap of thickness P, and (vi) a semi-infinite vacuum. In general, the various parts of the system may consist of different metals, including disordered substitutional alloys. The thickness of the left ferromagnetic slab may be extended to infinity.

2.1. Ab initio formulation

The system described above is formally partitioned into two subsystems, denoted left ($\mathcal{L}$) and right ($\mathcal{R}$). The left subsystem contains the non-magnetic substrate and the left ferromagnetic slab. The right subsystem consists of the spacer, the right ferromagnetic slab, the cap, and the vacuum. We employ the Lloyd formulation of the IEC combined with a spin-polarized surface Green function technique as based on the tight-binding linear muffin-tin orbital (TB-LMTO) method [10]. Considering M and M' as implicit, system specific parameters, the exchange coupling energy $\mathcal{E}_x(N, P)$ for a given set (N, P) of spacer and cap layers can be written as the difference between the configurationally averaged grandcanonical potentials of the antiferromagnetic and the ferromagnetic alignment of the magnetic slabs,

$$\mathcal{E}_x(N, P) = \frac{1}{\pi N_{\|}} \operatorname{Im} \sum_{\mathbf{k}_{\|}} \int_C f(z) \operatorname{tr}_L \ln \mathrm{M}(N, P; \mathbf{k}_{\|}, z) \, \mathrm{d}z. \tag{1}$$

Here $f(z)$ is the Fermi–Dirac distribution function and tr_L denotes the trace over angular momentum indices $L = (\ell m)$. In (1), the energy integration is performed along a contour in the upper half of the complex energy plane, the $\mathbf{k}_{\|}$-summation runs over the surface Brillouin zone (SBZ), and $N_{\|}$ is the number of sites in the two-dimensional lattice. The quantity $\mathrm{M}(N, P; \mathbf{k}_{\|}, z)$ is defined as

$$\mathrm{M} = (1 - A_0^{\uparrow} B_0^{\uparrow})^{-1}(1 - A_0^{\uparrow} B_0^{\downarrow})(1 - A_0^{\downarrow} B_0^{\downarrow})^{-1}(1 - A_0^{\downarrow} B_0^{\uparrow}), \tag{2}$$

where

$$A_0^{\sigma} = S^{\dagger}(\mathbf{k}_{\|}) \bar{\mathcal{G}}_{\mathcal{L}}^{\sigma}(M; \mathbf{k}_{\|}, z) S(\mathbf{k}_{\|}), \qquad B_0^{\sigma} = \bar{\mathcal{G}}_{\mathcal{R}}^{\sigma}(M', N, P; \mathbf{k}_{\|}, z). \tag{3}$$

In Eq. (3), the quantities $S^{\dagger}(\mathbf{k}_{\|})$ and $S(\mathbf{k}_{\|})$ are the screened structure constants which couple neighboring (principal) layers [11], $\bar{\mathcal{G}}_{\mathcal{L}}^{\sigma}(M; \mathbf{k}_{\|}, z)$ is the configurationally averaged surface Green function of the left magnetic subsystem, while $\bar{\mathcal{G}}_{\mathcal{R}}^{\sigma}(M', N, P; \mathbf{k}_{\|}, z)$ is its counterpart for the right magnetic subsystem which also contains the spacer, the cap, and the vacuum. Finally, σ denotes the spin index ($\sigma = \uparrow, \downarrow$). We refer the reader to [11] for an efficient evaluation of the configurationally averaged surface Green function in the framework of the present formalism. The use of a Green function formulation of the IEC is essential for describing randomness in the spacer, magnetic slabs, and the cap within the coherent potential approximation (CPA). The calculations are significantly simplified [12] by using the vertex-cancellation theorem [13]. It should be noted that due to the block tridiagonal form of the inverse of the Green function matrices, inherent to the TB-LMTO method, the evaluation of the surface Green functions, and hence the evaluation of the IEC itself, scales linearly with M, M', N and P.

2.2. Electron confinement in a non-magnetic cap

For a qualitative interpretation of the ab initio calculations it is convenient to summarize first the simple free-electron like model of the cap proposed in [9]. The expression for $\mathcal{E}_x(N, P)$ in the limit of large spacer and cap thicknesses, and for the simplest case when the cap and the spacer are formed by the same material, is given by

$$
\mathcal{E}_x(N, P) \propto \text{Im} \sum_\alpha \left[\frac{A_\alpha}{N^2} e^{iq_\alpha N} + \frac{B_\alpha}{(N+P)^2} e^{iq_\alpha(N+P)} \right]. \tag{4}
$$

Here, A_α and B_α are the (complex) amplitudes of the oscillations expressed in terms of reflection and transmission matrices and q_α is a stationary spanning wave vector of the bulk Fermi surface of the spacer (and the cap material) in the direction normal to the layers. The wave vector is measured in units of d^{-1}, where d is the distance between the atomic layers. There may be several such vectors labelled by the index α. The first term in Eq. (4) gives the IEC for an infinitely thick cap and oscillates with respect to the spacer thickness with a period $p_\alpha = 2\pi/q_\alpha$ and decays as N^{-2}. The second one is found to oscillate with the same period p_α but now with respect to both the spacer and the cap thickness, and to decay as $(N + P)^{-2}$. Generally, the second term is smaller than the first one, so that for a given spacer thickness N, the oscillations of the IEC with the cap thickness P often do not lead to changes of the sign of the coupling.

Let us now consider the more complicated case when the cap and the spacer are different materials. The expression for the IEC is then given by [9]

$$
\mathcal{E}_x(N, P) \propto \text{Im} \sum_\alpha \left[\frac{A_\alpha}{N^2} e^{iq_\alpha N} + \frac{C_\alpha}{(N + \lambda P)(N + \lambda' P)} e^{i(q_\alpha N + q'_\alpha P)} \right]. \tag{5}
$$

Here, q_α and q'_α are stationary spanning vectors of the bulk Fermi surface of the spacer and the cap, respectively, corresponding to the same in-plane wave vector $\mathbf{k}_{\|(\alpha)}$. If there is no such vector, the second term in Eq. (5) has to be dropped. Note, that unless $\mathbf{k}_{\|(\alpha)}$ corresponds to a high-symmetry point, a spanning vector q' of the cap material Fermi surface is unlikely to be stationary at one and the same $\mathbf{k}_{\|(\alpha)}$. A detailed discussion of this point is given in [4]. In the above equation, λ and λ' are related to the perpendicular Fermi velocity and the curvature radius of the Fermi surface of the spacer and of the cap, respectively.

As before, the first term in Eq. (5) corresponds to the contribution of an infinitely thick cap, and the second one to the correction due to electron confinement in the cap. The first term is of the same form as before, but the second one turns now out to be more complicated: it is no longer an oscillatory function of $N + P$ and the amplitude no longer decays as $(N + P)^{-2}$. An approximate $(N + P)^{-2}$ decay may still be obtained provided that the Fermi surfaces of the spacer and the cap material are not too different.

2.3. Details of calculations

Numerical studies were performed by assuming an ideal fcc(0 0 1) stacking for the substrate, the spacer, the magnetic slab, and the cap corresponding to the experimental lattice spacing of fcc Cu. In each case, the magnetic slabs consist of Co layers, the spacer and the substrate are formed by Cu layers while the cap consists either of Cu or Rh layers, or of random substitutional $Cu_{75}Zn_{25}$ alloy layers. We employ the frozen potential approximation and align bulk fcc-Cu, fcc-Co, fcc-Rh, and fcc-$Cu_{75}Zn_{25}$ to the Fermi energy of the substrate (for more details see [14]). The value of the dipole barrier between the cap and the vacuum was determined self-consistently for the case of a semi-infinite fcc(0 0 1) cap. Special care was devoted to the Brillouin zone and energy integrations (for more details see [10]). Since temperatures can obscure an analysis of the oscillation amplitudes [4,10], all calculations reported here refer to $T = 0\,\text{K}$.

2.4. Ab initio formulation of the free-electron model

The general formalism developed in Section 2.1 can be used to study a free-electron model which allows one to understand the effect of cap-layers on the properties of the IEC in a particularly transparent manner. We place empty spheres [15] at each lattice site of our ideal fcc(0 0 1) stack of layers and choose the system Fermi energy such that the corresponding Fermi vector is far enough from artificial zone boundaries introduced by the underlying lattice. The spacer-, cap-, and magnetic-bands differ from each other only by a rigid shift. There is only a single stationary vector at $\mathbf{k}_\| = 0$ that corresponds to the Fermi sphere diameter, but it can have a different value (leading to different periods) for the spacer and the cap. It should be noted that the solution of the free-electron model as obtained from an ab initio formulation is not limited to the asymptotic region as in other approaches but is valid also for systems with a thin spacer and/or a thin cap.

2.5. Analysis of results

In order to analyze the set of values $\mathcal{E}_x(N, P)$ where pairs (N, P) typically include $N, P \in (1, 50)$, we employ two different methods. In the first approach a direct representation of $\mathcal{E}_x(N, P)$ with respect to either N (spacer thickness) or P (cap thickness) is used by keeping the respective other variable implicit. The second method consists of a two-dimensional discrete Fourier transformation of $\mathcal{E}_x(N, P)$,

$$F(q_N, q_P) = \sum_{N=N_1}^{N_2} \sum_{P=P_1}^{P_2} (N + P)^2 [\mathcal{E}_x(N, P) - \mathcal{E}_1(N)] e^{i(q_N N + q_P P)}, \tag{6}$$

with respect to the so-called biased value $\mathcal{E}_1(N)$,

$$\mathcal{E}_1(N) = \mathcal{E}_x(N, \infty) = \lim_{P \to \infty} \mathcal{E}_x(N, P), \tag{7}$$

The prefactor $(N + P)^2$ in Eq. (6) is consistent with the asymptotic behavior discussed in Section 2.2. Strictly speaking, the prefactor $(N + P)^2$ in Eq. (6) is only correct when the spacer and cap are formed by the same material (see Eq. (4)), however, we shall deliberately use this prefactor also when the spacer and the cap correspond to different materials. The use of a large enough set of input data permits to exclude the preasymptotic region, e.g. by choosing $N, P \in (10, 50)$. In this way the periods of oscillations and their amplitudes as a function of both the spacer and the cap thickness can clearly be extracted.

3. Results and discussion

3.1. Free-electron model

Consider a system with a semi-infinite magnetic slab (left subsystem) and a five monolayer (ML) magnetic slab with a spacer and a cap (right subsystem) as motivated by comparison with a realistic Co/Cu/Co(0 0 1) system. In all cases considered, in the following the spacer and the magnetic majority bands coincide whereas the minority band is shifted upwards by 0.075 Ry. The Fermi energy is chosen to give the spacer spanning vector corresponding

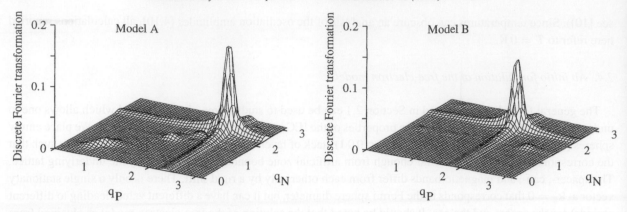

Fig. 1. Absolute values of the discrete two-dimensional Fourier transformation of $(N + P)^2 \mathcal{E}_2(N, P)$ $(10 \leq N, P \leq 50)$ with respect to the spacer and the cap thickness for the free-electron case described in the text: (a) Model A, and (b) Model B.

to oscillations with a period of about 2.8 MLs. The vacuum is again represented by a free-electron band shifted upwards by 1 Ry as compared to the spacer band and represents a perfectly reflecting barrier.

Three different geometries are assumed: (i) Model A: the cap band coincides with the spacer band, and the same is true for their spanning vectors; (ii) Model B: the cap band is shifted rigidly upwards by 0.0375 Ry with respect to the spacer band and its spanning vector is reduced in its size (the opposite is true for a downward shift); and (iii) Model C: the cap band is shifted upwards by 0.15 Ry such that its bottom lies above the Fermi level (the band is unoccupied and has thus no spanning vector at $\mathbf{k}_\parallel = 0$).

The absolute values of the discrete two-dimensional Fourier transformation of $(N + P)^2 \mathcal{E}_2(N, P)$ are presented in Fig. 1. The following conclusions can be drawn: (i) Model A (Fig. 1(a)): we observe a clearly pronounced peak situated at $q_N = q_P \approx 2.25$ which corresponds to oscillations with respect to the spacer and cap thickness with the same period $p_N = p_P \approx 2.8$ MLs, and, in accordance with the asymptotic expression, Eq. (4), with the functional dependence $\mathcal{E}_2(N, P) = \mathcal{E}_2(N + P)$. The amplitude of oscillations (the height of the peak) remains unchanged for different subsets (N, P) of $\mathcal{E}_2(N, P)$ used for the Fourier transformation which confirms the predicted $(N + P)^{-2}$ decay of the oscillations; (ii) Model B (Fig. 1(b)): the most important difference in comparison with the Model A is the shift of the peak of the discrete Fourier transformation to $q_P \approx 1.87$, however, with an unchanged value of q_N. This is consistent with a smaller Fermi sphere radius of the cap (the smaller spanning vector) and, consequently, with a weaker confinement. As a result, the amplitude of oscillations also decreases. It should be noted that for a downward shift of the cap-band relative to the spacer leads to a stronger confinement which in turn shifts the peak to larger values of q_P; (iii) Model C (not shown): no peak is seen because there is no spanning vector (the cap-band is not occupied). Consequently, the values of $\mathcal{E}_x(N, P)$ are identical for $P \geq 5$.

3.2. Co/Cu/Co(0 0 1)-trilayer with the Cu cap

Consider now a realistic Co/Cu/Co(0 0 1) system with two different geometrical arrangements for the Cu-cap, namely (i) magnetic slabs of different thicknesses $(M = \infty, M' = 5)$, and (ii) the case when both magnetic slabs have the same finite thickness $(M = M' = 1)$. In the limit of an infinite cap [10] in the former case the so-called long-period oscillations (LPO) with respect to the spacer thickness are almost suppressed, while in the latter case both, the SPO and the LPO, show up with comparable weights in the infinite cap limit [10]. The corresponding plots of two-dimensional discrete Fourier transformations of $(N + P)^2 \mathcal{E}_2(N, P)$ are presented in Figs. 2 and 3,

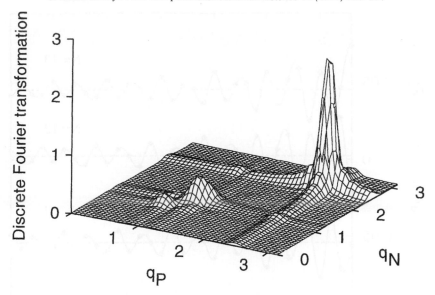

Fig. 2. Absolute values of the discrete two-dimensional Fourier transformation of $(N + P)^2 \mathcal{E}_2(N, P)$ $(10 \leq N, P \leq 50)$ with respect to the spacer and the cap thickness for the case of a Co/Cu/Co(0 0 1) trilayer ($M = \infty$, $M' = 5$) with a Cu-cap.

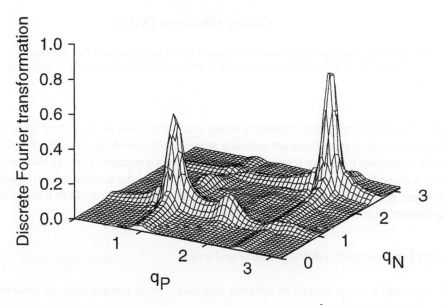

Fig. 3. Absolute values of the discrete two-dimensional Fourier transformation of $(N + P)^2 \mathcal{E}_2(N, P)$ $(10 \leq N, P \leq 50)$ with respect to the spacer and the cap thickness the case of a Co/Cu/Co(0 0 1) trilayer ($M = M' = 1$) with a Cu-cap.

respectively. The case $M = \infty$, $M' = 5$ is formally similar to the related free-electron case (Model A). We observe a pronounced peak at $q_N = q_P \approx 2.5$ which corresponds to the SPO period $p \approx 2.5$ MLs with respect to the spacer and the cap thickness. The case $M = M' = 1$ is different. Two prominent peaks located at $q_N = q_P \approx 1$ and at $q_N = q_P \approx 2.5$, which correspond to the LPO and SPO oscillations, respectively, are clearly seen.

By using a direct representation [7–9], for the case $M = \infty$, $M' = 5$ the IEC is shown in Fig. 4 for three chosen spacer thicknesses N. Fig. 4 illustrates an important feature of the IEC, namely, that the coupling can be

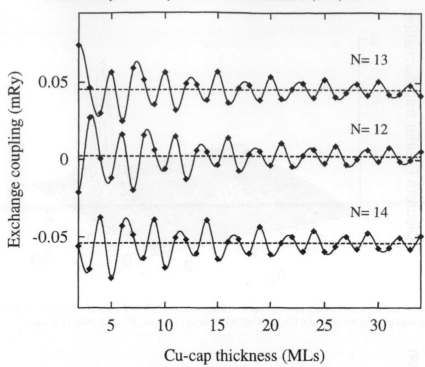

Fig. 4. Dependence of the exchange coupling (in mRy) on the thickness of the Cu-cap (diamonds) for a fixed spacer thickness of $N = 12$, 13, and 14 in the case of $M = \infty$, $M' = 5$. The lines serve as a guide to the eye. The dashed lines refer to the corresponding value for an infinite Cu-cap.

purely ferromagnetic or antiferromagnetic depending on the spacer thickness N. At a given spacer thickness N the oscillations with respect to the cap thickness are generally around a non-vanishing value, the so-called biased value. As illustrated in Fig. 5, another important feature of the IEC oscillations relative to the cap thickness is the decrease of their amplitudes with both the spacer- and cap-thickness. In particular, the amplitudes corresponding to one and the same cap thickness, progressively decrease with increasing spacer thickness N. Both features are in agreement with a recent experiment on a Co/Cu/Co(0 0 1) cap system [5].

3.3. Co/Cu/Co(0 0 1)-trilayer with a different spacer and a cap

The case with a cap and a spacer formed by different materials is more complicated. As an example the case of a Cu-spacer and a $Cu_{75}Zn_{25}$-cap is considered for which one expects that the spanning vectors of the spacer and cap materials are stationary approximately at the same $\mathbf{k}_{\parallel(\alpha)}$ because the Fermi surfaces are similar. This situation is related to Model B of the free-electron case with a downward shift of the cap-band because the alloying of Cu with Zn atoms increases the number of valence electrons and hence the size of the alloy Fermi surface. The IEC of $Co/Cu_{75}Zn_{25}/Co$ trilayers has been studied in a previous paper [12], in which oscillations with a period of about 3.1 MLs (i.e., $q \approx 2.0$) were found, that corresponds to an in-plane wave vector close to the one giving rise to the SPO oscillation for a pure Cu spacer. From the two-dimensional Fourier transform of $(N + P)^2 \mathcal{E}_2(N, P)$, shown in Fig. 6, a peak located at $(q_N \approx 2.5, \ q_P \approx 2.0)$ is clearly identified, which is in complete agreement with the predictions of the electron confinement model (Section 2.2).

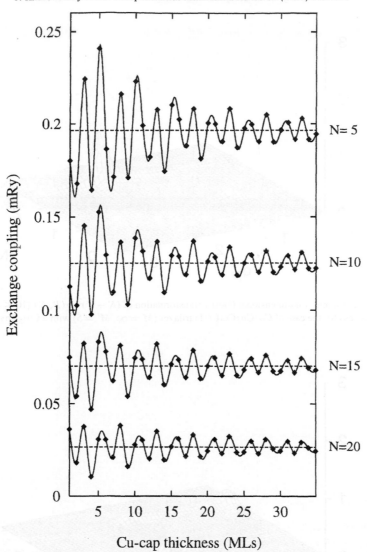

Fig. 5. Dependence of the exchange coupling (in mRy) on the thickness of the Cu-cap (diamonds) for a fixed spacer thickness of $N = 5$, 10, 15, and 20 ML in the case of $M = \infty$, $M' = 5$ as a function of the thickness of the Cu-cap. The lines serve as a guide to the eye. The dashed lines refer to the corresponding value for an infinite Cu-cap. An arbitrary shift is applied to each curve in order to prevent overlapping.

The other example is the case of a Cu-spacer and a Rh-cap for which one expects that the spanning vectors are stationary at different $\mathbf{k}_{\|(\alpha)}$. Indeed, inspection of the Rh Fermi surface [3] reveals that there is no stationary spanning vector in the vicinity of the $\mathbf{k}_{\|(\alpha)}$ that corresponds to the short period oscillation for a Cu spacer. One thus expects no oscillations with respect to the cap thickness in the asymptotic regime. This is indeed the case as can be seen from the two-dimensional Fourier transform of $(N + P)^2 \mathcal{E}_2(N, P)$ in Fig. 7 which in turn has to be compared to the case of a Cu-cap in Fig. 2. We see that the Rh-cap essentially suppresses the oscillations with respect to the cap-thickness.

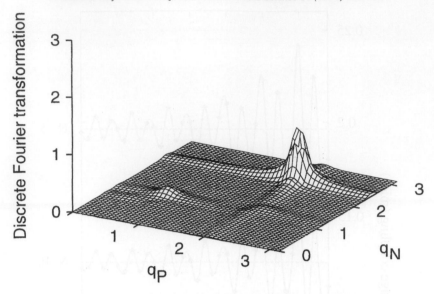

Fig. 6. Absolute values of the discrete two-dimensional Fourier transformation of $(N + P)^2 \mathcal{E}_2(P, N)$ ($10 \le N, P \le 50$) with respect to the spacer and the cap thickness for the case of Co/Cu/Co(0 0 1) trilayer ($M = \infty$, $M' = 5$) with a $Cu_{75}Zn_{25}$-cap.

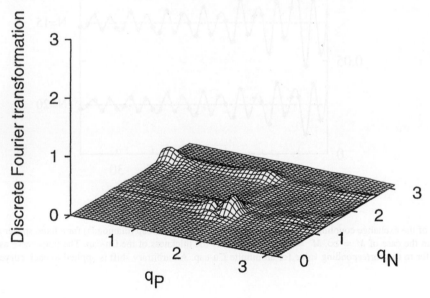

Fig. 7. Absolute values of the discrete two-dimensional Fourier transformation of $(N + P)^2 \mathcal{E}_2(P, N)$ ($10 \le N, P \le 50$) with respect to the spacer and the cap thickness for the case of Co/Cu/Co(0 0 1) trilayer ($M = \infty$, $M' = 5$) with a Cu-cap.

4. Conclusions

We have investigated systematically the effect of a finite cap on the interlayer exchange coupling using a free-electron model as well as ab initio calculations. The results confirm the basic predictions of the electron confinement model as well as available experimental data, namely (i) the periods of the oscillations of the IEC with respect to the cap thickness can be related to the Fermi surface of the cap material; (ii) the oscillations are around a biased value which is generally non-zero and depends on the spacer thickness; (iii) the coupling energies decrease asymptotically with the thickness of the spacer (N) and the cap (P) as $(N + P)^{-2}$, and (iv) the oscillations with respect to the cap thickness can be strongly suppressed if the spanning vectors of the spacer and cap material are stationary at different $\mathbf{k}_\parallel$-vectors.

Acknowledgements

This work is a part of activities of the Center for Computational Material Science sponsored by the Academy of Sciences of the Czech Republic. Financial support for this work was provided by the Grant Agency of the Czech Republic (Project No. 202/97/0598), the Project 'Scientific and Technological Cooperation between Germany and the Czech Republic', the Center for the Computational Materials Science in Vienna (GZ 45.384 and GZ 45.420), and the TMR Network 'Interface Magnetism' of the European Commission (Contract No. EMRX-CT96-0089).

References

[1] D.M. Edwards, J. Mathon, R.B. Muniz and M.S. Phan, Phys. Rev. Lett. 67 (1991) 493.
[2] P. Bruno, J. Magn. Magn. Mat. 121 (1993) 248.
[3] M.D. Stiles, Phys. Rev. B 48 (1993) 7238.
[4] P. Bruno, Phys. Rev. B 52 (1995) 411.
[5] S.N. Okuno and K. Inomata, J. Phys. Soc. Jpn. 64 (1995) 3631.
[6] J.J. de Vries, A.A. Schuldelaro, R. Jugblut, P.J.H. Bloemen, A. Reinders, J. Kohlhepp, R. Coehoorn and W.J.M. de Jonge, Phys. Rev. Lett. 75 (1995) 4306.
[7] A. Bounouh, P. Beauvillain, P. Bruno, C. Chappert, R. Mégy and P. Veillet, Europhys. Lett. 33 (1996) 315.
[8] J. Barnaś, Phys. Rev. B 54 (1996) 12 332.
[9] P. Bruno, J. Magn. Magn. Mater. 164 (1996) 27.
[10] J. Kudrnovský, V. Drchal, I. Turek and P. Weinberger, Phys. Rev. B 50 (1994) 16 105; V. Drchal, J. Kudrnovský, I. Turek and P. Weinberger, Phys. Rev. B 53 (1996) 15 036.
[11] J. Kudrnovský, B. Wenzien, V. Drchal and P. Weinberger, Phys. Rev. B 44 (1991) 4068.
[12] J. Kudrnovský, V. Drchal, P. Bruno, I. Turek and P. Weinberger, Phys. Rev. B 54 (1996) R3738.
[13] P. Bruno, J. Kudrnovský, V. Drchal and I. Turek, Phys. Rev. Lett. 76 (1996) 4254.
[14] J. Kudrnovský, V. Drchal, I. Turek, M. Šob and P. Weinberger, Phys. Rev. B 53 (1996) 5125.
[15] O.K. Andersen, O. Jepsen and D. Glötzel, in: Highlights of Condensed-Matter Theory, eds. F. Bassani, F. Fumi and M.P. Tosi (North-Holland, New York, 1985).

COMPUTATIONAL
MATERIALS
SCIENCE

Computational Materials Science 10 (1998) 198–204

Micromagnetic study of ultrathin magnetic films

Xiao Hu [a,*], Yoshiyuki Kawazoe [b]

[a] *National Research Institute for Metals, Sengen 1-2-1, Tsukuba 305, Japan*
[b] *Institute for Materials Research, Tohoku University, Sendai 980, Japan*

Abstract

The phenomenon of spin reorientations in ultrathin magnetic films is discussed. A micromagnetic theory is presented which reveals the competition between the in-plane shape anisotropy and the normal surface anisotropy through a finite exchange stiffness. For small surface anisotropy, two continuous transitions in spin orientation are derived as the film thickness is increased: first from the uniform, normal configuration to a nonuniform, canting configuration, and then to the uniform, in-plane configuration. This result is consistent with experimental observations. For large surface anisotropy, it is derived theoretically that only the first spin reorientation occurs and the canting configuration remains stable even at large thickness limit. Copyright © 1998 Elsevier Science B.V.

Keywords: Ultrathin magnetic film; Normal surface anisotropy; Spin-reorientation transition; Micromagnetics

1. Introduction

The stable configuration of magnetization in an ultrathin magnetic film coated by other materials, such as a Co film sandwiched by Au, changes with the film thickness: At low thicknesses the magnetization is normal, while at high thicknesses it is parallel, to the film plane. The in-plane magnetization configuration is favored by the shape anisotropy from dipolar interactions in the thin-film geometry, while the normal magnetization is achieved by a strong normal anisotropy localized at the surface of the magnetic film. The existence of a surface anisotropy in magnetic films coated by other materials was first discussed theoretically by Néel [1] based on the breaking of translational symmetry at the surface. Experimental evidence for the normal surface anisotropy was obtained by Grandmann and Müller [2] about three decades ago. Since the normal magnetization state can be used for high-density magnetic and magneto-optic recording [3], the spin-reorientation phenomena and the origin of the surface anisotropy are of considerable interests, both from academic and applied points of view [4–10]. Spin-reorientation transitions are also observed as the temperature is varied [11–16].

---

* Corresponding author. Tel.: +81 298 59 2627; fax: +81 298 59 2601; e-mail: xhu@nrim.go.jp.

2. Discrete model

Spin reorientations in ultrathin magnetic films occur as the result of competition between the two above-mentioned anisotropies via ferromagnetic exchange coupling among neighboring atomic layers. Therefore, a micromagnetic approach is well suited for modeling and understanding this phenomenon [17–21]. Since the transitions have been observed in magnetic films of several atomic layers, the discreteness of these systems in the direction of the film normal is important. We thus study the following free energy functional [19]:

$$\gamma = -Jm_s^2 \sum_{i=1}^{N-1} \cos(\varphi_i - \varphi_{i+1}) - K_v' \sum_{i=2}^{N-1} \sin^2 \varphi_i + K_s'(\sin^2 \varphi_1 + \sin^2 \varphi_N). \tag{1}$$

The first term covers the exchange coupling between the magnetization of adjacent atomic layers, where J is the exchange constant, m_s the saturation magnetization, and the orientation of magnetization φ is measured from the direction of the film normal. The second and third terms are for the volume anisotropy energy and the surface anisotropy energy. In the volume anisotropy term, the shape anisotropy part is assumed to dominate over the intrinsic part and thus $K_v' > 0$. We neglect domain structures across the film [21,22], since we want to concentrate on the nonuniformity in the normal direction. This approximation is sufficient in many cases, noticing the difference between the size of domains across the film, of order of 1 μm, and the film thickness, of order of 1 nm [6]. It is also assumed that, for simplicity, the demagnetization factor (involved in K_v'), the surface anisotropy K_s', and the magnetization are independent of the film thickness, an assumption not always accurate, true [4].

The stable magnetic configuration is determined using the variational technique. We fix the magnetic quantities and change the number of atomic layers. We find a spin-reorientation transition from the uniform, normal configuration to a nonuniform, canting configuration as the number of layers is increased from $N = 2$. Another transition is

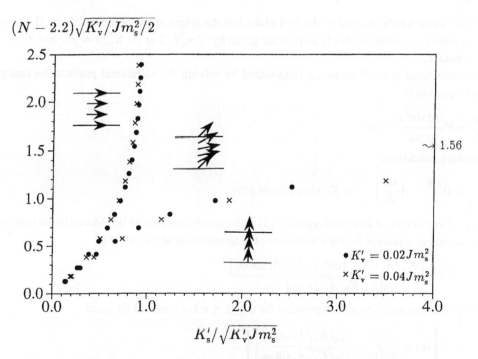

Fig. 1. Phase diagram for the stable magnetic configurations derived from the discrete model.

observed when N is increased further, where the nonuniform, canting configuration is converted into the uniform, in-plane configuration, provided the value of K_s' is small.

Since the critical number of atomic layers depends sensitively on the values of the anisotropies, we have tried scaling plotting. It is then found that by using two renormalized variables, $(N - \Delta N)\sqrt{K_v'/(Jm_s^2)}/2$ and $K_s'/\sqrt{K_v' Jm_s^2}$, and choosing $\Delta N = 2.2$, all the phase boundaries fall into two smooth curves, as shown in Fig. 1.

The scaling relations mentioned above are satisfied sufficiently in the whole region of the surface anisotropy $K_s'/\sqrt{K_v' Jm_s^2}$, while only for small volume anisotropies: $K_v'/(Jm_s^2) \leq 0.04$. Since for a real material like Fe, K_v' is about $0.003 Jm_s^2$, we expect the scaling behaviors shown in Fig. 1 should be observed experimentally. Meanwhile, this scaling plot provides a unified way to analyze experimental data on spin-reorientation transitions observed in different materials. It may also be used for plotting data at different temperatures where the magnetic quantities vary with temperature while fluctuation effects are not critical.

3. Continuum model

In order to understand the numerical results derived from the discrete model, we investigate a continuum model for the same system [17,18]. Continuum models have been used for the study of the magnetic structures in semi-infinite systems with surface anisotropy by several authors [23–25]. Consider a thin magnetic film of thickness $2a$: On the two surfaces there exist normal anisotropies K_s; within the film the shape anisotropy K_v is in the film plane; and the exchange stiffness A is ferromagnetic and finite. From the symmetry of the system, we consider only half of the system. The free energy functional is

$$
\gamma = \int_0^a \left[A \left(\frac{d\varphi}{dz} \right)^2 - K_v \sin^2 \varphi \right] dz + K_s \sin^2 \varphi(0), \tag{2}
$$

where the z axis is taken to be normal to the film plane and the origin at the bottom surface. The relations among the magnetic quantities in functionals (1) and (2) are given by: $Jm_s^2 \hat{a}/2 = A$, $K_v'/\hat{a} = K_v$ and $K_s' = K_s$, where $\hat{a}$ is the lattice constant.

The stable magnetization configuration is determined by solving the variational problem for energy functional (2). The Euler equation is

$$
2A \frac{d^2\varphi}{dz^2} + K_v \frac{d \sin^2 \varphi}{d\varphi} = 0 \tag{3}
$$

with two boundary conditions

$$
\left. \frac{d\varphi}{dz} \right|_{z=a} = 0, \quad A \left. \frac{d\varphi}{dz} \right|_{z=0} = K_s \sin \varphi(0) \cos \varphi(0). \tag{4}
$$

The problem of solving the differential equation (3) under conditions (4) can be reduced to the problem of solving the following nonlinear equation for the orientation of magnetization at $z = a$ [18]:

$$
\frac{K_s}{\sqrt{AK_v}} = \frac{\mathrm{sn}\left[a\sqrt{K_v/A}, \sin \varphi_a\right] \mathrm{dn}\left[a\sqrt{K_v/A}, \sin \varphi_a\right]}{\mathrm{cn}\left[a\sqrt{K_v/A}, \sin \varphi_a\right]} \tag{5}
$$

with $\varphi_a \equiv \varphi(a)$. The magnetization configuration for $0 \leq z \leq a$ is expressed by φ_a as

$$
\varphi(z) = \sin^{-1} \left\{ \sin \varphi_a \frac{\mathrm{cn}\left[(a - z)\sqrt{K_v/A}, \sin \varphi_a\right]}{\mathrm{dn}\left[(a - z)\sqrt{K_v/A}, \sin \varphi_a\right]} \right\}. \tag{6}
$$

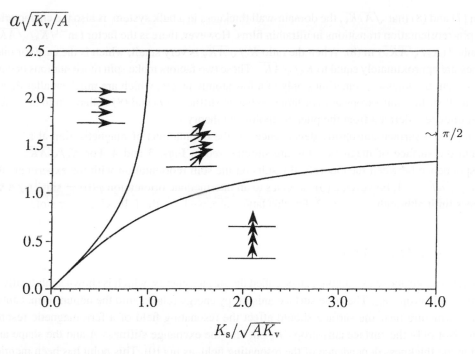

Fig. 2. Phase diagram for the stable magnetic configurations derived from the continuum model.

The stable configuration of the system should be determined by comparing the energies associated with the above magnetic configuration and the two configurations $\varphi = 0$ and $\varphi = \pi/2$, which are trivial solutions of (3) and (4). The stable configuration depends on the magnetic quantities as well as the film thickness, as shown in Fig. 2. The phase boundary between the nonuniform, canting configuration phase and the uniform, normal configuration phase is derived by putting $\varphi_a = 0$ in (5):

$$a_{c1} = \sqrt{A/K_v} \, \tan^{-1}\left(K_s/\sqrt{AK_v} \right). \tag{7}$$

The phase boundary between the phase of nonuniform, canting configuration and the phase of the uniform, in-plane configuration is derived by putting $\varphi_a = \pi/2$ in (5):

$$a_{c2} = \sqrt{A/K_v} \, \tanh^{-1}\left(K_s/\sqrt{AK_v} \right). \tag{8}$$

We find a good agreement between the scaled phase diagram Fig. 1 derived from the discrete model and the phase diagram Fig. 2 derived from the continuum model. This coincidence justifies the use of the continuum approach in the study of magnetic properties of ultrathin magnetic films of several atomic layers.

From Figs. 1 and 2, the film thickness in the continuum model should be evaluated from the number of atomic layers of the relevant lattice as

$$2a = (N - \Delta N)\hat{a} \tag{9}$$

with $\Delta N \simeq 2$. The physical meaning of ΔN becomes clear when comparing the two treatments of surface anisotropy in (1) and (2). In other words, the treatment of surface anisotropy in the continuum model (2) is correct only when the film thickness is evaluated as in (9).

We find in (7) and (8) that $\sqrt{A/K_{\mathrm{v}}}$, the domain-wall thickness in a bulk system, is also the characteristic length in describing spin-reorientation transitions in ultrathin films. However, there is the factor $\tan^{-1}(K_{\mathrm{s}}/\sqrt{AK_{\mathrm{v}}})$ in (7) and the factor $\tanh^{-1}(K_{\mathrm{s}}/\sqrt{AK_{\mathrm{v}}})$ in (8). When the ratio $K_{\mathrm{s}}/\sqrt{AK_{\mathrm{v}}}$ is very small, which is the case for many magnetic materials, they are approximately equal to $K_{\mathrm{s}}/\sqrt{AK_{\mathrm{v}}}$. These two factors make spin reorientations occur in magnetic films of ultra small thicknesses, sometimes only of a few atomic layers, which are much smaller than $\sqrt{A/K_{\mathrm{v}}}$. At the limit of small surface anisotropy and/or large exchange stiffness, (7) and (8) collapse into $a_{\mathrm{c}} = K_{\mathrm{s}}/K_{\mathrm{v}}$, which is the critical thickness derived from the phenomenological theory.

The thickness and surface-anisotropy dependences of the orientations of magnetization at $z = a$ and $z = 0$, the center and the surface of magnetic film, are summarized in Figs. 3 and 4. For $K_{\mathrm{s}}/\sqrt{AK_{\mathrm{v}}} < 1$, we find a direct correspondence between the theoretical results on the spin reorientation with the experimental observation in [6]. For $K_{\mathrm{s}}/\sqrt{AK_{\mathrm{v}}} > 1$, however, $\varphi(0)$ saturates to an intermediate orientation $\varphi(0) = \sin^{-1}(\sqrt{AK_{\mathrm{v}}}/K_{\mathrm{s}})$ at the large-thickness limit although $\varphi_a \to \pi/2$. For this latter case, we have the following asymptote of the free energy [18]:

$$\gamma \simeq -K_{\mathrm{v}}a + 2\sqrt{AK_{\mathrm{v}}} - AK_{\mathrm{v}}/K_{\mathrm{s}}. \tag{10}$$

The constant term in the above asymptote can be called the surface energy, which is shown analytically to be smaller than the surface anisotropy K_{s}. The large surface-anisotropy energy relaxes into the nonuniform, canting structure.

The canting structure near the surface should affect the resonating field of a ferromagnetic resonance (FMR) measurement. Not only the surface anisotropy K_{s} but also the exchange stiffness A and the shape anisotropy K_{v} are involved in the thickness dependence of the resonating field, as in (10). This point has been mentioned in [27].

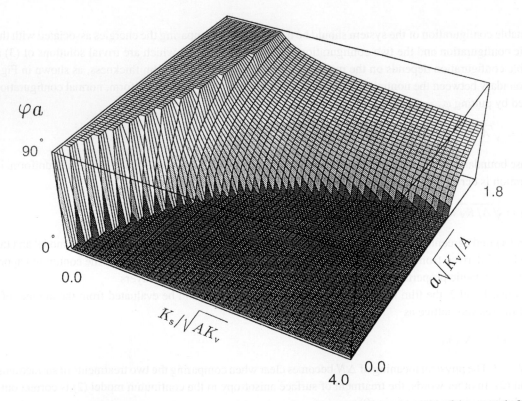

Fig. 3. Thickness and surface anisotropy dependences of the orientation of magnetization at the center of the magnetic film.

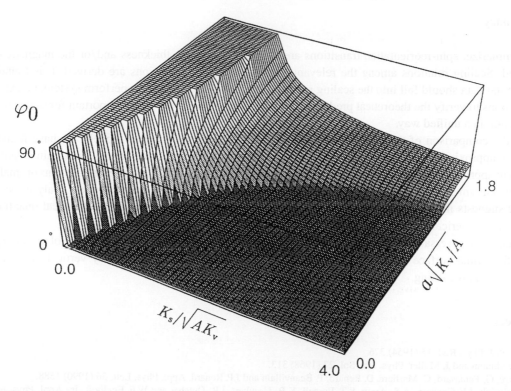

Fig. 4. Thickness and surface anisotropy dependences of the orientation of magnetization on the surface of the magnetic film.

4. Higher-order anisotropies

In many materials, higher-order terms in the expansions of the anisotropies are of comparable magnitudes with the dominant terms included in free energy functional (2), and thus should also be taken into account. Here we show briefly some results including the biquadratic anisotropies. The free energy functional is given as

$$\gamma = \int\limits_0^a \left[A \left(\frac{d\varphi}{dz} \right)^2 - K_{v1} \sin^2 \varphi - K_{v2} \sin^4 \varphi \right] dz + K_{s1} \sin^2 \varphi(0) + K_{s2} \sin^4 \varphi(0). \tag{11}$$

The critical thickness between the uniform, normal phase and the nonuniform, canting phase is given by

$$a_{c1} = \sqrt{A/K_{v1}} \, \tan^{-1} \left(K_{s1}/\sqrt{AK_{v1}} \right). \tag{12}$$

The critical thickness between the uniform, in-plane phase and the nonuniform, canting phase is

$$a_{c2} = \sqrt{\frac{A}{K_{v1} + 2K_{v2}}} \, \tanh^{-1} \frac{K_{s1} + 2K_{s2}}{\sqrt{A(K_{v1} + 2K_{v2})}}. \tag{13}$$

Both of these two expressions are similar to their counterparts in Section 3. Therefore, the model that includes only the dominant anisotropy terms can explain most of the properties of spin-reorientation transitions. However, the model including higher-order anisotropies predicts more complex classification of spin-reorientation transitions [8] since the case $a_{c1} > a_{c2}$ is possible, in principle.

5. Summary

To summarize, spin-reorientation transitions are revealed as the film thickness and/or the magnetic constants are varied. Scaling relations among the relevant quantities in these transitions are derived. It is found that the experimental data should fall into the scaling region. Therefore, it is expected to perform systematic experimental investigations to verify the theoretical predictions. The scaling relations should be important for understanding the phenomenon in a unified way.

From the comparison of the results derived from the continuum and discrete models, we have found that the continuum approximation is sufficient in the study of the spin-reorientation transitions in ultrathin magnetic films. Continuum approach is more reliable than expected generally, in studies of surface effect on systems of small relevant scales, such as small particles and thin films. The continuum approach makes mathematical analyses possible and therefore suggests many important properties such as scaling relations among various physical quantities, more directly than numerical calculations based on discrete models.

Magnetization reversals under external fields form another important problem in magnetism of ultrathin magnetic films with normal surface anisotropy, since they are the key process in high-density vertical magnetic and/or magneto-optical recording.

References

[1] L. Néel, J. Phys. Rad. 15 (1954) 376.
[2] U. Gradmann and J. Müller, Phys. Stat. Sol. 27 (1968) 313.
[3] J. Ferré, G. Pénissard, C. Marliere, D. Renard, P. Beanvillain and J.P. Renard, Appl. Phys. Lett. 56 (1990) 1588.
[4] B. Heinrich, J.F. Cochran, A.S. Arrott, S.T. Purcell, K.B. Urquhart, J.R. Dutcher and W.F. Egelhoff, Jr., Appl. Phys. A 49 (1989) 473.
[5] C. Chappert and P. Bruno, J. Appl. Phys. 64 (1988) 5736.
[6] R. Allenspach, M. Stampanoni and A. Bischof, Phys. Rev. Lett. 65 (1990) 3344.
[7] V. Grolier, J. Ferre, A. Maziewski, E. Stefanowicz and D. Renard, J. Appl. Phys. 73 (1993) 5939.
[8] H. Fritzsche, J. Kohlhepp, H.J. Elmers and U. Gradmann, Phys. Rev. B 49 (1994) 15665.
[9] I. Harada, O. Nagai and T. Nagamiya, Phys. Rev. B 16 (1977) 4882.
[10] K. Takanashi, S. Mitani, M. Sano, H. Fujimori, H. Nakajima and A. Osawa, Appl. Phys. Lett. 67 (1995) 1016; S. Mitani, K. Takanashi, H. Nakajima, K. Sato, R. Schreiber, P. Grünberg and H. Fujimori, J. Magn. Magn. Mater. 156 (1996) 7.
[11] J.J. Krebs, B.T. Jonker and G.A. Prinz, J. Appl. Phys. 63 (1988) 3467.
[12] M. Stampanoni, A. Vaterlaus, M. Aeschlimann and F. Meier, Phys. Rev. Lett. 59 (1987) 2483.
[13] D. Pescia and V. L. Pokrovsky, Phys. Rev. Lett. 65 (1990) 2599.
[14] D.P. Pappas, K.P. Kamper and H. Hopster, Phys. Rev. Lett. 64 (1990) 3179.
[15] A. Berger, A.W. Pang and H. Hopster, J. Magn. Magn. Mater. 137 (1994) L1.
[16] S.D. Bader, D.Q. Li and Z. Qui, J. Appl. Phys. 76 (1994) 6419.
[17] A. Thiaville and A. Fert, J. Magn. Magn. Mater. 113 (1992) 161.
[18] X. Hu and Y. Kawazoe, Phys. Rev. B 51 (1995) 311.
[19] X. Hu and Y. Kawazoe, J. Appl. Phys. 79 (1996) 5842.
[20] X. Hu, T. Yorozu, Y. Kawazoe, S. Ohnuki and N. Ohta, IEEE Trans. Magn. 29 (1993) 3790; X. Hu and Y. Kawazoe, Phys. Rev. B 49 (1994) 3294; J. Appl. Phys. 75 (1994) 6486.
[21] R.-B. Tao, X. Hu and Y. Kawazoe, Phys. Rev. B 52 (1995) 6178.
[22] Y. Yafet and E.M. Gyorgy, Phys. Rev. B 38 (1988) 9145.
[23] D.L. Mills, Phys. Rev. B 39 (1989) 12306.
[24] R.C. O'Handley and J.P. Woods, Phys. Rev. B 42 (1990) 6568.
[25] A. Aharoni, Phys. Rev. B 47 (1993) 8296.
[26] N.D. Mermin and H. Wagner, Phys. Rev. Lett. 17 (1966) 1133.
[27] G.T. Rado, Phys. Rev. B 26 (1982) 295.

ELSEVIER

Computational Materials Science 10 (1998) 205–210

COMPUTATIONAL
MATERIALS
SCIENCE

Magnetization processes in submicronic Co dots studied by means of micromagnetic calculations

R. Ferré

Institut de Physique et Chimie des Matériaux de Strasbourg, 23 rue du Loess, 67037 Strasbourg Cedex, France

Abstract

The magnetization nucleation mechanisms taking place in "perfect" submicronic cobalt dots obtained by means of three-dimensional micromagnetic simulations are shown. Attention is focused on the study of the transition from parallel-to-plane to perpendicular-to-plane magnetization occurring as the thickness of the dot is increased. Simulations show different nucleation and magnetization reversal mechanisms as thickness varies. For a 35 nm dot, nucleation of stripe domains takes place. Stripe domains evolve with the decrease of the in-plane applied field towards a configuration of "bended stripe" domains as a result of the need for a lateral flux closure. The same nucleation mechanism has been found for thicker dots and is responsible for the appearance of diagonal domains for high thicknesses. For 25 nm thick dots no stripe domains are present and a full in-plane magnetization configuration has been obtained for which flux closure takes place with the formation of two distorted vortices at both sides of the dot. Copyright © 1998 Elsevier Science B.V.

Keywords: Micromagnetics; Nanomagnets; Magnetization processes

1. Introduction

Recently, we have reported on the domain structure of square magnetic dots patterned out of an epitaxial hcp cobalt film with thicknesses varying from 250 to 1500 Å and lateral sizes of 0.5 μm which had been fully characterized by means of magnetic force microscopy (MFM) and magnetization measurements [1]. Although domain size and orientation can be derived from domain theory, a detailed understanding of the domain nucleation and the domain evolution with applied field cannot be obtained with simple arguments and a micromagnetic approach of the problem should be considered. Micromagnetic theory, first developed by Brown, allows for an analytical description of nucleation mechanisms in isolated magnetic

particles as well as in thin films [2–4]. Nevertheless, the general micromagnetic equations are highly non-linear and therefore the resolution of problems other than nucleation requires the use of numerical simulations [5–10]. The starting point of all these numerical calculations is the requirement to use a discrete form for the Gibbs free energy (G) and a "dynamics" or way the simulated system has to evolve in phase space. This dynamics can be either given by an equation of motion (of the Landau–Lifshitz type [11]), by a master equation (as is the case for Monte Carlo–Kawasaki dynamics [12]) or simply by an energy minimization procedure. All the simulations presented here have been performed using an algorithm simulating the Landau–Lifshitz equation (more details on numerical simulation procedure are given in [13]).

Physically, the discretization of the micromagnetic equations is equivalent to the assumption that the magnetization varies smoothly on a certain length scale given by the size of the mesh (d). For the sake of physical meaning of simulations, d must be small enough to guarantee smooth variations of M between adjacent elements. It has been shown that d must be, at least, 6–7 times smaller than the domain wall thickness (λ_w) and 3–4 times smaller than the exchange coherence length (λ_{ex}) which are both related to material characteristics (for Co $\lambda_w \approx 200$ Å and $\lambda_{ex} \approx 100$ Å) [13].

The physical parameters of the problem have been chosen according to the values known for cobalt films with perpendicular anisotropy, i.e. $M_s = 1400$ emu/cm^3, $K = 5 \times 10^6$ erg/cm^3, $A = 1.8 \times 10^{-6}$ erg/cm, and the non-dimensional damping constant α has been taken to be unity as a matter of computation simplification [14]. On these bases we have calculated the magnetization equilibrium configurations for 0.45 μm wide cobalt dots with thicknesses ranging from 166 to 1000 Å. The size of the cubic finite element (d) has been taken equal to 35 Å, leading to three-dimensional grid sizes of $125 \times 125 \times N$, where N ranges between 6 and 14 in order to study dots with thicknesses ranging from 20 to 50 nm. Our particular interest in this thickness range is motivated by the fact that, according to the observations and in agreement with Kittel's domain theory, for thicknesses below 30 nm the magnetostatic (or demagnetizing energy) overcomes the crystal anisotropy energy. This leads to an in-plane orientation of the magnetization. Contrarily, for thicknesses above 30 nm the crystal anisotropy energy dominates and the magnetization tends to orient along the normal to the dot's plane. It is in this transition region from in-plane to out-of-plane oriented domains that the equilibrium domain pattern obtained after in-plane demagnetization changes from a "stripe" pattern to a concentric domain "pattern" [15,16].

The aim of the simulations presented here is the understanding of the origin of these different domain patterns and to give general answers to domain pattern formation in square dots. Since for small thicknesses the magnetostatic energy overcomes the crystal anisotropy energy, all the results obtained for small thicknesses are qualitatively applicable to the explanation of domain pattern fromation in soft (permalloy, iron, etc.) magnetic dots (see [17] for instance).

2. Simulation procedure and results

Since we are interested in domain formation after an in-plane field is applied, the simulation procedure employed is the following. Starting the simulation at in-plane saturation (very large values of the in-plane applied field) the magnetic field is slowly reduced (the field speed is about 4×10^7 Oe/s) while recording the hysteresis loop.

Using this simulation procedure for a Co dot 35 nm thick and 0.45 μm large it has been possible to obtain the nucleation mechanism (the "way" the magnetization of the system deviates from the uniform magnetization). The nucleation process is shown in Fig. 1, where we report the equilibrium magnetization configurations for decreasing values of the external magnetic field (the gray contrast corresponds to magnetization pointing up or down along the direction perpendicular to the plane of the figure). For this thickness (and also for thicker systems) the nucleation mechanism found consists of the homogeneous nucleation of out-of-plane up and down domains beginning at the two sides of the dot oriented normal to the initial orientation of the magnetization. A reduction of the applied field results then in a propagation inwards of the nucleated domain configuration leading, when the applied field is small enough, to the formation of "bent" stripe domains. The magnetization inside the domains being fully oriented along the direction perpendicular to the plane of the dot, the domain walls separating them are, as expected, of the Néel–Bloch–Néel type (or vortex domain walls). Thus, for a 35 nm thick Co dot the most favorable nucleation mechanism results in the formation of a "strong stripe" domain pattern.

Reducing the thickness of the dot from 35 to 25 nm results in a completely different nucleation mechanism. Contrary to what has been shown above, the first deviation from the uniform magnetization

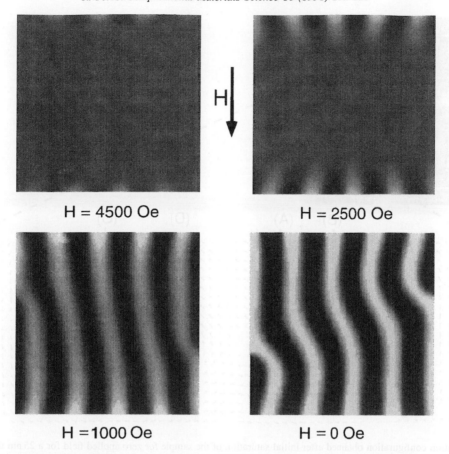

H = 4500 Oe   H = 2500 Oe

H = 1000 Oe   H = 0 Oe

Fig. 1. Nucleation mechanism obtained for a square Co dot 35 nm thick and 0.45 μm large. The magnetic field is applied in the plane of the dot. The sequences shown here correspond to four equilibrium states obtained after in-plane saturation of the magnetization at different fields. The gray tones show the intensity of the perpendicular-to-plane component of the magnetization going from a negative magnetization pointing inwards (white) to a positive magnetization pointing towards the reader (black). Notice the formation of stripe domains at zero applied field.

configuration consists in an in-plane rotation of the magnetization. In this case the magnetostatic energy dominates over the crystal anisotropy energy and the magnetization tends to remain in-plane. Then the only way to reduce the internal energy, and particularly the magnetostatic energy, is the lateral closure of the magnetic flux. The magnetization configuration so obtained evolves with the reduction of the applied field towards a particular magnetization configuration resulting from the penetration of one vortex at each side of the dot (see Fig. 2(a)). A detailed look at this magnetization configuration shows that each vortex (inset) present at zero field has a characteristic fine structure. One single vortex would contain a "singular" line at the vortex center. In this case this line becomes a plane and the center of the vortex spreads out forming a line containing a cross-tie wall (Fig. 2(b) A), two vortices (Fig. 2(b) B and C) carrying opposite sign magnetostatic charge and a quasi-wall (Fig. 2(b) D). This magnetization structure is reminiscent of that obtained for permalloy films with in-plane orientation of the magnetization and is the consequence of the forced stretching of the vortex core when entering at both sides. Important is to notice here that, the observed presence of the quasi-wall is surely related to the geometrical arrangement of the vortices inside the dot. In another system with another geometrical characteristics (different width,

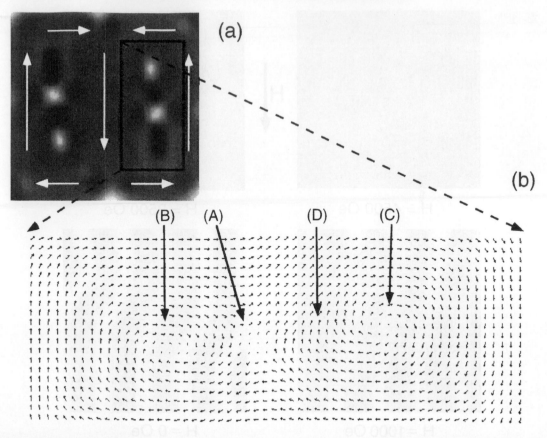

Fig. 2. Magnetization configuration obtained after initial saturation of the sample for zero applied field for a 25 nm thick and 0.45 μm large Co dot. In (a) we show a complete view of the magnetization configuration in which the appearance of two elongated regions constituting the core of two magnetization vortices can be noticed. Gray levels have the same meaning as described for Fig. 1 and the arrows show the orientation of the magnetization. In (b) we show a close up view of one of these vortex cores. In this figure we can notice that the vortex core is formed by a cross-tie wall (A), two vortices (B) and (C) and a quasi-wall (D).

for instance), this quasi-wall would not exist at all or give rise to another pair of vortices.

Finally, for intermediate thicknesses (we present in Fig. 3 the case of a 28 nm thick dot) we find that nucleation begins as for thicker dots with the formation of out-of-plane up and down domains at the two sides of the dot oriented normal to the initial orientation of the magnetization. Nevertheless, in this case, the relative high magnetostatic energy does not permit a full out-of-plane orientation of the magnetization and an in-plane component remains. Thus, the need for an in-plane closure of the magnetic flux makes the magnetization to suffer an in-plane rotation (buckling) that gives rise to the formation of two "bubble domains" (or vortices) at zero field. In this case, magnetization

reversal is achieved as the two magnetic bubbles cross the dot.

3. Discussion

We have presented here first simulations of the micromagnetic properties of perfect Co dots made using a regular cubic mesh. The results obtained show a crucial influence of the magnetostatic energy on the nucleation mechanism as well as on the equilibrium zero-field magnetization configuration. Thus, for thin systems the strong magnetostatic energy only allows for in-plane rotations of the magnetization vector. Contrarily, for thick systems the strong crystal

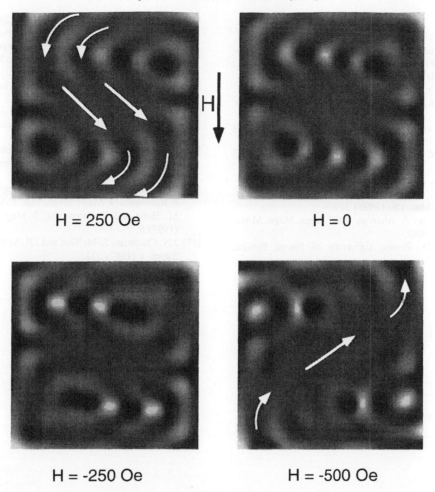

H = -250 Oe   H = -500 Oe

Fig. 3. Magnetization reversal mechanism obtained in the intermediate case, i.e. a Co dot 28 nm thick and 0.45 μm large. The arrows denote the direction of the in-plane component of the magnetization. Notice that the reversal is triggered by the dynamics of two bubble domains that cross the dot from one side to the other.

anisotropy will force the magnetization to nucleate domains pointing along the out-of-plane direction.

The simulated systems and also the mesh employed present a high symmetry. Thus, all the simulations presented here have been made adding a fluctuating term (white gaussian noise with zero time-average) in order to allow for a symmetry breaking. Even so, the magnetization configurations obtained here and so the nucleation mechanisms described present a high degree of symmetry. Since in "real life" the magnetic systems we deal with are not perfect (nonhomogeneous, polycrystalline, etc.) new calculations have been undertaken to compare all the preceding simulations with new ones on systems with a weak disorder. These new simulations will permit us to see the robustness of the magnetization processes presented here and to confirm (or discard) them as the "real" nucleation mechanisms for "actual" perfect systems.

Acknowledgements

I would like to thank the IDRIS and the CNCPST for the permission to use their computing facilities and their technical support to the practical realization of this work.

References

[1] M. Hehn, S. Padovani, K. Ounadjela and J.P. Bucher, Phys. Rev. B 54 (1996) 3428.

[2] W.F. Brown, Jr., J. Appl. Phys. 30 (1959) 62S.

[3] W.F. Brown, Jr., Phys. Rev. 124 (1961) 1348.

[4] W.F. Brown, Jr., Micromagnetics (Wiley/Interscience, New York, 1963).

[5] D.R. Fredkin and T.R. Koehler, J. Appl. Phys. 67 (1990) 5544.

[6] Y. Nakatani, Y. Uesaka and N. Hayashi, J. Appl. Phys. 28 (1989) 2485.

[7] M.E. Schabes, J. Magn. Magn. Mat. 95 (1991) 249.

[8] R. Fischer, T. Schrefl, H. Kronmüller and J. Fidler, J. Magn. Magn. Mat. 150 (1995) 329.

[9] M.E. Schabes and A. Aharoni, IEEE Trans. Magn. MAG-23 (6) (1987) 3882.

[10] R. Ferré, Ph.D. Thesis, University of Joseph Fourier, Grenoble (1995).

[11] T.L. Gilbert, Phys. Rev. 100 (1955) 1243 (abstract only; a full report can be found in Armour Research Foundation Project No. A059, Supplementary Report, 1 May, 1956); L.D. Landau and E.M. Lifshitz, Phys. Z. Sowjet. 8 (1935) 153.

[12] K. Binder, ed., Monte Carlo Methods in Statistical Physics (Springer, Heidelberg, 1979).

[13] R. Ferré, Comput. Phys. Comm. 105 (1997) 169.

[14] Y. Nakatani, N. Hayashi, Y. Uesaka and H. Fujiwara, Jpn. J. Appl. Phys. 33 (1994) 6546.

[15] M. Hehn, K. Ounadjela, J.P. Bucher, F. Rousseaux, D. Decanini, B. Bartenlian and C. Chappert, Science 272 (1996) 1782.

[16] M. Hehn, R. Ferre, K. Ounadjela, J.P. Bucher and F. Rousseaux, J. Magn. Magn. Mat. 165 (1997) 5; R. Ferre, M. Hehn and K. Ounadjela, J. Magn. Magn. Mat. 165 (1997) 9.

[17] J.N. Chapman, S. McVitie and I.R. McFadyen, Scan. Micr. Suppl. 1 (1987) 221.

ELSEVIER

Computational Materials Science 10 (1998) 211–216

COMPUTATIONAL
MATERIALS
SCIENCE

Periodic Anderson model for the description of noncollinear magnetic structure in low-dimensional 3d-systems

V.M. Uzdin [a,*], N.S. Yartseva [b]

[a] *Saint-Petersburg State University, CAPE, V.O. 14 linia, 29, St. Petersburg, 199178, Russian Federation*
[b] *Institute of Metal Physics, GSP-170, Ekaterinburg, 620219, Russian Federation*

Abstract

Distribution of magnetic moments in the low-dimensional metallic structures has been studied theoretically on the basis of periodic Anderson model. Calculation of noncollinear magnetic order was performed in the Hartree–Fock approximation using tight binding real space recursion method. Iteration process includes self-consistent determination of population numbers for the electrons with different directions of the magnetic moments at given atom relatively to the fixed axis. Energies of all states corresponding to the different directions of magnetic moments at the atom under consideration have been calculated, and the state with minimal energy being accepted for the next step.

Analytical transformations based on the generalised "zeros and poles method" were performed for the Green function that allows to avoid some time-consuming numerical procedures. It gives the possibility to develop efficient algorithm for the calculation of noncollinear magnetic structure of complex space nonhomogeneous systems.

Calculations performed for the parameters corresponding to Fe and Cr show the qualitatively different dependencies of the magnetic moment magnitude and the energies of d-electrons on the angles, which define the direction of magnetic moments. Copyright © 1998 Elsevier Science B.V.

Keywords: Model Hamiltonians; Noncollinear magnetism; 3d-metals

Magnetic structure of low-dimensional metallic systems (LDMS) has been recently studied quite intensively. It is connected with a number of new physical phenomena discovered in these structures as well as with development of new technologies which allow to create systems with well-controlled parameters. For most of the new phenomena the magnetic structure on atomic scale plays a crucial role. Imperfections such as surface roughness and interdiffusion lead to the breakdown of space homogeneity. It makes the theoretical description of real systems under experimental investigation extremely complex and time-consuming problem. So far ab initio calculations of such complex systems are out of the possibilities of modern computers. That is why the approach of tight binding model Hamiltonians such as Hubbard-like model (HM) and periodic Anderson model (PAM) is used for this purpose.

* Corresponding author. Present address: Institut fuer Angewandte Physik, Universitaet Duesseldorf, Universitaetsstrasse 1, D-40225 Duesseldorf, Germany. E-mail: uzdin@moss.pu.ru.

HM has been applied to describe the distribution of magnetic moments on the stepped Cr surface. The results allow a rationalisation of the apparently contradictory observations obtained by spin-resolved and angle-and-energy resolved photoemission [1]. In [2] for a Fe (Cr) monolayer deposited on a vicinal Cr (Fe) substrate different self-consistent solutions were obtained. This is important for the description of experimental results of samples prepared under similar but not identical conditions.

PAM has been used to calculate the magnetic properties of Cr atoms impurity in the Fe matrix [3] and Fe clusters embedded near Cr surface, Fe/Cr interface [4], and pinhole defects in Fe/Cr superlattices [5]. Special algorithm was suggested for the modelling of rough surfaces and interfaces with consequent self-consistent calculation of obtained nonideal structure [6].

However, all these theories were developed only for the description of collinear magnetism. Whereas the distribution of magnetic moments with directions in LDMS is very important to understand the physical nature of phenomena in these systems. For example, the mechanisms of noncollinear exchange coupling and giant magnetoresistance in the Fe/Cr and Fe/Si multilayers are unknown, despite the essential efforts undertaken in this direction. So, the development of theoretical scheme for the description of complex nonhomogeneous systems taking into account the noncollinear magnetic structure is of great significance.

Variants of such a theory based on Hubbard Hamiltonian in mean-field approximation were developed in [7] for description of magnetism in transition metals at finite temperature and in [8] for investigation of helical spin density wave state in fcc iron. Modification of the theory was used to study the dependence of exchange coupling in Fe/Cr superlattices as a function of the relative orientation of the magnetic moments at the centre of two adjacent Fe layers [9] and for description of magnetic structure of Mn monolayer on Fe substrate [10].

In this work on the basis of PAM we develop the approach for the self-consistent calculation of noncollinear distribution of the magnetic moments in complex space nonhomogeneous systems. PAM assumes the existence of two bands, one of which corresponds to the quasilocalised d-band and another one to the itinerant s-electrons [3–6,11,12]. The s–d coupling on the site is presupposed to be more stronger than d–s–d interaction of d-electron on different sites. In this case, at first, one should construct resonant d-states and only after that to introduce the electron hopping between different states. d-electron energies have a finite width which is determined by s–d interaction

$$\Gamma_i = \mathrm{Im} \sum_{sk} \frac{V_{i,sk} V_{sk,i}}{\omega - \varepsilon_{sk}},$$

where $V_{i,sk}$ is the s–d hybridisation, ε_{sk} the energy of itinerant s-electrons and i is the coordinate of the site for localised d-electron.

For the description of noncollinear magnetism the Hamiltonian of PAM can be rewritten either in terms of spin quantisation along a global z-axis which is the same for all atoms or using a local spin-quantisation axis along local magnetic moment at site. We will use here the first possibility. In this case after Hartree–Fock approximation in the PAM Hamiltonian one can obtain together with intersite hopping without change of spin-projection V the hopping on site with spin inversion $V^{\uparrow\downarrow}$.

It is schematically depicted in Fig. 1, where every site is shown as two circles corresponding to the up and down spin projection, the solid lines show the usual intersite hopping, the hopping on the site is shown by dash-lines. Parameters $V_{ll}^{\uparrow\downarrow}$ and $V_{ll}^{\downarrow\uparrow}$ depend on the magnitude of magnetic moment on given site as well as on its direction

$$V_{ll}^{\uparrow\downarrow} = (V_{ll}^{\downarrow\uparrow})^* = -e^{-i\varphi_l} \sin\theta_l \frac{U_l M_l}{2}.$$

Here U_l and M_l are Coulomb integral and magnetic moment on l-site correspondingly; θ_l and φ_l are polar angles which define the magnetic moments direction.

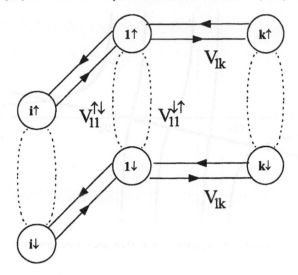

Fig. 1. Schematic representation of hopping in PAM. Solid lines correspond to the intersite hopping without change of spin projection, dash-lines correspond to the hopping on site with the change of spin projection.

Using the recursive method in real space, when the d–d interaction is taken into account inside of first coordination sphere of given atom, one can obtain expression for the d-electron Green function

$$g_{ll}^{\uparrow\uparrow} = \frac{1}{\omega - E_l^{\uparrow} - \sigma_{ll}^{\uparrow\uparrow}},$$

where $E_l^{\uparrow}$ is the energy of d-electron with up spin projection on z-axis, mass operator $\sigma_{ll}^{\uparrow\uparrow}$ is defined as

$$\sigma_{ll}^{\uparrow\uparrow} = Z_l^{-1} \left\{ \frac{U_l^2 M_l^2}{4} \sin^2\theta_l \prod_{i=1}^{8} D_i + \sum_{j=1}^{8} \left(\frac{U_l U_j M_l M_j}{2} \sin\theta_l \sin\theta_j \cos(\varphi_l - \varphi_j) - a_1^{\downarrow} a_j^{\downarrow} - V^2 \right) V^2 \prod_{i\neq j}^{8} D_i \right.$$
$$\left. - V^2 \sum_{j=1}^{8} \sum_{i<j}^{8} \left(a_i^{\downarrow} a_j^{\uparrow} + a_j^{\downarrow} a_i^{\uparrow} - \frac{U_i U_j}{2} M_i M_j \sin\theta_i \sin\theta_j \cos(\varphi_i - \varphi_j) \right) \prod_{i'\neq i,j}^{8} D_{i'} \right\},$$

and following notifications were introduced:

$$Z_l = \left(a_l^{\downarrow} - \sum_{j=1}^{8} \frac{a_j^{\uparrow} V^2}{D_j} \right) \prod_{i=1}^{8} D_i, \quad D_i = a_i^{\uparrow} a_i^{\downarrow} - \frac{U_i^2 M_i^2}{4} \sin^2\theta_i, \quad a_i^{\uparrow} = \omega - E_i^{\uparrow}.$$

For the numerical calculations we used the modification of "zeros and poles" method, which allows to determine very effectively the poles of mass operator and Green function and to avoid time-consuming numerical integration of density of d-electron states in the process of self-consistency.

Fig. 2 illustrates the graphical solution of equation $Z_l = 0$ for the determination of mass operator poles. It is easy to see that all poles x_i are separated from each other, so that $x_i \in (\omega_{i-1}, \omega_i)$, where the ω_i are the roots in ascending order of equation $D_j = 0$ for different j. All x_i can be obtained by means of bisection of the interval (ω_{i-1}, ω_i) and choosing the interval where the function $Z_l(\omega)$ has different signs at the ends of interval. After calculation of mass operator poles one can rewrite mass operator as a sum of simple fraction so that the denominator of Green

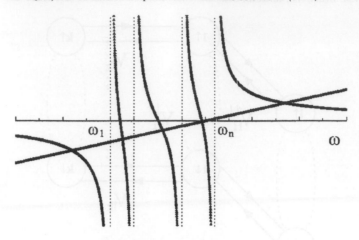

Fig. 2. Graphical solution of the equation $Z_l = 0$. Dash-lines show zeros of D_i for different i.

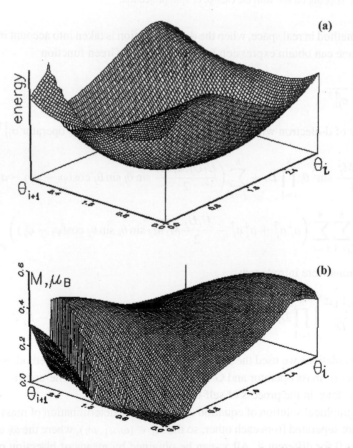

Fig. 3. Energy of d-electrons on i-site (a) and its magnetic moment M (b) as a function of angles θ_i and θ_{i+1} between axis of quantisation and magnetic moments on i and $i + 1$-layer, $\theta_{i-1} = \pi/2$, $\varphi = 0$ for all atoms. All other parameters correspond to the antiferromagnetic Cr in collinear case.

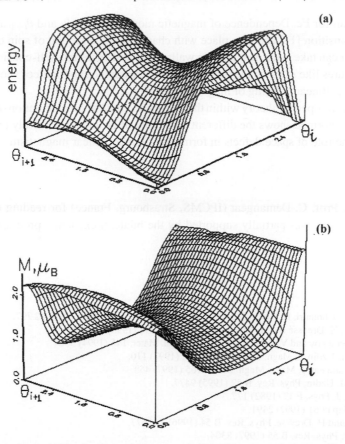

Fig. 4. The same as in Fig. 3 with parameters correspond to ferromagnetic Fe in collinear case.

function will have the same structure as Z_l. Green function roots $y_i(l)$ can be obtained after that by the same way. So the d-electron Green function takes the form

$$g_{ll}^{\uparrow\uparrow} = \sum_i \frac{C_i(l)}{\omega - y_i(l)},$$

and the population number can be found analytically without numerical calculations and it is essentially reduced the time of calculations.

Magnetic moment M_l and occupation number N_l on each site have been determined self-consistently for the different angles θ_l and φ_l, and the state with minimal energy is saved for the next iteration. The magnitude of the moments and their direction for other atoms are kept constant during the self-consistency on a given site.

The dependence of the energy and of the magnetic moments on the polar angles θ_l between quantisation axis and direction of magnetic moment depends essentially on the parameters of the model. As an example, we calculated the magnetic moments of atom in the layer i as a function of θ_i and θ_{i+1} when for the previous layer $i-1$ the value $\theta_{i-1} = \pi/2$ is taken. In Fig. 3 all φ were taken zero and all other parameters correspond to the antiferromagnetic chromium in collinear case. Fig. 4 displays the results where the same calculations were performed for the ferromagnetic iron. The shape of the surfaces in Figs. 3 and 4 differs essentially. First of all there is a minimum of energy in Fig. 3 for θ_i and θ_{i+1} not equal to 0 or π. In Fig. 4 on the contrary the dependence of energy on angles looks like saddle. So, minimal energy will be for $\theta_i = 0$ or $\theta_i = \pi$. It means that for Cr the formation of noncollinear structure can be much more probable than for Fe. Dependence of magnetic moment on the θ_i and θ_{i+1} appears to be complex for Cr, so that first phase transition [13] can take place with change of the direction of spin of the nearest neighbours. Note that such behaviour can take place near domain walls in the bulk material. Self-consistent calculations of space nonhomogeneous structures like steps on the surface and pinholes in the superlattices show that such defects of the structure lead to the noncollinear ordering of magnetic moments.

In summary we have developed the theory within the framework of PAM for calculation of noncollinear magnetism in LDMS. Calculation performed shows the different sensitivity of Fe and Cr relatively to perturbation of collinear structure and displays the role of space defects in formation of noncollinear magnetism.

It is a pleasure to thank Prof. C. Demangeat (IPCMS, Strasbourg, France) for reading our manuscript and for a stimulating discussion. Work was partially supported by the bilateral exchange programme between France and Russia (CNRS-RAS).

References

[1] A. Vega, L.C. Balbas, A. Chouairi, H. Dreyssé and C. Demangeat, Phys. Rev. B 49 (1994) 12 797.
[2] A. Vega, C. Demangeat, H. Dreyssé and A. Chouairi, Phys. Rev. B 51 (1995) 11 546.
[3] V.N. Gittsovich, V.G. Semenov and V.M. Uzdin, J. Magn. Magn. Mater. 146 (1994) 3564.
[4] M.S. Borczuch and V.M. Uzdin, J. Magn. Magn. Mater. 172 (1997) 110.
[5] V.M. Uzdin and C. Demangeat, J. Magn. Magn. Mater. 165 (1997) 458.
[6] A.K. Kazansky and V.M. Uzdin, Phys. Rev. B 52 (1995) 9477.
[7] M.V. You and V. Heine, J. Phys. F 12 (1982) 177.
[8] K. Hirai, J. Phys. Soc. Japan 61 (1992) 2491.
[9] M. Freyss, D. Stoeffler and H. Dreyssé, Phys. Rev. B 54 (1996) R12677.
[10] D. Spisak and J. Hafner, Phys. Rev. B 55 (1997) 8304.
[11] Y. Teraoka, Progr. Theor. Phys. Suppl. N 101 (1990) 181.
[12] C. Demangeat and V.M. Uzdin, J. Magn. Magn. Mater. 156 (1996) 202.
[13] A.K. Kazansky, A.S. Kondratyev and V.M. Uzdin, Phys. Met. Metall. 69 (1990) 53.

ELSEVIER   Computational Materials Science 10 (1998) 217–220

COMPUTATIONAL
MATERIALS
SCIENCE

Improved angular convergence in noncollinear magnetic orders calculations

Daniel C.A. Stoeffler *, Clara C. Cornea

IPCMS – Gemme, 23, rue du Loess, F-67037 Strasbourg, France

Abstract

The method we use for determining the noncollinear orders in metallic superlattices is presented. The electronic structure is obtained within a semi-empirical tight binding framework and the real space recursion technique which allows to study nonsymmetric systems. We focus our attention on the iterative procedure for reaching the angular self-consistency which is extremely slow to converge. We show that the usual mixing scheme is unsuited (because the output angle is directly equal to the next input) and we give an example of possible speed increase by extrapolating the angles resulting from mixing over a few iterations. Copyright © 1998 Elsevier Science B.V.

Keywords: Noncollinear magnetism; Convergence acceleration

1. Introduction

Noncollinear magnetism in the electronic structure framework is a relatively new research area. However, most of the work has been done fixing the direction of each local quantization axis and determining the length of the magnetic moments for these angular frozen noncollinear states. Spin spiral [1] or nearly helicoidal [2] magnetic configurations are illustrations of such studies. It is only during these last years that the angular degree of freedom has been included in the self-consistency resulting from the increase of computer capabilities. The major computing time increase comes not from a higher complexity of the electronic structure description (larger matrix, complex Hamiltonian, three components of the magnetic

moment vector) but from the very slow angular relaxation during the iterative calculation. This is mainly due to the small torque acting on each local moment at the end of the calculation for magnetic configurations nearly converged. In this paper, we illustrate this behaviour in the case of metallic superlattices and show that the convergence becomes extremely slow when a mixing scheme is used. In the second part, we give a very simple method allowing a speed up of the convergence of more than 3 by extrapolation of the angular variations.

2. Method of calculation for noncollinear magnetism

We use a tight binding Hamiltonian which can be written as the sum of a band H_{band} and an exchange

* Corresponding author. Tel.: +33 3 88 10 70 65; fax: +33 3 88 10 72 49; e-mail: daniel@zooropa.u-strasbg.fr.

Hamiltonian with:

$$H_{band} = \sum_{\substack{i,l,m \\ i',l',m'}} |i; l; m\rangle [(\varepsilon_{i,l,m}^0 + U_{il}\Delta N_{i,l})$$

$$\times \delta_{i,i'}\delta_{l,l'}\delta_{m,m'} + t_{i',l',m'}^{i,l,m}(1 - \delta_{i,i'})]$$

$$\times \langle i'; l'; m'| \begin{bmatrix} 1 & 0 \\ 0 & 1 \end{bmatrix}, \tag{1}$$

$$H_{exc} = \sum_{i,l,m} \left(-\frac{1}{2}I_{il}M_{il}\right)|i; l; m\rangle\langle i; l; m|$$

$$\times \begin{bmatrix} \cos\theta & e^{-i\varphi_i}\sin\theta_i \\ e^{i\varphi_i}\sin\theta_i & -\cos\theta_i \end{bmatrix}, \tag{2}$$

where $\varepsilon_{i,l,m}^0$ is the energy level of the site i for the symmetry (l, m); $U_{i,l}\Delta N_{i,l}$ is the intrasite Coulomb contribution from the local charge transfer $\Delta N_{i,l}$, $U_{i,l}$ being the intrasite effective Coulomb parameter; $t_{i',l',m'}^{i,l,m}$ is the so-called hopping integral between orbitals of site i for the symmetry (l, m) and of site i' for the symmetry (l', m'); I_{il} is the intrasite effective exchange parameter and M_{il} the local magnetic moment whose direction is given by the two angles (θ_i, φ_i) in the usual spherical representation; the $[2 \times 2]$ matrix represents the spin part. Such a Hamiltonian allows the determination of the electronic structure for a set of vectorial magnetic moments.

In our case, we determine the electronic structure with the real space recursion technique which allows the determination of the local density of states projected on an arbitrary local quantization axis by rotating the initial element of the recursion basis. We use a cluster containing approximately 5000 atoms for determining 24 levels in the continuous fraction for each "s", "p", "d" symmetry, each spin and each quantization axis. The resulting magnetic moment vector $M_i = M_r\hat{u}_r + M_\theta\hat{u}_\theta + M_\varphi\hat{u}_\varphi$ is obtained by the successive calculation of three components M_r, M_θ, M_φ in the local spherical basis $(\hat{u}_r, \hat{u}_\theta, \hat{u}_\varphi)_i$. We use an iterative method to obtain the angular self-consistent solution meaning that the output vector must be the same as the input one. In our case, the angular self-consistency is achieved when the "d" part of the components M_θ and M_φ perpendicular to the local quantization axis are equal to zero (smaller than 5×10^{-5}). This is equivalent to the assumption that the local quantization axis

is given by the direction of the "d" local magnetic moment. More details about the self-consistency are given in another paper of these proceedings [3].

Finally, the tight-binding parameters are either obtained from the literature [4] or determined in order to reproduce satisfactorily the ab initio results of a preliminary study in the collinear case [5].

3. Angular self-consistency

In a simple mixing scheme, the next input variable is chosen between the input and the output ones with a mixing factor x corresponding to the weight of the

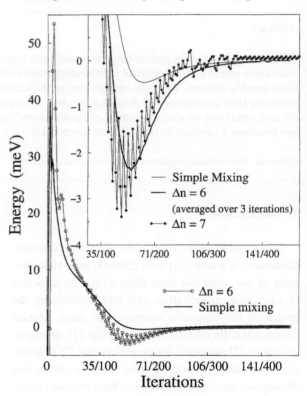

Fig. 1. Variations of the total energy during the iterations relative to the converged value obtained with a simple complete mixing scheme (thick solid line) and with the *mixing-extrapolation* method for $\Delta n = 6$ (open circles) for Co_5Mn_7 superlattices. The insert shows the comparison between the two previous converging cases and the nonconverging variations for $\Delta n = 7$; the variations for $\Delta n = 6$ have been averaged over three iterations. The scale of the horizontal axis has been renormalized by the speed up: the smallest (largest) axis labels correspond (respectively) to the accelerated (normal) calculation.

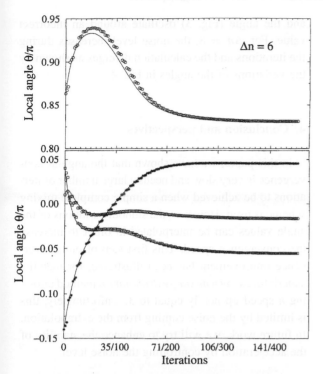

Fig. 2. Variations of the local θ angle on selected sites obtained with a simple complete mixing scheme (solid line) and with the *mixing-extrapolation* method for $\Delta n = 7$ (open symbols) for Co_5Mn_7 superlattices. The scale of the horizontal axis has been renormalized by the speed up: the smallest (largest) axis labels correspond (respectively) to the accelerated (normal) calculation.

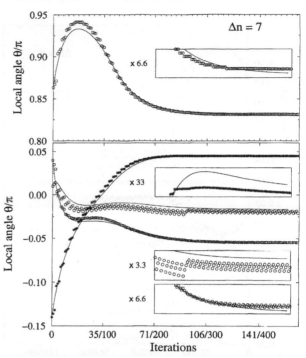

Fig. 3. Variations of the local θ angle on the same selected sites as in Fig. 2 obtained with a simple complete mixing scheme (solid line) and with the *mixing-extrapolation* method for $\Delta n = 7$ (open symbols) for Co_5Mn_7 superlattices. The inserts correspond to a magnification for the last iterations. The scale of the horizontal axis has been renormalized similar to Figs. 1 and 2.

output values, $1 - x$ being the weight of the input values. This is the case for the charge transfers (with $x = 0.1$–0.2) and the magnitude of the magnetic moments (with $x = 0.5$–1) and can be the case for the local quantization axis (the input and the output magnetic moments vectors defining a plane and the new input is in the same plane between these two vectors). However, the angular relaxation is so slow that a complete mixing can be used: the next input direction of the quantization axis is directly the output one. This corresponds to the fastest variations for a mixing scheme. Unfortunately, even with a complete mixing the convergence needs to much iterations. When the number of atoms in the cell is large, like in superlattices, it is possible that different noncollinear solutions are obtained. We have then to compare their energies which gives the most stable solution [3].

The figures of this paper are illustrations of the calculations of the magnetic properties for Co_5Mn_7 superlattices with $\varphi_i = 0$ [3,5] and show the total energy (Fig. 1) and the angles on selected sites (Figs. 2–4) during the iterations n. With the simple mixing, the convergence is obtained after $\tilde{n} = 454$ iterations. Fig. 2 shows clearly that the angles vary very smoothly during the calculations. This is why, it is reasonable to assume that they follow regular curves which can be interpolated. This has been done using a mixing-extrapolation method which consists in (i) calculating three points $\{\theta_i(n), \theta_i(n + 1), \theta_i(n + 2)\}$ with the usual complete mixing, (ii) interpolating these three angle values as a function of the iteration index n with a quadratic polynomial $\tilde{\theta}_i(n)$, and (iii) extrapolating up to a large iteration index jumping a few iterations Δn. The next input is then given by

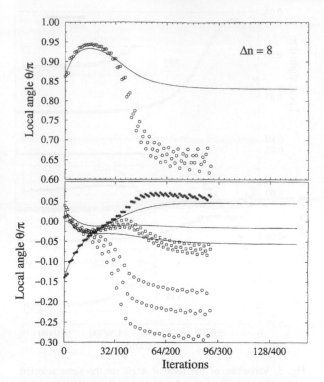

Fig. 4. Variations of the local θ angle on selected sites obtained with a simple complete mixing scheme (solid line) and with the *mixing-extrapolation* method for (open symbols) for Co_5Mn_7 superlattices. The scale of the horizontal axis has been renormalized similar to Figs. 1 and 2.

$\tilde{\theta}_i(n + 2 + \Delta n)$. This interpolation is repeated until the convergence is achieved. In principle, if the interpolating function follows the regular curve, the convergence speed up is equal to $1 + \Delta n/3$. For the situation presented in Figs. 1 and 2 the convergence is obtained after 161 iterations with $\Delta n = 6$: the speed up is then equal to 2.8 representing 93% of the best one. This shows the validity of the *mixing-extrapolation* technique.

Of course, the speed up increases when Δn is augmented. However, the extrapolation introduces a little noise which increases with Δn and the calculation will become unstable for large Δn values. This is illustrated by Figs. 3 and 4 for $\Delta n = 8$ and $\Delta n = 7$. For, $\Delta n = 7$ the noise level is nearly constant so that the calculation does not converge but the energies (Fig. 1)

and the angle (Fig. 3) oscillate around an incorrect value. For $\Delta n = 8$, the noise level increases during the iterations and the calculation diverges as shown by the variations of the angles in Fig. 4.

4. Conclusion and perspectives

In this paper, we have shown that the angular convergence is very slow and needs a large number of iterations to be achieved when a simple complete mixing scheme is used. However, the smooth variations of the angle values can be interpolated in order to increase the convergence speed. This first step of our convergence improvement has been illustrated, for metallic superlattices, by a *mixing-extrapolation* method allowing a speed up nearly equal to 3. Unfortunately, this is limited by the noise coming from the extrapolation. In future work, we will try to enhance the stability of the acceleration by controlling the noise level.

Acknowledgements

The calculations have been mainly realized on the T3E parallel computer at the Institut du Développement et des Ressources en Informatique Scientifique (IDRIS) of the CNRS and on the SP2 parallel computer of the Centre National Universitaire Sud de Calcul (CNUSC).

References

[1] L.M. Sandratskii and J. Kubler, Phys. Rev. B 47 (1993) 4854.

[2] D. Stoeffler and F. Gautier, J. Magn. Magn. Mater. 121 (1993) 259–265.

[3] C.C. Cornea and D. Stoeffler, these proceedings.

[4] O.K. Andersen, O. Jepsen and D. Gloetzel, in: Highlights of Condensed Mater Theory, eds. F. Bassani, F. Fumi and M.P. Tosi (North-Holland, Amsterdam, 1985) p. 59.

[5] C.C. Cornea-Borodi, D.C.A. Stoeffler and F. Gautier, J. Magn. Magn. Mater. 165 (1997) 450–453; D.C.A. Stoeffler, to be published.

ELSEVIER

Computational Materials Science 10 (1998) 221–224

COMPUTATIONAL
MATERIALS
SCIENCE

On the validity of two-current model for systems with strongly spin-dependent disorder

A. Vernes [a], H. Ebert [a,*], John Banhart [b]

[a] *Institute for Physical Chemistry, University of Munich, Theresienstr. 37, 80333 Munich, Germany*
[b] *Fraunhofer-Institute for Applied Materials Research, Lesumer Heerstr. 36, 28717 Bremen, Germany*

Abstract

The resistivities of the ferromagnetic alloy systems Fe–Ni and Co–Ni were studied in detail by application of the Kubo–Greenwood formalism. The electronic structure of the randomly disordered ferromagnetic alloys was computed by use of the spin-polarized Korringa–Kohn–Rostoker coherent potential approximation (KKR-CPA) method. Two sets of calculations were carried out: one fully relativistic and another one based on the two-current model. The former one will judge whether the two, not directly coupled spin-subsystems, could lead to the spontaneous magnetoresistance anisotropy, as it is supposed within the two-current model. Furthermore, all the results obtained are compared with the experimental data. We found that the two-current model calculations yield spin-resolved resistivities more polarized than could be expected from the experiments. Finally they lead to a much lower total resistivity than the relativistic calculations, showing that the scattering processes between the spin-systems are of crucial importance. Copyright © 1998 Elsevier Science B.V.

Keywords: Two-current model; Disordered alloys; Kubo–Greenwood formalism; Magneto-resistance anisotropy; Relativistic scattering theory

1. Introduction

The *two-current model* [1] assumes that the majority and minority electron spin systems contribute independently to electronic conduction giving rise to two spin-dependent partial resistivities $\rho^\uparrow$ and $\rho^\downarrow$, respectively. Mechanisms which change the spin direction of an electron are either assumed to vanish or to be weak [2]. In the latter case one introduces phenomenologic correction terms which describe the effect of spin mixing. Possible spin-mixing mechanisms are scattering by magnons, electron–electron interaction and spin–orbit interaction.

Recently it became possible to calculate the resistivity of ferromagnetic alloys using the first principles Kubo–Greenwood formalism based on solutions of the Dirac equation [3]. This way all relativistic effects, as e.g. the spin–orbit interaction, are taken rigorously into account. By manipulating the Dirac equation, it could be shown explicitly that spin–orbit interactions cause the magnetoresistance anisotropy [4]. This rigorous, relativistic approach is used in the present work to supply proper theoretical reference data for calculations of the isotropic residual resistivity for the binary alloy systems Fe–Ni and Co–Ni based on the two-current model.

* Corresponding author. Tel.: +89-2394-4642; fax: +89-2394-248; e-mail: he@gaia.phys.chemie.uni-muenchen.de.

2. Simple and extended two-current model

Without any spin-flip processes the two sub-band resistivities add like those of two parallel resistors and the resulting average isotropic resistivity is simply given by

$$\bar{\rho} = \left(\frac{1}{\rho^\downarrow} + \frac{1}{\rho^\uparrow} \right)^{-1}. \tag{1}$$

To account for spin mixing, a parameter $\rho^{\uparrow\downarrow}$ describing the rate of spin-flip transitions is introduced [5–7]:

$$\bar{\rho} = \frac{\rho^\downarrow \rho^\uparrow + \rho^{\uparrow\downarrow}(\rho^\downarrow + \rho^\uparrow)}{\rho^\downarrow + \rho^\uparrow + 4\rho^{\uparrow\downarrow}}. \tag{2}$$

Within the two-current model, the difference $\Delta\rho$ between the resistivities in parallel and perpendicular to the spontaneous magnetization depends on the ratio $\alpha = \rho^\downarrow / \rho^\uparrow$. So, if again spin-mixing effects are neglected [7]

$$\frac{\Delta\rho}{\bar{\rho}} = \gamma \left(\frac{\rho^\downarrow}{\rho^\uparrow} - 1 \right) = \gamma \, (\alpha - 1) , \tag{3}$$

where $\Delta\rho/\bar{\rho}$ is the so-called spontaneous magneto-resistance anisotropy (SMA) ratio and γ is a measure for the momentum transfer between the two spin systems due to spin–orbit coupling.

For the case that spin mixing has to be considered, the relation contains the spin-flip parameter $\rho^{\uparrow\downarrow}$ [7]:

$$\frac{\Delta\rho}{\bar{\rho}} = \gamma \frac{(\rho^\downarrow - \rho^\uparrow)\rho^\downarrow}{\rho^\uparrow\rho^\downarrow + \rho^{\uparrow\downarrow}(\rho^\uparrow + \rho^\downarrow)}. \tag{4}$$

3. Results and discussion

3.1. Spin-dependent disorder within the two-current model

In a first set of calculations the Kubo–Greenwood formalism was used in combination with the two-current model. This means that conductivity calculations were performed separately for the minority and the majority sub-bands of each alloy. Accordingly, the

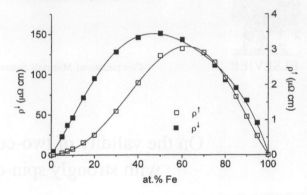

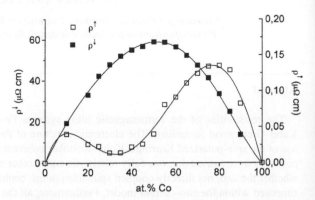

Fig. 1. Sub-band resistivities for the alloy systems Fe–Ni (a) and Co–Ni (b) calculated assuming the two-current model. Open symbols: majority resistivity $\rho^\uparrow$ (see right scale), full symbols: minority resistivity $\rho^\downarrow$ (left scale).

two sub-bands possess a different electronic structure having only the Fermi energy in common. As it can be seen in Fig. 1, the partial resistivities $\rho^\uparrow$ and $\rho^\downarrow$ obtained are quite different. While $\rho^\uparrow$ takes very low values – especially for Co–Ni – $\rho^\downarrow$ is found to be quite large. Consequently, the corresponding ratio $\alpha = \rho^\downarrow / \rho^\uparrow$ is also rather high, reaching values of up to 370 for Fe–Ni and 3800 for Co–Ni. These high values are in strong contrast to those determined experimentally ($\alpha \approx 20, \ldots, 30$).

The reason for these very different partial resistivities is that, the conduction electrons effectively seem to see a strongly different degree of disorder, i.e. loosely spoken there is a strongly spin-dependent disorder present.

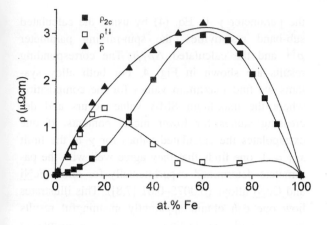

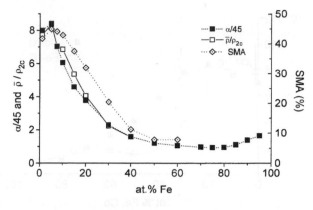

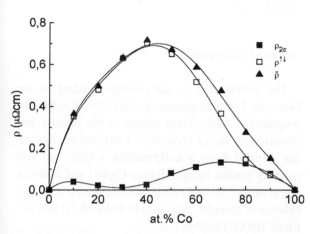

Fig. 2. Resistivities for the alloy systems Fe–Ni (a) and Co–Ni (b): triangles: relativistic resistivities $\bar{\rho}$, full squares: two-current model resistivities ρ_{2c}, open squares: spin-mixing parameter $\rho^{\uparrow\downarrow}$.

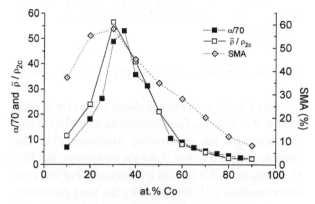

Fig. 3. Quantities related to deviations from two-current model for Fe–Ni (a) and Co–Ni (b). Squares: parameter α/C, open squares: ratio $\bar{\rho}/\rho_{2c}$, diamonds: $\Delta\rho/\bar{\rho}$ (SMA [3]). For further details see text.

3.2. Comparison with fully relativistic results

The total resistivity calculated within the framework of the two-current model (in the following $\bar{\rho}_{2c}$) has to be distinguished from the resistivity $\bar{\rho} = (2\rho_{\perp} + \rho_{\parallel})/3$ which is calculated fully relativistically. The resulting resistivities $\bar{\rho}_{2c}$ and $\bar{\rho}$ are compared in Fig. 2 with one another. Obviously, $\bar{\rho}$ is always larger than the two-current model resistivity $\bar{\rho}_{2c}$ for both alloy systems. This is much more pronounced for Co–Ni than for Fe–Ni. In the former case, it was found that the two-current model resistivity is wrong up to two orders of magnitude with respect of $\bar{\rho}$ calculated fully relativistically.

Comparing the ratio $\bar{\rho}/\rho_{2c}$ and α, one finds that these are almost proportional to one another

$$\frac{\rho}{\rho_{2c}} \approx C\frac{\rho^{\downarrow}}{\rho^{\uparrow}} = C\alpha. \tag{5}$$

But as is demonstrated in Fig. 3, C must be set to 45 for Fe–Ni and to 70 for Co–Ni. Thus the more the two-current model results deviate from the proper results the more the resistivity of the two sub-bands differs, i.e. the higher the ratio α is. In conclusion, the strongly spin-dependent disorder in the two alloy systems Fe–Ni and Co–Ni is responsible for the failure of the two-current.

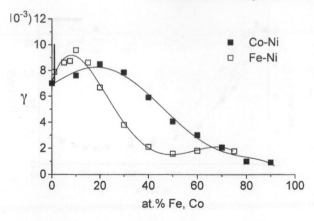

Fig. 4. Parameter γ calculated from Eqs. (2) and (4). Vertical bar: experimental values for the dilute limit of both alloy systems [7,9].

3.3. Spin–orbit related parameters $\rho^{\uparrow\downarrow}$ and γ, SMA

Using Eq. (2) with the calculated values for $\rho^{\uparrow}$ and $\rho^{\downarrow}$ together with the fully relativistic resistivity $\bar{\rho}$ one can determine the spin-mixing parameter $\rho^{\uparrow\downarrow}$. The result is included in Fig. 2 for the two alloys. For Fe–Ni values up to 1.3 $\mu\Omega$ cm are obtained. For low iron concentrations $\rho^{\uparrow\downarrow}$ is essentially the total resistivity, i.e. the most effective scattering for majority electrons seems to be via spin–orbit interaction into the minority sub-band. For Co–Ni the situation is even more dramatic: the spin–orbit interaction induced scattering that is deduced from Eq. (2) dominates over the entire composition range. For Ni diluted with Fe and Co, experimental values for $\rho^{\uparrow\downarrow}$ at $T = 0$ K were found between 0.2 and 0.35 $\mu\Omega$ cm [8], having the same order of magnitude as the calculated values for comparable Fe or Co contents.

The spontaneous magnetoresistance anisotropy (SMA) ratio $\Delta\rho/\bar{\rho}$ calculated relativistically (for Fe–Ni, see [3]) is compared to the ratios $\bar{\rho}/\rho_{2c}$ and α/C in Fig. 3. Apparently there is a close correlation between SMA and α/C, especially for Fe–Ni – as one would expect on the basis of Eq. (3).

If one accepts Eqs. (2) and (4) as valid for situations with strong spin-mixing, one can determine

the parameter γ in Eq. (4) by using the calculated sub-band resistivities, the spin-mixing parameter $\rho^{\uparrow\downarrow}$ and the calculated $\Delta\rho/\bar{\rho}$. The corresponding results are shown in Fig. 4. For both alloy systems we find maximum values for the composition where the maximum SMA value occurs and decreasing values for lower nickel contents. If one extrapolates the calculated values for γ to the limit $x_{Ni} = 1$, one finds that they agree well with the parameters determined experimentally for dilute FeNi and CoNi alloys (0.0075–0.01 [7,8]). This illustrates how one can obtain apparently meaningful results by repeatedly neglecting the crucial spin-mixing contribution.

Acknowledgements

The authors like to acknowledge funding by the Deutsche Forschungsgemeinschaft (DFG) within the programme *"Relativistic effects in the physics and chemistry of heavy elements"*. Furthermore, this paper resulted from a collaboration within, and was partially funded by, the Human Capital and Mobility Network on "Ab initio (from electronic structure) calculation of complex processes in materials" (Contract: ERBCHRXCT930369).

References

[1] N.F. Mott, Adv. Phys. 13 (1964) 325.
[2] I.A. Campbell and A. Fert, in: Ferromagnetic Materials, ed. E.P. Wohlfahrth, Vol. 3 (North-Holland, Amsterdam, 1982) p. 751.
[3] J. Banhart and H. Ebert, Europhys. Lett. 32 (1995) 517.
[4] J. Banhart, A. Vernes and H. Ebert, Solid State Commun. 98 (1996) 129.
[5] A. Fert, J. Phys. C 2 (1969) 1784.
[6] A. Fert, Phys. Rev. Lett. 21 (1968) 1190.
[7] I.A. Campbell, A. Fert and O. Jaoul, J. Phys. C 1 (1970) Suppl., S95.
[8] O. Jaoul and I.A. Campbell, J. Phys. F 5 (1975) L69.
[9] O. Jaoul, I.A. Campbell and A. Fert, J. Magn. Magn. Mat. 5 (1977) 23.

Computational Materials Science 10 (1998) 225–229

COMPUTATIONAL
MATERIALS
SCIENCE

Percolation mechanism for colossal magnetoresistance

Paul J.M. Bastiaansen *, Hubert J.F. Knops

Institute for Theoretical Physics, University of Nijmegen, PO Box 9010, 6500 GL Nijmegen, Netherlands

Abstract

We propose a new mechanism to explain colossal magnetoresistance. The explanation assumes that the materials displaying colossal magnetoresistance are halfmetallic and proposes that the effect is a critical phenomenon, which is intimately connected with the ferromagnetic-to-paramagnetic phase transition present in these materials. The proposed mechanism is a percolation mechanism; the behavior of the resistance is described using a resistor network. An analysis of the percolation phase diagram and Monte Carlo calculations on the resistor network show a full qualitative correspondence with the experimentally observed features of colossal magnetoresistance. Copyright © 1998 Elsevier Science B.V.

Keywords: Magnetic materials; Electrical conductivity; Critical phenomena

Recently there has been a lot of interest in the magnetism and transport properties of rare earth manganese perovskites [1–10], such as $La_{1-x}Ca_xMnO_3$. The manganites having $0.2 < x < 0.5$ are ferromagnets, and their resistivity depends strongly on temperature and external magnetic field. Both the resistance and the magnetoresistance peak at or near the Curie temperature; and in particular an external magnetic field causes the resistivity to drop several orders of magnitude. Hence the name colossal magnetoresistance (CMR) for this phenomenon.

The manganese perovskites are of a mixed valence type; if only Mn^{3+} or Mn^{4+} are present the material is insulating and antiferromagnetic, and no CMR is observed. It is known for a long time that the double exchange interaction between pairs of Mn^{3+} and Mn^{4+} is responsible for the ferromagnetic and metallic properties of the perovskites [11]. In this picture,

a dependence of the conductance on the spin direction of the charge carriers is already present. More recently band structure calulations [12] using the local spin-density approximation indicated an effective halfmetallic behavior, and in similar calculations adopting the generalized gradient approximation the halfmetallic character emerged more clearly [13]. We will use this halfmetallic behavior as a starting assumption of our mechanism for CMR.

The typical temperature dependence displayed in the measurements on CMR shows that the peak in the resistance occurs at or close to the Curie temperature T_C. We regard this as a strong indication that CMR is a critical phenomenon. The Curie temperature is a critical point, and the occurrence of the peak in the resistance at this temperature strongly suggests that the behavior of the resistance is intimately connected with the occurrence of the critical, long range correlations that are present at this point. The mechanism we want to propose treats CMR as a critical phenomenon, and will be shown to explain all qualitative features

---

* Corresponding author. Tel: +31-24-3652849; fax: +31-24-3652120; e-mail: paulb@tvs.kun.nl.

of CMR as seen from experiments. We do not aim at a reproduction of the actual numerical values of peak heights or critical temperatures, as they will be strongly material-dependent. The concept of universality in critical phenomena, however, assures that the universal features such as critical exponents should be reproduced correctly even by a simple model that only incorporates the basic mechanism that we want to propose.

The halfmetallic character of the CMR materials implies that the band structure of the electrons depends on the relative orientation of their spin with respect to the magnetization of the material. Electrons with spin parallel to the local magnetization are conducting, electrons with spin antiparallel are insulating. The conductance of the material thus depends on the structure of the domains with different directions of magnetization. As suggested in [9], this explains the observation that the resistance increases with temperature, as more domain walls occur with higher temperatures, thereby hampering the percolation of charge carriers. It is less clear that this effect also explains the decrease in resistance above T_C, but we will see that it does, so there is no need to invoke another mechanism (e.g., magnetic polarons) to explain the behavior above T_C.

Magnetic phase transitions such as displayed in the CMR materials are described with lattice models with a Hamiltonian, depending on temperature and magnetic field, that determines the probability distribution of the domains with different directions of the dipole moments $\mathbf{m}(\mathbf{r})$. Realistic models typically are of the Heisenberg type, with an anisotropy term that incorporates the lattice structure and eventually breaks the full rotational freedom of $\mathbf{m}(\mathbf{r})$.

The system described with such a Heisenberg Hamiltonian should be dressed with a resistance definition that incorporates the halfmetallic character. The resistance must therefore depend on the local direction of the magnetic moments $\mathbf{m}(\mathbf{r})$. Electrons having their spin parallel to $\mathbf{m}(\mathbf{r})$ experience a low resistance, electrons with spin antiparallel feel resistances that are orders of magnitude higher. The question as to the overall resistance now results in a percolation-type problem. If the domains with a certain direction of

$\mathbf{m}(\mathbf{r})$ form a percolating cluster through the material, the net resistance is low, whereas it will be high when there is no such cluster. The precise value of the resistance of course depends on the precise definition of the model and the local resistances, and will thus be material-dependent. The qualitative behavior of the resistance, however, can be extracted exclusively from the percolation properties of the model.

We will illustrate the mechanism using a model where the directions of magnetic moments $\mathbf{m}(\mathbf{r})$ are confined to be up or down. The resulting model is the Ising model. In this simple case, electrons with spin up are conducting only in domains where the Ising moments $\mathbf{m}(\mathbf{r})$ are up, and vice versa for spin down. The net resistance that is predicted by this model is low when one or both of the directions of $\mathbf{m}(\mathbf{r})$ form a percolating cluster, and is high when there is no percolation for both directions.

The percolation phase diagram of this model depends on the dimensionality and on the definition of which bonds are considered percolating. As the CMR materials have a low resistance above the Curie temperature, there have to be percolating clusters present in this regime. This is the case only when percolating bonds are defined such that different clusters can cross each other. The generic phase diagram [14,15] for this case is shown in Fig. 1. In general, the percolation threshold T_p lies below or at T_C.

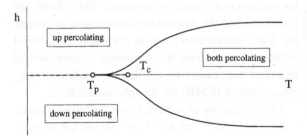

Fig. 1. The generic phase diagram for percolation in the Ising model. T is temperature, h is magnetic field. Indicated are the regions where one or both of the directions of $\mathbf{m}(\mathbf{r})$ are percolating. In general, the percolation threshold T_p lies below T_C, but in the limiting case of a two-dimensional lattice with next-nearest neighbor percolating bonds, the two temperatures coincide. The thick solid line is a critical percolation line in the universality class of random percolation and the same is expected for the endpoint of the line. The dashed line is a first-order transition line.

From the phase diagram in Fig. 1 and from the observation that the percolation threshold lies just below or at the Curie temperature, we extract the qualitative predictions for the overall resistance of the model and compare those with the experimental data. To make these predictions more explicit, we performed Monte Carlo calculations on a specific model as well. We chose the simple case of a two-dimensional model where the directions of the local moments $\mathbf{m}(\mathbf{r})$ are confined to be 'up' or 'down'. The Hamiltonian, governing the statistical distribution of the domains with different directions of $\mathbf{m}(\mathbf{r})$, is in this case

$$H = -J \sum_{\langle \mathbf{r}, \mathbf{r}' \rangle} \mathbf{m}(\mathbf{r}) \cdot \mathbf{m}(\mathbf{r}') - h \sum_{\mathbf{r}} \mathbf{m}(\mathbf{r}), \qquad (1)$$

where the coupling J measures the inverse temperature, h is the magnetic field pointing in the positive z direction, and the double summation is over nearest neighbor lattice sites $\mathbf{r}$ and $\mathbf{r}'$.

The definition of the resistance we used incorporates the simplest possible case when different clusters are allowed to cross. That means that we place resistors between each pair of magnetic moments $\mathbf{m}(\mathbf{r})$ that

are nearest or next-nearest neighbors. In this particular case, the percolation threshold T_p and the Curie temperature T_C happen to coincide [15]. The resistance definition incorporates the halfmetallic character in the following way: the charge carriers with spin up are conducting in the domains having $\mathbf{m}(\mathbf{r})$ pointing in the positive z-direction, the charge carriers with spin down are conducting in domains with $\mathbf{m}(\mathbf{r})$ pointing along $-z$. In the other cases, that is, on the isolating bonds, the resistances are infinite. This is not too bad an approximation; the resistance on the isolating bonds in realistic models will be finite, albeit orders of magnitude higher than on the conducting bonds. The effect of this will be that the phase transition is somewhat smeared out.

We performed MC calculations on this system using the standard Metropolis algorithm. The conductance is calculated using the multigrid method [16].

Now we consider the following features of CMR, as observed from experiments, and compare them with the results of our simulations:

(1) The peak in the resistance in CMR experiments is observed close to or at the Curie temperature T_C.

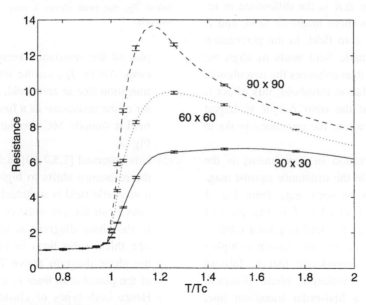

Fig. 2. Resistance as a function of temperature for different system sizes at zero field. The system sizes are indicated in the plot, the resistance is scaled such that the minimal resistance (all bonds having $R = 1$) is unity. The lines are guides to the eye. For larger systems, the peak shifts to T_C, becomes infinitely high but remains asymmetric. For $T < T_C$ there is a rapid convergence to the minimal (unit) resistance.

According to the phase diagram, the percolation threshold T_P lies, for all directions of $\mathbf{m(r)}$, close to or at T_C. That means that clusters with a low resistance are very sparse at that point, if present at all, yielding a high overall resistance. In fact, in any other domain of the phase diagram the resistance is much lower, as at least one of the directions of $\mathbf{m(r)}$ lies well into its percolating regime.

(2) The shape of the curve of the resistance against temperature is experimentally observed to be asymmetric [1,7,8,10]. At temperatures below T_p the resistance drops much more rapidly than for temperatures above T_p. Again this follows from the phase diagram, as below the threshold T_p one of the directions of $\mathbf{m(r)}$ is well into its percolating regime, whereas above T_p all directions remain close to their percolation line (the thick, solid line in Fig. 1). Our MC simulations confirm this: we calculated the resistance for temperatures around T_C and for several system sizes, yielding the plot in Fig. 2.

(3) Experiments show that, for all temperatures, switching on a magnetic field lowers the resistance. This is exhibited in the magnetoresistance ratio $\Delta R/R$, where ΔR is the difference in resistance with and without magnetic field, and R is the resistance at zero field. In the percolation language, the magnetic field tends to align the moments $\mathbf{m(r)}$ and thus enhances the percolation. Close to the percolation threshold, this effect is strongest, such that the ratio $\Delta R/R$ is largest at the temperature where the resistance peaks in zero field.

(4) A particular observation in experiments is the shape of the curve in the resistance against magnetic field. As can be seen, e.g., from Fig. 1 in [4], Fig. 5 in [5], and Fig. 5 in [6], plots of the resistance versus the field display a cusp at temperatures below T_C but are smooth at higher temperatures. This remarkable feature follows directly from the percolation phase diagram: below T_p there is a first-order transition line. Moving away from this line in both directions of the field immediately enhances percolation and hence decreases the resistance, meaning that a

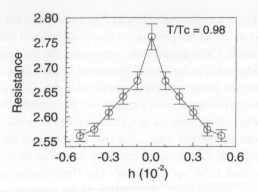

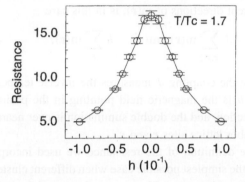

Fig. 3. The resistance as a function of the external magnetic field for temperatures just below T_C (upper plot) and above T_C (lower plot). The system has 60×60 spins. For temperatures below T_C, the peak shows a cusp, whereas above T_C it is smooth.

plot of the resistance versus the field shows a cusp. Above T_C, on the other hand, there is no transition line at zero field, and hence the behavior of the resistance as a function of the magnetic field is smooth. MC simulations confirm this; see Fig. 3.

(5) It is observed [1,5,7,10] in CMR that the peak in the resistance shifts to higher temperature when a magnetic field is switched on. This directly follows from the presence of the percolation lines in the phase diagram at temperatures above T_p (the thick, solid lines in Fig. 1). The region in the phase diagram above T_p shows the vicinity of the percolation lines at a small magnetic field. Hence both types of clusters are close to their percolation threshold. Below this temperature, however, the magnetic field pushes one of the directions $\mathbf{m(r)}$ well into its percolating regime,

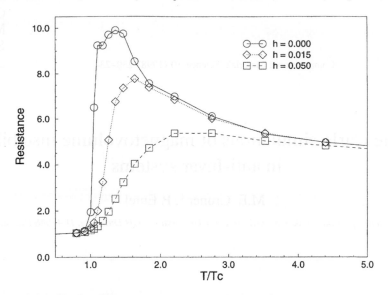

Fig. 4. The resistance around $T = T_C$ for different external magnetic fields h, for a system of 60×60 spins. The peak in the resistance is seen to decrease and shifts to higher temperatures.

thereby immediately lowering the resistance. MC simulations confirm this; see Fig. 4.

The above comparison with experiments shows that the qualitative features of CMR are reproduced by a single and simple mechanism, above as well below the Curie temperature, with as well as without magnetic field. On the other hand, the model we used for illustrating the mechanism cannot, regarding its extreme simplicity, be expected to reproduce precise, numerical values of involved temperatures, fields, and resistances. The concept of universality in statistical physics, however, assures that certain so-called universal features of the mechanism, such as critical exponents and scaling functions, do not depend on the details of the system but only on general features like symmetries and dimensionality. Quantitative tests of the mechanism thus should focus on these universal features, in particular on the critical exponent describing the behavior of the resistance in the neighborhood of the Curie temperature. Work on this is in progress.

We thank Rob de Groot and Peter de Boer for drawing our attention to the problem of CMR and for enlightening discussions; Alan Sokal for providing us with the code for the resistance calculations, which saved us a lot of work and Erik Luijten for discussions on the Monte Carlo part.

References

[1] R.M. Kusters, J. Singleton, D.A. Keen, R. McGreevy, W. Hayes, Physica B 155 (1989) 362.
[2] K. Chahara, T. Ohno, M. Kasai, Y. Kozono, Appl. Phys. Lett. 63 (1993) 1990; S. Jin, T.H. Tiefel, M. McCormack, R.A. Fastnacht, R. Ramesh, L.H. Chen, Science 264 (1994) 413.
[3] R. von Helmolt, J. Wecker, B. Holzapfel, L. Schultz, K. Samwer, Phys. Rev. Lett. 71 (1993) 2331.
[4] M. McCormack, S. Jin, T.H. Tiefel, R.M. Fleming, J.M. Phillips, R. Ramesh, Appl. Phys. Lett. 64 (1994) 3045.
[5] R. von Helmolt, J. Wecker, K. Samwer, L. Haupt, K. Bärner, J. Appl. Phys. 76 (1994) 6925.
[6] S. Jin, M. McCormack, T.H. Tiefel, R. Ramesh, J. Appl. Phys. 76 (1994) 6929.
[7] P. Schiffer, A.P. Ramirez, W. Bao, S.-W. Cheong, Phys. Rev. Lett. 75 (1995) 3336.
[8] G.-Q. Gong, C. Canedy, G. Xiao, J.Z. Sun, A. Gupta, W.J. Gallagher, Appl. Phys. Lett. 67 (1995) 1783.
[9] H.L. Ju, J. Gopalakrishnan, J.L. Peng, Qi Li, G.C. Xiong, T. Venkatesan, R.L. Greene, Phys. Rev. B 51 (1995) 6143.
[10] Y. Shimakawa, Y. Kubo, T. Manako, Nature 379 (1996) 53.
[11] C. Zener, Phys. Rev. 82 (1951) 403; P.-G. de Gennes, Phys. Rev. 118 (1960) 141.
[12] W.E. Pickett, D.J. Singh, Phys. Rev. B 53 (1996) 1146.
[13] P.K. de Boer, H. van Leuken, R.A. de Groot, T. Rojo, G.E. Barberis, Solid State Commun. 102 (1997) 621.
[14] H. Müller-Krumbhaar, Phys. Lett. A 50 (1974) 27.
[15] P. Bastiaansen, H. Knops, J. Phys. A 30 (1997) 1791.
[16] R.G. Edwards, J. Goodman, A.D. Sokal, Phys. Rev. Lett. 61 (1988) 1333; J. Goodman, A.D. Sokal, Phys. Rev. D 40 (1989) 2035.

ELSEVIER

Computational Materials Science 10 (1998) 230–234

COMPUTATIONAL
MATERIALS
SCIENCE

Monte Carlo simulations of magnetovolume instabilities in anti-Invar systems

M.E. Gruner*, P. Entel

Theoretische Tieftemperaturphysik, Gerhard-Mercator-Universität – GH Duisburg, D-47048 Duisburg, Germany

Abstract

We perform constant pressure Monte Carlo simulations of a spin-analogous model which describes coupled spatial and magnetic degrees of freedom on an fcc lattice. Our calculations qualitatively reproduce magnetovolume effects observed in some rare earth manganese compounds, especially in the anti-Invar material YMn_2. These are a sudden collapse of the magnetic moment which is connected with a huge volume change, and a largely enhanced thermal expansion coefficient. Copyright © 1998 Elsevier Science B.V.

Keywords: Anti-Invar effect; High moment–low moment transitions; Rare earth transitions metal compounds

1. Introduction

The observation that some transition metal alloys and complexes have an anomalously low or high thermal expansion coefficient is widely referred to as the Invar or anti-Invar effect, respectively. Since the discovery of Fe–Ni Invar one century ago, uncounted investigations in this topic have taken place (for recent reviews see [1]). A major advance has been achieved by ab initio band structure calculations (e.g. [2,3]). KKR-CPA calculations of $Fe_{65}Ni_{35}$ [3] reveal two nearly degenerate minima on the binding surface, i.e. the energy as a function of lattice constant and mean magnetic moment. One at a large lattice constant and a large magnetic moment (high moment or HM state) and another at a smaller lattice constant and a smaller

or vanishing magnetic moment (low moment or LM state). In this case, thermal excitation of the LM state can lead to a compensation of the volume expansion of the material. This scenario reminds of the phenomenological $2\text{-}\gamma$-states model [4] introduced by Weiss in order to explain Fe–Ni Invar. However, most ab initio calculations are performed for zero temperature only. So the comparison with experimental data is hampered by the fact that the extrapolation of these results to finite temperatures is connected with severe methodological problems. So far mainly continuous Ginzburg–Landau like spin models in combination with a Gaussian fluctuation theory have been used to achieve this goal [5]. Another approach is to use a Weiss-type model with its parameters adapted to ab initio results. This has been done for the classical Invar system Fe–Ni, and although we neglected the itinerant character of magnetism by using localized spins, we were able to reproduce qualitatively the major magnetovolume effects of this alloy [6].

---

* Corresponding author. Tel.: +49-203-379-3564; fax: +49-203-379-2965; e-mail: me@thp.uni-duisburg.de.

0927-0256/98/$19.00 Copyright © 1998 Elsevier Science B.V. All rights reserved
PII S0927-0256(97)00145-6

However, it has still to be proven, whether this approach can be applied to other materials, especially to anti-Invar systems. One of the most prominent anti-Invar materials is the Laves phase compound YMn_2. It occurs in the C15 structure with a frustrated antiferromagnetic (AF) spin order in the Mn sublattice. Around $T_N \approx 100$ K YMn_2 undergoes a first order phase transition into a paramagnetic state, accompanied by a huge volume contraction of 5% and a large thermal hysteresis of 30 K. Above T_N a large thermal expansion coefficient of 50×10^{-6} K^{-1} is encountered [7]. LMTO calculations reveal a nonmagnetic solution at a lower volume with a slightly higher energy than the AF ground state [8].

Within the scope of this work, we will show that for a suitable choice of parameters the large abrupt volume change at T_N as well as the enhanced thermal expansion of YMn_2 can be qualitatively reproduced by a Weiss-type model assuming a HM–LM instability of the Mn atom.

2. The model

Our model Hamiltonian consists of a magnetic part H_m describing the magnetic properties of the system and a vibrational part H_v responsible for the magnetovolume coupling:

$$H = \underbrace{D \sum_i S_i^2 + J \sum_{\langle i,k \rangle} S_i S_k}_{H_m} + \underbrace{\sum_{\langle i,k \rangle} U(r_{ik}, S_i, S_k)}_{H_v}.$$

(1)

H_m is a spin-1 Ising Hamiltonian, also known as Blume–Capel model. We identify the spin states $S_i = 0$ with the atomic LM state, whereas $S_i = \pm 1$ refers to the HM state. D denotes the crystal field, which imposes an energetic separation between HM and LM states; J is the magnetic exchange constant. The brackets indicate a summation over nearest neighbors. We did not consider a distance dependence of J so far, since we believe its influence to be small compared to the effect observed. Furthermore, compressible frustrated Ising antiferromagnets show additional effects like tetragonal lattice distortions

[9], that are not covered by experimental data in our case. The second part H_v introduces pair interactions between the atoms, which depend on their distance r_{ik} and on their spin states. If both the interacting atoms are in the HM state, we employ HM potentials that are characterized by a larger equilibrium distance than the LM potentials we use otherwise. So a system with a considerable concentration of LM atoms is supposed to have a smaller lattice constant than a pure HM system. For the sake of simplicity we choose Lennard–Jones interactions:

$$U(r_{ik}, S_i, S_k)$$
$$= \begin{cases} 4\epsilon_L \left(\left(\dfrac{d_L}{r_{ik}} \right)^{12} - \left(\dfrac{d_L}{r_{ik}} \right)^6 \right), & S_i S_k = 0, \\[2ex] 4\epsilon_H \left(\left(\dfrac{d_H}{r_{ik}} \right)^{12} - \left(\dfrac{d_H}{r_{ik}} \right)^6 \right), & S_i S_k \neq 0, \end{cases}$$

(2)

where $\epsilon_{L,H}$ denote the energy at the equilibrium neighbor distances $r_{ik} = 2^{1/6} d_{L,H}$ for LM and HM potentials, respectively.

3. The simulation

Our Monte Carlo routine consists of two local and one global update steps. For each lattice site we choose a new spin state $S_i \in \{0, \pm 1\}$ using the Metropolis criterion. Afterwards a new trial position is elected out of a cube around the old position. Again, the new position is accepted with the probability $\max(1, \exp(-\beta \Delta H))$, where β is the inverse temperature. The size of the cube is given by the condition that about half of the propositions are to be accepted in order to improve convergence. When all atoms have been updated, the volume of the complete lattice is adapted by another Metropolis step, except that now the quantity $\Delta \mathcal{H} = \Delta H - N k_B T \ln(V'/V)$ has to be considered, which accounts for the difference in translational entropy caused by the change of the volume from V to V' (for more information on constant pressure Monte Carlo methods see [10]). Our calculations were performed on a $8^3 \times 4 = 2048$ site fcc lattice with periodic boundary conditions at constant pressure $P = 0$. 80 000 lattice sweeps were performed for each temperature. The first 40 000 lattice sweeps were used to allow the system to reach equilibrium, afterwards every 10th sweep data were gathered for the statistics.

Although YMn₂ occurs in the C15 structure, we chose instead an fcc lattice in our simulations. This allows us to omit the Y species, which we do not expect to play an important role in the thermodynamics of the HM–LM transition. Furthermore we can easily relate our results to previous investigations [6]. Since the fcc Ising antiferromagnet has also a frustrated ground state spin structure as it is assumed for YMn₂, simulating an fcc structure is not a serious restriction. We used $d_H = 2.432$ Å and $d_L = 2.397$ Å for the interatomic spacing between the Mn atoms in AF and paramagnetic YMn₂. We chose $\epsilon_L = \epsilon_H = 30.86$ mRy in order to achieve a reasonable elastic behavior. The exchange constant was set to $J = 0.459$ mRy corresponding to a Néel temperature $T_N = 126.5$ K [11]. The crystal field D was varied in the range $D = 0, \ldots, 3J$. We examined the energy per lattice site $\langle e \rangle$, the mean magnetic moment $\langle S^2 \rangle = \langle 1/N \sum_i S_i^2 \rangle$, the atomic volume $\langle V \rangle$ and the linear thermal expansion coefficient $\alpha = 1/(3V) \, dV/dT$ (brackets denote thermal averaging). Additionally, for several values of D, we approximated ground state energy and average magnetic moment as a function of the lattice constant by exponentially cooling down the system to $T = 4$ K at constant volume.

4. Results

For $D < 2J$, the ground state of the system is an AF HM state as can be seen from simple energetic considerations concerning Hamiltonian (1). Correspondingly, for $D > 2J$, the ground state turns into a nonmagnetic LM state with a smaller volume. This is also verified by the plots of low temperature energy and magnetic moment vs. lattice constant shown in Fig. 1.

Disallowing the LM spin state ($D = -\infty$), around $T_N = 127$ K a first order phase transition occurs from a type 1 AF state to a paramagnetic state. This transition can be observed as a finite jump in the

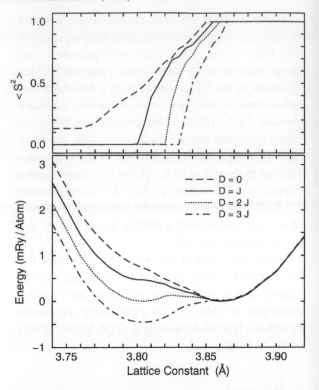

Fig. 1. Energy per lattice site and average magnetic moment at $T = 4$ K as a function of the lattice constant.

energy per lattice site (Fig. 2). Enlarging D allows thermal excitations of the $S = 0$ spin state, leading initially to a continuous decrease of HM atoms with increasing temperature, which appears in our results for $D = 0$ (Fig. 3). Due to the magnetovolume coupling introduced by H_v, this is connected with a continuous decrease of the volume and, correspondingly, a reduction of the thermal expansion coefficient, as depicted by Fig. 4. The AF phase transition is hardly affected by the dilution of the magnetic system and remains of first order; only T_N is slightly reduced. We now find a finite jump in $\langle e \rangle$ at $T_N = 124$ K. An investigation of the probability distribution (not shown) of the internal energy at T_N exhibits the two maxima structure that is typical for first order phase transitions.

For $D = J$, we find a sudden collapse of the magnetic moment at $T_s = 127$ K which is connected with an abrupt decrease of the volume of about 3%. The magnetic phase transition takes place a few Kelvin

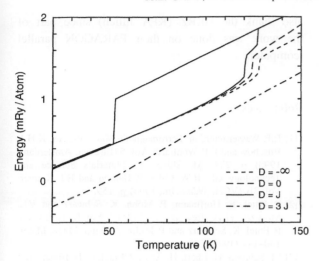

Fig. 2. Energy per lattice site for various values of D as a function of the temperature.

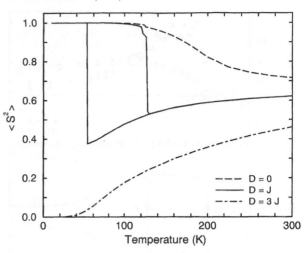

Fig. 3. Temperature dependence of the average magnetic moment for various values of D.

below at $T_N = 120$ K. This situation is reflected in a double jump of the internal energy. Upon cooling down, a large thermal hysteresis is encountered: The system has to be cooled down to $T'_N = 53$ K before it jumps back into the AF HM state. Responsible for this hysteresis is that lattice strain imposes a large free energy barrier between HM and LM states which is unlikely to be surmounted by our constant pressure Monte Carlo algorithm. However, the large energy barrier is suppressed in Fig. 1, since at constant volume the system is forced into a mixture of two coexisting HM and LM phases. Similar first order transitions with a large thermal hysteresis have also been observed in simulations of ferromagnetic systems; for a detailed discussion we refer to [6]. Above T_s the mean magnetic moment is considerably lower than the high temperature limit $\langle S^2 \rangle = 2/3$ of a spin-1 Ising model. So with increasing temperature more atoms will be excited to the HM state, leading to an enhanced thermal expansion. This process is further encouraged with increasing temperature, because the displacement of the atoms due to lattice vibrations and the difference between HM and LM neighbor distances become comparable. This levels out the energetic difference between $S_i = 0$ and $S_i = +1$ states introduced by H_v. For the same reason, discontinuous magnetovolume effects do not occur for smaller values of D: At higher

temperatures, the HM and LM minima of the free energy are smeared out by lattice vibrations.

Above $D = 2J$, the ground state of the system is nonmagnetic. So HM–LM transitions as well as magnetic phase transitions do not appear. However, we observe an enhanced thermal expansion coefficient, due to excitation of HM atoms.

5. Discussion

For $D = J$, our model shows magnetovolume effects similar to the properties of YMn$_2$ mentioned in Section 1: An abrupt collapse of the mean magnetic moment connected with a decrease of the volume of several percent and a largely enhanced thermal expansion coefficient. Lowering D produces continuous magnetovolume effects. This resembles the situation in Y(Mn$_{1-x}$Al$_x$)$_2$, where small concentrations of Al destroy the first order transition and anti-Invar behavior [12]. With increasing Al content the magnetovolume effects gradually vanish. This is due to the enlargement of the lattice constant by the Al atoms which makes it energetically more difficult for the Mn atoms to switch to the LM state. On the other hand, substituting a few percent of the Y atoms with the smaller Sc has the opposite effect: The Mn atoms are

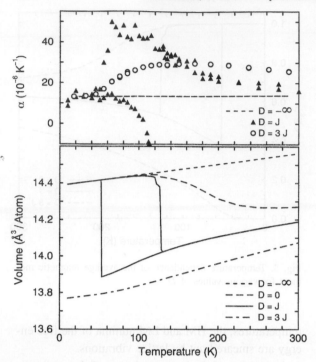

Fig. 4. Temperature dependence of the volume and the linear thermal expansion coefficient for various values of D.

nonmagnetic at $T = 0$, the Mn moment increases with temperature, leading to an enhanced thermal expansion coefficient [13]. This is the situation we observed for $D > 2J$.

Acknowledgements

This work has been supported by the Deutsche Forschungsgemeinschaft through the SFB 166. We also thank the HLRZ (KFA Jülich) since part of the work was done on their PARAGON parallel computer.

References

[1] E.F. Wassermann, in: Ferromagnetic Materials, eds. K.H.J. Buschow and E.P. Wohlfarth, Vol. 5 (Elsevier, Amsterdam, 1990) p. 237; M. Shiga, in: Materials Science and Technology, eds. R.W. Cahn, P. Haasen and E.J. Kramer, Vol. 3B (VCH, Weinheim, 1994) p. 159.

[2] P. Entel, E. Hoffmann, P. Mohn, K. Schwarz and V.L. Moruzzi, Phys. Rev. B 47 (1993) 8706; E. Hoffmann, P. Entel, K. Schwarz and P. Mohn, J. Magn. Magn. Mater. 140–144 (1995) 237.

[3] M. Schröter, H. Ebert, H. Akai, P. Entel, E. Hoffmann and G.G. Reddy, Phys. Rev. B 52 (1995) 188.

[4] R.J. Weiss, Proc. Roy. Phys. Soc. (London) 82 (1963) 281.

[5] P. Mohn, K. Schwarz and D. Wagner, Phys. Rev. B 43 (1991) 3318; M. Uhl and J. Kübler, Phys. Rev. Lett. 77 (1996) 334, and references herein.

[6] M.E. Gruner, R. Meyer and P. Entel, Europ. Phys. J. B, to be published.

[7] M. Shiga, Physica B 149 (1988) 293; R. Ballou, J. Deportes, R. Lemaire, Y. Nakamura and B. Ouladdiaf, J. Magn. Magn. Mater. 70 (1987) 129, and references herein.

[8] S. Asano and S. Ishida, J. Magn. Magn. Mater. 70 (1987) 39.

[9] L. Gu, B. Chakraborty, P.L. Garrido, M. Phani and J.L. Lebowitz, Phys. Rev. B 53 (1996) 11 985.

[10] M.P. Allen and D.J. Tildesley, Computer Simulation of Liquids (Clarendon Press, Oxford, 1987).

[11] D.F. Styer, Phys. Rev. B 32 (1985) 393.

[12] M. Shiga, H. Wada, H. Nakamura, K. Yoshimura and Y. Nakamura, J. Phys. F 17 (1987) 1781.

[13] M. Shiga, H. Wada, Y. Nakamura, J. Deportes, B. Ouladdiaf and K.R.A. Ziebeck, J. Phys. Soc. Jpn. 57 (1988) 3141.

COMPUTATIONAL
MATERIALS
SCIENCE

Computational Materials Science 10 (1998) 235–239

Preliminary investigations of weak non-adiabatic effects in materials from simulations on model clusters

Didier Mathieu [a],*, Philippe Martin [b]

[a] *CEA – Le Ripault, DXPL/CPX, Laboratoire d'Ingénierie Moléculaire, BP16, 37260 Monts, France*
[b] *CISI, Aéropôle, Im. Rafale B, 1 rue Charles Lindbergh, 44340 Nantes-Bouguenais, France*

Abstract

According to earlier work, excited electronic states, despite their very small concentrations in wide-gap systems, might be involved in the initiation of energetic materials. To get further insight into the non-adiabatic electronic behaviour under intense mechanical perturbations, a simple approach is suggested, using finite clusters as models for materials. Preliminary calculations are carried out for isolated clusters undergoing vibrations. Electronic populations in excited states obtained from these simulations are much larger than expected on the basis of statistical mechanics. Therefore, an extension of the present scheme to the canonical NVT ensemble would be desirable. Copyright © 1998 Elsevier Science B.V.

Keywords: Non-adiabatic simulations; Molecular dynamics; Impact sensitivity

1. Introduction

Impact sensitivity is a long-standing problem in the field of energetic materials. Progress is hampered by the formidable problems associated with probing at the molecular level such fast physico-chemical phenomena in a condensed system. On the other hand, the complexity of the processes involved precludes realistic simulations on real systems [1]. Therefore, further insights could be gained from computer experiments on simple models. Previous experience in other fields shows that such approaches can reveal interesting relationships between macroscopic properties and atomic or molecular features [2].

These relationships are very useful to build empirical models, similar to those employed so far to estimate impact sensitivities of energetic materials [3]. Indeed, for such a predictive scheme to be used safely, every contribution to initiation must be well understood and included through descriptors associated with it. On the other hand, in some attempts to rationalize observed impact sensitivities, it was suggested that excited electronic states should be considered [4]. Recently, the role of excited states has been revisited in order to explain a set of observations not consistent with standard theories, taking into account the closure of the band gap at very high densities [5]. However, because extreme conditions are required for such a metallization, it can hardly be invoked with regard to the initiation process. Nonetheless, the number of electrons in the conduction band can be enhanced as soon as the gap is reduced, as a result of non-adiabatic

* Corresponding author. Tel.: +33 2 47 34 41 85; fax: +33 2 47 34 51 42; e-mail: mathieu@ripault.cea.fr.

processes. In other words, a complete closure of the band gap is not necessary.

Should excited states play a major role in initiation, it is clear that impact sensitivity would directly depend on parameters associated with non-adiabatic transitions, whereas only static parameters such as the density of states or the band gap energy E_g at equilibrium are considered in searches for relationships between electronic features and impact sensitivity [1,6]. Therefore, further insight into non-adiabatic effects and their dependence on the characteristics of the materials would be welcome to elaborate better models of impact sensitivity.

In view of such investigations, a computer code to simulate the weakly non-adiabatic electron dynamics above the ground state Born–Oppenheimer surface resulting from a mechanical perturbation is currently being implemented in our group. Recently, several approaches have been proposed to simulate the electron dynamics including non-adiabatic transitions [7,8]. Useful semi-classical methods for condensed systems are reviewed by Coker [9]. With respect to established surface-hopping methods [9], the present model has to be more efficient, because every valence electron of the material should be dealt with. On the other hand, in view of the wide gap of energetic materials, the drift of the non-adiabatic wave function above the ground state potential energy surface remains small in any case: taking advantage of this situation affords a dramatic simplification because only trajectories on the ground state energy surface need to be considered.

A detailed discussion of the physics of the non-adiabatic electron dynamics is beyond the scope of this article. In Section 2, a general view of the model is provided, with an emphasis on the many approximations involved. Then, in Section 3, preliminary simulations are carried out for vibrationally excited atomic clusters, showing the fluctuations of the electronic population in the conduction band due to the coupling between the electronic and ionic subsystems. Moreover, rough estimates show that the average population in the conduction band is much larger than expected on the basis of statistical mechanics.

2. Overview of the model

The model is conceptually very simple. As in other semi-classical approaches [9,10], we consider classical nuclei surrounded by quantum electrons. The time-dependent behaviour of the ions is described by the Newton equation

$$M\ddot{X} = -\nabla V_{NN} + F_{e \to N}, \tag{1}$$

where X stands for ionic coordinates, M the matrix with the atomic masses, V_{NN} the internuclear repulsion potential and $F_{e \to N}$ the force on the ions arising from the electrons. In the adiabatic case, the electronic wave function is supposed to be, at any time, one of the eigenstates Ψ_k of the electronic hamiltonian H. In contrast, allowing for non-adiabatic effects requires to solve the time-dependent Schrödinger equation (TDSE) for the dynamic wave function Ψ. The lowest Hartree–Fock orbitals ϕ_k are used to build the ground state Ψ_0. The orbital approximation is also used to represent the dynamic wave function Ψ, using so-called dynamic orbitals φ_μ. Both sets of orbitals ϕ_k and φ_μ are derived from equations involving an effective one-electron hamiltonian. For the time being, the latter is described at the EHT semi-empirical level [11]. The initial conditions are such that the electrons start from the ground state: $\varphi_\nu(t=0) = \phi_k(t=0)$.

To solve the Newton equation (1), there remains to specify the forces $F_{e \to N}$ exerted on the ions by the electrons. This can be done e.g. using an empirical scaling of ionic velocities [9] or a generalized Hellmann–Feynman theorem [10]. However, as stated in Section 1, the non-adiabatic contribution to $F_{e \to N}$ can be neglected in the present case (validation of this approximation against simulations using velocities scaling will be presented elsewhere). This makes possible to employ an empirical potential for any small drift of Ψ from the ground state.

In principle, electronic populations in the eigenstates Ψ_k could be obtained at any time from the squared amplitudes of $\langle \Psi_k | \Psi \rangle$, but this would involve accurate descriptions of excited states obtained through interaction of many excited configurations. In the present context, since we are not interested in the instantaneous excited states, it is in fact more

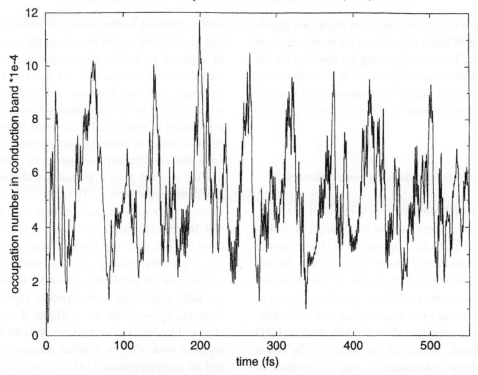

Fig. 1. Fluctuations of the average occupation number n_C in the conduction band: simulation results for a homoatomic cluster (five atoms) with vibrational interatomic frequency $1500\,cm^{-1}$ and band gap $2\,eV$. Other relevant parameters are the atomic mass (6 amu) and the ratio $D/r = 3$ of the equilibrium interatomic distance to the atomic orbital radius. Starting with every electron in the valence band ($n_C = 0$) this figure illustrates an equilibration stage (i.e. a simulation with no impact) at 300 K.

natural and much simpler to define the fraction of electrons promoted into the conduction band under the influence of the perturbation. To this aim, the electron carried by each occupied dynamic orbital φ_μ is shared between the static orbitals ϕ_k taking advantage of the normalization of the N dynamic orbitals and the closure relation for the set of static orbitals. Since

$$N = \sum_\mu \langle \varphi_\mu | \varphi_\mu \rangle = \sum_{\mu,k} \langle \varphi_\mu | \phi_k \rangle \langle \phi_k | \varphi_\mu \rangle$$

$$= \sum_k \left(\sum_\mu |\langle \phi_k | \varphi_\mu \rangle|^2 \right), \tag{2}$$

the natural choice to define the population in the eigenstate ϕ_k is

$$n_k = \sum_\mu |\langle \phi_k | \varphi_u \rangle|^2. \tag{3}$$

This looks like a Mulliken analysis. However, while the latter shares the electrons of the molecular

orbitals (MO) between the atomic orbitals (AO), in the present scheme the electrons of the dynamic orbitals are shared between the static ones (actually the MO). Because the differences between orbitals in a given energy band stem only from the wave vector, it is more convenient to consider the band as a whole. Thus, we define the average occupation number of electronic levels in the valence and conduction bands, respectively,

$$n_V = \frac{1}{N} \sum_{k \in VB} n_k$$

and

$$n_C = \frac{1}{N} \sum_{k \in CB} n_k = 1 - n_V. \tag{4}$$

The populations n_V or n_C allow to assess how relevant non adiabatic effects might be. Whenever n_C is significantly increased after a perturbation, it is reasonable to assume that excited states might be involved in the response of the material. This requires a significant increase of ionic velocities with respect to their thermal values. Indeed, expressing the equation for the dynamic orbitals in the basis set of the eigenstates ϕ_k, it is easy to realize that the transitions arise because of non-adiabatic coupling coefficients involving ionic velocities [9].

3. Preliminary calculations on small clusters

In view of preliminary investigations of the electronic response to very intense mechanical perturbations such as impacts or shockwaves, 1D linear atomic clusters are considered. Such 1D systems proved valuable models in previous studies of the electronic structure of materials [12]. They provide convenient tools for preliminary test calculations and validation. Presently, our computer code deals only with such 1D clusters as simple models of the material. The atoms treated quantum mechanically can be embedded within classical buffer atoms, in order to get a realistic description of the ions dynamic at reduced cost. In contrast to buffer atoms, quantum atoms carry gaussian atomic orbitals in 1D, used as a basis set to represent static and dynamic orbitals ϕ_k and φ_μ of the model cluster. In the simulations reported here, s and p gaussian AOs are used to describe valence and conduction states, respectively. Any simulation is started from the equilibrium configuration, with initial velocities selected from a Boltzmann distribution at the specified temperature T. To simulate the compression wave, a flyer plate is used, in line with previous 2D and 3D simulations [13]. This flyer plate is launched from the left onto the model cluster with an impact velocity V. Preliminary calculations have been carried out in view of validating the code. Results obtained are fully consistent with physical expectations. For instance, $n_C \to 0$ as $V \to 0$ and $T \to 0$. Results obtained under impact will be presented elsewhere, together with a more detailed account of the model and its implementation [14].

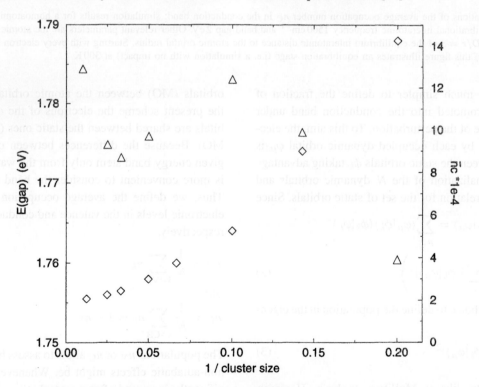

Fig. 2. Convergence against cluster size of E_g (diamonds) and $\bar{n}_C$ (triangles) for atomic clusters. Parameters of the atoms are the same as in Fig. 1.

In this paper, we focus on the preliminary equilibration phase before the impact is applied. The fluctuations of the average occupation number n_C in the conduction band during this stage are illustrated in Fig. 1. They exhibit two typical timescales, the fast one being associated with the electrons and the other with ionic motions. A time averaged excited state population, denoted $\bar{n}_C$, is obtained from a time average of n_C over sufficiently long simulations. In practice, simulations were carried out at $T = 300\,\text{K}$ for several vibrational periods of the model clusters, in order to get rid of the transient drift away from the initial equilibrium state. Then, $\bar{n}_C$ was roughly estimated as an average over one or two vibrational periods. Values obtained for $\bar{n}_C$ and the gap energy E_g are plotted in Fig. 2 against the inverse cluster size $1/N_a$.

While E_g can be easily extrapolated to the limit $N_a \to \infty$, this is less true for $\bar{n}_C$, presumably because of insufficient sampling for the time average. However, the magnitude of $\bar{n}_C$ remains in any case $\sim 10^{-4}$. In view of the band gap value $E_g = 2\,\text{eV}$ selected, this is clearly much larger than thermal populations derived for real materials from statistical mechanics ($\sim 10^{-15}$ according to the Fermi–Dirac statistics). If values consistent with the thermal equilibrium electronic distribution at $300\,\text{K}$ were required, the present scheme should be extended to the canonical ensemble, i.e. the influence of a thermal bath on the system should be taken into account.

On the other hand, Fig. 2 suggests that $\bar{n}_C$ increases with the cluster size: this is consistent with the band gap reduction associated with increased electronic delocalization, which appears clearly in the figure. Therefore, the model should allow to study enhanced electronic excitations in high density regions. The systematic overestimation of $\bar{n}_C$ may not preclude a priori some qualitative conclusions to be drawn from comparative calculations on several model clusters.

Acknowledgements

Discussions with A. Llor, J.-P. La Hargue, Ph. Simonetti and Ph. Martin (CEA-Saclay, author of Ref. [12]) are gratefully acknowledged.

References

[1] A.B. Kunz, Phys. Rev. B 53 (1996) 9733.
[2] D.D. Vvedenski, S. Crampin, M.E. Eberahrt and J.M. MacLaren, Contemp. Phys. 31 (1995) 73.
[3] H. Nefati, J.-M. Cense and J.-J. Legendre, J. Chem. Inf. Comp. Sci. 36 (1996) 804.
[4] A. Delpuech and J. Cherville, Propellants Explos. 3 (1978) 169.
[5] J.J. Gilman, Chem. Propulsion Inf. Agency 589 (1992) 379.
[6] A.B. Kunz, Mat. Res. Soc. Symp. 418 (1996) 287.
[7] E. Deumens, A. Diz, R. Longo and Y. Öhrn, Rev. Modern Phys. 66 (1994) 917.
[8] V.S. Filinov, Mol. Phys. 88 (1996) 1517.
[9] D.F. Coker, in: Computer Simulation in Chemical Physics, eds. M.P. Allen and D.J. Tildesley (Kluwer Academic Publishers, Dordrecht, 1993).
[10] R.E. Allen, Phys. Rev. B 50 (1994) 18 629.
[11] J.-L. Rivail, Eléments de Chimie Quantique à l'usage des chimistes (Interéditions/CNRS, Paris, 1989).
[12] P. Martin, J. Phys. B 29 (1996) L635.
[13] C.T. White, J.J.C. Barrett, J.W. Mintmire, M.L. Elert and D.H. Robertson, Mat. Res. Soc. Symp. Proc. 418 (1996) 277.
[14] D. Mathieu and P. Martin, in preparation.

ELSEVIER

COMPUTATIONAL
MATERIALS
SCIENCE

Computational Materials Science 10 (1998) 240–244

The electronic structure of colossal magnetoresistive manganites

P.K. de Boer *, R.A. de Groot

ESM, Faculty of Sciences, Toernooiveld 1, 6525 ED, Nijmegen, Netherlands

Abstract

The ferromagnetic manganites $A_{1-x}B_x MnO_3$ (A a trivalent element and B divalent) have been the subject of intensive study in the past few years. These manganites exhibit colossal magnetoresistance (CMR), i.e. their resistance can drop several orders of magnitude under influence of an external magnetic field. Electronic structure calculations show that these manganites are half-metallic: they are conducting for one spin direction exclusively. The possible relation between half-metallic magnetism and colossal magnetoresistance is discussed. Copyright © 1998 Elsevier Science B.V.

There is considerable interest in large negative magnetoresistance. Early developments were in the area of metallic multilayers [1]. These multilayers consist of alternate stacking of ferromagnetic and non-magnetic metals. The exchange coupling between the ferromagnetic layers takes place through the non-magnetic layers. It can be tuned by the variation of the thickness of the non-magnetic layer to be (weakly) antiferromagnetic.

The direction of the magnetization of the magnetic layers can be forced to be parallel by an external magnetic field. A large reduction in the resistance results. This phenomenon was called "giant" magnetoresistance. While the details of the explanation of the giant magnetoresistance is still an area of active research, it is worthwile remarking here that the asymmetry of the electronic structure for the two spin directions at the Fermi energy for the magnetic layer is a crucial ingredient in any theory. The largest asymmetry possible

is found in the so-called half-metallic ferromagnets, systems which are metallic for one spin direction exclusively [2].

Considerable larger negative magnetoresistances are found in the quaternary manganites $A_{1-x}B_x MnO_3$ where A stands for a trivalent metal (La, rare earth) and B is a divalent metal (Ca, Sr, Ba) [3]. Magnetoresistances of $10^6\%$ [4] and higher have been reported ("colossal" magnetoresistance). This paper reports on the electronic structure of one of these materials and the possible relation with the colossal magnetoresistance.

We focus on the compound $La_{1-x}Ca_x MnO_3$, which has been studied extensively experimentally. The magnetic ordering of $La_{1-x}Ca_x MnO_3$ at low temperatures depends on the doping. $La_{1-x}Ca_x MnO_3$ is an antiferromagnetic insulator for $x < 0.2$ and $x > 0.5$, while it is a ferromagnet for $0.2 < x < 0.5$. In and near the ferromagnetic regime the colossal magnetoresistances occur most strongly. Manganites with other elements A and B sometimes have a more complex phase diagram. But in general they all have insulating end points $x = 0$ and $x = 1$ (though not always

* Corresponding author. Tel.: +31-243652810; fax: +31-243652120; e-mail: peterdb@tvs.kun.nl.

antiferromagnetic but sometimes with a canted spin structure) and a ferromagnetic intermediate regime.

The crystal structure of colossal magnetoresistive manganites is basically a simple perovskite. However, depending on the specific elements A and B the perovskite structure can be distorted, leading to a wide variation in reported crystal structures such as cubic, orthorhombic, tetragonal and rhombohedral. For instance, $CaMnO_3$ is cubic while $LaMnO_3$ has a strongly distorted pervoskite structure with orthorhombic symmetry. In the intermediate region the structure is either an undistorted (cubic) or slightly distorted (orthorhombic) perovskite.

A special case is the $x = 0.5$ compound, for several reasons. First of all, it has a complex temperature dependent magnetic behaviour [5]. $La_{0.5}Ca_{0.5}MnO_3$ orders ferromagnetically at 220–230 K, but the magnetic ordering changes to antiferromagnetic at 150–180 K. Second, the FM–AFM transition is accompanied by a structural transition. There is some controversy about the exact crystal structure of $La_{0.5}Ca_{0.5}MnO_3$. Some authors reported a cubic crystal [4], others an orthorhombic one [5], the distortion of the perovskite structure being small in the FM phase and somewhat bigger in the AFM phase. And third, for this special value of x charge ordering can occur.

We now present the calculations on $La_{1-x}Ca_xMnO_3$. The doping of La by Ca can – in a calculation – only be performed by considering a unit cell consisting of several formula units $LaMnO_3$ and replacing one or more La atoms by Ca atoms. For general values of the doping x this results in very large unit cells and elaborate calculations. Therefore we concentrate on $La_{0.5}Ca_{0.5}MnO_3$. Keeping in mind the controversy about the crystal structure of this compound the crystal structure is assumed to be cubic. The substitution of La by Ca was performed by taking two cubes of the perovskite $LaMnO_3$ above each other and replacing one La atom by a Ca atom, resulting in a tetragonal unit cell with $c = 2a$. Later we will discuss the influence of the crystal structure on the electronic structure in manganites.

The calculations were performed with the full potential LAPW method [6], i.e., no shape approx-

imations to the potential were assumed. The LAPW method is based upon the density functional theory (DFT), which is in principle exact but needs an approximation to the exchange–correlation energy to become numerically feasible. The generalized gradient approximation (GGA) [7] was used, but a comparison is made with standard calculations within the local density approximation (LDA) [8], in order to indicate the small but important difference in the results between the two approximations in this system. For details on the calculation we refer to [9].

The total density of states (DOS) of ferromagnetic $La_{0.5}Ca_{0.5}MnO_3$ is shown in Fig. 1. The zero energy is the Fermi energy. The bands between 7.5 and 1.5 eV below the Fermi energy have primarily oxygen $2p$ character. Around $1 - 1.5$ eV below the Fermi energy lie the spin up Mn t_{2g} bands. The Fermi energy crosses a broad spin up band of manganese e_g orbitals strongly mixed with oxygen 2p orbitals. There are no spin down states at the Fermi energy, since the Fermi energy lies in a gap of 1.5 eV between the oxygen 2p states and the minority conduction band. This means that $La_{0.5}Ca_{0.5}MnO_3$ is half-metallic: metallic for the majority spin electrons but semiconducting for the minority electrons.

We now emphasize the importance of GGA in this system. An LDA calculation leads to a similar picture concerning the position of the oxygen 2p-states and the manganese 3d-states. The most important difference is the minority conduction band which lies slightly lower in energy in the LDA case as compared with the GGA picture. As a consequence the gap between the oxygen 2p-states and the minority conduction band has decreased to 1.4 eV, and, more importantly, the Fermi energy lies now in the conduction band. This means that the LDA calculations do not result in a half-metallic electronic structure, but just in a metallic one. Previous LDA calculations also showed this metallic picture of the manganites, which was called "nearly" half-metallic [10]. Further improvements like self-interaction correction (SIC) could enhance the half-metallic character, because SIC leads possibly to energetically lower occupied states, an effect which is stronger when more localized the bands are. In the manganites this would lead to a larger band gap and

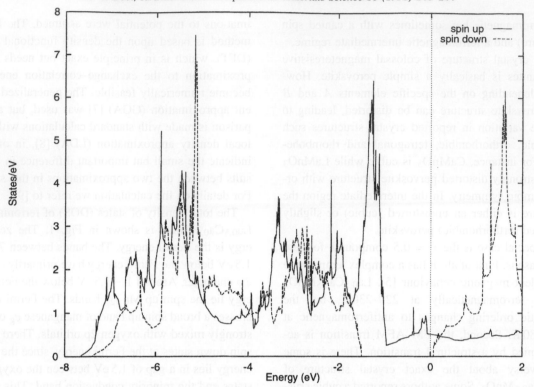

Fig. 1. Total DOS of ferromagnetic La$_{0.5}$Ca$_{0.5}$MnO$_3$ calculated within GGA.

a Fermi energy well below the minority conduction band.

At low temperature La$_{0.5}$Ca$_{0.5}$MnO$_3$ is antiferromagnetic. We calculated A-type AFM La$_{0.5}$Ca$_{0.5}$MnO$_3$ in the same crystal structure as in the FM phase. The total energy of the AFM phase is lower in energy (9 meV per unit cell) than the FM total energy. Though the A-type antiferromagnetic structure is not of the experimentally found type, it is clear that the FM phase is not the ground state, in agreement with experiment.

The total DOS (Fig. 2) shows that AFM La$_{0.5}$Ca$_{0.5}$MnO$_3$ is metallic, in disagreement with the experimentally known transport properties, which indicate that it is insulating. Several aspects can influence the conductivity here. First, while the crystal structure in the FM phase is a perfect or only slightly distorted perovskite, it is more distorted in the AFM phase. Second, the AFM phase is in fact of the complex CE-type [11]. Third, it is well known that at $x = 0.5$ in

the AFM phase of $A_{1-x}B_x$MnO$_3$ charge ordering can occur [12].

It is not unreasonable to expect that ferromagnetic manganites with $x < 0.5$ are also half-metallic. Previously published calculations revealed the electronic structure of $x = \frac{1}{3}$ and $x = \frac{1}{4}$ compounds of La$_{1-x}$Ca$_x$MnO$_3$. They showed ordinary metallic behaviour, but these calculations were performed within LDA. With the Fermi energy lying just at the bottom of the minority conduction band it is expected that improvements over LDA will give truly half-metallic properties. Half-metallic magnetism could give an explanation for the colossal magnetoresistance, as we will see later.

We now address the question of the origin of the half-metallic behaviour in these manganites and the comparison with earlier half-metallic systems. The explanation of the half-metallic behaviour in the first compounds recognized as such (NiMnSb) [2] depends on both the crystal structure and the chemical

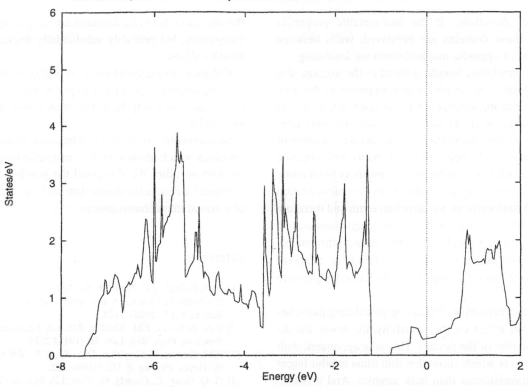

Fig. 2. Total DOS of A-type antiferromagnetic $La_{0.5}Ca_{0.5}MnO_3$.

composition. The crystal structure is pseudo-isostructural with zincblende, while the minority spin direction is iso-electronic with GaAs. The bandstructure, gap, bonding, etc., for the minority spin direction of NiMnSb is typical that of a III–V semiconductor, while for the majority spin the compound behaves like an alloy. Thus the half-metallic properties are quite subtle, depending on both the structure and the composition.

A quite distinct category may exist in the limit of narrow band width materials. In the case of an exchange and crystal field splitting, etc., in access of the band width one can expect half-metallic magnetism to occur frequently. However, whether systems in this limit still conduct or are Mott insulators is questionable. Recent work on the archetype in this category, Fe_3O_4 [13], contradicts earlier predictions of half-metallic properties [14].

A third category of half-metallic materials is in the area of strong magnetic ionic compounds. Strong magnetism is defined by a situation that a hypothetical increase in the exchange splitting does not lead to an increase in magnetic moment, i.e., the majority sub-shell is full or the minority empty. An example of a strong magnetic metal is Ni. But of course Ni shows hardly any spin polarization in the conduction because the mobile 4s, 4p electrons are hardly polarized. But exactly these itinerant electrons are transferred to the electronegative species if one forms an ionic compound, filling up a p-shell and thus disabling these electrons to conduct. Half-metallic behaviour results. The archetype of the third category of half-metallic magnetism is CrO_2 [15], but also the manganites fall into this category.

We now address the possible relation between half-metallic magnetism and colossal magnetoresistance. At zero temperature a half-metallic ferromagnet is uniformly magnetized in one direction. The majority electrons take care of good conduction. At non-zero temperature domains are formed with magnetizations in other directions. If the half-metallic properties within these domains are preserved, walls between domains of opposite magnetization are insulating.

The correlation length, defined as the average size of domains with magnetization opposite to the total magnetization, increases if the temperature increases and diverges at T_C. Therefore, majority electrons have, with increasing temperature, a decreasing number of percolating paths between the domains with opposite magnetization, which causes the resistivity to increase.

An external magnetic field forces the magnetization to point uniformly in one direction again and therefore lowers the resistivity, resulting in magnetoresistance. Statistical mechanical calculations have confirmed this explanation and have even reproduced the temperature dependent behaviour above the Curie temperature [16].

The mechanism of decreasing percolating paths has a stronger effect on the resistivity, the lower the dimensionality of the system. This is in agreement with experiments which show that thin films exhibit larger magnetoresistances than bulk samples. And it is also in agreement with experiments on layered manganite perovskites $(A, B)_{n+1}Mn_nO_{3n+1}$, which are built by stacking of n-layers of $(A, B)MnO_3$, separated by insulating AO layers. The conduction occurs primarily within the pseudo-two-dimensional n-layers and indeed are the highest CMR values found for the lowest n [17].

The occurrence of colossal magnetoresistance in half-metallic ferromagnetic Heusler alloys like NiMnSb has not been reported, to our knowledge, and is also not expected. There are two reasons for this. First, the charge carriers in NiMnSb have primarily Sb-5p character, which shows a large spin–orbit interaction. Therefore spin-flip scattering in the interface between domains of opposite magnetization would break down the mechanism of half-metallic magnetism causing CMR. And second, the charge carriers in the Heusler alloys are almost free electron like and therefore much more delocalized than the charge carriers in the manganites. As a consequence

the interfaces between domains are very sharp in the manganites, but probably substantially thicker in the Heusler alloys.

Colossal magnetoresistance in CrO_2, however, is not improbable a priori but experiments on magnetoresistance in CrO_2 have not been reported to our knowledge.

In conclusion, we presented electronic structure calculations which show that ferromagnetic manganites are half-metallic. We discussed the possible relation between half-metallic magnetism and the occurrence of colossal magnetoresistance.

References

[1] M.N. Baibich, J.M. Broto, A. Fert, F.N. Van Dau, F. Petroff, P. Eitenne, G. Creuzet, A. Friederich and J. Chazelas, Phys. Rev. Lett. 61 (1988) 2472.
[2] R.A. de Groot, F.M. Mueller, P.G. van Engen and K.H.J. Buschow, Phys. Rev. Lett. 50 (1983) 2024.
[3] R.M. Kusters, J. Singleton, D.A. Keen, R. McGreevy and W. Hayes, Physica B 155 (1989) 362.
[4] G.-Q. Gong, C. Canedy, G. Xiao, J.Z. Sun, A. Gupta and W.J. Gallagher, Appl. Phys. Lett. 67 (1995) 1783.
[5] P.G. Radaelli, D.E. Cox, M. Marezio, S.-W. Cheong, P.E. Schiffer and A.P. Ramirez, Phys. Rev. Lett. 75 (1995) 4488.
[6] P. Blaha, K. Schwarz, P. Dufek and R. Augustyn, WIEN95, Technical University of Vienna (1995) (Improved and updated Unix version of the original copy-righted WIEN-code, which was published by P. Blaha, K. Schwarz, P. Sorantin and S.B. Trickey, Comput. Phys. Commun. 59 (1990) 399.)
[7] J.P. Perdew, J.A. Chevary, S.H. Vosko, K.A. Jackson, M.R. Pederson, D.J. Singh and C. Fiolhais, Phys. Rev. B 46 (1992) 6671.
[8] J.P. Perdew and Y. Wang, Phys. Rev. B 45 (1992) 13 244.
[9] P.K. de Boer, H. van Leuken, R.A. de Groot, T. Rojo and G.E. Barberis, Solid State Commun. 102 (1997) 621.
[10] W.E. Pickett and D.J. Singh, Phys. Rev. B 53 (1996) 1146.
[11] E.O. Wollan and W.C. Koehler, Phys. Rev. 100 (1955) 545.
[12] C.N.R. Rao and A.K. Cheetham, Science 276 (1997) 911.
[13] V.I. Anisomov, F. Aryasetiawan and A.I. Lichtenstein, J. Phys.: Condens. Mater 9 (1997) 767.
[14] A. Yanase and K. Siratori, J. Phys. Soc. 53 (1984) 312.
[15] K. Schwarz, J. Phys. F 16 (1986) L211.
[16] P.J.M. Bastiaansen and H.J.F. Knops, unpublished; Comput. Mater. Sci. (1997), these proceeding.
[17] Y. Moritomo, A. Asamitsu, H. Kuwahara and Y. Tokura, Nature 380 (1996) 141.

ELSEVIER

Computational Materials Science 10 (1998) 245–248

COMPUTATIONAL
MATERIALS
SCIENCE

Non-collinear magnetism and exchange couplings in FeCo/Mn superlattices

Clara C. Cornea *, Daniel C.A. Stoeffler

I.P.C.M.S - Gemme (U.M.R. 46 du C.N.R.S.), 23 rue du Loess, 67037 Strasbourg, France

Abstract

We present the main results of a systematic study of the magnetic properties of Fe_xCo_{1-x}/Mn_n for various Fe concentrations x and Mn thicknesses n. We show that the magnetic order in the Mn spacer changes from collinear to non-collinear when the Fe concentration decreases. This behaviour is discussed in relation with a bulk 'canted' magnetic state nearly degenerate with the collinear AF order. The origin of the exchange of stability between these two magnetic states is ascribed to a stronger collinear character of the Fe/Mn interface than the Co/Mn one. © 1998 Elsevier Science B.V.

Keywords: Non-collinear magnetism; Metallic superlattices; Interlayer couplings

1. Introduction

It has been shown recently that FeCo/Mn multi-layers present very interesting magnetic properties related to non-collinear orders in the Mn spacer. First, a large biquadratic coupling has been obtained [1] and it has been shown that the coupling energy follows the Slonczewski parabolic law [2]. Second, a net magnetic contribution of the antiferromagnetic Mn spacer has been found [3,4] which makes an angle of 23° with the total FeCo magnetization. These properties have been ascribed to interfacial roughness [2]. However, the used models are based on micromagnetism approaches. The aim of the pre-

sent work is to study these systems from the electronic structure viewpoint.

Such a complete electronic structure study of Fe_xCo_{1-x}/Mn_n superlattices is under progress since more than two years in our group within the tight binding framework using the real space recursion technique [5–7]. We have developed a new code for massively parallel computers in order to determine self consistently the magnetic moments vectors distribution in superlattices. The method and technical details are given in two other papers in these proceedings [8,9]. In this paper, we present a summary of the latest achievements of this study. We show that the magnetic behaviour of thin Mn spacers is highly sensitive to the Fe concentration of the FeCo ferromagnetic layer: from collinear antiferromagnetic for pure Fe, it becomes non-collinear for pure Co. This result is discussed in relation with bulk 'canted' states and interfacial non-collinear character of the Mn.

* Corresponding author. Tel.: +33-388107065; fax: +33-388107249; e-mail: clara@taranis.u-strasbg.fr.

2. Non-collinear magnetic behaviour of the Mn spacer

A preliminary study of the bulk Mn [8,9] has shown that the antiferromagnetic (AF) collinear order is the most stable state as compared to helicoidal or spin spiral magnetic configurations. However, it as been pointed out that a 'canted' state corresponding to a spin spiral with the direction of the magnetic moments in the i-th plane equal to $\theta_i = \theta \cong 0.4\pi$ and $\varphi_i = i \cdot \pi$ is nearly degenerate with the AF order. In the case of superlattices under study in this paper, where the angular degree of freedom is restricted to θ ($\varphi_i = 0$), this 'canted' state plays an important role. Indeed, it corresponds to the magnetic configuration which is the most similar to the AF one when all magnetic moments are in the (xOz) plane. However, the energy difference of approxi-

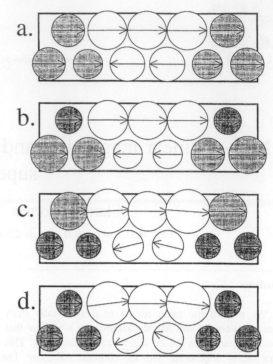

Fig. 2. Magnetic moments distributions for Fe_xCo_{1-x}/Mn_5 superlattices for (a) $x = 1$, (b) $x = 1/2$ with a pure Fe monolayer at the interface, (c) $x = 1/2$ with a pure Co monolayer at the interface and (d) $x = 0$. Each magnetic moment is represented by an arrow giving its direction and the diameter of the circle is equal to its length. The Mn are in white, the Co atoms in dark grey and the Fe atoms in light grey. Only the magnetic moments on the atoms in the unit cell are shown.

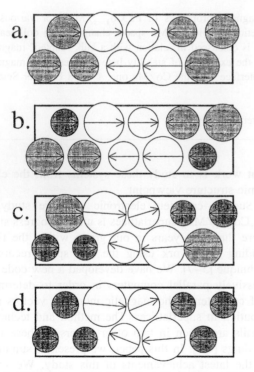

Fig. 1. Magnetic moments distributions for Fe_xCo_{1-x}/Mn_4 superlattices for (a) $x = 1$, (b) $x = 1/2$ with a pure Fe monolayer at the interface, (c) $x = 1/2$ with a pure Co monolayer at the interface and (d) $x = 0$. Each magnetic moment is represented by an arrow giving its direction and the diameter of the circle is equal to its length. The Mn are in white, the Co atoms in dark grey and the Fe atoms in light grey. Only the magnetic moments on the atoms in the unit cell are shown.

mately 1 meV/atom between both states should give collinear ground states for the superlattices.

Surprisingly, the ground state in not always collinear depending on the Fe concentration of the FeCo ferromagnetic layers. Figs. 1 and 2 show the magnetic moments distributions for Fe_xCo_{1-x}/Mn_4 and Fe_xCo_{1-x}/Mn_5 superlattices for the expected ground state i.e. for an antiferromagnetic interlayer arrangement when $n = 4$ and for a ferromagnetic interlayer arrangement when $n = 5$. For a pure Fe ferromagnetic layer (Fig. 1a and Fig. 2a) or for $x = 1/2$ with a pure Fe interfacial monolayer (Fig. 1b and Fig. 2b), the magnetic order is obtained collinear. The coupling between the Fe and Mn interfacial magnetic moments is parallel and the Mn moments of the second atomic layer from the interface are reduced as compared to the others in agreement with ab initio calculations of these collinear states [5–7]. For $x = 1/2$ with a pure Co interfacial monolayer (Fig. 1c and Fig. 2c) or for a pure Co ferromagnetic layer (Fig. 1d and Fig. 2d), two solutions are obtained: the collinear one with very similar magnetic moments distributions as for the two previous cases and a non-collinear one shown by the figures. For $n = 4$ (Fig. 1), the canted character of the non-collinear state in the Mn spacer can be clearly seen mainly on the two inner atomic layers. This state is less pronounced for $n = 5$ (Fig. 2) where the inner Mn magnetic moments is exactly collinear with the ferromagnetic moments by symmetry.

3. Interlayer coupling energy

We have calculated the total energy as a function of the angle $\Delta\theta$ between the inner magnetic moments (whose directions are fixed) of successive FeCo layers. This allows the determination of the interlayer magnetic couplings between the FeCo layers through the Mn spacer as a function of $\Delta\theta$ when

the Fe concentration and the Mn thicknesses vary. This situation is similar to the case where an external magnetic field is applied maintaining the relative angle $\Delta\theta$ constant. The results are reported in Fig. 3 for the same four cases as for Figs. 1 and 2. The energy curves are very similar and roughly follow parabolic laws. However, for pure ferromagnetic layers, the energy variations are approximately two times smaller than the ones obtained for FeCo alloys. They are, for these alloys, nearly the same whatever the element at the interface is. For $x = 1$ and $x = 1/2$ with a pure Fe interfacial monolayer, the energy is the lowest for $\Delta\theta = 180°$ when $n = 4$ and for $\Delta\theta = 0°$ when $n = 5$. For $x = 1/2$ with a pure Co interfacial monolayer, the collinear solution is nearly degenerate in energy with the non-collinear states presented in Figs. 1 and 2. However, only the collinear solution is stable from the angular viewpoint. For $x = 0$ (pure Co ferromagnetic layer), the energy minimum is obtained for $\Delta\theta \cong 150°$ when $n = 4$ and for $\Delta\theta \cong 25°$ when $n = 5$. However, as mentioned previously [8,9], these minima do not correspond to angular stable solutions because the perpendicular compo-

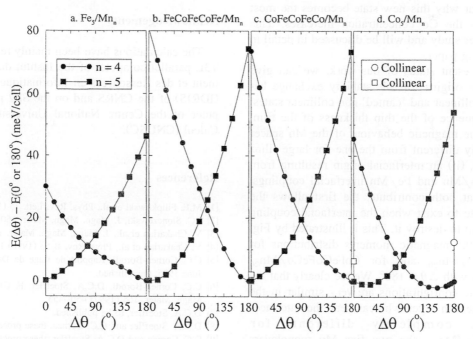

Fig. 3. Total energy of Fe_xCo_{1-x}/Mn_n superlattices as a function of $\Delta\theta$ ($n = 4$ circles and $n = 5$ squares) for (a) $x = 1$, (b) $x = 1/2$ with a pure Fe monolayer at the interface, (c) $x = 1/2$ with a pure Co monolayer at the interface and (d) $x = 0$. The energy of the collinear situation are given by the open symbols.

nent $M_{d,\theta}$ on the sites for which the angles are fixed is not equal to zero. As a consequence, the angular stable states correspond to $\Delta\theta \cong 175°$ when $n = 4$ and to $\Delta\theta \cong 3°$ when $n = 5$. Finally, these states are clearly more stable than the collinear solution with a surprising large difference (more than 12 meV per cell corresponding to 3 meV per Mn atom when $n = 4$).

4. Discussion

To summarise, we have shown that for $x = 1$ and $x = 1/2$, the most stable state is obtained for $\Delta\theta = 180°$ when $n = 4$ and for $\Delta\theta = 0°$ when $n = 5$ and is collinear (even if other solutions are obtained for $x = 1/2$ with a pure Co interfacial monolayer presenting non-collinear Mn configurations). We have also shown that for $x = 0$, the most stable state corresponds to $\Delta\theta \cong 175°$ when $n = 4$ and to $\Delta\theta \cong 3°$ when $n = 5$ with a significant non-collinear order in the Mn spacer. This result can be qualitatively understood from the existence of a bulk 'canted' state nearly degenerate with the AF order. However, it is not clear why this new state becomes the most stable when the Co concentration increases. This point is under study and will be discussed in detail in a forthcoming paper.

At the present status of our work, we can give two possible origins for this stability exchange between AF collinear and 'canted' non-collinear states: (i) a consequence of the thin thickness of the film, for which the magnetic behaviour of the Mn spacer is completely different from the one for large films or bulk and, (ii) an interfacial origin resulting from different Co/Mn and Fe/Mn interfacial couplings. It seems that both contribute: the first allows the 'canted' state to exist when the interfacial coupling does not tend to destroy it. This is illustrated by Fig. 4 showing the magnetic moments distributions for $FeCoFeCoFe/Mn_{20}$ and for $CoFeCoFeCo/Mn_{20}$ superlattices with $\Delta\theta = 180°$. We see clearly that the two magnetic configurations are very similar in the centre of the Mn layer whereas near the interfaces they are completely different: for $FeCoFeCoFe/Mn_{20}$, the two first Mn monolayers

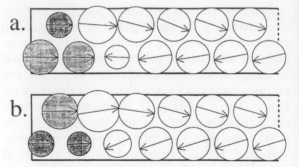

Fig. 4. Magnetic moments distributions for $Fe_{0.5}Co_{0.5}/Mn_{20}$ superlattices for (a) a pure Fe monolayer at the interface, (b) a pure Co monolayer at the interface. Each magnetic moment is represented by an arrow giving its direction and the diameter of the circle is equal to its length. The Mn are in white, the Co atoms in dark grey and the Fe atoms in light grey. Only the magnetic moments on the atoms in half of the unit cell are shown.

present a nearly collinear magnetic configuration. This shows that the Fe/Mn interface presents a higher level of collinearity than the Co/Mn one. This point will be extensively studied in the future.

Acknowledgements

The calculations have been mainly realised on the T3E parallel computer of the Institut du Développement et des Ressources en Informatique Scientifique (IDRIS) of the CNRS and on the SP2 parallel computer of the Centre National Universitaire Sud de Calcul (CNUSC).

References

[1] M.E. Filipkowski et al., Phys. Rev. Lett. 75 (1995) 1847.
[2] J.C. Slonczewski, J. Magn. Magn. Mater. 150 (1995) 13.
[3] V. Chakarian et al., J. Magn. Magn. Mater. 156 (1996) 265.
[4] V. Chakarian et al., Phys. Rev. B 53 (1996) 11313.
[5] C.C. Cornea-Borodi, Rapport de stage de DEA, Strasbourg, June 1995, unpublished.
[6] C.C. Cornea-Borodi, D.C.A. Stoeffler, F. Gautier, J. Magn. Magn. Mater. 165 (1997) 450–453.
[7] D.C.A. Stoeffler, to be published.
[8] D.C.A. Stoeffler and C.C. Cornea, these proceedings.
[9] C.C. Cornea and D.C.A. Stoeffler, these proceedings.

ELSEVIER

Computational Materials Science 10 (1998) 249–254

COMPUTATIONAL
MATERIALS
SCIENCE

Determination of magnetic moment vectors distributions in FeCo/Mn superlattices

Clara C. Cornea *, Daniel C.A. Stoeffler

I.P.C.M.S - Gemme (U.M.R. 46 du C.N.R.S.), 23 rue du Loess, 67037 Strasbourg, France

Abstract

We present in details a method for determining non-collinear orders in metallic superlattices and the interlayer magnetic couplings between ferromagnetic layers separated by antiferromagnetic spacers. A semiempirical tight binding framework and the real space recursion technique is used to obtain the electronic structure of non-symmetric systems. We focus on the angular self-consistency and we discuss the angular stability of the solutions obtained by fixing the magnetic moment direction of some atoms. We give a few examples of calculations for bulk Mn and FeCo/Mn superlattices and discuss the validity of the results. © 1998 Elsevier Science B.V.

Keywords: Non-collinear magnetism; Metallic superlattices

1. Introduction

It has been shown recently that FeCo/Mn multi-layers present very interesting magnetic properties related to non-collinear orders in the Mn spacers (large biquadratic couplings [1], net magnetic contribution of the antiferromagnetic Mn spacer [2,3]). A complete electronic structure study of Fe_xCo_{1-x}/Mn_n superlattices is under progress since two years in our group within the tight binding framework using the real space recursion technique [4–6]. In this paper, we focus mainly on technical aspects of the angular convergence compared to magnetic collinear calculations. We use a method combining (i) a mixing scheme between output and input values and (ii) an extrapolation of the variations during these few 'mixed' iterations (*mixing-extrapolation* technique) for improving the solving of our problems [7]. We illustrate our purpose with various calculations of bulk and superlattices magnetic properties and discuss the stability of the results in terms of perpendicular magnetic moments contributions and total energies.

2. Method of calculation

2.1. Hamiltonian and electronic structure

In the tight binding approximation, the total Hamiltonian can be written as the sum of a band

* Corresponding author. Tel.: +33-388107065; fax: +33-388107249; e-mail: clara@taranis.u-strasbg.fr.

0927-0256/98/$19.00 © 1998 Elsevier Science B.V. All rights reserved.
PII S 0927-0256(97)00149-3

H_{band} and an exchange H_{exc} term with (the $[2 \times 2]$ matrix represents the spin part):

$$H_{band} = \sum_{\substack{i,l,m \\ i',l',m'}} |i; l; m\rangle$$

$$\times \left[\left(\varepsilon_{i,l,m}^0 + U_{i,l} \Delta N_{i,l} \right) \delta_{i,i'} \delta_{l,l'} \delta_{m,m'} \right.$$

$$\left. + t_{i',l',m'}^{i,l,m} (1 - \delta_{i,i'}) \right] \langle i'; l'; m' \begin{bmatrix} 1 & 0 \\ 0 & 1 \end{bmatrix}$$

$$(1)$$

$$H_{exc} = \sum_{i,l,m} \left(-\frac{1}{2} I_{il} M_{il} \right) |i; l; m\rangle \langle i; l; m|$$

$$\times \begin{bmatrix} \cos \theta_i & e^{-i\varphi_i} \sin \theta_i \\ e^{i\varphi_i} \sin \theta_i & -\cos \theta_i \end{bmatrix} \quad (2)$$

where $\varepsilon_{i,l,m}^0$ is the energy level of site i for symmetry (l, m); $U_{i,l} \Delta N_{i,l}$ is the intrasite Coulomb contribution due to the local charge transfer $\Delta N_{i,l} = N_{i,l} - N_{i,l}^0$, $U_{i,l}$ being the effective intrasite Coulomb parameter; $t_{i',l',m'}^{i,l,m}$ is the so-called hopping integral; I_{il} is the effective intrasite exchange parameter and M_{il} the local magnetic moment whose direction is given by the two angles (θ_i, φ_i) in the spherical representation (Fig. 1). The electronic structure can be determined with such a Hamiltonian for a set of vectorial magnetic moments assuming that the local s, p and d moments are aligned.

In the present work, we use the real space recursion technique which allows the determination of local density of states (LDOS) $n_{i,l}^\sigma(\varepsilon)$ projected on

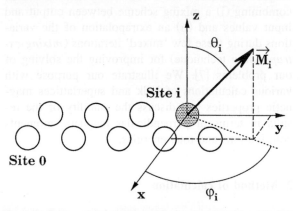

Fig. 1. Definition of the direction of the local quantisation axis for the site i given by the angles θ_i and φ_i. The local magnetic moment $\vec{M}_i$ is parallel to this direction and corresponds to $M_i \cdot \hat{u}_r$ in the local spherical basis $(\hat{u}_r, \hat{u}_\theta, \hat{u}_\varphi)_i$.

arbitrary local quantisation axis by rotating the initial recursion vector. The cluster we use contains approximately 5000 atoms and we determine 24 levels in the continuous fraction for each s, p, d symmetry, each spin and each quantisation axis. The charge transfers ΔN_i and the magnetic moments M_i on site i are obtained by integration of these densities of states up to the Fermi level.

2.2. Collinear magnetism

For a collinear situation, we determine self-consistently the sets $\{\Delta N_{i,l}\}$ and $\{M_{il}\}$ of charge transfers and magnetic moments on each unequivalent site using an iterative method starting from an initial sets $\{\Delta N_{i,l}(0)\}$ and $\{M_{il}(0)\}$. This method is the best one for solving the self-consistent solutions for such complex systems. At step p of the iteration, the electronic structure is calculated with the following input energy levels for the spin σ:

$$\varepsilon_{i,l,m,\sigma}(p) = \varepsilon_{i,l,m}^0 + U_{i,l} \Delta N_{i,l}(p, \text{input})$$

$$- \sigma \frac{I_{i,l}}{2} M_{i,l}(p, \text{input}) \quad (3)$$

and the Fermi level ε_F is obtained by requiring the global neutrality in the unit cell.

The output charge transfers and magnetic moments are equal to:

$$\Delta N_{i,l}(p, \text{output}) = \int_{-\infty}^{\varepsilon_F} \sum_\sigma n_{i,l}^\sigma(\varepsilon) \, d\varepsilon - N_{i,l}^0,$$

$$M_{i,l}(p, \text{output}) = \int_{-\infty}^{\varepsilon_F} \left[n_{i,l}^+(\varepsilon) - n_{i,l}^-(\varepsilon) \right] d\varepsilon$$

$$(4)$$

The new input sets at step $p + 1$ are obtained by mixing input and output at step p driven by the mixing factor x:

$$M_{i,l}(p + 1, \text{input})$$

$$= x M_{i,l}(p, \text{output}) + (1 - x) M_{i,l}(p, \text{input})$$

$$(5a)$$

$$\Delta N_{i,l}(p + 1, \text{input})$$

$$= x \Delta N_{i,l}(p, \text{output}) + (1 - x) \Delta N_{i,l}(p, \text{input})$$

$$(5b)$$

The choice of the mixing factor results from a compromise between convergence celerity (x nearly equal to 1) and divergence breaking ($x < 1$). Usually, the convergence is reached easier for the magnetic moments than for the charge transfers: we use then different mixing factors. For example, the collinear calculations of this work have been done using $x = 1$ or 0.5 for the magnetic moments and $x = 0.05$ to 0.2 for the charges. Self-consistency is assumed to be achieved at step $\tilde{p}$ when:

$$\max_{(i,l)} |M_{i,l}(\tilde{p}) - M_{i,l}(\tilde{p} - 1)| < \varepsilon,$$

$$\max_{(i,l)} |\Delta N_{i,l}(\tilde{p}) - \Delta N_{i,l}(\tilde{p} - 1)| < \varepsilon \qquad (6a)$$

$$|E_{\text{tot}}(\tilde{p} + 1) - E_{\text{tot}}(\tilde{p})| < \varepsilon \qquad (6b)$$

for a given ε. In our calculations we set ε equal at least to 10^{-5}.

2.3. Non-collinear magnetism

For a non-collinear magnetic situation we use a similar iterative method. The charges are obtained exactly in the same way as for collinear situations. However, the angular degree of freedom needs to be restricted to the 'd' component of the moment. Effectively, in the general case, the magnetic part of the calculation starts with a set of 's', 'p' and 'd' magnetic moments $M_{s,i}$, $M_{p,i}$, $M_{d,i}$ and directions of each local quantisation axis characterised by the spherical angles (θ_i, φ_i), the local magnetic moment being the sum $M_i = M_{s,i} + M_{p,i} + M_{d,i}$ and aligned with the local quantisation axis (Fig. 1). We determine the 3 components of the output *magnetic moment vector* $(M_r, M_\theta, M_\varphi)_i$ by aligning successively the first element of the recursion basis with the 3 perpendicular directions given by the local spherical basis $(\hat{u}_r, \hat{u}_\theta, \hat{u}_\varphi)_i$. The direction of the local magnetic moment $\vec{M}_i = M_r \hat{u}_r + M_\theta \hat{u}_\theta + M_\varphi \hat{u}_\varphi$ gives the direction of the output local quantisation axis. The problem comes from the fact that each component is the sum of 's', 'p' and 'd' magnetic moments and that this symmetry decomposition is lost during the determination of the input values $M_{s,i}$, $M_{p,i}$, $M_{d,i}$ for the next step. Indeed, the input magnetic moment magnitude is equal to:

$$M_i = \Big[\big(M_{s,r} + M_{p,r} + M_{d,r} \big)^2$$
$$+ \big(M_{s,\theta} + M_{p,\theta} + M_{d,\theta} \big)^2$$
$$+ \big(M_{s,\varphi} + M_{p,\varphi} + M_{d,\varphi} \big)^2 \Big]^{1/2} \qquad (7)$$

which can not be rewritten as $M_i = M_{s,i} + M_{p,i} + M_{d,i}$. As a consequence, we set I_s and I_p to zero so that the values of $M_{s,i}$ and $M_{p,i}$ do no more play any role and we restrict the angular self consistency to the 'd' symmetry. This means that the new local quantisation axis is given by the direction of the 'd' magnetic moment vector $\vec{M}_{d,i} = M_{d,r} \hat{u}_r + M_{d,\theta} \hat{u}_\theta + M_{d,\varphi} \hat{u}_\varphi$. When self-consistency is achieved, the local 'd' exchange field $\vec{H}_{ex,d,i}$ must be aligned with the local 'd' magnetic moment $\vec{M}_{d,i}$: this corresponds to having zero values for the $M_{d,\theta}$ and $M_{d,\varphi}$ components. The solution is then obtained for:

$$\max_i \{ |M_{d,r}(\tilde{p}) - M_{d,r}(\tilde{p} - 1)| \} < \varepsilon$$

and

$$\max_i \{ |M_{d,\theta}(\tilde{p})|, |M_{d,\varphi}(\tilde{p})| \} < \varepsilon \qquad (8)$$

Finally, in order to accelerate the angular convergence, we use a combination of usual mixing and an extrapolation of the variations of each local angle [7].

3. Angular stability

Because we do not include the spin–orbit contribution in the hamiltonian, a global rotation can be applied without modifying the results and the calculation could never converge if all moments rotate similarly during the iterations. Consequently, we fix the direction of at least one magnetic moment — the directions of all others will be defined relative to the fixed direction — and we do not require the angular self-consistency for this magnetic moment. Particular magnetic orders can then be built by fixing the directions of a few magnetic moments. However, these orders are not necessarily stable from the angular point of view i.e. $M_{d,\theta}$ and $M_{d,\varphi}$ are not equal to zero. We give two illustrations of such cases.

3.1. Bulk Mn

Helicoidal states along the [001] direction, satisfying $\theta_i = i \cdot \theta$ and $\varphi_i = i \cdot \varphi$ where θ_i and φ_i are the angles defining the direction of all magnetic moments in the same i-th (001) plane, are stable from the angular view point because the contribution to $\vec{H}_{ex,d,i}$ of each magnetic moment is compensated exactly by another one. The energy allows then to compare different situations as a function of (θ, φ):

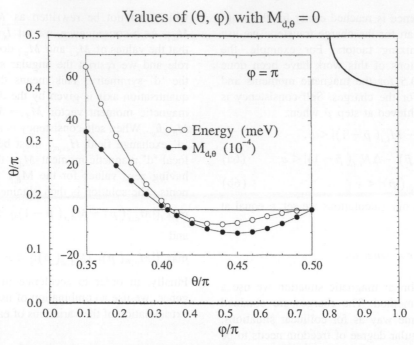

Fig. 2. Values of (θ, φ) for which $M_{d,\theta} = 0$ (thick solid lines) corresponding to angular stable spin spiral magnetic states along the [001] direction. The insert gives the energy relative to the AF order ($\theta = \pi/2$, $\varphi = \pi$) and $M_{d,\theta} \times 10^4$ for $\varphi = \pi$ as a function of θ; two spin spiral states are stable ($M_{d,\theta}$) for $\theta = \pi$ and $\theta \cong 0.4\pi$ but the first one has the lowest energy.

we have found that the antiferromagnetic (AF) order is the most stable. Spin spiral states along the same [001] direction, satisfying $\theta_i = \theta$ and $\varphi_i = i \cdot \varphi$, do not have $M_{d,\theta} = 0$ for all θ values. Effectively, it is only for $\theta = 0$ and $\theta = \pi/2$ that $M_{d,\theta}$ is necessarily equal to zero. Surprisingly, we have found another set of (θ, φ) values for which $M_{d,\theta} = 0$. Fig. 2 shows this new set which represents an arc roughly centred on $(\theta = \pi/2, \varphi = \pi)$. The energy most stable spin spiral state of this new set is obtained for $\varphi = \pi$ and $\theta \cong 0.4\pi$. It corresponds to a 'canted' magnetic state where the angle between successive magnetic moments is nearly equal to 0.8π and with

a net magnetisation of $0.7\mu_B$/atom. However, as shown in the inset of Fig. 2, the collinear AF order remains the most stable one by approximately 1 meV/atom.

3.2. FeCo/Mn superlattices

In bulk, the directions of all magnetic moments are fixed and only the magnitude of the moments and the charge transfers have to be determined self-consistently. In superlattices, the aim is to determine the interlayer coupling energy between FeCo ferromagnetic layers separated by a Mn antiferromagnetic

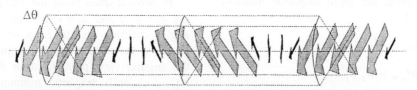

Fig. 3. Schematic representation of the non-collinear magnetic arrangements considered for the calculation of the interlayer magnetic couplings as a function of the angle $\Delta\theta$ between successive inner ferromagnetic atomic layer magnetic moments.

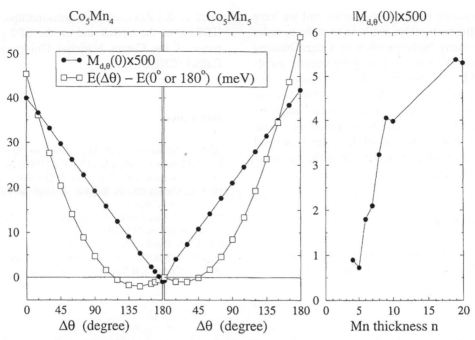

Fig. 4. Total energy (open squares) and $M_{d,\theta}$ values on site 0 for which the angle is fixed (filled circles) for Co_5Mn_4 and Co_5Mn_5 superlattices as a function of $\Delta\theta$. The right graph shows the variation of $|M_{d,\theta}(0)|\times 500$ for Co_5Mn_n superlattices as a function of n.

spacer as a function of the relative orientation of successive FeCo magnetisations. For this study we restrict the angular variation to the angle θ by fixing φ to 0 and we take into account the symmetry of the superlattice relative to the (001) plane at the centre of the Mn spacer. We fix the directions of the inner magnetic moments of each FeCo layer (site 0) and their relative angle is set to $\Delta\theta$ as shown in Fig. 3. All other moments can freely rotate and the local angle is determined up to satisfying the angular self-consistency (Eq. (8)). Such situations correspond to magnetic configurations of the superlattices when an external magnetic field is applied maintaining $\Delta\theta$ fixed. Angular stable states are then given by $M_{d,\theta}$ on site 0 equal to zero. The study of the magnetic orders and interlayer couplings for various Fe concentrations in the FeCo layers is presented in another paper of these proceedings [8] and we discuss only the results obtained for Co_5Mn_n superlattices for $n = 4$ and 5.

Surprisingly, even for $\Delta\theta = 180°$ with $n = 4$ and $\Delta\theta = 0°$ with $n = 5$, the magnetic order is non-collinear whereas it was expected to be collinear (a collinear solution is obtained only when the starting order is collinear, but it is found a few meV less stable than the non-collinear solution [8]). However, it does not correspond to an angular stable situation for a zero external magnetic field as shown by Fig. 4 because $M_{d,\theta}$ is not exactly equal to zero; the angular stable solution is obtained for a small tilt (approximately 5°) of $\Delta\theta$. This shows that the magnetic order is intrinsically non-collinear in such superlattices even for perfect interfaces. For large n values, the Mn is found ordered like the 'canted' state obtained in bulk. As a consequence, the torque applied on the fixed magnetic moment (site 0) increases with n as shown by Fig. 4. The stabilisation of the 'canted' state finds its origin in the interfacial coupling as exhibited by the fact that it disappears for a pure Fe ferromagnetic layer [8].

4. Conclusion and perspectives

In this paper we have detailed the method for determining non-collinear magnetic orders from the electronic structure of metallic systems and we have focussed on the restrictions we have to make in order to solve the many body problem in a tight binding approximation. We have discussed the angular stability of bulk Mn spin spirals and non-collinear states in FeCo/Mn superlattices. We have exhibited new stable situations in relation with 'canted' magnetic states in the Mn layer. The main limitation comes from the fact that the values of $M_{d,\theta}$ are small and need a high level of convergence in order to determine them accurately.

Acknowledgements

The calculations have been mainly realised on the T3E parallel computer of the Institut du Développement et des Ressources en Informatique Scientifique (IDRIS) of the CNRS and on the SP2 parallel computer of the Centre National Universitaire Sud de Calcul (CNUSC).

References

[1] M.E. Filipkowski et al., Phys. Rev. Lett. 75 (1995) 1847.
[2] V. Chakarian et al., J.M.M.M. 156 (1996) 265.
[3] V. Chakarian et al., Phys. Rev. B 53 (1996) 11313.
[4] C.C. Cornea-Borodi, Rapport de stage de DEA, Strasbourg, June 1995, unpublished.
[5] C.C. Cornea-Borodi, D.C.A. Stoeffler, F. Gautier, J. Magn. Magn. Mater. 165 (1997) 450–453.
[6] D.C.A. Stoeffler, to be published.
[7] D.C.A. Stoeffler and C.C. Cornea, these proceedings.
[8] C.C. Cornea and D.C.A. Stoeffler, these proceedings.

ELSEVIER

Computational Materials Science 10 (1998) 255–259

COMPUTATIONAL
MATERIALS
SCIENCE

Magnetic structure of nonideal Fe/Cr interface

N.S. Yartseva [a], V.M. Uzdin [b,*], C. Demangeat [c]

[a] *Institute of Metal Physics, GSP-170, Ekaterinburg, 620219, Russian Federation*
[b] *Saint-Petersburg State University, CAPE, V.O. 14 linia, 29, St. Petersburg, 199178, Russian Federation*
[c] *IPCMS, 23, rue du Loess, 67037 Strasbourg, France*

Abstract

Different kinds of space defects on the background of ideal smooth interface in Fe/Cr multilayers, such as steps, embedded clusters, pinholes, random interface roughness, are investigated within model Hamiltonian approach. Calculations of magnetic structure are performed in the periodic Anderson model (PAM) in mean field approximation by recursion method in the real space.

The distribution of magnetic moments obtained for two sets of interfaces (with different roughness) shows strong sensitivity of magnetic structure to the interface roughness and interdiffusion. Application of the theory for interpretation of Mössbauer spectra of Fe/Cr multilayers is discussed. Spectra obtained for three perpendicular orientations of the sample relatively to the incident beam allowed to avoid the problem of nonrandom distribution of magnetic moment direction in the multilayers.

Information about the angular dependence of hyperfine fields on the Fe atoms in Fe/Cr superlattices obtained by means of Mössbauer spectroscopy technique is discussed within the framework of the theory of noncollinear magnetic structure. Copyright © 1998 Elsevier Science B.V.

Keywords: Roughness; Interdiffusion; Electronic structure; Mössbauer spectroscopy

Fe/Cr interface magnetism is one of the challenging topics of modern science of low-dimensional structures. It attracts a number of experimental tools and theoretical approaches [1], but up to now the electronic and magnetic structure at the interface region remains partly unknown. One of the reasons is connected with the formation of defects such as embedded clusters, steps, and pinholes near the interface in the layered system. The space structure of this region appears to be very complex and it crucially influences the physical properties. For example, Heinrich et al. [2] have shown that bilinear exchange coupling of Fe/Cr mul-

tilayers can be changed by factor of 5 by varying the substrate temperature during the growth of the first atomic layer. So, the development of the theory for description of different kinds of space defects near the interface is very important for the adequate interpretation of the experiments in Fe/Cr systems.

In this paper we discuss the results of calculation of nonideal Fe/Cr interface and its application to the explanation of Mössbauer spectra of such a system.

Note that taking into account complex space defects for the description of real interface magnetic structure forbid the use of ab initio-like theories due to huge increasing of numerical calculation which is out of possibility of modern computers. In this case tight binding models like PAM or Hubbard model (HM)

---

* Corresponding author. Present address: Inst. fur Angewandte Physik, Universitat Dusseldorf, Universitatsstrasse 1, D-40225 Dusseldorf, Germany. E-mail: uzdin@moss.pu.ru.

remain so far the main tool for the description of non-ideal rough interface under experimental investigation.

On the basis of such theories the different kinds of defects were investigated. Calculation of magnetic moments on the stepped vicinal Cr surface and Fe/Cr stepped interface [3,4] revealed reasonable agreement of the magnetic moment distribution obtained in PAM and HM. Computation of magnetic moments perturbation of Fe atoms due to Cr impurities [4] and Fe clusters embedded into the Cr matrix (both in the bulk and near the surface or interface) [5] have revealed important role of defects of magnetic structure for interpretation of hyperfine fields (hff) distribution given by Mössbauer spectroscopy. In particular, the deviation from additivity in the perturbation of magnetic moment of Fe atoms by Cr impurities and dependence of magnetic moments of Fe clusters in the Cr matrix on the distance from the surface and interface, that are usually not considered in the spectra treatment, are discussed.

In [6] within PAM the distribution of magnetic moments in Fe/Cr/Fe sandwiches with pinhole defects has been studied. The distribution of the Fe moments inside pinholes appears to be similar to the ones in Fe clusters embedded into a Cr matrix. Such defects also should be taken into account especially for interpretation of experiments on the Fe/Cr superlattices with short period.

However, all mentioned papers give only magnetic structure around special kinds of defects and cannot be compared with experimental distributions of obtained on samples with a number of different defects.

Modelling of the magnetic properties of the Fe/Cr interface [7] on the basis of the special "epitaxy algorithm" within the framework of PAM allows to obtain the distribution of magnetic moments for the random structure with different kinds of defects. Self-consistent calculation of magnetic moments gave reasonable explanation of Turtur's and Bayreuther experiment [8]. However, for this purpose it was not necessary to calculate the distribution of magnetic moments, because only total magnetic moment of the sample was measured. On the contrary, for interpretation of Mössbauer data it is necessary to find the number of atoms with given magnitude of

magnetic moments, because it is the distribution that can be compared with one for hff. Below we will represent such a data for two variants of modelling of epitaxy process.

Detailed description of epitaxy algorithm was done in [6,7]. The prism of $8 \times 8 \times 18$ was filled with atoms Fe and Cr using random procedure. The base of this prism in our calculation comprises 8×8 elementary cells; its height is 18 levels. Outside of the prism the structure is repeated periodically. Two variants of algorithms A and B give the surfaces and interfaces with different roughness. A-algorithm gives relatively smooth interface without pores. The width of the interface region, where there are both atoms Fe and Cr do not exceed four layers. B-algorithm produces more irregular interfaces with some hollows and interface mixing of Fe and Cr atoms in the 7–10 layers. Self-consistent calculations of magnetic moments were performed for sets of 20 interfaces constructed by A or B variant of epitaxy algorithm when at first 64×6 atoms of Fe were distributed in the prism and after that $64 \times \xi$ atoms of Cr were "sputtered" on Fe substrate. Covering parameter ξ in our calculation was taken from 0 to 2.5 with a step of 0.5. As a result we obtain the structure where low layers (12–18) are filled preferably by Fe atoms. After that follows "Cr" slab with thickness depending on the covering parameter ξ. Upper layers (with small number 1, 2 . . .) are empty.

In Fig. 1 the distribution of magnetic moments on Fe atoms in the interface regions for A-type interfaces is depicted. 11, 12, and 13 layers are interface regions where there are both Fe and Cr atoms. All the sites lower than 13 layer are fully filled by Fe and all the sites in the layer higher than 11 are filled by Cr atoms or empty. The number of Fe atoms at every interface layer is shown in Fig. 1. The figure demonstrates the dynamics of change of magnetic moments on Fe atoms with Cr covering.

For $\xi = 0$ near the surface one can find the enhancement of magnetic moment up to 2.5–2.6 μ_B per site. Small Cr covering ($\xi = 0.5$) leads to the decrease of the moments. However, this decreasing differs for each layer. For Fe atoms in 13-layer most atoms have the moment which corresponds to bulk Fe, and only part of the atoms which have Cr among

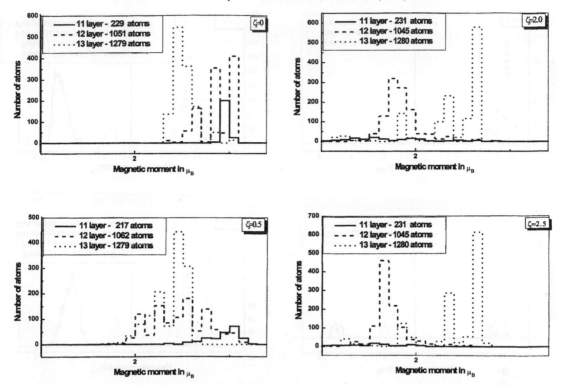

Fig. 1. Distribution of the magnetic moments on Fe atoms near Cr/Fe interface for Cr overlayer on Fe. ξ is the Cr-covering parameter. Random interface is produced by A-algorithm "epitaxy". In the box of figure the number of Fe atoms in every interface layer are shown.

their nearest neighbours decrease their moment. For 12-layer, small Cr covering leads to the very wide distribution of the moments, because in this layer stay surface atoms with enhanced moment, which do not have Cr neighbours, as well as atoms with several Cr neighbours with reduced moments. Most of Fe atoms of the 11-layer conserve their large moments because Cr transfer through this layer and they stay at the surface on the top of the sample.

When the Cr covering increases, the moments of 13-layer slightly grow. Similar behaviour for Fe atoms which have no Cr atoms in its nearest neighbours but only among second neighbours takes place for the inner atoms in clusters [5,9] and pinholes [6]. The 12-layer for the larger Cr covering has reduced magnetic moments because the most of these atoms have several neighbouring Cr atoms. As for 11-layer, one can find the change of direction of magnetic moments on some sites for higher Cr covering due to Fe–Cr exchange interaction.

For more rough interface of B-type the general behaviour of magnetic moments stays almost the same but the distribution of moments on every layer becomes much more wide.

Special interest concerns the distribution of magnetic moments of Cr atoms and its dependence on the roughness of the interface. In Figs. 2 and 3 such distribution for A-type and B-type surfaces correspondingly are shown. One can see the transition from layered antiferromagnetic structure to random structure with increasing of interface roughness. Despite the large (more than $2\mu_B$) local magnetic moment of subsurface Cr atoms the average moment of these layers appear to be almost zero. So, not only the existence of steps on the surface but also the distribution of the moments on the random rough interface can explain the contradiction between experimentally measured zero total moment and enhanced local moments of Cr overlayers on Fe substrate [9].

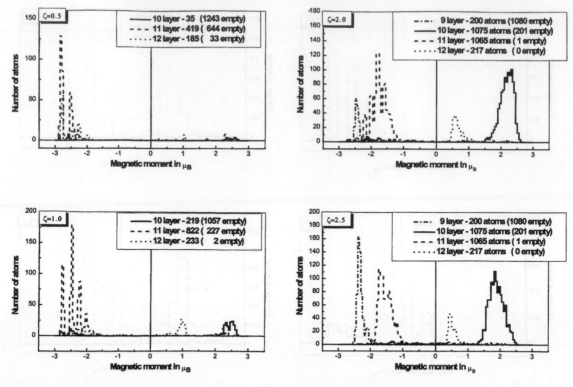

Fig. 2. Distribution of the magnetic moments on Cr atoms near interface Cr/Fe for Cr overlayer on Fe. ξ is the Cr-covering parameter. Random interface is produced by A-algorithm "epitaxy". In the box of figure the number of Cr atoms in every interface layer are shown.

Our results for the distribution of magnetic moments in Cr overlayer on Fe can be used for interpretation of Mössbauer spectra [10,11] if as a first approach we consider that hff is proportional to the local magnetic moments on Fe atoms.

There are however two circumstances which have to be taken into account for this interpretation:

(1) If for the bulk materials assumption about proportionality between magnetic moment and hff is usual, for low-dimensional metallic systems it is true only for hff which is connected with polarisation of inner shells. Conductivity band gives direct contact contribution to the hff field which have to be taken into account separately.

(2) In the layered structures the distribution of direction of magnetic moments is not random. It leads to the dependence of Mössbauer lines amplitude on the orientation of the sample relatively to the incident beam.

A second problem can be avoided experimentally if one takes three spectra obtained for perpendicular directions of γ-radiation incidence instead of one. After composition of these three spectra one can obtain free-texture spectrum which does not depend on the distribution of direction of magnetic moments in the sample [10].

From another side these three spectra can give an additional information about the direction of magnetic moments in the sample. For its theoretical interpretation it would be useful to develop the theory for the description of noncollinear structure in the low-dimensional metallic systems. Variant of such a theory based on the PAM gives the possibility for such calculations. Note, however, that our preliminary calculations show that Fe atoms prefer to keep their moments parallel or antiparallel and only in the neighbourhood of the space defects (like step or embedded cluster touching the interface) noncollinear ordering

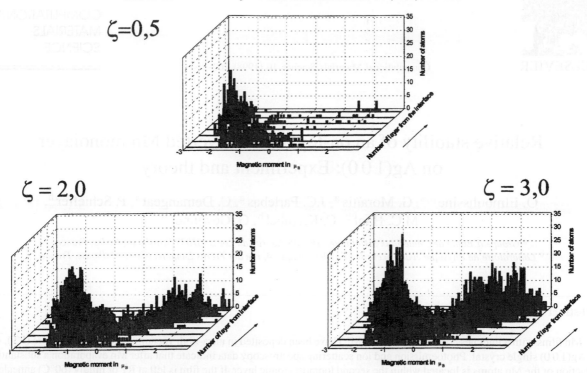

Fig. 3. Distribution of the magnetic moments on Cr atoms near interface Cr/Fe for Cr overlayer on Fe. ξ is the Cr-covering parameter. Random interface is produced by B-algorithm "epitaxy".

can appear. Cr magnetic moments seem to be much more flexible and change their direction more easily.

Hff distribution obtained from these spectra can be compared with magnetic moments to clarify the influence of the interface roughness on the magnetic structure. Existence of the correlation in the position of satellites in Mössbauer spectra and maximum in the magnetic moment distribution [11] show that the contribution of 4s-band does not hide this information.

Work was partially supported by the bilateral exchange program between France and Russia (CNRS-RAS).

References

[1] Workshop on Fe/Cr Interface Magnetism, Strasbourg, 2 3 June 1996, Satellite Meeting of E-MRS Symp. E. Book of Abstracts.

[2] B. Heinrich, J.F. Cochran, D. Venus, K. Tofland, D. Atland, S. Govorkov and K. Myrtle, J. Appl. Phys. 79 (1996) 4518.

[3] C. Demangeat and V.M. Uzdin, J. Magn. Magn. Mater. 156 (1996) 202.

[4] V.N. Gittsovich, V.G. Semenov and V.M. Uzdin, J. Magn. Magn. Mater. 146 (1994) 3564.

[5] M.S. Borczuch and V.M. Uzdin, J. Magn. Magn. Mater., in press.

[6] V.M. Uzdin and C. Demangeat, J. Magn. Magn. Mater. 165 (1997) 458.

[7] A.K. Kazansky and V.M. Uzdin, Phys. Rev. B 52 (1995) 9477.

[8] C. Turtur and G. Bayreuther, Phys. Rev. Lett. 72 (1994) 1557.

[9] A. Vega, L.C. Balbas, A. Chouairi, H. Dreyssé and C. Demangeat, Phys. Rev. B 49 (1994) 12797.

[10] V.N. Gittsovich, V.I. Minin, L.N. Romashev, V.G. Semenov, V.V. Ustinov and V.M. Uzdin, Proc. 10th Int. Conf. on Hyperfine Interactions. Book of Abstracts, Leuven, Belgium, 28 August–1 September (1995) P230-TH.

[11] V.N. Gittsovich, V.I. Minin, L.N. Romashev, V.G. Semenov, V.V. Ustinov and V.M. Uzdin, unpublished.

ELSEVIER

Computational Materials Science 10 (1998) 260–264

Relative stability of an on-top and an inverted Mn monolayer on Ag(1 0 0): Experiment and theory

O. Elmouhssine [a,*], G. Moraïtis [a], J.C. Parlebas [a], C. Demangeat [a], P. Schieffer [b],
M.C. Hanf [b], C. Krembel [b], G. Gewinner [b]

[a] *Groupe d'Etude des Matériaux Métalliques, IPCMS, 23, rue du Loess, 67037 Strasbourg Cedex, France*
[b] *Laboratoire de Physique et de Spectroscopie Electronique, 4, rue des Frères Lumière, 68093 Mulhouse, France*

Abstract

Mn films with a thickness of one monolayer (ML) have been deposited in ultra-high vacuum, at room temperature (RT), on a Ag(1 0 0) single crystal. Photoemission and ion scattering spectroscopy data indicate that after Mn evaporation a substantial fraction of the Mn atoms is located within the second topmost atomic layer. If the film is left at RT or mildly (60°C) annealed, the Mn atoms tend to exchange further with Ag atoms, increasing the Mn concentration in the second atomic plane. Eventually the second atomic plane of the sample is constituted mainly by Mn atoms, whereas the first atomic layer is almost a pure Ag plane, i.e. an inverted atomic Mn ML is formed. To directly compare experiment and theory, we have performed ab initio electronic band structure calculations on the energetic stability of 1ML of Mn on top of Ag(1 0 0) versus 1 ML of Mn covered by one Ag atomic plane. The case of a 2 ML-thick MnAg alloy is also investigated. The relation between magnetism and stability of Mn films is discussed. Copyright © 1998 Elsevier Science B.V.

Much work has been devoted in the last years to the growth and magnetic properties of thin transition metal films deposited on noble metals [1,2]. Interesting magnetic effects are expected when a species grows on a substrate with a crystallographic structure that is different from its bulk one. In this respect, we have grown Mn ultra-thin films on a Ag(1 0 0) substrate and analyzed the resulting overlayers by ion scattering spectroscopy (ISS) and X-ray photoelectron diffraction (XPD). We show that deposition of 1 ML of Mn on Ag(1 0 0) at RT gives rise to a superficial alloy, where the Mn atoms are confined within the first and the second (but not in deeper) atomic layers. Anneal-

ing the film or depositing the Mn at a higher temperature (however $< 75°C$) causes the Mn atoms initially in the first layer to occupy sites in the second layer. To complete this structural study we have performed ab initio electronic band structure calculations. The experimental and theoretical results are in good agreement and show evidence of an inverted Mn ML, i.e. 1 ML of Mn covered by one Ag plane.

The Ag(1 0 0) single crystal has been mechanically and chemically polished, and then cleaned in ultra-high vacuum by Ar^+-sputtering and annealing cycles. The Mn films were evaporated onto the Ag substrate kept at different temperatures (from RT to 170°C) and at a rate of ~ 0.3 ML/min (1 ML corresponds to the Ag(1 0 0) surface atomic density). The Mn thickness was calibrated by means of a quartz balance and the

---

* Corresponding author. Tel.: +33 03 88 10 70 84; fax: +33 03 88 10 72 49; e-mail: omar@nantos.u-strasbg.fr.

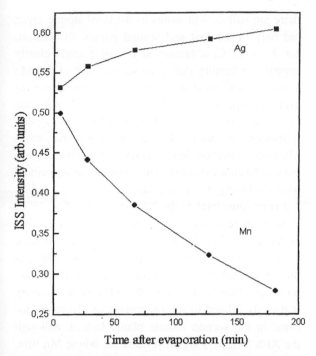

Fig. 1. Evolution of the Ag and Mn ion scattering spectroscopy peaks area as a function of time elapsed after deposition of 1 ML of Mn at RT on Ag(1 0 0).

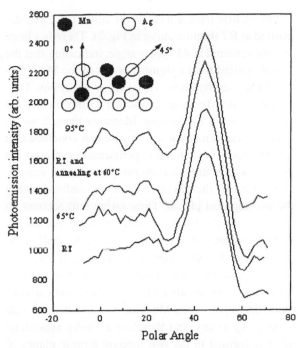

Fig. 2. Angular distributions of the $Mn2p_{3/2}$ core level intensity for 1 ML of Mn after deposition at RT 65°C, 95°C, and after deposition at RT followed by annealing at 60°C. The data are taken in the Ag(0 1 0) plane. The inset corresponds to the crystallographic configuration of the Mn film deposited at 95°C.

intensity of the $Mn2p_{3/2}$ core level photoemission signal. Every ISS spectrum (taken with 1 keV He^+ ion beam) exhibited two well-marked features corresponding to He^+ ions scattered from Ag and Mn atoms located in the surface, and whose area is proportional to the Ag and Mn concentration in the first atomic plane. Indeed, because of our experimental geometry (described in [3]), the contribution to the ISS signal of atoms belonging to the second atomic plane can be neglected.

Fig. 1 presents the ISS area of the Ag and Mn peaks in the ISS spectra for 1 ML of Mn deposited at RT as a function of elapsed time after Mn deposition. The Ag (Mn) ISS intensity has been normalized to the signal corresponding to the clean Ag(1 0 0) substrate (a thick Mn overlayer). We can see in Fig. 1 that immediately after Mn deposition there are only 0.5 ML of Mn in the surface of the film, and that about 0.5 ML of Ag are detected. Moreover, there is a Ag enrichment of the surface as a function of time (0.6 ML are present after 180 min), and the amount of Mn in the

first atomic plane tends clearly to decrease. Note that for intervals higher than 100 min, only the Ag signal augmentation is significant, as the Mn signal height is partly modified by selective adsorption of impurities (essentially oxygen) on the Mn atoms. At this stage, we conclude that the Mn growth is not layer by layer, as in this case the topmost plane would be constituted by Mn atoms only, and that the film is unstable at RT.

Fig. 2 displays the $Mn2p_{3/2}$ core level intensity modulations as a function of emission angle for 1 ML of Mn deposited at different temperatures: at RT 65°C, 95°C, and at RT followed by annealing at 60°C. All the data were taken at RT and after at least 100 min after Mn evaporation. Note that data recording is long (more than 1 h) because of the low intensity of the $Mn2p_{3/2}$ signal (only 1 ML has been deposited), thus it is not possible to probe by XPD the instability effects for these samples.

Let us first have a look at the curve for 1 ML deposited at RT (bottom curve in Fig. 2). There is a large feature centered at 45° polar angle, indicating that the signal emitted from a significant part of the Mn is forward focussed by above-lying Mn or Ag atoms. Thus a part of the Mn atoms is located within the second atomic plane of the sample. Moreover there is no intensity enhancement at 0°, thus we can conclude that no Mn atoms (within the experimental precision) are located within the third atomic plane of the sample. Consequently, the Mn atoms must be confined within the two topmost planes. Thus, taking into account on the one hand the ISS data, which indicate that the Ag surface concentration increases with time, and on the other hand the XPD curves, meaning that there are Mn atoms located within the second atomic plane at about 100–160 min after Mn deposition, one can conclude that a part of the Mn adatoms is exchanged with surface Ag atoms, and therefore a MnAg superficial alloy is formed in the two topmost atomic planes of the sample. As the $Mn2p_{3/2}$ intensity enhancement is found at 45° polar angle data, and according to our LEED measurements [4], this alloy is substitutional, i.e. the Mn atoms occupy Ag sites on the Ag fcc lattice. Moreover, the value of Mn concentration within the first plane immediately after deposition (0.5 ML, although 1 ML has been deposited) indicates that alloy formation takes place as soon as deposition starts. After the film has been deposited, the process of incorporating Mn atoms in the second plane keeps going, as indicated by the ISS signals evolution. In other words, the Mn films prefer to grow as an inverted ML (i.e. Mn atoms covered by a Ag plane), rather than an on-top Mn ML or Mn bilayers.

In order to study the influence of temperature on the film structure, 1 ML films have been deposited at different substrate temperatures, and then cooled to RT, or deposited at RT and then annealed at higher temperatures for about 250 min, from 60°C to 170°C. The $Mn2p_{3/2}$ modulations are shown in Fig. 2 for deposition at 65°C and 95°C, and annealing at 60°C. We found that for $T < 75°C$, the overall shape of the XPD spectrum is not modified when compared to the bottom curve of the figure: there is only one well-marked peak at 45° polar angle, indicating that

there are still no Mn atoms in the third atomic layer, but only in the first and second planes. In contrast, for $T \geq 75°C$, a feature at 0° polar angle clearly appears, indicating that a substantial part of the Mn atoms is now located in the third atomic plane, or even in deeper planes. This means that bulk diffusion occurs, and that a more dilute bulk Mn_xAg_{1-x} alloy is formed (see inset in Fig. 2). In other words, when Mn is deposited on Ag(1 0 0) the stable structure is a bulk substitutional alloy. This result is in agreement with the MnAg phase diagram.

Let us come back to the XPD curves for $T < 75°C$. Clearly the feature at 45° is higher when Mn is evaporated at 65°C or deposited at RT and then annealed at 60°C than for a simple deposition at RT. Now, in the present simple case where the atomic chains are no longer than two atoms, the XPD peak intensity is directly related to the proportion of Mn atoms located in the second atomic plane. Indeed, although the XPS $Mn2p_{3/2}$ signal reflects the whole Mn film, only the Mn atoms embedded in the second layer have their emission forward focussed by the first plane, i.e. only these atoms contribute to the peak at 45°. Consequently, the higher substrate temperature during or after deposition causes a larger amount of Mn atoms to occupy sites within the second plane. A calculation of the contrast C in the XPD curves leads to a more quantitative description of the temperature influence. The contrast C is given by $C = (I_{max} - I_{min})/((I_{max} + I_{min})/2)$, where I_{max} is the height of the XPD peak at 45°, and I_{min} the minimum in the XPD curve before the 45° maximum. $I_{max} - I_{min}$ is determined after subtraction of a linear background. For depositions at RT and 65°C, the values of C are 60% and 63%, respectively. When the film is grown at RT and then heated at 60°C, C is even higher (70%). Thus we can conclude that the mechanisms implied in the inverted layer formation (during and after Mn evaporation), i.e. surface diffusion and atomic place exchange, are thermally activated. Moreover, the dependence of C with temperature suggests that the activation energy increases with the Mn concentration in the second atomic plane: at the beginning of deposition, Mn incorporation in the Ag surface is very easy, but becomes slower and slower when the second atomic plane is enriched with Mn;

probably the Mn atoms diffuse on the surface before to get finally exchanged with Ag atoms, or Ag atoms diffuse from a step edge above Mn atoms [5]. As a result, a significant increase of the Mn proportion in the second layer can only be observed by increasing the annealing time. This explains why the contrast C is higher for the annealing at 60°C (about 250 min) than for the deposit (that takes 4 min) at 65°C and cooling at RT.

Finally, we tried to estimate the Mn concentration in the second plane by relating the ISS to the XPD data. Fig. 1 (growth at RT) indicates that 0.6 ML of Ag and 0.4 ML of Mn are present at the surface after 180 min. As 1 ML of Mn has been deposited, there are ~0.6 ML of Mn in the second layer, which corresponds to a contrast C of 60% in Fig. 2. Therefore, for an annealing at 60°C, where $C = 70\%$, we find ~0.7 ML of Mn in the second plane. A completely buried Mn ML would correspond to 1 ML in the second plane. However, as the annealing temperature is limited (< 75°C) to avoid bulk diffusion, the time needed to reach a full inverted ML is too long. In fact, at 60°C, we have a kinetically hindered situation. Moreover, ISS data indicate that contamination of the film cannot be neglected, when the annealing times are quite long, i.e. several hours. Impurities atoms adsorbed on Mn atoms prevent the latter to be incorporated in the second atomic plane.

In the next step of this work, we performed electronic band structure calculations to study the relative stability of Mn/Ag(1 0 0) films in three structural situations, which are competing in the experimental process: (1) Mn (1ML)/Ag(1 0 0), (2) Mn_1Ag_1 (2 ML) alloy/Ag(1 0 0) and (3) Ag (1ML)/Mn (1ML)/Ag (1 0 0). The calculations are based on the density-functional theory in the local spin density approximation. We used the local exchange–correlation potential of von Barth and Hedin [6]. The equations are solved using the tight–binding linear muffin-tin orbital method [7,8] in the atomic-sphere approximation. The Mn/Ag(1 0 0) system is modelled by a nine-layer (1 0 0) slab consisting of a seven-layer Ag(1 0 0) film and 1 ML Mn on each side of the film, with two atoms per layer, hence allowing for the description of a $c(2 \times 2)$ structure. In the situation (2) (resp. (3)), we only inverted half (resp. all) Mn atoms

Table 1

Energy differences between the three structural situations (in mRy/Mn atom) ((1) Mn (1 ML)/Ag(1 0 0), (2) MnAg alloy/Ag(1 0 0), (3) Ag (1 ML)/Mn (1 ML)/Ag(1 0 0)) for the nonmagnetic (NM), ferromagnetic (FM) and antiferromagnetic (AFM) states

	$\Delta E_{(2)-(1)}$ (mRy/Mn atom)	$\Delta E_{(3)-(2)}$ (mRy/Mn atom)	$\Delta E_{(3)-(1)}$ (mRy/Mn atom)
NM	−8.50	−37.25	−45.75
FM	−10.50	−12.75	−23.25
AFM	−7.75	−11.25	−19.00

with half (resp. all) interface Ag atoms. All nonmagnetic and magnetic calculations are performed taking the $c(2 \times 2)$ unit cell into account. The effect of magnetism on the stability of the Mn films is investigated by performing calculations for nonmagnetic (NM) as well as ferromagnetic (FM) and antiferromagnetic (AFM) films.

The results of the energy differences between the above structural situations are reported in Table 1. First, we compare the total energy $E_{(1)}$ of 1 ML of Mn on top of Ag(1 0 0) with the total energy $E_{(2)}$ of 2 ML-thick MnAg alloy on the Ag(1 0 0) substrate. The energy difference $\Delta E_{(2)-(1)} = E_{(2)} - E_{(1)}$ is negative for nonmagnetic as well as for magnetic states, meaning that the alloy formation is energetically preferred to 1 ML of Mn overlayer on Ag(1 0 0). This result is in agreement with the experimental observation of the alloy formation in the two topmost layers of the sample, immediately after Mn deposition. Subsequently, we compare the total energy $E_{(2)}$ of 2 ML-thick MnAg alloy on the Ag(1 0 0) substrate with the total energy $E_{(3)}$ of 1 ML of Mn on Ag(1 0 0) covered by one Ag layer. This serves as a test for the stability of the MnAg alloy against the complete wetting of the Mn by the substrate atoms. The negative values of $\Delta E_{(3)-(2)}$ mean that the inverted Mn monolayer is the energetically preferred state, which confirms our experimental findings.

Complementary, we also reported in Table 1 the energy difference $\Delta E_{(3)-(1)} = E_{(3)} - E_{(1)}$ between the inverted Mn monolayer situation (3) and the situation (1) where 1 ML of Mn is formed on top of Ag(1 0 0). Since it is obvious from the above discussion that the inverted Mn monolayer situation is the energetically preferred state, this energy difference is given in order to estimate the effect of magnetism on the stability of the Mn films. In this way, we compare the $\Delta E_{(3)-(1)}$ values of the magnetic states (FM and AFM) with its NM state value. The effect of magnetism appears with a large reduction of $\Delta E_{(3)-(1)}$ by about 50%, which means that magnetism acts against interdiffusion and tends to stabilize the Mn monolayer film on the Ag(1 0 0) substrate. The magnetic energy gain for the FM state (22.5 mRy per Mn atom) is lower than for the AFM one (26.75 mRy per Mn atom), indicating that the strength of the effect of magnetism on the stability of the magnetic films depends on the nature of their magnetic structure. The main effect, however, remains the formation of large magnetic moments on Mn atoms ($\sim 4\mu_B$).

To summarize, we have shown that Mn monolayer on-top of Ag(1 0 0) is unstable at RT. A MnAg superficial alloy is formed in the two topmost atomic layers, immediately after deposition. Moreover, the Mn incorporation in the second plane is thermally activated, and an inverted Mn layer tends to be formed by mild annealing. These findings are confirmed by our electronic band structure calculations, which show that the inverted Mn layer constitutes the energetically preferred state as compared to a MnAg alloy formation or to Mn monolayer on top of Ag(1 0 0). Moreover, the calculations show that magnetism acts against interdiffusion and tends to stabilize the Mn monolayer film on top of Ag(1 0 0), but not to the extent to prevent the formation of an inverted Mn layer at RT or above.

References

[1] S. Blügel, Phys. Rev. Lett. 68 (1992) 851.
[2] J.A.C. Bland and B. Heinrich, Ultrathin Magnetic Structures (Springer, Berlin, 1994).
[3] P. Schieffer, C. Krembel, M.C. Hanf and G. Gewinner, Phys. Rev. B 55 (1997) 13 884.
[4] P. Schieffer, C. Krembel, M.C. Hanf, D. Bolmont and G. Gewinner, Solid State Commun. 97 (1996) 757.
[5] P. Schieffer, M.C. Hanf, C. Krembel, M.H. Tuilier, G. Gewinner and D. Chandesris, Surf. Rev. and Lett. (1997), in press.
[6] U. von Barth and Hedin, J. Phys. C 5 (1972) 1629.
[7] O.K. Andersen and O. Jepsen, Phys. Rev. Lett. 53 (1984) 2571.
[8] O.K. Andersen, Z. Pawlowska and O. Jepsen, Phys. Rev. B 34 (1986) 5253.

ELSEVIER

Computational Materials Science 10 (1998) 265–268

COMPUTATIONAL
MATERIALS
SCIENCE

Magnetism of epitaxial Ru and Rh monolayers on graphite

P. Krüger *, J.C. Parlebas, G. Moraitis, C. Demangeat

IPCMS (UMR 46 CNRS), 23, rue du Loess, 67037 Strasbourg Cédex, France

Abstract

Recently, Pfandzelter et al. (1995) reported the first observation of monolayer ferromagnetism of a 4d metal, namely in a Ru monolayer grown on graphite. Using the tight-binding linear-muffin-tin-orbital (TB-LMTO) method we have calculated the electronic and magnetic structure of epitaxial Ru and Rh monolayers on graphite with the experimentally determined atomic density. Monolayers of the other 4d elements were found to be non-magnetic already in the free-standing limit. The magnetic structure of the Ru and Rh monolayers is studied as a function of metal–graphite interlayer distance h. They become magnetic at $h = 4.5$ a.u. (Ru) and $h = 4.8$ a.u. (Rh) in a first-order transition. In the assumed $p(2 \times 2)$ super-structure, the moments on the "hollow" site atoms are up to four times bigger than those on the "on-top" site atoms. For $h > 5.4$ a.u. (Ru) and $h > 5.1$ a.u. (Rh) the site dependence vanishes and the moments of the free monolayers are approximately reached ($1.9 \, \mu_B$ and $1.2 \, \mu_B$, respectively). Copyright © 1998 Elsevier Science B.V.

Keywords: 2D ferromagnetism; 4d transition metals; Graphite surface

Two-dimensional ferromagnetism in 4d transition metal (TM) monolayers (MLs) deposited on non-magnetic substrates has been predicted theoretically already, several years ago, for Rh and Ru MLs on the Ag(0 0 1) or Au(0 0 1) surfaces [1]. In experiment, however, long range ferromagnetic order has never been observed in these systems, which is probably due to strong diffusion of the TM adatoms into the noble metal substrate [2,3]. Graphite has been suggested as an alternative substrate, because TM atoms diffuse much less into it. The graphite C(0 0 0 1) surface is furthermore known to be very flat and it has, like the noble metals, only a small band overlap with the TM d-bands. Pfandzelter et al. [4] studied the growth and magnetic properties of Ru MLs on the

graphite surface. They found that Ru grew laterally on C(0 0 0 1) until a homogeneous ML was formed. By spin-polarized secondary electron emission spectroscopy they observed bidimensional ferromagnetism of a 4d TM for the first time. Very recently, Chen et al. [5] calculated the spin-polarized electronic structure of Ru, Rh, and Pd MLs on the graphite surface using the full potential linearized augmented plane wave method. They considered different epitaxial structures where the interatomic distances between TM atoms were either about 10% smaller or 60% bigger than in the bulk structures. The magnetic moments of the free MLs depended strongly on the chosen structure, i.e. on the atomic density of the ML. For the adsorbed MLs at equilibrium interfacial distance, only in the most dilute structure, where the atomic density in the ML is 2.5 times lower than in a dense ML in bulk, small moments of $0.28 \, \mu_B$ and $0.24 \, \mu_B$

* Corresponding author. Tel.: +33 388 107005; fax: +33 388 107249; e-mail: Kruger@belenus.u-strasbg.fr.

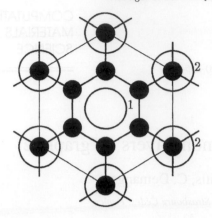

Fig. 1. Top view of the epitaxial structure assumed in the calculation. Small filled circles represent C atoms of the graphite surface; big open circles represent transition metal atoms a distance h above the surface. There are one "hollow" site adatom (1) and two "on-top" site adatoms (2) per $2D$ unit cell, which is indicated by thin lines. Thick lines symbolize the σ bonds in graphite.

could survive for, respectively, Ru and Rh. The moments were furthermore found to be very sensitive to changes in the interfacial distance.

In this article we present calculations on the electronic and magnetic structure of pseudomorphically adsorbed 4d TM MLs on the graphite surface by means of the tight-binding linear-muffin-tin-orbital (TB-LMTO) method within the atomic sphere approximation (ASA) [6]. We assume a $p(2 \times 2)$ superstructure where the adsorbed TM ML forms a simple hexagonal lattice with nearest neighbour distance twice that of graphite. These hexagonal MLs are slightly expanded dense planes of the bulk structures. The mismatch compared to a (0001) plane of hcp Ru or to a (111) plane of fcc Rh is 6%. The calculations have been carried out in the high symmetry structure depicted in Fig. 1, which is the simplest epitaxial structure compatible with the experimental data in [4]. In this geometry, the $2D$ unit cell contains eight C atoms at the graphite surface and three TM adatoms, one of them being at a "hollow" adsorption site (1) and the two others at the "on-top" adsorption sites (2). We have used a repeated slab geometry. The graphite surface is modelled by a symmetric bilayer (AA stacking of MLs rather than AB as in Bernal graphite). On either side of the graphite bilayer we

have put one TM ML in $p(2 \times 2)$ structure. For the in-plane interatomic distance and the interlayer distance in graphite, the experimental values have been taken (2.68 and 6.32 a.u., respectively). The interfacial distance h between TM ML and graphite ML is varied in the calculations. Three layers of empty spheres are added to separate two slabs in the three-dimensional supercell and a few more empty spheres are inserted in the interstitial region between the graphite layers and at the metal–graphite interface. The insertion of empty spheres is necessary in order to keep the overlap sufficiently small ($< 20\%$) between the space-filling muffin-tin spheres used in the ASA [7]. The empty spheres contain no ion core but the valence states are developed in the same manner as in the atomic spheres. When we vary h, only the radii of the empty spheres between TM and graphite are adjusted according to the variation of the interfacial volume. The radii of all other atomic and empty spheres are kept constant. By this procedure we have tried to minimize the unphysical influence on the results as a function of h which is induced by different empty sphere positions and radii. The k-space integrations are carried out with the tetrahedron method using 38 irreducible k points in the 1/24 wedge of the first Brillouin zone (point group 6/mmm).

Previously, we studied free-standing MLs with the same structure as the adsorbed ones considered here [8]. We found that the Ru and Rh MLs are ferromagnetic with moments of $1.9 \mu_B$ and $1.2 \mu_B$, respectively. The rest of the 4d series was found to be non-magnetic. That only Ru and Rh are ferromagnetic could be understood simply by comparing the densities of states (DOS). The DOS around the Fermi energy for these two elements was with about 6 states/eV/atom more than twice bigger than that of the other 4d metals. So they have a much bigger tendency for ferromagnetism according to the Stoner criterion (assuming that the Stoner parameter does not vary much within the series). The adsorption of TMs on a non-magnetic substrate usually leads to a reduction of the magnetic moments because of stronger d-electron delocalization due to the adsorbate–substrate hybridization. For the adsorbed MLs we therefore expect only Ru and Rh to be possibly magnetic. For

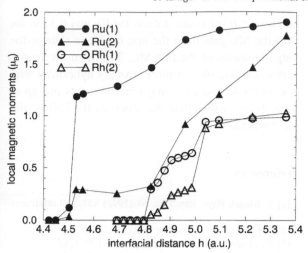

Fig. 2. Local magnetic moments of the adsorbed Ru and Rh monolayer as a function of interfacial distance. Full (open) symbols for Ru (Rh) moments. Circles for the hollow adsorption site (1) and triangles for an on-top adsorption site (2).

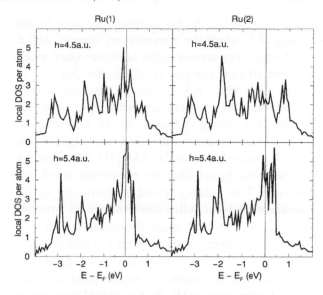

Fig. 3. Local densities of states of the non-magnetic state on the Ru(1) and Ru(2) atoms for two different interfacial distances h.

these two elements we have studied the magnetic structure of the MLs as a function of interfacial distance h in the interesting range from the onset of magnetism to a distance where the free-standing ML limit is approximately reached as far as the magnetic moments are concerned. The local magnetic moments for the two different adsorption sites are shown in Fig. 2. The Ru ML becomes magnetic at $h = 4.5$ a.u. The moment of the atom at the hollow site (1) jumps to $1.2\,\mu_B$ and the moment of the atom at the on-top site (2) to merely $0.3\,\mu_B$. They then remain approximately constant until about $h = 4.8$ a.u. before they smoothly converge to the free ML limit ($\mu = 1.9\,\mu_B$). The discontinuous magnetic transition at 4.5 a.u. can be understood by analysing the local DOS (in the paramagnetic state) which is shown in Fig. 3. At the transition ($h = 4.5$ a.u.) we find a relatively small local DOS at the Fermi energy (E_F) of about 3 states/eV/atom for Ru(1) and 2 states/eV/atom for Ru(2). The Ru(1) local DOS shows large peaks next to E_F on both sides. Such a structure of the DOS, where E_F falls in a relative gap between two peaks, is typical for a first-order magnetic transition. Krasko [9] showed that the ferromagnetic transition is of first order if the quantity $\chi \equiv n'' - 3(n')^2/n$ is positive and of second order otherwise. Here n, n', and n''

are, respectively, the DOS, its first and second energy derivatives, evaluated at E_F. For two high peaks below and above E_F that can be approximated around E_F by narrow parabola we get a particularly large value of χ and thus a first-order transition with a big jump in the magnetic moment. This analysis, which was derived for the total magnetization of a bulk system and the total DOS, should still approximately hold when it is applied to a local magnetic moment if we consider the corresponding local DOS. The local DOS on Ru(2) does not exhibit very large peaks near E_F for $h = 4.5$ a.u. This shows that it is more strongly hybridized with the graphite surface than the Ru(1) atom. We interpret its small magnetic moment of about $0.3\,\mu_B$ as induced by the large moments of the neighbouring Ru(1) atoms. The strong site dependence of the moments was already seen in a previous tight-binding calculation and is related to the directional character of the hybridization between graphite π- and TM d-states [10]. At $h = 5.4$ a.u. large peaks around E_F have developed in the local DOS of both sites leading to big moments close to the free ML value of $1.9\,\mu_B$.

In the Rh ML the moments are considerably smaller than in the Ru ML, but the overall dependence of the moments on the interfacial distance is quite similar. The main differences are: the onset of magnetism occurs at 0.3 a.u. higher interfacial distance h; the range of h, where the moments vary strongly with h and where they depend on the adsorption site is much smaller (0.3 a.u. compared to 0.9 a.u. for Ru).

In the ASA, energy changes associated with anisotropic deformations are badly reproduced, because they strongly depend on the non-spherical contributions to the potential and the charge density, which are neglected in the ASA [11]. Therefore we cannot reliably calculate the equilibrium interfacial distance. This quantity should be determined by experiment or by a full-potential electronic structure calculation. The equilibrium distances of the Ru MLs in the three structures investigated by Chen et al. [5] lie between 3.2 and 4.0 a.u., and thus a good deal below 4.5 a.u., the magnetic transition point we found here. Furthermore, a ML in the epitaxial structure of Fig. 1 is likely to show some corrugation between type (1) and type (2) atoms since the chemical environments are different. This might change the local magnetic moments slightly. After all, more structural information is needed to answer the question if dense, homogeneous Ru or Rh MLs on graphite are magnetic.

In summary, we have investigated the electronic and magnetic structure of dense epitaxial Ru and Rh monolayers on the graphite surface as a function of metal–graphite interlayer distance h. In a small interval of h (of width 0.9 a.u. for Ru and 0.3 a.u. for Rh) the ML gets from the non-magnetic state to the large moments of the free ML. The atoms at the on-top adsorption site are more strongly hybridized with the substrate and their magnetic moments are up to four times smaller than the atoms at the hollow site in this range of h.

References

[1] S. Blügel, Phys. Rev. Lett. 68 (1992) 851, and references therein.
[2] I. Turek et al., Phys. Rev. Lett. 74 (1995) 2551.
[3] W. Hergert et al., Surf. Rev. Lett. 2 (1995) 203.
[4] R. Pfandzelter, G. Steierl and C. Rau, Phys. Rev. Lett. 74 (1995) 3467.
[5] L. Chen, R. Wu, N. Kioussis and J.R. Blanco, J. Appl. Phys. 81 (1997) 4161.
[6] O.K. Andersen, O. Jepsen and D. Glötzel, in: Highlights of Condensed Matter Theory, eds. F. Bassami, F. Fumi and M.P. Tosi (North-Holland, Amsterdam, 1985) p. 59.
[7] O.K. Andersen, A.V. Postnikov and S. Yu. Savrasov, Mat. Res. Symp. Proc. 253 (Materials Research Society) (1992) 37.
[8] P. Krüger, A. Rakotomahevitra, G. Moraitis, J.C. Parlebas and C. Demangeat, Physica B 237–238 (1997) 278.
[9] G.L. Krasko, Phys. Rev. B 36 (1987) 8565.
[10] P. Krüger, C. Demangeat, J.C. Parlebas and A. Mokrani, Material Science and Engineering B 37 (1996) 242.
[11] M. Methfessel, C.O. Rodriguez and O.K. Andersen, Phys. Rev. B 40 (1989) 2009.

ELSEVIER

Computational Materials Science 10 (1998) 269–272

COMPUTATIONAL
MATERIALS
SCIENCE

Magnetic properties simulations of Co/Ru interfaces

K. Rahmouni, S. Zoll, N. Persat, D. Stoeffler, A. Dinia *

IPCMS-GEMM, UMR 46 CNRS-ULP, 23 rue du Loess, 67037 Strasbourg, France

Abstract

In plane-electronic transport in UHV Co(32 Å)/Ru(x Å)/Co(32 Å) sandwiches is analyzed with particular attention paid to the role of interface mixing on the moment and the giant magnetoresistance (GMR) effect. The resistivity measurements show that the GMR is very small in these sandwiches due to the presence of intermixed CoRu alloy at the interfaces. The magnetic profile of this mixed interface has been determined by ab initio DFT-LDA calculations which show two different regions. Magnetic and non-magnetic regions which are centers of spin-dependent and strong spin-independent scattering, respectively, reduce the flow of polarized electrons from one ferromagnetic layer to the other. This intermixed interface has been successfully included in our GMR simulation using a semi-classical description of the two spin channel currents to reproduce the experimental results. Copyright © 1998 Elsevier Science B.V.

Keywords: Magnetoresistance; Interface; Intermixing; Simulation; Co/Ru

1. Introduction

The giant magnetoresistance (GMR) recently observed in several magnetic superlattices such as Fe/Cr [1] and Co/Cu [2] has attracted much attention as one of the novel magnetotransport phenomena. It has been clearly established that the magnetoresistance (MR) is due to the spin-dependent scattering (SDS) of electrons in the magnetic/non-magnetic layers. Depending on the used metals, the SDS may have an interfacial [3] or bulk [4] origin, or both. The bulk SDS is a property of the ferromagnetic element, while interfacial SDS is a consequence of the combination of magnetic and non-magnetic metals at the interfaces. In both cases the spin asymmetry of the scattering coefficients α is assumed to be $\alpha\lambda^{\uparrow}/\lambda^{\downarrow}$ where $\lambda^{\uparrow}(\lambda^{\downarrow})$

is the mean free path (MFP) of spin$^{\uparrow}$ (spin$^{\downarrow}$) electrons. For systems with interfacial SDS, such as Fe/Cr multilayers, the structure of the interface plays a dominant role. In particular interface roughness increases the GMR in Fe/Cr multilayers [5]. However in the present study we show an opposite tendency where the roughness (or compositional mixing) of the interface reduces the amplitude of the GMR.

The aim of this work is to study the effect of interfaces on the magnetic and transport properties. The GMR results of [100 Å Ru/32 Å Co/x Å Ru/32 Å Co/24 Å Ru] sandwiches prepared by ultra high vacuum evaporation on mica substrates [6] are reported. These show a small GMR effect in this system in spite of the predicted large effect by several authors [7,8]. To understand the origin of the GMR reduction, the moment profile at the interfaces is determined by ab initio DFT-LDA calculations of the magnetization with intermixed interfaces. They show that the moment of

* Corresponding author. Tel.: +33 3 88 10 70 67, fax: +33 3 88 10 72 49; e-mail: aziz@atlas.u-strasbg.fr.

the Co atoms located at the interfaces decreases continuously with increasing Ru concentration and disappears completely for high Ru concentrations. This intermixed interface is taken into account in our GMR simulation based on the semi-classical model of Camley and Barnas. The results of the simulation reproduce very well the experimental data with realistic physical parameters.

2. Magnetic moment profile

The NMR analysis of the spectra performed on the Co/Ru sandwiches have shown that the interfaces are intermixed along three monolayers and accounted for in terms of dead magnetic layers to about 1 monolayer per interface [9]. This is in agreement with the results of the variation of the saturation magnetic moment surface density of Co, $M_s t_{Co}$, as a function of cobalt thickness (t_{Co}) as shown in Fig. 1 for mica/100 Å Ru/x Å Co/50 Å Ru single magnetic layer. Indeed the magnetization decreases linearly with decreasing t_{Co} and intercepts the abscissa at a thickness of 4 Å. To confirm this experimental result, calculations of the moment profile for Co/Ru superlattice with intermixed interfaces has been performed. We used the first principles augmented spherical waves (ASW) [10] method to determine the magnetic moments on each

site from the electronic structure. The cell contains 32 equivalent atoms (four atoms per in-plane cell) and the band structure was computed with 300 k-points in the irreducible part of the Brillouin zone. The accuracy on the magnetic moment is estimated to be better than 0.01 μ_B [11]. To provide a realistic comparison with the experimental data, we used the mixed interfaces on the basis of the linear NMR profile [9] with the following structure:

$$Co/Co/Co/Co_{0.75}Ru_{0.25}/Co_{0.5}Ru_{0.5}/$$
$$Co_{0.25}Ru_{0.75}/Ru/Ru.$$

Fig. 2 shows the evolution of the magnetic moment for all atoms in the stack as a function of their position. The horizontal dashed line corresponds to the bulk Co magnetic value. The Co moment value for the three pure Co atomic layers is approximately close to the bulk value. However for the first, second and third mixed layers we observe a reduction in the Co moment, which is approximately of the order of 12%, 40% and 95% for the Co layers with 25%, 50% and 75% Ru, respectively. As a consequence, the last monolayer with no magnetic moment does not contribute to the SDS and may induce spin-flip scattering. Thus the intermixed layers decrease the flow of electrons across the spacer, effectively isolating the

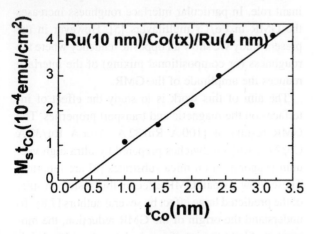

Fig. 1. Variation of the measured saturation magnetization per unit surface of Co, $M_s t_{Co}$, with Co layers thickness t_{Co} for mica/100 Å Ru/x Å Co/50 Å Ru single magnetic layer at $T = 300$ K.

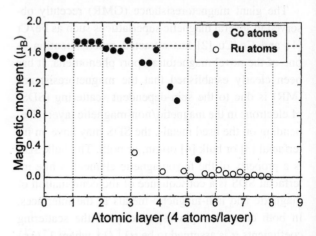

Fig. 2. Evolution of the magnetic moment as a function of the position of the Co atoms in the following stack: Co/Co/Co/Co$_{0.75}$Ru$_{0.25}$/Co$_{0.5}$Ru$_{0.5}$/Co$_{0.25}$Ru$_{0.75}$/Ru/Ru from ab initio calculations. The dashed line corresponds to the bulk Co moment value.

ferromagnetic layer from each other and reduce the effect of the GMR as it will be developed in the next section.

3. Magnetoresistance simulations

In this section, we present the results of the MR simulations for Co/Ru sandwiches with flat and intermixed interfaces. To describe the transport phenomena we used the classical model of Camley and Barnas [12] based on Boltzmann kinetic equation. The conductivity for both antiparallel and parallel configurations is obtained by adding the contributions of the spin-up and the spin-down electrons, calculated separately. This corresponds to the two currents model, which provides a good description of electron transport in magnetic 3d metals. In this approach, the bulk scattering processes are taken into account by spin-dependent relaxation times in the ferromagnetic layers. The interface roughness is included by some specularity coefficients which enter appropriate boundary conditions. The present model is an extension of the Barnas model, where calculations of the MR incorporate several layers simultaneously. In this case the mixed interface is represented by a 6 Å thick layer with a higher resistivity and a different scattering coefficient parameter than for bulk layers. To simplify the problem, we have assumed that the specular reflection coefficients $R^\uparrow$ and $R^\downarrow$ are sufficiently small to be neglected, i.e. $R^\uparrow = R^\downarrow = 0$. This assumption is reasonable if the potential step at the interfaces is relatively small. Since the conductivity of both materials is comparable and the interlayer thickness is smaller than the ferromagnetic thickness [12], we can reduce the number of independent parameters to only 5: $T^\uparrow$, $T^\downarrow$, Nb, λ_{Co} and λ_{Ru}; where $T^{\uparrow(\downarrow)}$ are the transmission coefficients depending on the interface potentials, $Nb = \lambda^\downarrow/\lambda^\uparrow$ (MFP ratio) which describes the spin-asymmetry in scattering inside the ferromagnetic films and $\lambda_{Co(Ru)}$ ($\lambda = \frac{1}{2}(\lambda^\downarrow + \lambda^\uparrow)$) is the MFP for the Co(Ru) layer. Due to the omitted reflection coefficients, the interface diffusive scattering parameters are determined by $D^{\downarrow(\uparrow)}$ where $D^{\downarrow(\uparrow)} = 1 - T^{\downarrow(\uparrow)}$. The parameter $N_S = D^\uparrow/D^\downarrow$ describes the spin asym-

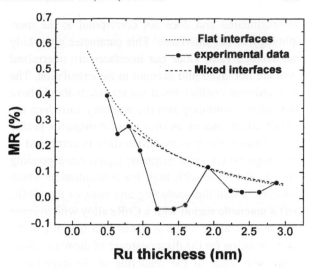

Fig. 3. Variation in the magnetoresistance as a function of the Ru spacer layer thickness at $T = 300$ K for [100 Å Ru/32 Å Co/x Å Ru/32 Å Co/24 Å Ru] sandwiches. The dotted and dashed lines correspond to simulations using the Camley and Barnas model [12] with the following structure: [100 Å Ru/Ru(Co)/32 Å Co/Co(Ru)/x Å Ru/Ru(Co)/32 Å Co/Co(Ru)/24 Å Ru] in which Co(Ru) and Ru(Co) take the thickness 0 Å (flat interface) and 6 Å (mixed interfaces), respectively.

metry of the diffusive scattering of electrons at the interfaces.

To elucidate the magnetic behavior of Co/Ru interfaces, we first fit MR data of a symmetrical sandwich, with the following composition: [100 Å Ru/32 Å Co/x Å Ru/32 Å Co/24 Å Ru] where interfaces are considered to be flat. The result of the simulation (dotted line) is reported in Fig. 3 which reproduces well the experimental data. These results have been obtained with the following parameters: the mean free path MFP = $\frac{1}{2}(\lambda^\downarrow + \lambda^\uparrow)$ are of 30 Å and 25 Å in Co and Ru, respectively, and the bulk scattering coefficient $N_b = 0.11$. These values are in agreement with Bruce experiments which use the GMR to measure the mean free paths [13]. However the transmission parameters which describe the interface behavior and which depend specially on the potentials at the interfaces give the values $T^\uparrow = 0.47$ and $T^\downarrow = 0.9$. These values give for a diffusive scattering parameter $N_S = 5.3$ which is close to the value found for Fe/Cr interfaces. This N_S parameter is certainly overestimated and does not correspond to the morphology of our interfaces. This parameter gives only an average value since our interfaces are intermixed and are not taken into account in the calculation. The transmission coefficients at the interfaces should have a smaller asymmetry and the tendency to reduce the GMR effect. Indeed, as shown in the magnetic profile calculations, the intermixed interface is composed of two regions: (i) a non-magnetic region corresponding to a CoRu alloy with high Ru concentration which induces a spin-flip scattering and reduces the GMR, (ii) a magnetic region with a CoRu alloy with average Ru concentration which is lower than the critical Ru concentration for the disappearence of the magnetism. This also leads to the reduction of the asymmetry between the spin-up and spin-down and consequently to the reduction of the GMR. In this magnetic region we can also expect another effect if the interfaces are asymmetrical. Indeed in the case where the Ru concentration is different within the interfaces, we can obtain for a certain concentration range an α parameter lower than 1 for one interface while it is larger than 1 for the second interface inducing inverse GMR [6]. For all these reasons we have decided to include mixed interfaces in our calculations as a thin layer with 6 Å thickness using the following structure: [97 Å Ru/Ru(Co)/26 Å Co/Co (Ru)/x Å Ru/Ru(Co)/26 Å Co/Co(Ru)/21 Å Ru] where Co(Ru) and Ru(Co) are CoRu (RuCo) alloys with few atomic percent Ru in Co. To simplify the problem we have taken the same electron MFP of about 15 Å for this intermixed layer and the same MFP for Co and Ru layers as used for flat interfaces. As shown in Fig. 3 (dashed line) the best agreement between simulation and the experimental results is obtained with the following parameters: (i) the scattering coefficients for Co(Ru) and Ru(Co) are 0.5 and 1.1, respectively, which are in good agreement with Itoh and Inoue calculations [9] of the density of states for Co alloyed with Ru as

impurity ($\alpha = \rho^{\downarrow}/\rho^{\uparrow} = 0.4$) and Ru alloyed with Co as impurity ($\alpha = 1.2$); (ii) the transmission coefficients are $T^{\uparrow} = 0.64$ and $T^{\downarrow} = 0.9$ for Co/CoRu interfaces and $T^{\uparrow} = 0.9$ and $T^{\downarrow} = 0.9$ for CoRu/Ru interfaces. The transmission coefficients for the Co/CoRu interfaces lead to an $N_S = 3.6$ which is more realistic than the previous one. We can conclude that by including mixed interfaces in our simulations, the GMR could be well fitted by the classical model of Camley and Barnas with realistic physical parameters.

In summary, we have demonstrated that our model correctly describes the majority of the experimental data. In particular we have seen that diffused interfaces in Co/Ru systems decrease the amplitude of the MR effect.

References

[1] M.N. Baibich, J.M. Broto, A. Fert, F. Nguyen Van Dau, F. Petroff, P. Etienne, G. Creuzet, A. Friederich and J. Chazelas, Phys. Rev. Lett. 61 (1988) 2472.
[2] S.S.P. Parkin, R. Bhadra and K.P. Roche, Phys. Lett. 66 (1991) 2152.
[3] P.M. Levy, S. Zhang and A. Fert, Phys. Rev. Lett. 65 (1990) 1643.
[4] B. Dieny et al., Phys Rev B 45 (1992) 806.
[5] E.F. Fullerton et al., Phys. Rev. Lett. 68 (1992) 859.
[6] A. Dinia, K. Rahmouni, D. Stoeffler, N. Persat, K. Ounadjela and H.A.M. van den Berg, Mater. Res. Soc. Symp. Proc. (1997).
[7] H. Itoh, J. Inoue and S. Maekawa, Phys. Rev. B 47 (1993) 5809.
[8] I.A. Campbell and A. Fert, Ferromagnetic Mater. 3 (1982) 9.
[9] J.P. Schillé, A. Michel, C. Mény, E. Beaurepaire, V. Pierron-Bohnes and P. Panissod, submitted.
[10] A.R. Williams, J. Kubler and C.D. Gelatt, Jr., Phys. Rev. B 19 (1979) 6094.
[11] D. Stoeffler and K. Ounadjela, Phys. Rev. B 49 (1994) 299.
[12] R.E. Camley and J. Barnas, Phys. Rev. Lett. 63 (1989) 664; J. Barnas and A. Fuss, Phys. Rev. B 42 (1990) 8110.
[13] B.A. Gurney, V.S. Speriosu, Phys. Rev. Lett. 71 (1993) 4023.

ELSEVIER

Computational Materials Science 10 (1998) 273–277

Complex magnetic behavior at the surface of B2 ordered FeCr alloy

F. Amalou [a], H. Bouzar [a], M. Benakki [a], A. Mokrani [b], C. Demangeat [c,*], G. Moraïtis [c]

[a] *Institut de Physique, Université de Tizi-Ouzou, 15000 Tizi-Ouzou, Algeria*
[b] *LPME, EA1153 DS4, 2, Rue de la Houssinière, 44072 Nantes, France*
[c] *IPCMS-GEMME, 23 rue du Loess, 67037 Strasbourg, France*

Abstract

Bouzar et al. (1997) have recently investigated the surface of B2 FeCr alloy. In all cases of crystal growth $\langle 0\,0\,1\rangle$, $\langle 1\,1\,0\rangle$ and $\langle 1\,1\,1\rangle$ they found the local polarization at the surface to be antiferromagnetically coupled with the subsurface layer in contrast to parallel coupling between Fe and Cr in bulk FeCr. In order to assert these results, we have investigated the local polarization of Fe at the $(0\,0\,1)$ surface of this alloy with a tight-binding linear muffin tin orbitals (TB-LMTO) model. Using general gradient approximation with Langreth–Mehl–Hu functional for $p(1 \times 1)$ and $c(2 \times 2)$ configurations we found the local polarization at the Fe surface layer to be antiferromagnetically coupled with the subsurface Cr layer and high magnetic moments compared to the bulk values. Copyright © 1998 Elsevier Science B.V.

Keywords: Surface magnetism; Fe/Cr alloy

1. Introduction

Several studies of FeCr alloy have been performed in these last few years. This system is of interest not only because of the possibility of growing it layer by layer with molecular beam epitaxy (MBE) but also because it may be viewed as alternating $(0\,0\,1)$ layers of Fe and Cr, thus forming the low thickness limit in the Fe/Cr/Fe multilayer family. Using the general potential linearized augmented plane wave (LAPW) method, Singh [1] has observed that shear elastic constants are found to be positive, indicating metastability of the B2 structure. He has also found that ferromagnetic Fe monolayers are ferromagnetically aligned with the adjacent Cr layers. This result

is opposite to what is observed in multilayers [2–4]. For $Fe_n Cr_m$ multilayers (with $n > 2$) Singh observed an antiparallel coupling at the interfacial Fe and Cr moments. Moroni and Jarlborg [5], using a linearized muffin tin orbitals (LMTO) calculations within local density approximation (LDA), have also found Fe and Cr to be ferromagnetically coupled. The same result has been reported by Moraïtis et al. [6] using a tight-binding linearized muffin tin orbitals (TB-LMTO) method [7]. In their calculation [6] a parallel coupling is found to be the ground state while an antiparallel configuration occurs when the lattice parameter is increased by 5%.

Bouzar et al. [8] have recently performed semi-empirical calculations on B2 FeCr surfaces at low Miller indices. Using a self-consistent tight-binding real space model within the unrestricted Hartree–Fock approximation to the Hubbard Hamiltonian, combined

* Corresponding author. Tel.: (33) 3.88.10.70.75; fax: (33) 3.88.10.72.49; e-mail: claude@belenus.u-strasbg.fr.

with the recursion method, they have investigated all possible solutions starting with parallel configuration. Self-consistency procedure leads to a spin-flip of the surface magnetic moments. As proposed by Bouzar et al. this configuration of the surface for B2 FeCr seems to be the ground state. In this work we will shed some light on this problem using a self-consistent TB-LMTO method [7] in general gradient approximation (GGA) [9,10]. We have considered two configurations (Fig. 1) for the surface: (i) ferromagnetic $p(1 \times 1)\uparrow$ and $p(1 \times 1)\downarrow$, (ii) in plane antiferromagnetic $c(2 \times 2)$, and we have performed a total energy calculation in order to study the relative stability of these two solutions.

2. Calculation model

To be able to compare between the total energies of the two $p(1 \times 1)$ and $c(2 \times 2)$ configurations we have chosen the same unit cell in real space as shown

in Fig. 1. The same number of k-points in the irreducible Brillouin zone for the two configurations has been used. In our TB-LMTO approach we used the general gradient approximation with the Langreth–Mehl–Hu [9] and Perdew–Wang [10] functional. In our calculations, the surface consists of *slabs* that are superposition of alternating Fe and Cr monolayers separated by five layers of empty spheres. It has been found that this number (5) is sufficient to obtain well-separated noninteracting slabs, that is charge vanishing in the central layer of empty spheres and no dispersion along the z-axis direction. Empty spheres consist of pseudo-atoms with no core states. They are located alternatively at the Fe and Cr sites. Their role is double: (i) reproduce the symmetry along the z-axis broken for a semi-infinite system, thus allowing us to calculate electronic structure using a method operating in the k-space, (ii) break the bonds for the atoms of the top layer of the slab thus creating the surface. First we have computed the total energy versus lattice parameter for bulk B2 FeCr

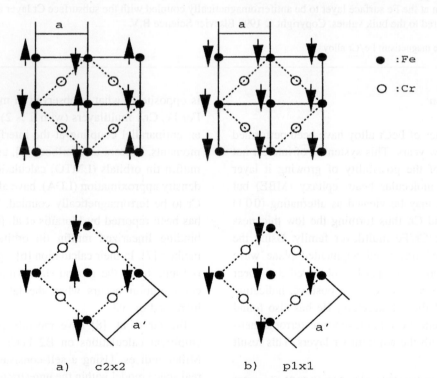

● : Fe

○ : Cr

Fig. 1. Unit cells (dashed lines) for (a) $c(2 \times 2)$ and (b) $p(1 \times 1)\downarrow$ configurations, for the surface of B2 FeCr with Fe top layer (filled circles), Cr atoms (hollow circles) are located on the adjacent subsurface layer, $a' = \sqrt{2}a$.

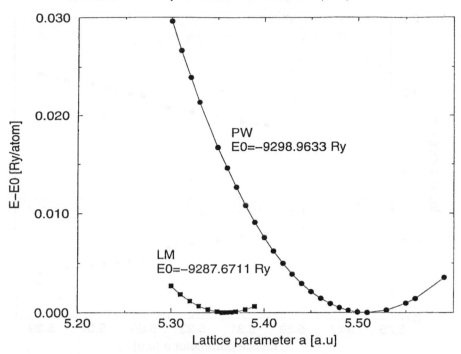

Fig. 2. Total energy versus lattice parameter around the ground state (E_0) for Langreth–Mehl–Hu (squares) and Perdew–Wang (circles) functional for B2 FeCr alloy.

alloy (Fig. 2) for Langreth–Mehl–Hu and Perdew–Wang functional. The equilibrium lattice constant in bcc magnetic B2 FeCr is found to be 5.36 a.u. for Langreth–Mehl–Hu functional while it is 5.51 a.u. for Perdew–Wang. We have also reported corresponding magnetic moment versus lattice parameter. For Langreth–Mehl–Hu functional (Fig. 3), the magnetic moments vary slowly over the entire range of the lattice parameter around the ground state and shows parallel coupling between Fe and Cr magnetic moments. The same result has been reported by Singh [1]. For Perdew–Wang functional, the same behavior is shown until lattice parameter becomes ≥ 5.53 a.u. in which case an antiparallel coupling occurs.

3. Results and discussion

Starting with $p(1 \times 1) \uparrow$ configuration, self-consistency procedure leads to a spin-flip transition giving rise to $p(1 \times 1) \downarrow$ one. For $p(1 \times 1) \downarrow$

converged configuration we observe (Fig. 4) that the magnetic moment on the surface, corresponding to Fe2 atoms on the top layer of the *slab*, is antiparallel to the subsurface layers. Magnetic moment on the surface is found to be slightly enhanced: $\mu_{Fe2} = -2.35\mu_B$ as compared to the bulk value $\mu_{Fe} = 1.51\mu_B$. For Cr subsurface layer $\mu_{Cr1} = 0.92\mu_B$. The same results are obtained even when we start the self-consistency procedure with $p(1 \times 1)\downarrow$. These results are in qualitatively good agreement with those obtained by Bouzar et al. [8]. For $c(2 \times 2)$ configuration using a $6 \times 6 \times 2$ points in the Brillouin zone, we obtain a less stable state than $p(1 \times 1)\downarrow$ with a difference in energy of 0.06 Ry which is an important difference compared to LDA calculations reported by different authors. This large difference can be due either to the reduced number of k-points or to the general gradient approximation used in our calculations which was also shown to overestimate lattice parameters corresponding to the ground state. For 3d transition metals Bagno et al. [11] have shown

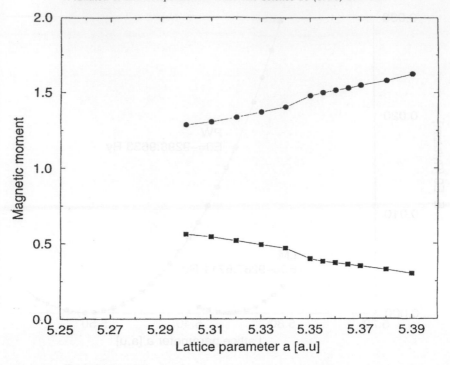

Fig. 3. Magnetic moments versus lattice parameter with Langreth–Mehl–Hu functional for Fe (circles) and Cr (squares) for B2 FeCr around the minimum of energy, i.e. $a = 5.36$ a.u.

that gradient corrections (in particular Perdew–Wang [12,13] (PW) version) improve the total energies. Subsequent works have shown that the Perdew–Wang potential leads to an overestimation of the lattice constants and compressibility in $4d$ and $5d$ transition metals (TMs). Moreover, using the PW potential, even if they obtain the right ground state for iron, the computed structural energy difference between the bcc FM and fcc NM phase is of about 14 mRy, and is three times larger than the experimental results [14].

4. Conclusion and outlook

Using a tight binding LMTO method we have investigated the local polarization at the surface of B2 FeCr alloy. We have considered the two $p(1 \times 1)$ and $c(2 \times 2)$ configurations. Our preliminary results show that for Fe at the surface, the $p(1 \times 1)\downarrow$ configuration seems to be more stable but one cannot deem about this since our calculations were performed using

a reduced number of k-points in the Brillouin zone. A large energy difference was obtained between the two $p(1 \times 1)\downarrow$ and $c(2 \times 2)$ configurations. This may be related to the general gradient approximation which has also overestimated the lattice parameter corresponding to the ground state compared to LDA calculations of Moraïtis et al. [6]. Actually we are computing these configurations using more k-points ($18 \times 18 \times 6$). We are also considering the case of Cr top layer. In further works we will study the different configurations (1 1 0) and (1 1 1) of the surface. We will also consider the *full potential* extension to the problem which will permit us to calculate more accurately the electronic structure at the surface, and to perform relaxation calculations.

Acknowledgements

This work was made possible through a cooperative program between France and Algeria (93 MEN 222). FA would like to thank the IPCMS-GEMME

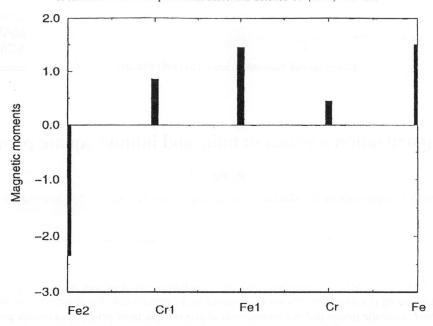

Fig. 4. Magnetic moments at the surface of (0 0 1) B2 FeCr alloy, Fe2 being the iron top layer and Cr1 the subsurface chromium layer.

group for their kind hospitality. This work was partly supported by the European Community Human Capital and Mobility Programme through Contract No. CHRX-CT93-0369.

References

[1] D.J. Singh, J. Appl. Phys. 76 (1994) 6688.
[2] D. Stoeffler and F. Gautier, Phys. Rev. B 44 (1991) 10 389.
[3] A. Vega et al., J. Appl. Phys. 69 (1991) 4544.
[4] F. Herman, J. Sticht and M. Van Schilgaarde, J. Appl. Phys. 69 (1991) 4789.
[5] E.G. Moroni and T.J. Jarlborg, Phys. Rev. B 47 (1993) 3255.
[6] G. Moraïtis, M.A. Khan, H. Dreyssé and C. Demangeat, JMMM 156 (1996) 250.
[7] O.K. Andersen and O. Jepsen, Phys. Rev. Lett. 53 (1984) 2571.
[8] H. Bouzar et al., Surf. Sci. 381 (1997) 117.
[9] D.C. Langreth and M.J. Mehl, Phys. Rev. Lett. 47 (1981) 446.
[10] J.P. Perdew et al., Phys. Rev. B 46 (1992) 6671.
[11] P. Bagno, O. Jepsen and O. Gunnarsson, Phys. Rev. B 40 (1989) 1997.
[12] J.P. Perdew, Phys. Rev. Lett. 55 (1985) 1655.
[13] J.P. Perdew, Phys. Rev. B 33 (1986) 8822.
[14] G.L. Krasko and G.B. Olson, Phys. Rev. B 40 (1989) 11 536.

ELSEVIER

Computational Materials Science 10 (1998) 278–282

COMPUTATIONAL
MATERIALS
SCIENCE

Magnetization reversal in finite and infinite square prisms

R. Ferré [1]

Institut de Physique et Chimie des Matériaux de Strasbourg, 23, rue du Loess, F-67037 Strasbourg, France

Abstract

We present first 3D micromagnetic calculations on finite as well as infinite square prisms. We have studied the magnetization reversal mechanisms taking place in these systems as a function of the prism width. We observe strong differences between reversal mechanisms in infinite prisms and nucleation/reversal processes in finite prisms. For infinite prisms buckling and curling are the modes that have been obtained in the whole width range studied. Substantial differences have been observed between magnetization reversal in finite prisms and inifinite prisms. While for small widths the magnetization reversal takes place following a buckling-like deformation strongly influenced by the extremities, for large widths, domain nucleation and wall propagation has been found to be the ad hoc reversal mechanism after a curling like nucleation. From stability arguments we have determined the critical domain (or activation volume) in low anisotropy systems. Both the switching fields and the activation volume obtained from simulations compare favorably with measurements on Ni electrodeposited nanowires. Copyright © 1998 Elsevier Science B.V.

Keywords: Micromagnetics; Magnetization processes

1. Introduction

Pioneer studies of the magnetization processes in elongated particles in the light of micromagnetic theory [1] were carried out in the mid-1960s on single crystal bars (whiskers) prepared by evaporation techniques [2]. Early experiments such as those carried out by Luborski [3] and Luborski and Morelock [4] on ensembles of iron whiskers showed a relatively poor general agreement between the measured coercive fields as a function of the radius and theoretical results for the nucleation fields in a cylinder with infinite length (infinite cylinder model). These pioneering experiments also indicated the enormous

difficulty to interpret the experimental results obtained for ferromagnetic bars of finite length in terms of simple micromagnetic models such as the infinite cylinder [5]. Recently, electrodeposited Ni nanowires, with small crystal anisotropy, have been the object of single particle magnetization reversal studies using microSQUIDS and magnetic force microscopy [6,7]. Statistical studies of thermally activated magnetization reversal in Ni wires, with diameters ranging between 40 and 100 nm, have shown that the reversal process occurs as the result of the nucleation of a single domain with a volume 200 times smaller than the volume of the entire wire [6]. So, it seems clear now, that, even for small crystal anisotropy, the reversal mechanism for real elongated particles (with finite length) cannot be described in terms of "coherent" modes such as "curling" and "buckling" but seems to

[1] Tel.: 33 3 88 10 70 81; fax: 33 3 88 10 72 49; e-mail: ricardo@taranis.u-strasbg.fr.

be well described in terms of nucleation–propagation reversal mechanisms. Nevertheless, despite the non-coherent character of the magnetization reversal mechanism found, the measured angular dependence of the switching field for Ni wires seems to be in good agreement with that found for the 'curling' mode in infinite cylinders, although the fitted value for the exchange length (about 17 nm) is smaller than that found in the literature (about 21 nm) ($\lambda_{ex} = \sqrt{A/M_s^2}$, with A the exchange stiffness constant and M_s the saturation magnetization). Thus, although the infinite cylinder model gives a first approximation to the nucleation mechanisms and the switching fields of an elongated particle, the deviations between the experimental values and those predicted for the infinite cylinder show that we should avoid the use of simple models to describe correctly the magnetization reversal mechanisms (and so, the switching field values) occurring in elongated particles, even for ferromagnetic systems with low crystal anisotropy.

In order to check the way the infinite hypothesis (in the infinite cylinder model for example) affects the reversal mechanism and consequently, the values of the switching fields we have undertaken micromagnetic simulations for infinite and finite systems. The simulation algorithm employed here is based on the solution of the Brown micromagnetic equations [1] and permits us to simulate both finite and infinite systems provided we use periodic boundary conditions (PBC). This calculation method has been extensively discussed elsewhere [8].

2. The infinite cylinder and the infinite square prism

The first step of the work we present here has been the solution of a well-known example with well-defined magnetization curve and analytical values of the nucleation fields [9]. This case is that of the infinite cylinder. It is well known that for infinite cylinders with a magnetic field applied parallel to the axis of the cylinder, only two nucleation (departure from uniform configuration parallel to cylinder axis) modes are possible [9]: buckling for small radii and

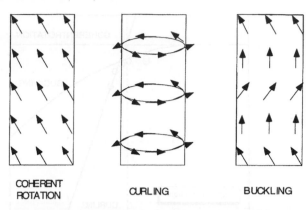

Fig. 1. Diagram showing the nucleation/reversal mechanisms existing in infinite cylinders.

curling for large radii [9] (see Fig. 1). As the radius of the cylinder goes to zero, the wavelength of the oscillation of the magnetization for buckling diverges and the reversal mechanism approaches the coherent rotation [9].

The calculations presented here have been performed in cylinders with their circular section divided into 154 square cells and a length period divided into 128 cells (we make use of PBC along the cylinder length). The cylinder diameter has been chosen for the characteristic length of magnetization variations (the exchange length, defined as $\lambda_{ex} = \sqrt{A/M_s}$) to remain at least three times larger than the elementary cell size.

In agreement with the analytical results we have found only curling and buckling to be the nucleation mechanisms. The results obtained for the reduced nucleation fields $h_n = H_n/2\pi M_s$ and the comparison with analytical calculations [9] are summarized in Fig. 2. The obtained results show a very good agreement between the calculations and the analytical results for large values of the reduced radius S ($S = r/\lambda_{ex}$, where r is the radius of the cylinder) for which the curling mode has been found. For smaller S, the calculations reproduce the cross-over between curling and buckling, the values obtained for the buckling nucleation mechanism being slightly smaller than those obtained by Frei et al. [9] (it should be noticed here that the calculations for buckling by Frei et al. [9] are only approximate calculations and they do not give

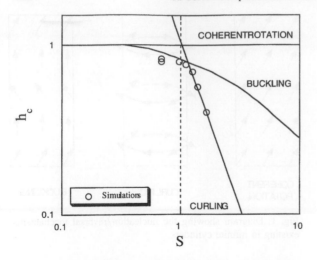

Fig. 2. Nucleation fields for the three reversal mechanisms described in Fig. 1 (from [2]) together with the values obtained from simulations. Notice that for $S < 1$ the simulated values strongly deviate from the analytical values (due to the use of PBC). The two points for $S = 0.7$ correspond to two different simulated cylinder periods.

the real value for the nucleation field but an overestimation). Thus, with the numerical calculations the transition from buckling to curling can be well described. Contrarily, since the calculations consider the magnetization inside the cylinder to be periodic with reduced period $L = 12.8$ ($L = 1/\lambda_{ex}$, where l is the simulated period). Since the buckling wavelength λ for an infinite cylinder does not necessarily coincide with L/n (where n is an integer), the values obtained here for the buckling nucleation mechanism are an approximation of the exact values, which is the more correct when the smaller λ is (i.e., for large diameters [9]). Thus, since the period of the oscillations of the magnetization for buckling diverges as S decreases from 1 to 0, the numerical calculations are unable to describe correctly the behavior for extremely thin cylinders. This has been evidenced by calculations for infinite cylinders with $S = 0.7$ and with two different reduced period values $L = 12.8$ and $L = 25.6$. As shown in Fig. 2 the nucleation fields obtained in the two cases are different, the one obtained for a longer value of L being higher than that obtained for shorter L and being also closer to the analytical value obtained by Frei et al. [9].

These calculations have allowed to validate the employed simulation method and to define accurately the validity limits of the PBC approach.

The second case we have examined is the infinite square prism. For this system, contrary to the infinite cylinder, no analytical values of the nucleation fields exist and only upper and lower bounds for the nucleation fields have been published prior to this article [10]. These bounds are given by the relationships

$$h_n \sigma \pi S^{-2} \quad \text{with } \tfrac{1}{4} < \sigma < \tfrac{1}{2}, \tag{1}$$

for curling and

$$h_n = \rho/2\pi \quad \text{with } 4 < \rho < 2\pi \tag{2}$$

for coherent rotation. In this case no values for the nucleation mode analog to buckling exist.

Making use of PBC we have determined that also curling and buckling are the reversal mechanisms occurring in infinite prisms. We have also determined the angular variation of the nucleation fields for large square prisms (reduced width $S = d/\lambda_{ex}$ larger than one) and for D close to one. The angular variation of the nucleation field has allowed to obtain approximate values for the two constants σ and ρ (see for example [11]). The values obtained are $\sigma = 0.4$ and $\rho = 5.98$ lying between the upper and lower bounds in (1) and (2).

3. The finite square prism

Since actual elongated ferromagnetic particles [2–4,6,7] are finite, we have studied the changes in switching field values and in reversal mechanisms related to the finiteness of these systems. We have studied elongated particles with square cross-section and with high aspect ratio (length to width ratio) AR = 25. The values of the reduced width S of the calculated prisms range between 0.8 and 4.2. The results obtained can be summarized as follows:

(i) For the entire width range studied we have obtained nearly square hysteresis loops.
(ii) For large reduced width ($S > 1.7$) the simulated switching fields h_c are close to the infinite prism and the infinite cylinder values (Fig. 3).

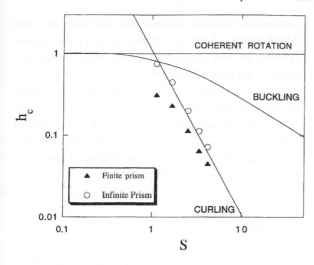

Fig. 3. Reduced values of the switching field as a function of the reduced radius for finite and infinite prisms with aspect ratio AR = 25 and with the applied field along their axis.

(iii) For very thin bars ($S < 1.7$), the switching fields h_c deviate strongly from the values obtained for the infinite prism and the infinite cylinder (Fig. 3).

Dynamical micromagnetic simulations have allowed to calculate the evolution with time of the magnetization configuration. In Fig. 4 we show the transient magnetization configurations near one of the extremities of a prism with $S = 1.1$. The first phase of the magnetization reversal consists in the nucleation of a reversed domain at one extremity of the wire (Figs. 4(a) and (b)). This first nucleation is accompanied by a new domain nucleation in the neighboring region, activated by the change in stray field produced by the first domain nucleation (Figs. 4(c) and (d)). The domain nucleation "avalanche" continues until the whole magnetization of the wire has been reversed. The estimated size of the initially reversed domain is of about $(32 \, \text{nm})^3$.

The most important result extracted from this nucleation mechanism derives from the analysis of the stability of the initially nucleated domain. We have taken the non-equilibrium magnetization configuration in Fig. 4(a) for $h = h_c$ and we have changed "instantaneously" the field to a value smaller than h_c but pointing in the same direction. After that, the stability of the artificially created new magnetization state was

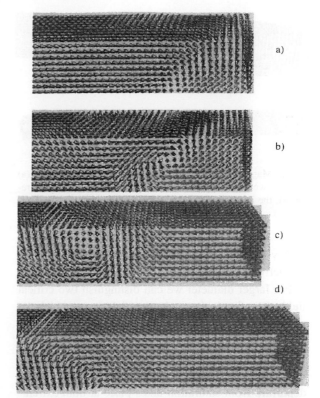

Fig. 4. Transient configurations for a finite prism with aspect ratio AR = 25 and with reduced width $S = 1.1$. The magnetization reversal takes place as the consequence of a domain nucleation at one of the extremities of the prism. The nucleated domain initiates a "cascade" domain nucleation until the total magnetization of the prism is reversed. It is important to notice that magnetization reversal begins at one of the ends of the prism.

checked. In all the cases explored, we have found that this configuration was unstable and that the reversal progressed until the total magnetization of the wire was reversed.

Thus, this analysis reveals that the nucleation of a domain of size around $32 \, \text{nm}^3$ at one extremity of the prism is always accompanied by the complete reversal of the magnetization and so the calculated domain volume gives an upper bound for the activation volume. This theoretical value is in agreement with that derived from a recent statistical analysis of the magnetization reversal in a single Ni wire [6]. This agreement suggests the validity of the reversal mechanism obtained from the numerical simulations or at least the

(a)

(b)

Fig. 5. Magnetization reversal in a large ($S = 2.5$) prism. In (a) and (b) we show two distinct phases of the magnetization reversal, the initial reversed domain formation, and the following wall displacement.

ability of the simulations to give a correct value for the activation volume.

For larger diameter we can distinguish two critical fields. The first is the nucleation field h_n, at which the magnetization of the system deviates from saturation. The second is the switching field h_c, at which an irreversible magnetization jump occurs. Dynamical simulations of the magnetization reversal have shown that departure from saturation takes place at both extremities of the wire according to the curling mode while far from the extremities the magnetization remains oriented along the wire axis. After curling nucleation at both extremities, reducing the applied field results in a propagation of the nucleated vortices inwards until $h = h_c$, where the magnetization of the wire reverses completely, according to a domain nucleation and domain wall propagation mechanism (Figs. 5(a) and (b)). As shown in Fig. 3 the size dependence of the calculated switching fields follows an S^{-2} law, the same as curling (in finite and infinite systems).

4. Conclusions

We have presented here calculations of the nucleation fields and the reversal mechanisms for finite and infinite elongated systems (cylinders and prisms). We have shown that the calculations reproduce the analytical values of the nucleation fields for the infinite cylinder as well as give a precise value for the nucleation fields in infinite prisms for which only upper and lower bounds were known. We have calculated the different nucleation-reversal mechanisms in finite systems and an upper bound for the activation volume in finite systems. Both nucleation fields and the calculated activation volume are in good agreement with recent experimental results [6,11].

References

[1] W.F. Brown, Magnetostatic Principles in Ferromagnetism (North-Holland, Amsterdam, 1962); W.F. Brown, Micromagnetics (Wiley, Interscience, New York, 1951).

[2] C.R. Morelock, Acta Met. 10 (1962) 161; S.S. Brenner, Acta Met. 4 (1956) 62; F.E. Luborsky, T.O. Paine and L.I. Mendelsohn, Powder Met. Bull. 4 (1959) 57.

[3] F.E. Luborsky, J. Appl. Phys. Suppl. 32 (1961) 171S.

[4] F.E. Luborsky and C.R. Morelock, J. Appl. Phys. 35 (1964) 2055.

[5] A. Aharoni, Coherent and incoherent magnetisation processes in non-interacting particles, in: Proc. NATO ASI on Magnetic Hysteresis in Novel Magnetic Materials, 1996, to appear.

[6] W. Wernsdorfer, B. Doudin, D. Mailly, K. Hasselbach, A. Benoit, J. Meier, J.-Ph. Ansermet and B. Barbara, Phys. Rev. Lett. 77 (1996) 1873.

[7] M. Lederman, R. O'Barr and S. Schultz, IEEE Trans. Mag. 31 (1995) 3793; R. O'Barr, M. Lederman, S. Schultz, W. Xu, A. Scherer and J. Tonucci, J. Appl. Phys. 79 (1996) 5303.

[8] R. Ferré, Comp. Phys. Comm. 105 (1997) 169.

[9] A. Aharoni and S. Shtrikman, Phys. Rev. 109 (1958) 1522; E.H. Frei, S. Shtrikman and D. Treves, Phys. Rev. 106 (1957) 446.

[10] W.F. Brown, Jr., J. Appl. Phys. 33 (1962) 3026; A. Aharoni, Phys. Stat. Sol. 16 (1966) 3.

[11] R. Ferré, K. Ounadjela, J.M. George, L. Piraux and S. Dubois, Phys. Rev. B, accepted.

ELSEVIER

Computational Materials Science 10 (1998) 283–286

COMPUTATIONAL
MATERIALS
SCIENCE

Monte Carlo simulations of magnetic properties in multilayers

L. Veiller, D. Ledue *, J. Teillet

GMP Magnétisme et Applications, UMR 6634 CNRS–Université de Rouen, 76821 Mont-Saint-Aignan Cedex, France

Abstract

A Monte Carlo method has been used to simulate Heisenberg multilayer systems ($L \times L \times 4P$) consisting of alternating P ferromagnetic layers A and B with antiferromagnetic interface coupling J_{AB}. Finite-size effects on the specific heat and magnetisation thermal variation for two kinds of boundary conditions at the top and bottom planes are investigated. In particular, our Monte Carlo data evidence that the specific heat exhibits two peaks and a single phase transition occurs at the temperature which corresponds to the location of the high temperature peak (as $L \to \infty$). Copyright © 1998 Elsevier Science B.V.

Keywords: Heisenberg model; Monte Carlo simulation; Magnetic multilayers

1. Introduction

In recent years, some bilayer systems made up of alternating layers of transition metal (TM) and rare earth (RE) atoms have been extensively studied for reasons of scientific interest and technological applications [1,2]. In particular, Tb/Fe ferrimagnetic multilayers with small layer thicknesses which exhibit large perpendicular magnetic anisotropy have potential application as magneto-optical recording media [3,4]. However, most of the theoretical works related to magnetic multilayers are restricted to simple Ising or Heisenberg systems consisting of identical spin ions and very few numerical or analytical studies dealing with RE/TM magnetic properties have been published. For example, mean-field and effective-field theories have been used to study magnetic

properties of RE/TM multilayers [2,5,6], but very few Monte Carlo studies deal with RE/TM multilayer systems.

In this article we examine, by Monte Carlo simulations, the temperature dependence of the specific heat and magnetisation of a magnetic multilayer system consisting of two alternating ferromagnetic materials (A and B) with different bulk properties. Finite-size effects in the case of two kinds of boundary conditions have been investigated.

2. Model and formulation

We consider a multilayer system in which P ferromagnetic layers of materials A and B alternate. For reason of simplicity, we restrict our study to simple cubic structure. Each layer is made up of four atomic planes which are $L \times L$ in cross section. Then, the total number of atoms is $N = 4 \times P \times L^2$.

* Corresponding author. Tel.: +33.(0)2.35.14.68.77; fax: +33.(0)2.35.14.66.52; e-mail: denis.ledue@univ-rouen.fr.

The Hamiltonian for this system is given by

$$H = -\sum_{\langle i,j \rangle} J_{ij} S_i S_j,$$

where S_i is a classical Heisenberg spin and the sum is taken over nearest-neighbour pairs of spins. J_{ij} denotes nearest-neighbour exchange interaction and is assumed to be $J_{AA}(J_{BB})$ between A (B) atoms and J_{AB} between different atoms at the interfaces. The exchange parameters are defined in temperature units.

In order to relate our results to real systems, such as RE/TM multilayers, we make the assumption that A and B atoms are transition metal (iron) and rare earth (terbium) like atoms, respectively. Then, the magnetic moments for A and B atoms are related to the spin momentum S_A and the total angular momentum J_B, respectively, by

$$m_A = -g_A \mu_B S_A,$$

$$m_B = -g_B \mu_B J_B = \frac{-g_B}{g_B - 1} \mu_B S_B,$$

where we assume that Landé factors $g_A = 2$ and $g_B = \frac{3}{2}$, $S_A = 1$, $S_B = 3$ and $J_B = 6$. We have considered $J_{AB} < 0$, $J_{AA} > 0$ and $J_{BB} > 0$. The exchange interactions J_{AA} and J_{BB} have been estimated by Monte Carlo simulations on pure monoatomic A and B systems, respectively, so that the maximum of the specific heat is located at $T_C^{iron} \approx 1044$ K and $T_C^{terbium} \approx 220$ K. We found $J_{AA} \approx 780$ K and $J_{BB} \approx 18$ K. As usually considered for Fe–Tb alloy systems, we have taken $J_{BB} < |J_{AB}| < J_{AA}$, here $J_{AB} = -200$ K.

Simulations were performed using the importance-sampling Monte Carlo procedure based on the standard Metropolis algorithm [7] and updating the spin configuration in visiting atomic site randomly. Our data were obtained by repeating the calculations at several temperatures. In our simulations, the temperature is slowly decreased. At each temperature, the first 2×10^3 Monte Carlo steps (MCS) were discarded for equilibration before averaging over the next 78×10^3 MCS. The thermodynamic quantities of interest at each temperature are the total energy, $E(T) = -\langle \sum_{\langle i,j \rangle} J_{ij} S_i S_j \rangle$, the specific heat calculated from the fluctuations of the internal energy,

$C(T) = (\langle E^2 \rangle - \langle E \rangle^2)/(NkT^2)$, and the magnetisation per atom

$$m(T) = \frac{1}{N} \left[\left\langle \left| \sum_i m_i^x \right| \right\rangle^2 + \left\langle \left| \sum_i m_i^y \right| \right\rangle^2 + \left\langle \left| \sum_i m_i^z \right| \right\rangle^2 \right],$$

where $\langle \ \rangle$ is the average over the MCS, i.e. thermal average at temperature T.

Periodic boundary conditions have been applied in the plane of the layers while two kinds of boundary conditions have been used along the perpendicular direction at the top and bottom planes: free boundary conditions (FBC) with $P = 4$ to take account of free surfaces in real systems and periodic boundary conditions (PBC) with $P = 2$ for a bilayer system. For a given size N, each data point was averaged over three runs using different starting configurations and different random number sequences. In the work reported, N has been assigned various values from 576 to 28 800 spins.

3. Numerical results and discussion

The first aim of this paper is to study the size effects with FBC at the top and bottom planes ($P = 4$). The cross sections vary from $L = 6$ to $L = 24$.

Specific heat profiles versus temperature are shown in Fig. 1. The temperature dependence of the specific heat exhibits two peaks, at about 220 and 960 K (± 20 K), as previously seen for Ising models [6,8]. This could suggest that two phase transitions occur. Indeed, for $J_{AB} = 0$ K, the multilayer system is decomposed into four independent layers so the net specific heat is the sum of the A and B independent layers: each peak occurs near to the temperature at which a maximum would occur for each independent layer (220 K ↔ B layers, 960 K ↔ A layers) and the system undergoes two phase transitions. However, when $J_{AB} \neq 0$ K, the maximum for the high temperature peak (C_{max}^1) increases and this peak narrows as L increases (Fig. 1). On the contrary, the

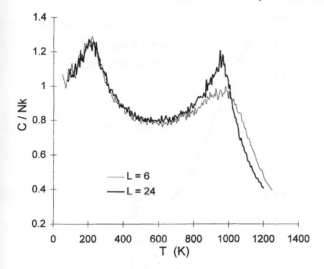

Fig. 1. Temperature dependence of the specific heat with free boundary conditions ($P = 4$) for different values of L. For reason of clarity, only curves for $L = 6$ and $L = 24$ are drawn.

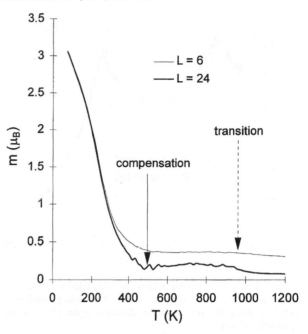

Fig. 2. Temperature dependence of the magnetisation per atom with free boundary conditions ($P = 4$) for different values of L. The continuous arrow indicates the compensation point for $L = 24$.

low temperature peak is roughly unchanged and thus reveals only an alteration of the short range order in B layers. Owing to these facts, the transition temperature T_C of the multilayer system can be determined from the location of the high temperature specific heat peak (as $L \to \infty$) and can be estimated to about 960 K in our system.

In Fig. 2, we have plotted the thermal variation of the magnetisation per spin (only results for $L = 6$ and $L = 24$ are drawn). Below the transition, the magnetisation per spin tends to the expected limit value of $3.5\mu_B$ which corresponds to the ferromagnetic ground state. As expected, finite-size effects are responsible for a tail above the transition where spontaneous magnetisation should be zero. Moreover, these finite-size effects prevent us from observing magnetic compensation for small sizes ($L < 16$) while a compensation point can be seen for large enough sizes ($L \geq 16$). In particular, this point is close to 500 K for the largest system ($L = 24$).

Secondly, we discuss the size effects in the case of PBC, at the top and bottom planes, on the specific heat and magnetisation ($P = 2$). The cross sections vary from $L = 10$ to $L = 60$.

The size effects on the two peaks are similar as for FBC, so the thermal variation of the specific heat for

different values of L is not shown. For comparison, we plot in Fig. 3 the thermal dependence of the specific heat for the two types of boundary conditions ($L = 24$). It can be seen that the location and the height of the high temperature peak seems to be independent of P and the boundary conditions. On the other hand, with PBC, the low temperature peak is shifted to higher temperatures due to a more rapid ordering of the B sites. Indeed, the B spins of the top plane have a more reduced *freedom* in the PBC system than in the FBC system because of the presence of magnetically ordered A planes in the neighbouring when $T < T_C$. This is also valid to explain that the compensation point exhibits a shift towards high temperatures ($T_{comp} \sim 620$ K) for the system with PBC while $T_{comp} \sim 500$ K for the system with FBC (Fig. 4). With PBC, the magnetisation of the sublattice A is more rapidly compensated by the magnetisation of the B layer. It should be noted that PBC allow the magnetisation curve to be more regular and, therefore, the position of the compensation temperature to be determined more precisely (Fig. 4).

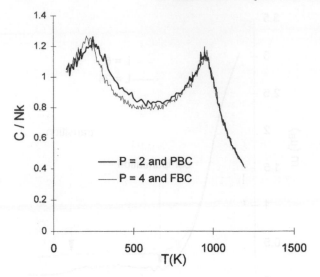

Fig. 3. Comparison of the temperature dependence of the specific heat for $L = 24$ with two kinds of boundary conditions applied along the perpendicular direction: open ($P = 4$) or periodic ($P = 2$).

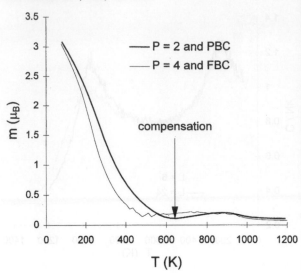

Fig. 4. Comparison of the temperature dependence of the magnetisation per atom for $L = 24$ with two kinds of boundary conditions applied along the perpendicular direction : open ($P = 4$) or periodic ($P = 2$). The arrow shows the location of the compensation point for the bilayer system with PBC.

In this paper, we have investigated the finite-size effects on the magnetisation and the specific heat for a Heisenberg multilayer system. Two kinds of boundary conditions have been applied along the perpendicular direction of the layers. Our results evidence that a single phase transition occurs in a Heisenberg multilayer with non-zero interface coupling and that the transition temperature is given by the location of the high temperature peak of the specific heat (as $L \rightarrow \infty$). The number of layers and the type of boundary conditions are ineffective on the location and the height of the high temperature peak. On the other hand, the effects of free surfaces are substantial concerning the low temperature peak and the determination of the compensation point. Then, it is more judicious to study magnetic properties of a bilayer with PBC and larger values of L for which Monte Carlo data will be less disperse in order to estimate more accurately the transition temperature and the compensation point.

Acknowledgements

This study was supported in part by the Conseil Régional de Haute-Normandie.

References

[1] S. Honda, T. Kimura and N. Nawate, J. Magn. Magn. Mater. 121 (1993) 144.
[2] S. Honda and M. Nawate, J. Magn. Magn. Mater. 136 (1994) 163.
[3] K. Yamauchi, K. Habu and N. Sato, J. Appl. Phys. 64 (1988) 5748.
[4] F. Richomme, J. Teillet, A. Fnidiki, P. Auric and Ph. Houdy, Phys. Rev. B 54 (1996) 416.
[5] T. Kaneyoshi and M. Jaščur, Physica A 203 (1994) 316.
[6] M. Jaščur and T. Kaneyoshi, J. Magn. Magn. Mater. 140–144 (1995) 488.
[7] N. Metropolis et al., J. Chem. Phys. 21 (1953) 1087.
[8] A.M. Ferrenberg and D.P. Landau, J. Appl. Phys. 70 (1991) 6215.

ELSEVIER

Computational Materials Science 10 (1998) 287–291

COMPUTATIONAL
MATERIALS
SCIENCE

Relation between the magnetic moment and the ionicity in NiO

D. Fristot, S. Doyen-Lang, J. Hugel *

Université de Metz-IPC-LSOM, 1 Bd Arago, 57078 Metz Cedex 3, France

Abstract

The magnetic moment and the ionicity have been investigated within the local density approximation. The simple LSDA results show close connection between the moment and the ionicity since the variations depend on the respective weights of the occupied oxygen 2p and nickel 3d states. The LSDA+U treatment allows the theoretical and experimental gap to match when U is regarded as a parameter. The values of U increase from 5.35 to 6.20 eV when the ionicity diminishes. Irrespective of the ionicity, the Hubbard energy U saturates the moment around 1.86 μ_B. Copyright © 1998 Elsevier Science B.V.

Keywords: Magnetically ordered materials; Electronic band structure; Electronic states

Nickel oxide is the classical prototype of the Mott–Hubbard insulators. It crystallizes in the NaCl structure and undergoes an antiferromagnetic phase transition at 525 K. The compound has been intensively studied since the standard band structure calculations failed to describe the experimental gap about 4 eV. The problem received satisfactory answers when corrections beyond the simple local density approximation (LDA) have been introduced. Various types of corrections leading to meaningful improvements have been applied to the monoxides: the generalized gradient approach [1,2], the self-interaction correction [3,4], the orbital polarization correction [5] and the LDA+U correction [6]. However, the first promising calculation applied to transition metal monoxides within a noncorrected spin polarized approach is due to Terakura et al. [7]. They obtained a small forbidden

gap in NiO and shoved the very sensitive dependence on the choice of the atomic radii of both the spin magnetic moment and the electron transfer from nickel to oxygen. In the present paper we do not focus on the soundness of the LDA corrections but propose an interpretation of the interrelation between the moment and the charge transfer in terms of the proportions of 2p functions within the valence bands. In contrast to the work of Terakura et al. [7] we fix the atomic potential radii but use various local basis sets which each generates a self-consistent solution for the moment and the ionicity. A numerical LCAO procedure derived from the discrete variational method developed by Zunger and Freeman [8] has been adopted. This method was previleged since it possesses some freedom in the choice of the local orbitals. The details of the derivation as well as the differences with the variational method can be found in [9]. The procedure is based on the resolution of the crystal Kohn–Sham equation and is briefly summarized in the following.

* Corresponding author. Tel.: +03 87 31 57 77; fax: +03 87 58 31 01; e-mail: hugel@ipc.univ-metz.fr.

0927-0256/98/$19.00 Copyright © 1998 Elsevier Science B.V. All rights reserved
PII S 0927-0256(97)00123-7

The self-consistent spin polarized scheme is organized in two stages. The first stage yields atomic-like functions obtained through the resolution of a local Kohn–Sham equation. The local potential depends upon two parameters, the muffin-tin radius and the constant potential outside V^0 whose role is simply to localize the functions. As a general result, a rise in the barrier height V^0 stresses the decay of the orbital tail while keeping the location of the nodes. The local functions serve in the basis expansion of the Bloch functions and have a significant effect on the crystal density. The second step determines the radial crystal density attached to each site together with the occupation numbers by solving numerically the LCAO Hamiltonian matrix. The occupations are fractionary and are used to redefine a new set of localized orbitals for the next iteration. Both steps are repeated until the desired convergence is reached. The originality of our approach is that the charge transfer amount can easily be modified through the additional external potential. This later potential does by no means appear in the crystal potential.

A first series of calculations have been performed within the simple LSDA approximation. The perfect ionicity has been assumed for the initial atomic configuration namely $1s^2 2s^2 2p^6 3s^2 3p^6 3d^8 4s^0 4p^0$ for the nickel ion and $1s^2 2s^2 2p^6$ for the oxygen ion. The potential spheres are allowed to overlap and their radii have been chosen equal to 2.65 and 2.80 a.u., respectively, for the metal and the oxygen. The convergence procedure is achieved after 30 iterations on average. The results are summarized in the following figures obtained by fixing the localizing potential on the metal site, respectively, to 1.2, 1.0 and 0.8 a.u., the oxygen potential ranging from 0.1 to 1.1 a.u. The magnetic moment is depicted in Fig. 1 and the ionicity in Fig. 2. The forbidden gap is almost stable since it results from the crystal field effect and amounts to 0.65 eV. Two facts emerge from the consideration of the LSDA results: both the moment and the ionicity are greatly dependent on the atomic like basic used; the magnetic moment and the ionicity vary in the same way.

A second series of calculations have been done with the same fixed values as before but by introducing the U correction proposed by Anisimov et al. [6]. The

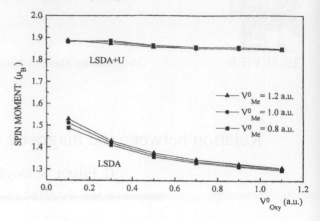

Fig. 1. Magnetic spin moment versus the localizing potential V^0 on the oxygen site. Both LSDA and LSDA+U results are reported.

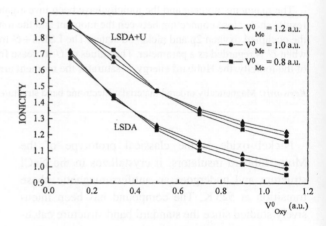

Fig. 2. Ionicity versus the localizing potential V^0 on the oxygen site. Both LSDA and LSDA+U results are reported.

values of U are adjusted so as to match with the experimental gap. The corrected magnetic moments and ionicities are presented for comparison on the same figures as the LSDA results. The dispersion of the Coulomb energy U is recorded in Fig. 3. The action of U appears clearly: it saturates the moment to constant a value of 1.86 μ_B and raises all the ionicities. The Hubbard interaction is growing in magnitude when the ionicity diminishes. Within our fixed parameters, the U values spread out over a limited interval ranging from 5.35 to 6.20 eV.

The interpretation of the results find their origin in the organization of the electronic states. It has been

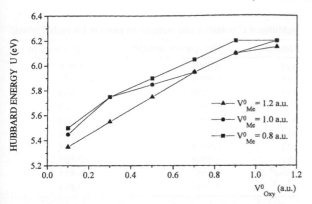

Fig. 3. Dispersion of the Coulomb energy U versus the localizing potential V^0 on the oxygen site.

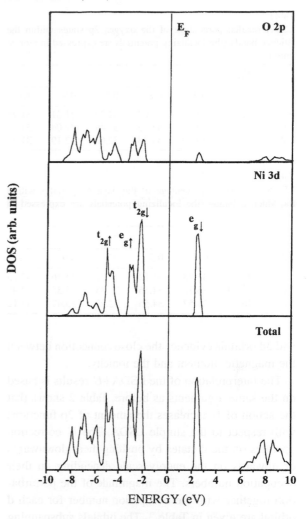

Fig. 4. The total and partial ground-state DOS within the LSDA+U calculations for V^0, respectively, equal to 0.3 and 1.2 a.u. for the oxygen and the metal.

checked that for all the present solutions the 2p oxygen bands stand below the 3d metal bands. The latter 3d bands are splitted according to the crystal field effect. Thus the successive band within increasing energy can be identified as 2p, $t_{2g\uparrow}$ and overlapping $e_{g\uparrow}$–$t_{2g\downarrow}$ occupied bands separated from an empty $e_{g\downarrow}$ band. As an example Fig. 4 shows the LSDA+U total and partial ground-state DOS for V^0, respectively, equal to 0.3 and 1.2 a.u. for the oxygen and the metal. The charge transfer between the two crystal constituents depend on the mixing rate between the occupied 2p and 3d states. Perfect ionicity corresponds to six p like states and eight d states below the Fermi level when the other very small contributions are neglected.

The variations of the moment and the ionicity depicted in Figs. 1 and 2 within the simple LSDA scheme is the result of the specific behavior of the oxygen ion. Two aspects have to be considered. Firstly it is well known that the oxygen 2p functions show a great adaptability in the sense that they fill all the space left by the other atoms in the crystal. The 2p functions with the largest spatial extent generate a doubly ionized ion whereas a compression of the 2p functions leads to a single ionized oxygen ion. These trends are easily simulated through the constant potential V^0 on the oxygen site by varying its values from 0.1 to 1.1 a.u. The ionicity is thus governed by the volume that is allotted to the 2p functions. The second point is to analyze how the variability of the oxygen charge transfer acts on the energy levels. According to the

results presented in Table 1 one observes a diminution of the 2p functions present in the valence bands when the 2p oxygen functions are progressively contracted via the action of the external V^0 potential. The reduction of the number of p levels decreases the ionicity. Since the total number of occupied states remains constant, a loss of p states is compensated by a gain of d states presently identified as minority spin levels. As a consequence the difference between the majority and minority metal states becomes smaller and lowers the magnetic spin moment. The interplay between the 2p

Table 1

LSDA results: percentage of the oxygen 2p states within the valence bands (the localizing potentials are expressed in atomic units)

V^0_{Met}	V^0_{Oxy}					
	0.1	0.3	0.5	0.7	0.9	1.1
0.8	35.47	34.27	33.02	32.30	31.81	31.45
1.0	35.67	34.22	33.15	32.49	32.04	31.70
1.2	35.79	34.19	33.23	32.63	32.19	31.87

Table 2

LSDA+U results: percentage of the oxygen 2p states within the valence bands (the localizing potentials are expressed in atomic units)

V^0_{Met}	V^0_{Oxy}					
	0.1	0.3	0.5	0.7	0.9	1.1
0.8	36.85	35.98	34.52	33.71	33.16	32.75
1.0	36.79	33.75	34.56	33.83	33.35	32.98
1.2	36.85	35.57	34.55	33.92	33.47	33.12

Table 3

Additional U potential and occupation number for each d orbital

Orbital	Occupation number	U potentiel in eV
xy↑	0.9996	−1.036
yz↑	0.9994	−1.036
zx↑	0.9996	−1.036
$3z^2$-r^2 ↑	0.9955	−1.020
x^2-y^2 ↑	0.9956	−1.020
xy↓	0.9988	−0.604
yz↓	0.9988	−0.604
zx↓	0.9988	−0.604
$3z^2$-r^2 ↓	0.0681	3.484
x^2-y^2 ↓	0.0685	3.484

and 3d orbitals evidence the close connection between the magnetic moment and the ionicity.

The interpretation of the LSDA+U results is based on the same arguments as before. Table 2 shows that the action of U increases the amount of 2p functions with respect to the simple LSDA. The U correction operates on the d states by pushing them downwards or upwards on the energy scale depending on their occupation numbers. The magnitude of the perturbation together with the occupation number for each d orbital are given in Table 3. The orbitals subspanning the $t_{2g\uparrow}$, $e_{g\uparrow}$ and the $t_{2g\downarrow}$ subbands are nearly filled whereas the $3z^2$-r^2 and the x^2-y^2 minority orbitals are practically empty. One deduces that the minority e_g orbitals initially located below the Fermi level have been moved above the highest occupied level. The missing 3d orbitals are substituted by 2p orbitals to maintain the conservation of the electron number. This process both enhances the presence of the 2p states within the valence bands and the imbalance between the majority and minority d spin. The greater number of 2p states in the valence band is at the origin of the simultaneous enhancement of the charge transfer and the spin moment.

For a transition metal ion the reduction of the ionicity is related to the gradual filling of its d shell. The repulsion strength felt by an electron when it jumps from one metal to another one depends on the number of electrons composing the cloud of the final site. U is commonly interpreted as the energy required to overcome the repulsive Coulomb forces for an additional electron put on a metal site. Following the interpretation one expetcs that U evolve to higher energies as the metal ion becomes progressively neutral. Fig. 4 shows that the variations of the calculated values are in agreement with the predicted behavior deduced from physical considerations.

The whole results find a unique explanation in terms of a competition between the 2p and 3d states below the Fermi energy. When one focuses on the agreement between the theoretical and experimental moments, various LSDA+U solutions can be retained. In fact, the available experimental moments range between 1.6 and 1.9 μ_B [10–12] and every calculated moment falling within the interval can be accepted as satisfactory. The moment, being an integrated quantity, is not very sensitive to the details of the electronic density of states. As a consequence the moment is not an absolute criterion to discriminate among the previous converged solutions. Other physical properties giving insight into the nature of the electronic states on the energy axis have to be taken into account.

Acknowledgements

The calculations have been performed thanks to computational facilities granted by the CIRIL.

References

[1] P. Dufek, P. Blaha, V. Sliwko and K.H. Schwarz, Phys. Rev. B 49 (1994) 10170.

[2] P. Dufek, P. Blaha and K.H. Schwarz, Phys. Rev. B 50 (1994) 7279.

[3] A. Svane and O. Gunnarsson, Phys. Rev. Lett. 65 (1990) 1148.

[4] Z. Szotek, W.M. Temmerman and H. Winter, Phys. Rev. B 47 (1993) 4029.

[5] M.R. Norman, Phys. Rev. B 44 (1991) 1364.

[6] V.I. Anisimov, J. Zaanen and O.K. Andersen, Phys. Rev. B 44 (1991) 9434.

[7] K. Terakura, T. Oguchi, A.R. Williams and J. Kübler, Phys. Rev. B 30 (1984) 4734.

[8] A. Zunger and A.J. Freeman, Phys. Rev. B 15 (1977) 4716.

[9] J. Hugel and M. Kamal, J. Phys.: Condens. Matter 9 (1997) 6473.

[10] B.E.F. Fender, A.J. Jacobson and F.A. Wegwood, J. Chem. Phys. 48 (1968) 990.

[11] H.A. Alperin, J. Phys. Soc. Japan Suppl. B 17 (1962) 12.

[12] A.K. Cheetham and D.A.O. Hope, Phys. Rev. B 27 (1983) 6964.

Computational Materials Science 10 (1998) 292–297

COMPUTATIONAL
MATERIALS
SCIENCE

Magnetic properties of molecular systems:
A nonlocal spin density functional study

I. Lado Touriño *, F. Tsobnang, A. Le Méhauté

Institut Supérieur des Matériaux du Mans, 44, Avenue F. A. Bartholdi, 72000 Le Mans, France

Abstract

A new class of molecular magnets based on aniline and aminonaphthalene sulfonic acid was previously reported. In the present work, a nonlocal density functional theory study was performed on simple molecular models, to determine the mechanism that stabilizes the parallel alignment of the spin component. The role played by transition metals on the magnetic properties of this kind of systems was also studied within this approximation. The results obtained show a high sensitivity of the magnetic coupling on the molecular geometry as well as on the nature and oxidation state of the metallic atom. Copyright © 1998 Elsevier Science B.V.

Keywords: Density functional theory; Molecular magnetism

1. Introduction

The question whether it is possible to create organic or molecular ferromagnets has been discussed since the 1950s [1]. An organic or molecular ferromagnet is a magnetic material made up of purely organic paramagnetic molecules or metallic atoms surrounded by an organic structure [2]. At the microscopic level, all the traditional ferromagnetic compounds contain metal atoms with d or f electrons as, for them, the exchange interaction is generally positive and arranges their spins in a parallel manner. On the contrary, for s and p electrons, the exchange interaction is generally negative and gives rise to an antiparallel arrangement of the electron spins. So, the problem is to find the way to achieve a ferromagnetic alignment on the

basis of s and p electrons. The interest is motivated by technological applications in various fields as molecular electronic, intelligent materials, electromagnetic compatibility, information storage, biomaterials, etc.

In order to create an organic ferromagnet, both, a building material (high-spin molecules) and a procedure to organize this material in a condensed phase in such a way that ferromagnetism is conserved, are needed. Important advances in this field are due to Higuchi [3] and Magata [4]. In 1987 Miller et al. [5] characterized a ferromagnetic transition in the organometallic salt decamethylferrocenium tetracyanoethenide, which provoked a great impulse in this field and soon, some other ferromagnetic salts were characterized [6,7].

We have previously reported a new class of molecular magnets based on aminonaphthalenesulfonic acid and aniline [8,9]. In the present work, we investigate

---

* Corresponding author. Tel.: +33 2 43214000; fax: +33 2 43214039; e-mail: ilado@ismans.univ-lemans.fr.

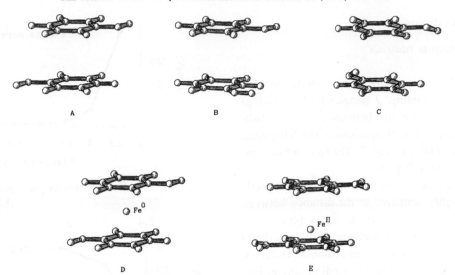

Fig. 1. Molecular models used for the study of magnetic properties of molecular systems; (A) pseudo-para; (B) pseudo-ortho; (C) pseudo-meta; (D) aniline radical-Fe^0(1/2)-aniline radical; (E) aniline molecule-Fe^{II}(1/4)-aniline molecule.

theoretically the magnetic coupling mechanism in a class of materials formed by encryption of transition metals, in sulfonic acid.

2. Model system and calculation method

As a first step we consider both a model system consisting of two aniline radicals with different relative angular NH positions (pseudo-ortho, para and meta) superposed in a parallel stack and a system in which an Fe (0, II or III oxidation state) or Cu (0 or II oxidation state) atom is placed between two parallel aniline radicals or molecules in pseudo-para position. Some of these models are shown in Fig. 1.

In a previous work [10] the two-radical system had been studied within the local spin density approximation (LSDA) [11] using the Dmol program [12]. In the present work, some of these calculations were repeated within the nonlocal spin density approximation (NLSDA) using the Becke's exchange functional together with the gradient-corrected correlation functional of Lee, Yang and Parr (BYLP). In this case, the exchange–correlation term depends not only on the value of the electron density at each point in space but

on its gradient. All the calculations were performed with DNP basis (double-numerical basis functions together with polarization functions), frozen inner-core orbitals and a density convergence criterion of 10^{-6} electron/$bohr^3$ for the energy and electronic density self-consistent field calculation.

In order to determine the magnetic coupling J, we performed two different and independent calculations on the two-radical system and its different configurations, which were spin constrained (singlet or triplet). Then the magnetic coupling J was obtained from the singlet–triplet splitting (E_S-E_T). This approach assumes that the operator S^2 commutes with the Hamiltonian and that the fundamental and first excited states have different multiplicity. In addition, the accuracy is numerically limited since J is obtained as the difference of two large numbers. In order to overcome this problem, some authors have proposed other methods, in which the relative energies of the low-lying states are calculated using the broken-symmetry formalism [13–15]. In spite of this weak point, surprisingly, this approach has led to satisfactory predictions and explanation of experimental results [16], and is in agreement with the results proposed by some other authors [17].

3. Results and discussion

3.1. Aniline–Aniline radicals

The first calculations were made to study the variation of the spin coupling, J (singlet–triplet splitting) as a function of the distance between the aniline radicals for the three different positions of the NH group. The results are shown in Fig. 2. The spin density distribution is shown in Fig. 3.

From these figures it can be seen that the magnetic coupling is highly sensitive to the distance between the two radicals and diminishes rapidly as this distance increases. Such a behavior is characteristic of a direct exchange coupling resulting from a combined effect of the Coulomb interaction and the Pauli principle. Ferromagnetic coupling ($J > 0$) occurs for pseudo-para and pseudo-ortho positions, while for the pseudo-meta position, antiferromagnetism ($J < 0$) is observed. The relative positions of the NH units control the magnetic coupling, in accordance with the McConnell mechanism [18]. All these results are consistent with the ones previously obtained from the LSDA, although an energy minimum has not been found within our approximation. This result is usually obtained when a gradient-corrected approximation is applied to weakly bonded systems. The reason for using the NLSDA is that this approximation must be applied to the study of structures of the type used in this work to obtain good results as it describes better the systems in which the charge is not distributed homogeneously.

3.2. Aniline–Metal–Aniline

All the calculations were performed for a fixed distance (d) between the aniline molecules of 3.5 Å, as this is the equilibrium distance found for polyaniline and also in our previous calculations within the LSDA.

The spin associated with the ground state as well as the symmetry of the Fe containing systems are shown in Table 1.

The spin for the Cu containing systems is always 1/2.

The spin of the free atoms and ions is shown in Table 2. It can be seen that the interaction of the

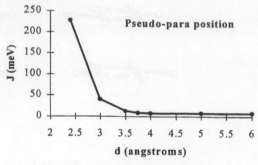

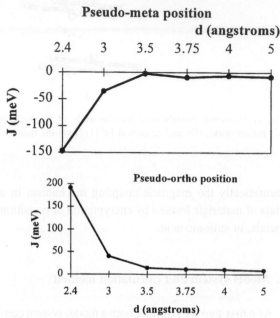

Fig. 2. Magnetic coupling J vs. aniline–aniline separation for the pseudo-para, -meta and -ortho positions.

Table 1
Spin associated with the ground state for the aniline-Fe-aniline systems

	1/2 (d)			1/4 (d)		
	Fe^0	Fe^{II}	Fe^{III}	Fe^0	Fe^{II}	Fe^{III}
Molecules	1	0	1/2	1	1	3/2
Symmetry	C2h	C2h	C2h	Cs	Cs	Cs
Radicals	0	0	1/2	1	1	3/2
Symmetry	Ci	Ci	Ci	C1	C1	C1

Table 2
Spin of free Fe and Cu atoms and ions

	Fe^0	Fe^{II}	Fe^{III}	Cu^0	Cu^{II}
Spin	2	2	5/2	1/2	1/2

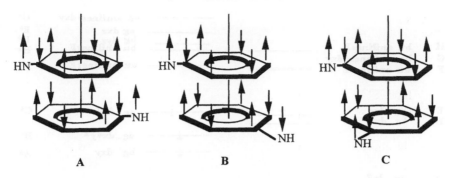

Fig. 3. Spin density distribution for the three different positions of the two radical system: (A) pseudo-para; (B) pseudo-meta; (C) pseudo-ortho.

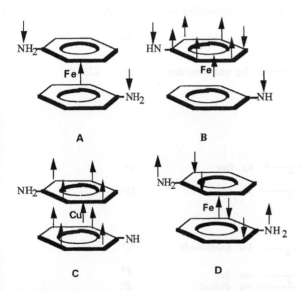

Fig. 4. Spin density distribution for: (A) aniline molecule-$Fe^0(1/2(d))$-aniline molecule; (B) aniline radical-$Fe^{III}(1/4(d))$-aniline radical; (C) aniline molecule-$Cu^0(1/2(d))$-aniline molecule; (D) aniline molecule-$Fe^{III}(1/2(d))$-aniline molecule.

metallic atoms with the aniline molecules and radicals modifies strongly its spin.

The spin density distribution for some of these systems is shown in Fig. 4. It can be observed that the spin value as well as the type of coupling between the anilines and metal atoms depend strongly on the nature, oxidation state and position of the latter.

It is worth noting that the spin value ($S = 0$) for the aniline radical-$Fe^{II}(1/2(d))$-aniline radical system is due to a global electronic effect rather than to the pairing of the electrons in the same molecular orbital.

The electrons are highly delocalized and the spin value cannot be attributed to any particular electron in the system. Such effect cannot be seen within the classical approaches used to calculate the spin of this kind of systems [19].

The use of symmetry properties makes easier the study of the interactions between the π the orbitals of the two rings and the valence orbitals of the metal atom. However, due to the great number of atomic orbitals used to expand the molecular orbitals, it is sometimes difficult to elucidate graphically the nature of the atomic orbitals which contribute to each molecular orbital. In addition, for some systems, the low degree of symmetry allows a complete mixing of these atomic orbitals as shown below.

In a first step, we calculated the energy level diagrams. In general, for similar molecular geometries, these energy level diagrams have the same qualitative features, but the relative order of the levels varies due to a variation in both the relatives energies of the metal atom and metallic ions and the relative magnitudes of the different interaction energies.

As an example, the energy level spectra obtained for the aniline molecule-$Fe^0((1/2)d)$-aniline molecule and aniline molecule-$Cu^0((1/2)d)$-aniline molecule systems are shown in Figs. 5 and 6. The separation between levels is only qualitative.

Two sets of single-particle wavefunctions, one for spin-up electrons and one for spin-down electrons, are represented. The "*" symbol is used to indicate the higher energy level. It is possible to observe that some of the Fe levels are largely split ($\Delta E_{EE^*} = 2.003$ eV

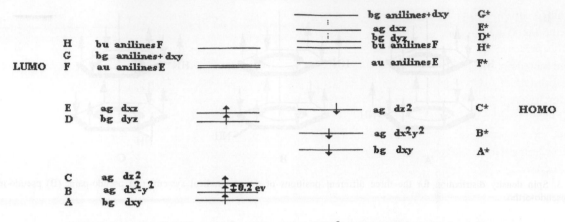

Fig. 5. Energy level diagram for the aniline molecule-Fe^0((1/2)d)-aniline molecule system.

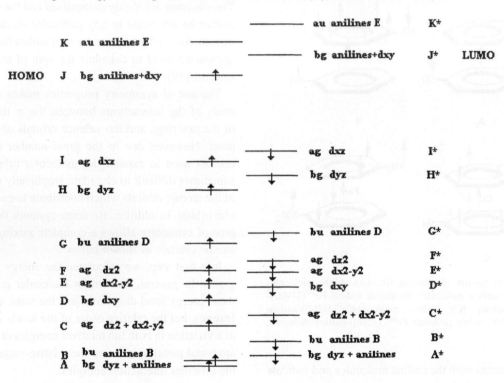

Fig. 6. Energy level diagram for the aniline molecule-Cu^0((1/2)d)-aniline molecule system.

(Fig. 5)). Such a result is quite usual for d atomic orbitals of magnetic atoms and supports the idea of a high degree of spin polarization for these systems. The mixing of the metal orbitals with the aniline orbitals allows for a complete delocalization of the electronic charge, which is quite a usual characteristic of this type of bond.

In this type of compounds, rings bind to metals via their π orbitals, by donating electrons into an empty metal d orbital. However, there is also another bond contribution: the donation of electron density from the metal to the ring into its π^* orbital. This back-donation into C–C antibonding levels leads to a weakening (lengthening) of the C–C bond. The C–C lengths obtained for the two molecule system with and without the metal are 1.41 and 1.38 Å respectively. Thus, in the presence of the metallic atom the bond distance increases. The HOMO (highest occupied molecular orbital) of the two aniline system has the right symmetry (Ag) for effective overlap with some occupied metal d orbitals (d_{z^2}, $d_{x^2-y^2}$, d_{xz}). On the other hand, the LUMO (lowest unoccupied molecular orbital) can overlap with the d_{xy} and d_{yz} orbitals of Bg symmetry.

4. Conclusions

Using a density functional theory (nonlocal spin density) approach as implemented in the Dmol program, we have investigated both the electronic structure and the magnetic properties of aniline-(Fe, Cu)-aniline systems.

We have shown that the magnetic coupling of the two-radical system depends strongly on the relative positions on the NH group. Thus, a triplet ground state for the pseudo-ortho and para configurations and a singlet ground state for the pseudo-meta configuration have been found.

For the metal containing systems, the magnetic behavior depends on the nature, oxidation state and position of the metal atom. These results show the possibility of creating molecular magnets by doping purely organic compounds with transition metal atoms and thus, the most interesting result was found for the aniline molecule-Cu^0(1/2(d))-aniline molecule system.

Pseudo ortho and meta configurations analysis as well as a more detailed study of the influence of the position of the metal on the magnetic coupling will be carried out in future studies.

Acknowledgements

Isabel Lado Touriño would like to thank the Spanish Ministerio de Educación y Cultura for their financial support.

References

[1] H.C. Longuet-Higgins, J. Chem. Phys. 24 (1950) 265.
[2] J.S. Miller, A.J. Epstein, Acc. Chem. Res. 23 (1988) 114.
[3] J. Higuchi, J. Chem. Phys. 38 (1963) 1237.
[4] N. Magata, Theor. Chim. Acta 10 (1968) 372.
[5] J.S. Miller, J.C. Calabrese, H. Rommelmann, S.R. Chittipeddi, J.H. Zhang, W.M. Reiff, A.J. Epstein, J. Am. Chem. Soc. 109 (1987) 769.
[6] W.E. Broderick, J.A. Thomson, E.P. Day, B.M. Hoffman, Science 249 (1990) 401.
[7] W.E. Broderick, J.A. Thomson, B.M. Hoffman, Inorg. Chem. 30 (1991) 2958
[8] S. Galaj, A. Le Méhauté, F. Tsobnang, D. Cottevieille, A. Léaustic, R. Clément, V. Cagan, M. Guyot, M.R. Soriano, J.C. Fayet, B. Villeret, L. Noirez, A. Périchaud, J. Magn. Magn. Mater. 140 (1995) 1445.
[9] S. Galaj, A. Le Méhauté, A.M. Tiller, E. Wimmer, Ext. Abstr., MRS Fall Meet., Boston, MA, USA, 1993.
[10] M.R. Soriano, F. Tsobnang, A. Le Méhauté, E. Wimmer, Synthetic Metals 76 (1996) 317.
[11] O.R. Jones, O. Gunnarson, Rev. Mod. Phys. 61 (1989) 3.
[12] B.J. Delley, Chem. Phys. 92 (1990) 508.
[13] H.K. Johnson, Adv. Quantum Chem. 7 (1973) 143.
[14] A.P. Ginsberg, J. Am. Chem. Soc. 102 (1980) 111.
[15] L. Noodleman, E.R. Davidson, Chem. Phys. 109 (1986) 131.
[16] K. Yoshino, H. Mizobuchi, H. Araki, T. Kawai, A. Sakamoto, Jpn. J. Appl. Phys. 33 (1994) 1624.
[17] C. Kollmar, O. Kahn, Acc. Chem. Res. 26 (1993) 259.
[18] O. Kahn, Molecular Magnetism, VCH Publishers, New York, 1993.
[19] H.M. McConnell, J. Chem. Phys. 39 (1963) 1910.
[20] C. Kollmar, O. Kahn, Acc. Chem. Res. 26 (1993) 259.
[21] M. Bochmann, Organometallics 2, Complexes with transition metal-carbon π-bonds, Oxford Chemistry Primers, New York, 1944, p. 47.

ELSEVIER

Computational Materials Science 10 (1998) 298–301

COMPUTATIONAL
MATERIALS
SCIENCE

AsNCa₃ at high pressure

P.R. Vansant [a,*], P.E. Van Camp [a], V.E. Van Doren [a], J.L. Martins [b,c]

[a] *Department of Physics, University of Antwerpen (RUCA), Groenenborgerlaan 171, B-2020 Antwerpen, Belgium*
[b] *INESC, Rua Alves Redol 9, P-1000 Lisboa Codex, Portugal*
[c] *Instituto Superior Tecnico, Avenida Rovisco Pais 1, P-1096 Lisboa, Portugal*

Abstract

A constant pressure optimization scheme is applied to the study of ternary calcium nitrides under pressure. The enthalpy is minimized with respect to the electronic configuration, the positions of the atoms and the cell metric (i.e. the lattice parameters). Symmetry corrections can be performed during the relaxation towards the equilibrium structure in order to be able to investigate the compound in a certain proposed symmetry. We obtain excellent agreement with experiment for the zero-pressure structural parameters of the cubic anti-perovskite structure $BiNCa_3$ and of the distorted anti-perovskite structures $AsNCa_3$ and $PNCa_3$. For $AsNCa_3$ the structural parameters, band gap energies, etc. are investigated as a function of the pressure. A new cubic phase is predicted to have a lower enthalpy than the orthorhombic phase for pressures above 59 GPa. Copyright © 1998 Elsevier Science B.V.

Keywords: Perovskite structure; Ternary nitrides; Structural optimization; Phase transition

1. Introduction

A structural optimization at constant pressure based on a variable-cell-shape scheme [1] is applied to the anti-perovskite structure $BiNCa_3$ (cubic $Pm3m$) and to the distorted anti-perovskite structures $AsNCa_3$ and $PNCa_3$ (orthorhombic $Pbnm$).

The optimizations are performed using a first-principles pseudopotential plane wave method which is based on the local density approximation (LDA), on the Ceperley–Alder correlation potential [2] as parametrized by Perdew and Zunger [3] and on the norm-conserving Troullier–Martins pseudopotentials [4]. The **k**-point sampling is performed by the

Monkhorst–Pack [5] scheme. The number of plane waves is determined by the energy cutoff (here 80 Ry) for the kinetic energy.

The minimization of the enthalpy $H = E + p_{ext}V$ is based on: (1) an iterative matrix diagonalization for the electronic SCF calculation, and (2) a conjugate gradient method using the forces on the atoms to determine the positions of the atoms, and using the stress tensor to obtain the metric g_{ij} (i.e. the lattice constants). For more details of the method we refer to [1]. It is in this recent reference that one explains e.g. the advantages of using the metric g_{ij} instead of the lattice parameters.

In order to be able to investigate the pressure dependence of the structural properties for a certain well-known symmetry, some precautions are taken to be able to impose this symmetry during the relaxation of

* Corresponding author. Tel.: +32.3.218.03.59; fax: +32.3-.218.03.18; e-mail: pvansant@ruca.ua.ac.be.

the structure [6]. However, all the possible transitions between different structures with the *same* symmetry are still accessible. Therefore, a minimization of the enthalpy as a function of *all* atomic positions is still performed at the end of each relaxation step. A check of the intermediate structure with respect to the symmetry reveals possible transitions which can be further investigated or skipped (when a certain symmetry is imposed).

2. The calcium nitrides

2.1. Ambient pressure properties

At ambient pressure we obtain theoretically that the structure of $AsNCa_3$ and $PNCa_3$ are distorted cubic anti-perovskite structures which have a primitive orthorhombic symmetry with spacegroup *Pbnm*. This orthorhombicity can be explained by considering the Ca_6N octaheder in the cubic $Pm3m$ structure (Fig. 1). A slight tilting of this octahedra results in the orthorhombic structure with atomic positions

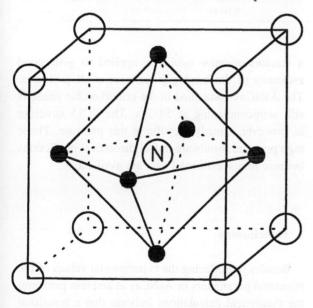

Fig. 1. The cubic anti-perovskite structure SC5 (five atoms per unit cell). The black dots represent the calcium ions, the open circles the arsenic ions and the circle with a "N" inside represents the nitrogen ion. The structure is built up with corner-shared Ca_6N octahedra.

as listed in Table 1. The distortion parameters $\Delta_1, \Delta_2, \delta, \gamma_1, \gamma_2, \lambda_1$ and λ_2 in this table are found to be small fractions of the coordinates. In Table 2 our calculated values of these distortion and lattice parameters are given in detail for the orthorhombic $AsNCa_3$ and $PNCa_3$ compounds. Experimentally (see [7,8]) the same orthorhombic structure was found and its parameters are also shown in Table 2 (for $PNCa_3$ the experimental distortion parameters were not available). An agreement of less than 1% between the theoretical and experimental values of all parameters is found. Also the lattice parameters of the $BiNCa_3$ compound are presented. No distortion parameters are shown because $BiNCa_3$ has an undistorted cubic ($Pm3m$) symmetry.

For the orthorhombic $PNCa_3$ the lattice constants are found to be smaller and the degree of distortion larger than in the case of the orthorhombic $AsNCa_3$. This corresponds to our expectations based on the atomic sizes. Indeed, the size of the As-atom is larger than the P-atom and, therefore, $AsNCa_3$ is less distorted. The Bi-atom is even so large that no distortions are present in the $BiNCa_3$ compound and, as a consequence, it has the cubic $Pm3m$ structure.

2.2. High pressure properties

The pressure dependence of the structure is shown in Fig. 2 by plotting the difference in enthalpy ($H - H_c$) between the investigated structure and the cubic $Pm3m$ perovskite structure (SC5) consisting of five atoms per unit cell as a function of the pressure. Through the solid dots which indicate the theoretical results we present quadratic fits given by: $0.5351 \times 10^{-1} - 0.1024 \times 10^{-2}p + 0.1370 \times 10^{-5}p^2$ for the SC15 structure and $-0.1065 \times 10^{-2} + 0.2730 \times 10^{-6}p - 0.3444 \times 10^{-6}p^2$ for the orthorhombic structure, where p indicates the pressure in the investigated region of 0–80 GPa. By increasing the pressure, the orthorhombic structure becomes more stable in comparison with the cubic SC5 structure and theoretically we also observe an increase of the orthorhombicity. However, starting from a pressure of about 59 GPa a new cubic phase SC15 (see Fig. 3) consisting of 15 atoms per unit cell is found to have a lower enthalpy.

Table 1

The structure *Pbnm* of AsNCa$_3$ found experimentally in [8]

Ca (1)	$\pm(\frac{3}{4}-\Delta_1, \frac{1}{4}+\Delta_2, \delta)$	$\pm(\frac{3}{4}-\Delta_1, \frac{1}{4}+\Delta_2, \frac{1}{2}-\delta)$
	$\pm(\frac{3}{4}+\Delta_1, \frac{3}{4}+\Delta_2, \delta)$	$\pm(\frac{3}{4}+\Delta_1, \frac{3}{4}+\Delta_2, \frac{1}{2}-\delta)$
Ca (2)	$\pm(\gamma_1, \frac{1}{2}-\gamma_2, \frac{1}{4})$	$\pm(\frac{1}{2}-\gamma_1, -\gamma_2, \frac{1}{4})$
N	$\frac{1}{2}, 0, 0$	$\frac{1}{2}, 0, \frac{1}{2}$
	$0, \frac{1}{2}, 0$	$0, \frac{1}{2}, \frac{1}{2}$
As	$\pm(-\lambda_1, \lambda_2, \frac{1}{4})$	$\pm(\frac{1}{2}-\lambda_1, \frac{1}{2}-\lambda_2, \frac{3}{4})$

Table 2

Experimental and theoretical results for the lattice and distortion parameters of AsNCa$_3$, PNCa$_3$ and BiNCa$_3$. For PNCa$_3$ no experimental data for the distortion parameters are available and in the case of the cubic BiNCa$_3$ they are not relevant

	AsNCa$_3$		PNCa$_3$		BiNCa$_3$	
	Experiment [8]	Theory	Experiment [7]	Theory	Experiment [7]	Theory
T (K)	15	0	305	0	305	0
a (Å)	6.716	6.720	6.709	6.707	4.888	4.862
b (Å)	6.711	6.715	6.658	6.659	4.888	4.862
c (Å)	9.520	9.526	9.452	9.451	4.888	4.862
Δ_1 (Å)	0.0329	0.0400	—	0.0464	—	—
Δ_2 (Å)	0.0321	0.0400	—	0.0459	—	—
δ (Å)	0.0209	0.0265	—	0.0396	—	—
γ_1 (Å)	0.0399	0.0510	—	0.0747	—	—
γ_2 (Å)	0.0048	0.0100	—	0.0220	—	—
λ_1 (Å)	0.0000	0.0032	—	0.0084	—	—
λ_2 (Å)	0.0170	0.0263	—	0.0449	—	—

This indicates a phase transition to this SC15 structure at a pressure of 59 GPa. The nitrogen atoms then obtain a higher coordination: *eight* calcium atoms as nearest neighbours instead of *six*.

In addition we calculate also the bandstructure of AsNCa$_3$ as a function of the pressure. For the orthorhombic AsNCa$_3$ structure a decrease of the band gap is found with increasing pressure. For the least squares fit $E_g(p) = a_0 + a_1 p + a_2 p^2$ the coefficients can be interpreted as the coefficients of a second order approximation to the band gap energy as a function of the pressure. The a_1 and a_2 are then recognized as the pressure derivatives $\partial E_g/\partial p$ and $\frac{1}{2}\partial^2 E_g/\partial p^2$, respectively, and a_0 as $E_g(0)$ the band gap energy at zero pressure. Their values are: $E_g(0) = 0.871$ eV, $\partial E_g/\partial p = -0.0249$ eV/GPa and $\frac{1}{2}\partial^2 E_g/\partial p^2 = 0.000173$ eV/GPa^2. Due to the fact that LDA-calculations give band gaps which are up to 50% smaller than the correct ones,

a scissors-operator could be applied to give good estimates of the band gap energies at all pressures. The AsNCa$_3$ compound in the orthorhombic phase is still semiconducting at 59 GPa. The SC15 structure is, however, already metallic at this pressure. Those high pressure results are in fact theoretical predictions because no experimental data are available.

3. Conclusions

Besides reproducing the experimental values of the structural parameters in AsNCa$_3$ at ambient pressure, the theoretical calculations indicate that a transition from the orthorhombic *Pbnm* phase to the new SC15 phase might occur at 59 GPa.

It should be noted that both, the earth-mineral MgSiO$_3$ (in the lower mantle of the earth), and

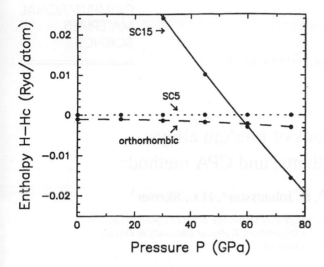

Fig. 2. The difference in enthalpy (Ry/atom) between the considered structure and the cubic SC5 structure is plotted as a function of the pressure (GPa). The solid dots result from the theoretical calculations and the curves are quadratic fits through those data points (see text). Around 59 GPa the SC15 structure obtains a lower enthalpy as the orthorhombic structure. Notice that the SC5 structure is slightly higher in enthalpy than the orthorhombic structure.

Al_2O_3 (having important applications in high pressure research) have an orthorhombic *Pbnm* perovskite structure at high pressures. The large similarities between the properties of the anti-perovskite $AsNCa_3$ and these perovskite structures indicate that this new SC15 phase could also have its analogue for both compounds, $MgSiO_3$ and Al_2O_3.

Acknowledgements

This work is supported by the Belgian NSF (FWO) under grant No. 9.0053.93 and partly by the HCM network "Ab initio Calculations of Complex Processes in Materials" grant No. ERBCHRXCT930369.

References

[1] I. Souza and J.L. Martins, Phys. Rev. B 55 (1997) 8733.
[2] D.M. Ceperley and B.J. Alder, Phys. Rev. Lett. 45 (1980) 566.
[3] J.P. Perdew and A. Zunger, Phys. Rev. B 23 (1981) 5048.
[4] N. Troullier and J.L. Martins, Phys. Rev. B 43 (1991) 1993–2006.
[5] H.J. Monkhorst and J.D. Pack, Phys. Rev. B 13 (1976) 5188.
[6] P.R. Vansant, P.E. Van Camp, V.E. Van Doren and J.L. Martins, to be published.
[7] M.Y. Chern, D.A. Vennos and F.J. Disalvo, J. Sol. Stat. Chem. 96 (1992) 415–425.
[8] M.Y. Chern, F.J. Disalvo, J.B. Parise and J.A. Goldstone, J. Sol. Stat. Chem. 96 (1992) 426–435.

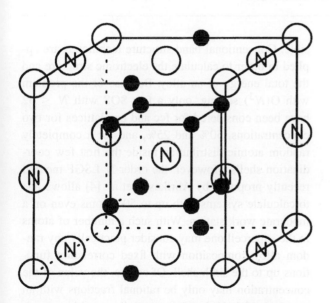

Fig. 3. The simple cubic SC15 structure. The black dots, open circles and circles with a "N" inside represent, resp., the calcium, arsenic and nitrogen ions. Notice the layered structure of edge-sharing Ca_8N cubes and cubes isolated by calcium ions.

ELSEVIER

Computational Materials Science 10 (1998) 302–305

COMPUTATIONAL
MATERIALS
SCIENCE

Total energy calculations of random alloys: Supercell, Connolly–Williams, and CPA methods

I.A. Abrikosov [a,*], A.V. Ruban [b], B. Johansson [a], H.L. Skriver [b]

[a] *Condensed Matter Theory Group, Physics Department, Uppsala University, S-75121 Uppsala, Sweden*
[b] *Center for Atomic-scale Materials Physics and Physics Department, Technical University of Denmark, DK-2800 Lyngby, Denmark*

Abstract

We compare the results of total energy calculations obtained by three most popular methods, the supercell method, the Connolly–Williams (CW) method, and the coherent potential approximation (CPA). As a reference we use the result for a supercell of 256 atoms obtained by the order-N locally self-consistent Green's function (LSGF) method. We show that for the particular case of a fcc Ag–Au alloy there is satisfactory agreement between the different techniques. Moreover, we demonstrate that in the framework of the CPA and the screened impurity model (SIM) one may obtain a reliable description of the total energy of a random alloy. Copyright © 1998 Elsevier Science B.V.

The supercell method, the Connolly–Williams (CW) method [1], and the methods based on the coherent potential approximation (CPA) [2] represent three different kinds of methods which are most frequently used in ab initio total energy calculations for completely random metallic alloys at the present time. They differ in the implementation of the averaging procedure used to treat substitutional disorder on a lattice and this determines their advantages and limitations.

Within the supercell approach an alloy is viewed as a big periodically repeated unit cell. In particular, in the framework of the so-called special quasirandom structures (SQS) method suggested by Zunger et al. [3] a random alloy is modeled by a supercell whose correlation functions match those of a system of interest for as high number of the first shells as possible. Conventional band structure techniques are applied in order to calculate the electronic structure and the total energy of an alloy. Because of the problem with $O(N^3)$ scaling, only a few SQS with $N \leq 32$ have been considered for fcc and bcc lattices for two concentrations, 50% and 25%, and for a completely random atomic distribution inside the first few coordination shells. However, the order-N LSGF method recently proposed by Abrikosov et al. [4] allows one to calculate systems with up to 500 atoms even on a moderate work station. With such a number of atoms in the supercell one may consider practically any random alloy composition with fixed correlation functions up to the sixth shell. Of course, the value of the concentration may only be rational fractions with an accuracy of $1/N$, but for many alloy problems such an accuracy is quite reasonable.

In the CW method the total energy is expanded over atomic distribution correlation functions and the

* Corresponding author. Tel.: +46 18 4713568; fax: +46 18 471 3524; e-mail: Igor.Abrikosov@fysik.uu.se.

procedure of determining the corresponding expansion coefficients, or interactions, is based on the calculation of a definite set of ordered alloys. Thus, the CW method should in principle work for an arbitrary system, and its main advantage is the possibility to estimate the accuracy of the expansion even in the case of a poor convergence. However, the CW method cannot be used in a study of some important physical phenomena connected, for instance, with Fermi surface effects and, moreover, it may give reliable results only for the systems with short-range interatomic interactions.

In contrast, the methods based on the CPA turn out to be very useful in these cases [5]. However, they give only a mean-field description of the electronic properties and atomic configurations in the alloy and their accuracy is unknown. Moreover, because of the single-site approximation they fail to determine the important Coulomb contribution to the total energy of an alloy coming from charge transfer effects. This is so because on the average the alloy is viewed as an ordered lattice of equivalent effective scatterers. Thus the local environment fluctuations existing in real alloys are completely neglected, and the spatial charge distribution around each atom in the alloy is unknown.

Recently, different models and forms of the Madelung energy for a completely random binary alloy $A_x B_{1-x}$ have been proposed. They include the screened impurity model (SIM) [5–7], the charge-correlated model [8] and its CPA version (cc-CPA) [9], the screened CPA (scr-CPA) [9], and the model based on the so-called "$q–V$ relation" [10]. However, as shown in [6] all except that of [10] may be reduced to the same formula with different prefactors:

$$E_M = -\beta e^2 x (1-x) \frac{(Q_A - Q_B)^2}{R_1}, \qquad (1)$$

where x is the concentration of the element A in an alloy, Q_A and Q_B are the net charges of the alloy components, R_1 is the radius of the first coordination shell and the prefactor β ranges from 0 to 1 in different models.

Unfortunately, β cannot be determined within the framework of CPA itself. Therefore, in order to obtain β it has been suggested in [7] to use other methods in the total energy calculations for random alloys which

do not rely on the single-site approximation. In particular, one can use the CW method with unrelaxed cluster interactions which allows one to calculate the total energy of completely random alloys on the ideal lattice as it is considered in the CPA method.

It turns out, for example, that for the fcc Cu–Au and Ni–Pt alloys system-dependent, but concentration and volume independent values of the prefactor in Eq. (1) may be found [7]. With this prefactor the total energies calculated in both ways, i.e. by the CW method and in the framework of the CPA, are consistent in the range of 1 mRy for different alloy compositions and lattice parameters. However, several questions remain. Firstly, there is a question whether Eq. (1) with β obtained by a comparison with other first-principle method really describes the Madelung energy of a random alloy or whether it compensates both kinds of errors coming from the unknown part of the Coulomb energy and the CPA itself. Secondly, the accuracy and reliability of the CW method itself has been so far established mainly in comparison with experiment which often could be misleading. Thirdly, it is interesting to see if different approaches to the random alloy problem, the CW, the supercell, and the CPA, lead to the same result at least in one system. Finally, one may investigate whether there exists a single value of the parameter β in Eq. (1) which gives satisfactory results in different systems.

To answer these questions one may find a system without any charge transfer (so that E_M defined in Eq. (1) vanishes) and with a short-range interatomic interaction leading to a good convergence of the CW total energy expansion and the supercell calculations with respect to its size [11]. Then one may compare the results obtained by different methods. We have found that these conditions are fulfilled for random fcc Ag–Au alloys. The net charges of the alloy components change their signs for different alloy compositions at definite Wigner–Seitz radii. Hence, a radius may be chosen where Q_{Ag} and Q_{Au} are practically zero and E_M therefore vanishes. At the same time the convergence of the CW interactions in this system is very good. We have used the CW method in two modifications. The first which we refer to as "5–5" is based on the 5-interaction expansion of the total energy up to the tetrahedra of the nearest neighbors in the fcc lattice on the basis of the total energies of five ordered alloys: two pure elements, A and B, the AB compound in the $L1_0$ structure and the A_3B and AB_3 compounds in the $L1_2$ structure. The second modification called "7–8" is based on the the 7-interaction expansion including that of the "5–5" modification plus two pair interactions between the second and third neighbors in the fcc lattice which have been determined on the basis of the eight different structures, including the five structures of the "5–5" modification, the A_3B, AB_3 compounds in the DO_{22} structure and AB in the structure Z2 (see [8]). We have found that the nearest-neighbor pair interactions for different volumes range from 500 to 900 K, the next nearest-neighbor pair interactions from to 2 to 20 K, the triangle interactions from 4 to 40 K and the tetrahedra interactions are less than 3 K. This is also reflected in a good convergence of the LSGF calculations with respect to the size of the local interaction zone. The later was used for the solution of the electronic structure problem for the supercells of different sizes. The calculations for the largest supercell consisting of 256 atoms are definitely converged with respect to all parameters and therefore they can be used as a reference point to all other calculations.

In Fig. 1 we present the results for the equiatomic Ag–Au random alloy. In this alloy the net charge of the components vanishes near $R_{WS} = 3.18$ and E_{Mad} is zero at this radius (see Fig. 1(a)). Hence, there is no contribution from Eq. (1) to the total energy in the CPA calculations. The agreement with the total energy for the supercell of 256 atoms is excellent. At the same time, the difference between the LSGF results and the results of the CW "7–8" scheme [8] is about 0.1 mRy/atom and has the same order of magnitude as the convergence of the CW method itself as may be judged from the difference between $E_{CW7–8}$ and $E_{CW5–5}$ (Fig. 1(a)). Thus, one may conclude that the errors introduced by the CPA are negligible if the alloy components are neighbors in the Periodic Table. This result is to be expected because the total energy is an integral quantity over the electronic spectrum and therefore it is not sensitive to the small details of the density of states which are in addition smeared out in alloys [2]. On the other hand, one may see in Fig. 1(a)

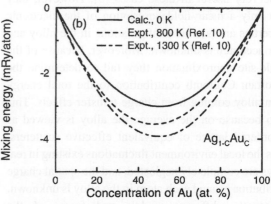

Fig. 1. Consistency between the supercell, the CPA and CW total energy calculations for random fcc $Ag_{50}Au_{50}$ alloy (a) and mixing energy in this system as a function of Au concentration (b). The supercell calculations for the 256 atomic system have been carried out by the LSGF method with the local interaction zone consisting of the central site and two shells of neighboring atoms. In (a) the Madelung contribution to the alloy total energy calculated within the SIM-CPA ($\beta = 0.6$) (full line) and for the supercell (long-dashed line) are presented. Also, the energies calculated by the CPA (circles, full line) and the CW in the "7–8" modification (squares, dashed line) and in the "5–5" modification (diamonds, dot–dashed line) methods are shown relative to one of the LSGF method as a function of Wigner–Seitz radius. In (b) calculated mixing energies (full line), as well as the experimental data at $T = 800$ K (dashed line) and 1300 K (dot–dashed line) [12] are shown.

that close to $R_{WS} = 2.9$ and 3.4 a.u. E_M becomes finite (for $\beta = 0.6$) and that its contribution to the total energy increases significantly. Nevertheless, the difference between the CPA and LSGF results remains negligible. This means that with some degree of accuracy Eq. (1) with this particular value of β describes the

Madelung energy of this particular alloy and does not just compensate for the difference between the CPA and the LSGF results. The last fact is clearly illustrated by a comparison of the Madelung energy of the $Ag_{50}Au_{50}$ alloy calculated by Eq. (1) and directly for a 256 atomic supercell, presented in Fig. 1(a). Finally, the difference between the LSGF and the CW results changes sign. At the same time the CW calculations definitely converge towards the LSGF result when increasing the number of input structures and calculated interaction parameters.

If we now compare the energy difference between the CPA or LSGF and the CW_{7-8} results with the mixing energy for the Ag–Au alloys presented in Fig. 1(b), we find that it is roughly 3–5% of the mixing enthalpy in this system which we consider as a satisfactory result. However the error of the CW_{5-5} calculations increases to 10%. The agreement with experiment is found to be good. Calculated results are expected to be higher than the experimental ones, because we did not account for any short-range order effects in this study. It is seen that with increasing temperature calculated and experimental results become close to each other.

In summary, we have shown that there is satisfactory agreement between results of three most popular techniques for the total energy calculations of random alloys, the supercell, the CPA, and the CW methods, at least in the Ag–Au system. However, the accuracy of the CW method depends strongly on the number of cluster interactions included into the expansion of the total energy. We also have demonstrated that in the framework of the CPA one may obtain a reliable description of the total energy of random alloys. This is important because the application of the CPA is relatively simple. On the other hand, there is no possibility to describe local environment effects (in particular, the Madelung contribution to the alloy total energy)

within this scheme in an exact manner and one needs to go beyond the single-site approximation for a better understanding of the properties of the random alloy.

We are grateful to the Swedish National Science Council for financial support. The support by the Swedish Materials Consortium #9 is acknowledged. Center for Atomic-scale Materials Physics is sponsored by the Danish National Research Foundation. The supercells used in this work were constructed using the computer code written by Dr. S.I. Simak.

References

[1] J.W.D. Connolly, A.R. Williams, Phys. Rev. B 27 (1983) 5169.

[2] J.S. Faulkner, Prog. Mater. Sci. 27 (1982) 1.

[3] A. Zunger, S.-H. Wei, L.G. Ferreira, J.E. Bernard, Phys. Rev. Lett. 65 (1990) 353.

[4] I.A. Abrikosov, A.M.N. Niklasson, S.I. Simak, B. Johansson, A.V. Ruban, H.L. Skriver, Phys. Rev. Lett. 76 (1996) 4203.

[5] I.A. Abrikosov, Yu.H. Vekilov, P.A. Korzhavyi, A.V. Ruban, L.E. Shilkrot, Solid State Commun. 83 (1992) 867.

[6] P.A. Korzhavyi, A.V. Ruban, I.A. Abrikosov, H.L. Skriver, Phys. Rev. B 51 (1995) 5773.

[7] A.V. Ruban, A.I. Abrikosov, H.L. Skriver, Phys. Rev. B 51 (1995) 12958.

[8] Z.W. Lu, S.-H. Wei, A. Zunger, S. Frota–Pessoa, L.G. Ferreira, Phys. Rev. B 44 (1991) 512.

[9] D.D. Johnson, F.J. Pinski, Phys. Rev. B 48 (1993) 11553.

[10] J.S. Faulkner, Y. Wang, G.M. Stocks, Phys. Rev. B 52 (1995) 17106; Phys. Rev. B 55 (1997) 7492.

[11] I.A. Abrikosov, S.I. Simak, B. Johansson, A.V. Ruban, H.L. Skriver, Phys. Rev. B 56 (1997) 9319.

[12] R. Hultgren, P.D. Desai, D.T. Hawkins, M. Gleiser, K.K. Kelley, Selected Values of Thermodynamic Properties of Binary Alloys, American Society for Metals, Ohio, 1973.

ELSEVIER

Computational Materials Science 10 (1998) 306–313

Electronic structure calculations of vacancies and their influence on materials properties

P.A. Sterne [a,b,*], J. van Ek [c], R.H. Howell [b]

[a] *Department of Physics, University of California, Davis, CA 95616, USA*
[b] *Lawrence Livermore National Laboratory, PO Box 808, L-407, Livermore, CA 94550, USA*
[c] *Department of Physics, Tulane University, New Orleans, LA 70118, USA*

Abstract

We provide two examples to illustrate how electronic structure calculations contribute to our understanding of vacancies and their role in determining materials properties. Diffusion and electromigration in aluminium are known to depend strongly on vacancies. Electronic structure calculations show that the vacancy–impurity interaction oscillates with distance, and this leads to an explanation for both the increased electromigration resistance and the slow impurity diffusion for copper in aluminium. Calculations of vacancies in plutonium have been used in conjunction with positron annihilation lifetime measurements to identify the presence of helium-filled vacancies. Helium stabilization of vacancies can provide the precursor for subsequent vacancy-related changes in materials properties. Copyright © 1998 Elsevier Science B.V.

1. Introduction

Vacancies contribute in a number of ways to the properties of materials. Their effects and interactions must be taken into account in describing a variety of processes and materials aging issues since they are intrinsic defects and their influence cannot be eliminated at finite temperatures. Vacancies play a crucial role in diffusion, the random motion of atoms in a solid, and in electromigration, the transport of atoms under the influence of electric fields and currents. Vacancies can also contribute significantly to resistivities in metals at elevated temperatures. Radiation damage in crystalline materials often

produces Frenkel defects, i.e. vacancy–interstitial pairs, and although the vast majority of these recombine rapidly, vacancies can still lead to changes in microstructure with subsequent changes in materials properties.

In this paper we consider the role played by vacancies in two different materials problems. The first example concerns the role of vacancies in diffusion and electromigration in dilute alloys, with particular attention to the case of Cu in Al. The variation of the impurity–vacancy interaction with distance is shown to be significant. The second example addresses the detection by positron annihilation of self-irradiation-induced vacancies in plutonium using positron annihilation. Theoretical calculations of positron lifetimes based on first principles electronic structure methods play an essential role in interpreting the experimental data in this case.

---

* Corresponding author. Lawrence Livermore National Laboratory, PO Box 808, L-407, Livermore, CA 94550, USA. Tel.: +1 510 422 2510; fax: +1 510 422 2510; e-mail: sterne1@llnl.gov.

Table 1
Experimental activation energies relative to Al, ΔQ_{exp}, for the diffusion of substitutional impurities in Al metal in the temperature range 600–900 K ($Q_{Al} = 1.36\,\text{eV}$)

Impurity	Sc	Ti	Cu	Mg	Si	Ag	Au
ΔQ (eV)	1.61	–	0.04	−0.01	−0.07	−0.15	−0.15

Positive values of ΔQ correspond to slow diffusers, i.e. slower than self-diffusion, and negative values correspond to fast diffusers.

2. Diffusion and electromigration

Bulk diffusion of substitutional impurities and self-diffusion in close packed fcc and hcp metals are mediated by vacancies [1,2]. Only the atoms adjacent to vacancies can change their lattice position at any given moment, so diffusion is strongly dependent on the vacancy formation energy which controls the number of thermal vacancies present in the system. Diffusion also depends on the migration energy, the amount of energy required to carry an atom over the energy barrier from its original lattice position into the vacancy site. Self-diffusion is generally described in terms of an activation energy Q_0 which combines these terms:

$$D = D_0 e^{-Q_0/kT}. \tag{1}$$

The diffusion of substitutional impurities can also be represented in this exponential form, but with a different activation energy [1–3]:

$$D_{imp} = D_2 e^{-Q_2/kT}. \tag{2}$$

It is customary to define the activation energy difference:

$$\Delta Q = Q_2 - Q_0 = \Delta E_h + E_{i-v} - C, \tag{3}$$

where ΔE_h is the difference in migration energies between bulk and impurity atoms, E_{i-v} is the impurity–vacancy interaction energy, and C is a correlation term which is typically small and has a slight temperature dependence.

The impurity–vacancy interaction enters into the activation energy difference in a very direct way [3,4] – impurities with negative impurity–vacancy interaction energies attract vacancies to nearest-neighbour sites thereby lowering the activation energy, while positive impurity–vacancy interactions indicate repulsion, and an increase in the activation energy. If we assume

that the migration energies are similar to those for the bulk atoms, then impurities with attractive impurity–vacancy interactions will tend to have lower activation energies than self-diffusion, and will therefore be fast diffusers. In contrast, those that repel vacancies will have higher activation energies, leading to slower diffusion rates than those observed for self-diffusion. Table 1 shows some fast and slow diffusers in Al, with corresponding values of ΔQ [5,6]. Note that Cu is a slow diffuser in Al, suggesting that Cu repels vacancies from its nearest-neighbour shell.

The kinetic five-frequency model is widely used to describe diffusion in fcc lattices [3]. In this model, frequencies are assigned to the various jumps that atoms can make into adjoining vacancy sites. The individual frequencies depend on the activation energies for individual atomic processes, so the impurity–vacancy interaction enters indirectly through the values of the frequencies for jumps that change the impurity-vacancy separation. The model effectively includes only nearest-neighbour impurity–vacancy interactions since it distinguishes only atomic jumps that involve impurities and vacancies in neighbouring sites either before or after the jump is completed. All other atomic jumps are represented in terms of a single "bulk" frequency, regardless of impurity–vacancy separation.

In cases where there is a difference in charge between the impurity and the host, the impurity–vacancy interaction has been estimated on the basis of electrostatic screening [1,7,8],

$$E_{i-v}(r) \sim \frac{e^{-qr}}{r}, \tag{4}$$

where q is the inverse screening length. This interaction is generally evaluated and used only at the nearest-neighbour distance. However, if it is taken to further distances, it clearly represents a monotonic variation of the impurity–vacancy interaction with distance. First principles electronic structure calculations can assess the applicability of these models to dilute alloys. By examining the impurity–vacancy interaction as a function of distance, it is possible to determine quantitatively if the nearest-neighbour term alone is sufficient for an accurate description of the impurity–vacancy interaction in diffusion, or if further neighbour terms should be included. It is also possible to assess the reliability of simple models like the statically screened electrostatic model for describing the interactions at longer distances, and if the interactions are indeed monotonic in nature.

Vacancies also play an important role in electromigration in which the random atomic diffusion is biased by an applied electric field leading to net mass transport in one direction. Electromigration in aluminium interconnects is one of the major failure mechanisms for microelectronic devices [9]. It is well known that the addition of a few atomic per cent of copper in aluminium leads to a significant increase in electromigration resistance, improving the operating lifetime of the microelectronic devices. The mechanism by which Cu increases electromigration resistance is unknown, however. There have been several suggestions, many of which are based on an assumed attractive interaction between the vacancies and the impurity Cu atom [10–12]. These models appear to be in contradiction with diffusion data since a repulsive impurity–vacancy interaction is required to account for the slow diffusion [5] of Cu in Al. This apparent contradiction has stimulated the present study of impurity–vacancy interactions in Al.

We have used a first principles self-consistent electronic structure approach to calculate the total energy as a function of impurity–vacancy separation for a variety of substitutional impurities in aluminium. The linear muffin tin orbital method together with the atomic sphere approximation [13] and the local density approximation [14] were used in a supercell geometry with 32 atoms, including an impurity atom and an empty sphere on the vacancy site. The all-electron calculations included fully relativistic core states and scalar-relativistic valence states. The basis set included s, p and d orbitals on all sites, and the Brillouin zone integration was performed with a

mesh of 125 k-points. Convergence in k-point sampling was checked by performing additional calculations with 216 and 343 k-points. The most significant approximation in this approach is the neglect of lattice relaxation around the vacancy and impurity sites. The implications of this approximation for our results will be addressed below.

The calculations were performed with the impurity atom at nearest-neighbour, second-neighbour, third-neighbour and fourth-neighbour positions with respect to the vacancy in the same unit cell. The total energies were converged to 1 meV or better in all cases. Seven different impurity atoms were considered to ensure reliability in the observed trends – Sc, Ti, Cu, Mg, Si, Ag and Au. We observe a pronounced oscillation in the total energy with impurity–vacancy separation for all seven impurities. The amplitude of this oscillation is numerically significant, and is unaffected when we increase the number of k-points, indicating that it is not an artefact of the calculation.

The oscillatory form of the total energy suggests that the impurity–vacancy interaction itself may oscillate with distance. The well-known Friedel oscillations [15] in the charge density around an impurity provide a natural source of an oscillating interaction in dilute alloys. The Friedel oscillations arise due to the metallic screening of the impurity and vacancy potentials by the host electrons, which in the case of Al have a significant free-electron character.

In order to check if our results are consistent with a Friedel-like mechanism, we have fitted our total energies to an oscillatory form [15]. We assume that the interaction between any given vacancy and impurity can be written as

$$\epsilon(R_i) = A \frac{\cos(BR_i + \phi)}{R_i^3}, \qquad (5)$$

where R_i is the distance to an impurity in the ith shell around the vacancy and A, B and ϕ are determined from the fitting procedure. This functional form should be valid for distances longer than screening length, which is less than the nearest-neighbour distance in Al. The calculated total energies include multiple impurity–vacancy interactions due to the supercell used in the calculations, so the fitting procedure

Table 2
Parameter B obtained from fitting the total energies using Eqs. (5) and (6)

Impurity	Sc	Ti	Cu	Mg	Si	Ag	Au
B (a_0^{-1})	1.63	1.71	1.77	1.75	1.56	1.62	1.69
ΔE_{21} (eV)	−0.19	−0.28	−0.18	−0.04	0.02	−0.08	−0.02

For comparison, the Al free-electron value of B is $1.85a_0^{-1}$. Also listed are the differences in energy between second-neighbour and nearest-neighbour impurity–vacancy configurations, $\Delta E_{21} = \epsilon(R_2) - \epsilon(R_1)$ as obtained from the fitted parameters.

must take this into account in determining the parameters A, B and ϕ. We write the total energy for a given configuration as

$$E_i = C + \sum_j n_{ij}\, \epsilon(R_j), \qquad (6)$$

where the total energies E_i for the configuration with the impurity in shell i around the vacancy is put equal to the sum over all the impurity–vacancy terms together with a constant background energy C. The constant C is eliminated analytically, and the fitting procedure is performed using a standard non-linear fitting procedure [16] to yield the parameters A, B and ϕ. If the oscillations truly arise from a Friedel-like form, the value of B will be close to twice the Fermi wave vector for Al, $2k_F = 1.85a_0^{-1}$

Table 2 shows that the values of B resulting from the fit are close to twice the free-electron Fermi wave vector in Al for all seven impurities considered here, as expected for a Friedel-based mechanism. In fact, the values of B fall slightly below the free-electron estimate, which may be due to an effective reduction in k_F due to Umklapp processes in the crystal. Nevertheless, they are all within 4–16% of the free-electron value, which strongly suggests that the observed oscillations in total energy originate from a Friedel-like mechanism.

Fig. 1 shows the impurity–vacancy interaction energy as a function of separation based on the fitted values obtained from the total energy calculations. The positions of the various atomic shells around the vacancy are indicated. All of the curves show a pronounced oscillatory form, although the amplitude is small in the case of Mg. There is clearly a strong repulsion at the nearest neighbour site for Sc, Cu and Ti, indicating that they are likely to be slow diffusers in Al. This is consistent with experiment, which shows

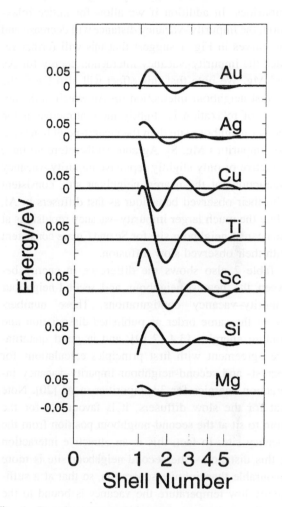

Fig. 1. Impurity–vacancy interaction energy versus impurity–vacancy separation for various impurities in Al. Distances from the vacancy to the first through fifth shell of surrounding atoms in an fcc lattice are indicated. The curves represent the oscillating part of Eq. (5) while zero interaction energy corresponds to the Al-vacancy interaction in Al metal.

that both Cu and Sc are slow diffusers. No experimental data are available for Ti, so this constitutes a theoretical prediction that Ti will be a slow diffuser in Al.

The Si and Au impurities show an attractive impurity–vacancy interaction at the nearest-neighbour site, and this is consistent with the observation that they are fast diffusers in Al. Mg and Ag show small repulsive impurity–vacancy nearest-neighbour interactions, so the contribution to the diffusion activation energy from the migration energy may be significant for these impurities. In addition if we allow for lattice relaxation, the impurity–vacancy distance will decrease and the curves in Fig. 1 suggest that this will further reduce the impurity–vacancy interaction energy for Ag and Mg. We note that this effect will also lower the nearest-neighbour interaction for Au which in the absence of relaxation is slightly more attractive at the second-neighbour site than the nearest-neighbour site. The impurities Mg, Si, Ag and Au therefore all have attractive or only slightly repulsive impurity–vacancy interactions at the nearest-neighbour site, consistent with their observed behaviour as fast diffusers in Al, while the much larger impurity–vacancy repulsions at the nearest-neighbour site for Sc and Cu are consistent with their observed slow diffusion.

Table 2 also shows the difference in energy between the nearest-neighbour and second-neighbour impurity–vacancy configurations. These numbers are of the same order as published dissociation and binding energies [3,4,17,19], and in good quantitative agreement with first principles calculations for nearest- and second-neighbour impurity–vacancy interaction energies for 3d impurities in Al [20]. Note that for the slow diffusers, it is favourable for the atom to sit at the second-neighbour position from the vacancy site. In fact, this is an attractive interaction at this distance – the second-neighbour site is more favourable than infinite separation, so that at a sufficiently low temperature the vacancy is bound to the impurity at the second-neighbour site.

The calculations confirm the validity of an oscillating Friedel-like form for the vacancy–impurity interaction, and clearly indicate that models relying on nearest-neighbour interactions alone are not sufficient to describe the impurity–vacancy interaction for all substitutional impurities. The traditional application of the kinetic model based only on nearest-neighbour interactions is very reasonable for impurities with attractive impurity–vacancy interactions, such as Si in Al and electropositive impurities in noble metals [1], where the simple screened-Coulomb electrostatic theory works well. However for systems with repulsive impurity–vacancy interactions, the oscillatory nature of the interaction, which has its origin in the very general Friedel oscillations, results in the possibility of an attractive impurity–vacancy interaction at the second-neighbour or some further neighbour site. This attraction, which can significantly affect the diffusion, is completely neglected in the widely used five-frequency kinetic model.

We noted in Section 1 that Cu in Al is a slow diffuser, implying a repulsive impurity–vacancy interaction. At the same time, the increase in Al electromigration resistance with Cu doping implies that the Cu impurity attracts vacancies, thereby inhibiting their participation in electromigration. This apparent contradiction is resolved by our results, which show that the impurity–vacancy interaction is repulsive at the nearest-neighbour site, but that there is a significant attraction at the second-neighbour site. While this attraction is not sufficient to permanently bind the vacancy to the Cu site, it may retard vacancy motion sufficiently to result in an increased electromigration resistance.

Lattice deformation around defects can significantly affect the total energies obtained from the electronic structure calculations. The relaxation energy for a vacancy in Al is in the range 0.1–0.2 eV [21,22]. This is comparable to the impurity-vacancy repulsion we calculate, so neglecting such a large energy contribution is clearly a source of concern. However, the relaxation around the vacancy will be similar in all configurations, and relatively independent of the impurity–vacancy separation. The impurity–vacancy interaction is only affected by *differences* in relaxation energies due to different configurations, and not by the size of the relaxation energy itself. For this reason, we have confidence that our results are not adversely affected by neglecting these relaxation energies. Model calculations indicate that relaxation energies vary by only about 0.015 eV depending on the impurity–vacancy separation [21]. This is much smaller than the dissociation energy for an impurity–vacancy pair for Sc,

Ti, Cu, and Au (Table 2). The effects of relaxation will be more important in cases where the amplitude A is small, such as Mg and Ag, and should then be included directly in the total energy calculations. In most other cases, however, relaxation will only have a small quantitative effect on the results. For this reason, our omission of relaxation is not as serious an approximation as it might at first appear, since, to first order, the different impurity–vacancy configurations have the same relaxation energy. In addition, the oscillatory nature of the impurity–vacancy interaction will not be affected significantly by lattice relaxation since it arises from the Friedel oscillations, a fundamental property of the electronic screening in Al.

The widely used five-frequency kinetic model [1–3] for diffusion is convenient since it provides a tractable model that facilitates comparison with experiment. In common with essentially all the analytic work on diffusion on fcc metals, it assumes that only the nearest-neighbour impurity–vacancy interaction energy is significant. We have shown here that for some impurities there may be a very strong nearest-neighbour repulsion together with a significant second-neighbour attraction. This can modify the diffusion characteristics of the impurity in a way that is not well described in the standard five-frequency model, and clearly calls for the development of a new model to account for the second-shell interactions. The calculated nearest-neighbour repulsion and second-neighbour attraction can also lend support to models of electromigration for Cu in Al which rely on an impurity–vacancy attraction to increase electromigration resistance.

3. Vacancies in plutonium studied by positron annihilation

Radioactive decay in plutonium results in the creation of a large number of vacancies, interstitials, and other defects. An understanding of the behaviour of these defects is important for predicting and understanding how materials properties evolve over time. Each radioactive decay creates a few thousand Frenkel pairs, but the majority of these recombine rapidly,

leaving a smaller fraction of uncompensated vacancies and interstitials that can diffuse and interact with other point and extended defects in the lattice.

Positron annihilation lifetime spectroscopy provides a sensitive probe of vacancies and open-volume defects in materials [23]. Positrons annihilate with electrons in the material with a lifetime that depends on the electron charge density in the vicinity of the positron. For defect-free crystals, the positron enters a Bloch-like extended state located primarily in the interstitial region where the repulsion with the similarly charged atomic nuclei is minimized. The positron then annihilates with a characteristic bulk lifetime determined by the electron charge density in this region. In materials with open-volume defects, the positron is naturally attracted to vacancies and voids, again due to repulsion with the atomic nuclei. The lifetime then differs from the bulk lifetime due to the different electron charge density in the vacancy. Vacancies and open volume defects typically have longer lifetimes than the defect-free material, since there are fewer electrons available for annihilation with the positron.

Theoretical calculations of positron lifetimes play a crucial role in interpreting experimental data [24]. Positron lifetimes reflect the local charge density around the positron, with a unique lifetime corresponding to each different positron state. When the experimental data provide a single lifetime component, indicating that the positrons all annihilate in a similar state, additional information is required to determine if the observed lifetime value corresponds to a bulk (defect-free) lifetime or a defect. In systems where several different defects may trap positrons, each with their own characteristic lifetime, it is often difficult to match the observed lifetime value with the appropriate defect.

Theoretical calculations of positron lifetimes are based on first-principles electronic structure methods [24–26] The electron–electron and electron–positron interactions are all treated within the local density approximation, as is the enhancement of the positron annihilation rate due to the mutual attraction of the electron and positron. The parameters for these interactions are obtained from many-body calculations [27,28] and are independent of measured positron lifetimes, so the method truly constitutes a parameter-free first-principles approach for calculating positron lifetimes. These lifetime calculations have been very successful in reproducing well established bulk lifetimes for many elemental metals [25,26,29], as well as vacancy lifetimes in metals, semiconductors [26] and oxides [30].

Positron lifetime measurements on aged plutonium samples produced a single component lifetime of 184 picoseconds (ps) [31], indicating that all the positrons annihilate from the same type of state; either all the positrons remained in extended bulk-states, or they all trapped in similar defect states. Positron lifetime calculations were performed using our first principles approach [25] based on the linear muffin tin orbital method. The resulting bulk lifetime was 139 ps, significantly lower than the measured lifetime. Supercell calculations for a monovacancy produced a converged positron lifetime of around 250 ps with a 64 atom supercell. This is significantly longer than the measured lifetime, indicating that bare monovacancies are not responsible for the observed lifetime. Divacancies and larger vacancy clusters and voids will have even longer lifetimes, corresponding to their lower electron charge densities compared to the monovacancy. The observed lifetime is therefore not consistent with a defect-free lifetime, or an empty vacancy, vacancy clusters, or larger voids.

Helium atoms are produced in plutonium due to alpha decay. There is a strong repulsion between helium atoms and metal atoms due to the closed He s-shell of electrons, so helium atoms preferentially occupy vacant sites in the lattice if these are available. We have calculated the positron lifetime in a Pu vacancy containing a He atom and find a lifetime value in the range 170–195 ps, depending on the position of the He atom. This is in excellent agreement with the measured lifetime value of 184 ps, leading to the conclusion that He-filled vacancies are present in significant numbers in this aged plutonium sample. Vacancy concentrations of a part per 10^3–10^4 are generally sufficient to trap all the positrons into vacancies, so the observation puts a lower limit of one part per 10^4 on the number of helium-filled vacancies in this aged sample.

Helium is known to have an important influence on radiation-induced swelling and subsequent embrittlement in other materials [32]. Helium atoms can stabilize vacancies and small voids, and lower the critical radius for cavity growth [33]. Helium stabilization of vacancies is therefore an important precursor for subsequent changes in materials properties, and the identification and observed evolution of these vacancies provides important information for subsequent modelling of aging mechanisms in irradiated materials.

4. Conclusions

We have used the sophisticated apparatus provided by electronic structure calculations to examine key issues associated with realistic materials problems. Calculations of the impurity–vacancy interactions in dilute aluminium alloys provide good agreement with observed trends in diffusion, and also indicate that attractive impurity–vacancy interactions at the second-neighbour lattice distance may be important for understanding diffusion and electromigration properties. First-principles-based positron lifetime calculations indicate that aged plutonium samples contain an appreciable concentration of helium-filled vacancies. In other materials like Al, similar observations have been important in understanding subsequent changes in materials properties due to aging.

Acknowledgements

This work was performed under the auspices of the US Department of Energy by Lawrence Livermore National Laboratory under contract number W-7405-ENG-48. The authors gratefully acknowledge the Louisiana Board of Regents.

References

[1] R.J. Borg, G.J. Dienes, An Introduction to Solid State Diffusion, Academic Press, San Diego, 1988.
[2] A.R. Allnatt, A.B. Lidiard, Atomic Transport in Solids, Cambridge University Press, Cambridge, 1993.

[3] A.D. Le Claire, J. Nuc. Mater. 69–70 (1978) 70 and references therein.

[4] M. Doyama, J. Nuc. Mater. 69–70 (1978) 350 and references therein.

[5] Landolt-Börnstein, New Series III/26, Springer, Berlin, 1990.

[6] S. Fujikawa, Sci. Eng. Light Metals RASELM-91, Tokyo, 1991, p. 959.

[7] D. Lazarus, Phys. Rev. 93 (1954) 973.

[8] A.D. Le Claire, Phil. Mag. 7 (1962) 141.

[9] A. Christou (Ed.), Electromigration and Electronic Device Degradation, Wiley, New York, 1994.

[10] R. Rosenberg, J. Vac. Sci. Tech. 9 (1971) 263.

[11] P.S. Ho, Phys. Rev. B 8 (1973) 4534.

[12] C. Kim, J.W. Morris, J. Appl. Phys. 73 (1993) 4885.

[13] O.K. Andersen, Phys. Rev. B 12 (1975) 3060.

[14] U. von Barth, L. Hedin, J. Phys. C 5 (1972) 1629.

[15] J. Friedel, Phil. Mag. 43 (1952) 153.

[16] Mathematica 2.2,Wolfram Research, Champaign, IL, 1993.

[17] U. Klemradt, B. Drittler, R. Zeller, P.H. Dederichs, Phys. Rev. Lett. 64 (1990) 2803.

[18] U. Klemradt, B. Drittler, T. Hoshino, R. Zeller, P.H. Dederichs, N. Stefanou, Phys. Rev. B 43 (1991) 9487.

[19] N.H. March, J. Nuc. Mater. 69–70 (1978) 490 and references therein.

[20] T. Hoshino, R. Zeller, P.H. Dederichs, Phys. Rev. B 53 (1996) 8971.

[21] P.S. Ho, R. Benedek, IBM J. Res. Dev. 18 (1974) 386.

[22] N. Chetty, M. Weinert, T.S. Rahman, J.W. Davenport, Phys. Rev. B 52 (1995) 6313.

[23] R.W. Siegel, in: P.G. Coleman, S.C. Sharma, L.M. Diana (Eds.), Positron Annihilation, North-Holland, Amsterdam, 1982, p. 351; K. Petersen, in: W. Brandt, A. Dupasquier (Eds.), Positron Solid-State Physics, North-Holland, Amsterdam, 1983, p. 298.

[24] M.J. Puska, R. Nieminen, Rev. Mod. Phys. 66 (1994) 841.

[25] P.A. Sterne, J.H. Kaiser, Phys. Rev. B 43 (1991) 13 892.

[26] B. Barbiellini, M.J. Puska, T. Korhonen, A. Harju, T. Torsti, R.M. Nieminen, Phys. Rev. B 53 (1996) 16 201.

[27] J. Arponen, E. Pajanne, Ann. Phys. 121 (1979) 343.

[28] L. Lantto, Phys. Rev. B 36 (1987) 5160.

[29] M. Puska, J. Phys.: Condens. Matter 3 (1991) 3455.

[30] W.D. Mosley, J.W. Dykes, P. Klavins, R.N. Shelton, P.A. Sterne, R.H. Howell, Phys. Rev. B 48 (1993) 611.

[31] C. Colmenares, R.H. Howell, D. Ancheta, T. Cowan, J. Hanafee, P. Sterne, UCID-20622-96, Lawrence Livermore National Laboratory (1996) p. 15.

[32] H. Ullmaier, Nuclear Fusion 24 (1984) 1039.

[33] L.K. Mansur, J. Nuc. Mater. 216 (1994) 97.

ELSEVIER

Computational Materials Science 10 (1998) 314–318

COMPUTATIONAL
MATERIALS
SCIENCE

Ab initio study of the chemical role of carbon within TiAl alloy system: Application to composite materials

S.F. Matar *, Y. Le Petitcorps, J. Etourneau

*Institut de Chimie de la Matière Condensée de Bordeaux-CNRS, Château Brivazac,
Avenue du Docteur Schweitzer, F33600 Pessac, France*

Abstract

Computations within the local spin density functional (LSDF) are carried out for the ternary Ti,Al,C system using the ASW (augmented spherical waves) method to address the chemical role of carbon within TiAl alloy whose electronic structure is firstly investigated. From the preliminary study of the hypothetical compounds formed by insertion and substitution we find carbon to preferably substitute for Al forming Ti-rich phases. The electronic structure of the actually forming carbide Ti_2AlC compound is then carried out and a role played by carbon is addressed through chemical bonding results using the so-called COOP (crystal orbital overlap populations) implemented for the first time in an LSDF method. Copyright © 1998 Elsevier Science B.V.

Keywords: LSDF; ASW; Carbide; Ti_2AlC

1. Introduction and method of calculation

Carbon-fibre reinforced titanium-based inter-metallics are of interest because their field of applications largely involves space and aeronautic industries. In recent years several works were devoted to their investigation both experimentally and theoretically [1,2]. From the latter point of view physical properties such as the change of brittleness can be addressed starting from the results of ab initio self-consistent calculations. For instance in TiAl, vanadium was found to enter substitutionally at Ti site leading to a reduced brittleness [3]. The Ti,Al,C phase diagram shows that carbon is not thermodynamically stable within TiAl and the chemical reaction between C and

TiAl may form TiC and Ti_2AlC [4]. This work is thus devoted to the investigation of TiAl and of Ti_2AlC with the objective to further understand the chemical role of carbon.

The electronic properties were calculated using the ab initio self-consistent ASW (augmented spherical waves) method based on local spin density functional (LSDF) [5]. The matrix elements were constructed involving solutions of the Schrödinger equation up to the secondary l quantum number: $l_{max} + 1$ where $l_{max} = 2$ for Ti, Al and C, i.e. with s, p, d and f ($l_{max} + 1$) basis set and $l_{max} = 1$ for empty spheres (ESs), i.e. with s, p and d ($l_{max} + 1$) basis set. Thus the basis set is limited and the contributions associated with $l_{max} + 1$ highest term should be small, i.e. the relevant charge should always be lower than 0.1 electron in order to ensure a convergence of the charges. Using the atomic sphere

* Corresponding author. E-mail: matar@cribx1.u-bordeaux.fr.

approximation (ASA) the ASW method assumes overlapping spheres centred on the atomic sites. The volume of all spheres is enforced to equal the cell volume. This is a good approximation for closely packed alloy systems but for loosely packed crystal structures such as that of Ti_2AlC, the empty space has to be represented by use of ESs which are pseudo-atoms with $Z = 0$ atomic number and no core states. Using a sufficiently large number of **k** points within the first Brillouin zone of the respective Bravais lattices, self-consistency was achieved with the following convergence criteria for charge and total energy: $\Delta Q < 10^{-8}$; $\Delta E < 10^{-8}$ Rydberg.

Further we discuss the chemical bonding based on the concept of crystal orbital overlap population (COOP). These are obtained from the expectation values of operators which consist of the non-diagonal elements of the overlap population matrix

$$c_{ni}^*(\mathbf{k})S_{ij}c_{nj}(\mathbf{k}) = c_{ni}^*(\mathbf{k})\langle\chi_{ki}(\mathbf{r})|\chi_{kj}(\mathbf{r})\rangle c_{nj}(\mathbf{k}),$$

where S_{ij} represents an element of the overlap matrix of the basis functions and the $c_{nj}(\mathbf{k})$ are the expansion coefficients entering the wave function of the nth band. Partial COOP coefficients $C_{ij}(E)$ are then obtained by integrating the above expression over the Brillouin zone:

$$C_{ij}(E) = C_{ji}(E) = 1/\Omega_{BZ}$$
$$\times \sum_n \int_{BZ} \mathrm{d}^3\mathbf{k}\,\mathrm{Re}\,\{c_{ni}^*(\mathbf{k})S_{ij}c_{nj}(\mathbf{k})\}\delta(E - \varepsilon_{nk})$$

(Dirac notation delta) which in a somewhat lax notation is often designated as the overlap-population-weighted-DOS. The total COOP are then evaluated as the sum over all non-diagonal elements, i.e.

$$C(E) = \sum_{ij\,(i \neq j)} C_{ji}(E).$$

For a detailed description and for significant examples we refer the reader to the recent work of Eyert [6]. Recently we implemented the COOP [7] in the ASW method with the objective to extract further information on the chemical bonding.

2. Results and discussion

Fig. 1 shows the site projected DOS (y-axis, eV^{-1}) in TiAl. The energy reference along the x-axis is taken with respect to Fermi level (E_F) within a reduced energy range $\{-6, 6\,eV\}$, i.e. excluding low lying C s-like states to make the presentation clear. Ti states are mainly of d character and Al DOS are mainly of s,p character exhibiting a free-electron like behaviour in the lower energy range $\{-6, -4\,eV\}$ of the valence band (VB). Due to the small filling of Ti d subshell Fermi level crosses the lower part of d states whose major contribution is found above E_F. The resemblance between the Ti and Al DOS in the energy range $\{-6, -1\,eV\}$ points to a covalent mixing between them in the lower energy part of Ti states. This is illustrated by the changes of electronic configurations of Ti and Al in the alloy system as with respect to the free atoms configurations. That is the initial valence electron configurations: Ti $4s^2\,4p^0\,3d^2\,4f^0$; Al $3s^2\,3p^1$ $3d^0\,4f^0$ become Ti $4s^{0.52}\,4p^{0.66}\,3d^{2.66}\,4f^{0.03}$; Al $3s^{1.11}\,3p^{1.69}\,3d^{0.29}\,4f^{0.03}$. The valence charge sums up to 14 electrons per unit cell when the multiplicity of the different atoms (2Ti, 2Al) is accounted for. The terms in italics correspond to charges in $l_{max} + 1$ defined in former section. From this the mixing between the states of Ti and Al thus results in a d-character brought into Al by Ti and a p-character brought into Ti by Al.

The character of the bonding can be further assessed by examining the total COOP (y-axis, eV^{-1}) given as an insert in Fig. 1 within the same energy range as the DOS to enable comparisons with them. Positive, negative and vanishingly small values refer to bonding, anti-bonding and non-bonding states, respectively. All the states below E_F and up to 1.5 eV are bonding thus stabilising the crystal lattice. A detailed analysis shows that the peak maximum dominating the COOP around $-2\,eV$ is relative to the interactions of the lower part of Ti d band with Al s, p; they are largely separated from their anti-bonding counterpart around 4 eV in the conduction band (CB) thus pointing to strong interactions. This is opposed to the COOP around E_F with exhibit much lower intensity. They are relevant to Ti–Ti interactions with in-plane d

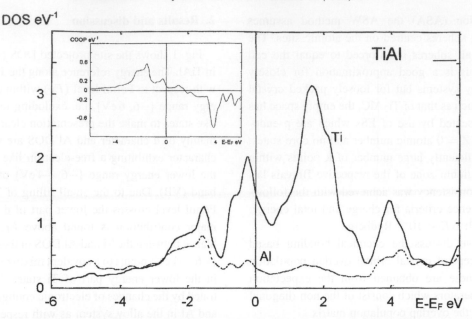

Fig. 1. Site projected densities of states in TiAl. The insert shows total COOP of TiAl.

orbitals. From this, one may expect the bonding characteristics which are dominated by Ti–Al interactions to be modified by contribution from foreign atoms such as carbon.

Preliminary studies were done within TiAl structure either by inserting or by substituting carbon. They are summarized in the following.
- Inserting carbon at the octahedral voids – otherwise occupied by ES – in the CuAu^I-type tetragonal structure of TiAl can be done either in the neighbourhood of Ti or Al. The self-consistent results show that charge transfer is from Ti and Al spheres to ES and C ones. There is however a larger charge carried by carbon when it is placed near Ti than in the neighbourhood of Al thus pointing to a relatively larger stabilisation of the lattice in such a configuration.
- TiAl contains two formula units per unit cell so that one carbon atom substituting for one Ti or Al atom in an ordered manner, yields two possible formulae: TiCAl$_2$ and Ti$_2$AlC. Whereas Ti$_2$AlC is thermodynamically stable, TiAl$_2$C is not [4]. The calculation of the COOP for both compounds confirms this thermodynamic trend. We find half

of the VB of TiCAl$_2$ bonding and the other half anti-bonding thus leading to destabilised lattice. On the contrary in Ti$_2$AlC all the COOP are bonding throughout the VB and even above E_F – just like in TiAl (Fig. 1). This composition can be hence expected to be more stable. From these results one can suggest that when carbon firstly enters the alloy lattice of TiAl interstitially, the tendency is to form Ti–C pairs leading to TiC precipitates in the close neighbourhood of the carbon fibre. Further diffusion of carbon through TiC is possible because this binary is always non-stoichiometric so that it hardly constitutes a diffusion barrier. This leads to the formation of the intermetallic system Ti$_2$AlC which we now study in its actual hexagonal structure [4].

The changes of the electron configuration in the Ti$_2$AlC carbide ($Z = 2$) upon self-consistency are as follows. Initial configurations: Ti: $4s^2$ $4p^0$ $3d^2$ $4f^0$; Al: $3s^2$ $3p^1$ $3d^0$ $4f^0$; C: $2s^2$ $2p^2$ $3d^0$ $4f^0$; ES: $1s^0$ $2p^0$ $3d^0$; final configurations: Ti: $4s^{0.37}$ $4p^{0.59}$ $3d^{2.33}$ $4f^{0.08}$; Al: $3s^{1.08}$ $3p^{0.82}$ $3d^{0.23}$ $4f^{0.08}$; C: $2s^{1.40}$ $2p^{3.05}$ $3d^{0.15}$ $4f^{0.04}$; ES: $1s^{0.50}$ $2p^{0.16}$ $3d^{0.04}$. The valence charge sums up to 30 electrons per inerac unit cell when accounting for the multiplicity of the different

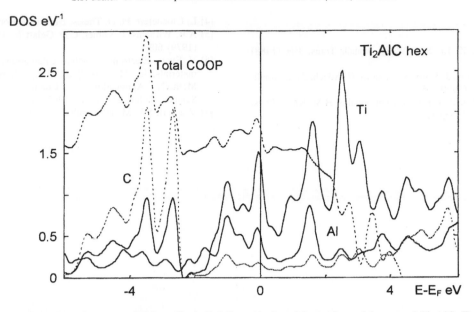

Fig. 2. Site projected densities of states (solid line: Ti; dashed line: Al; dotted line) of actual hexagonal Ti$_2$AlC. Total COOP are shown in dashed lines in the middle of the figure.

atoms (4Ti, 2Al, 2C, 4ES). Besides the general feature of an electron transfer Ti,Al→ C,ES on the one hand and the redistribution of electrons among the l states due to orbital mixing within a given atom on the other hand, the charge transfer into ES mainly arises from the s states of Ti and Al and there is a larger d chartacter of Al in the carbide as with respect to TiAl. Addressing this point should be enabled by the study of the DOS and the COOP.

Fig. 2 shows the site projected DOS of Ti$_2$AlC. Ti states show much larger structure towards the lower energies than in TiAl where they solely interact with Al s,p states – because of the extra interaction with carbon. This is resembled by the massifs between -6 and -2 eV which follow the skyline of carbon DOS. The d character of Al observed in the charge transfers above is resembled by the Ti–Al interaction which mainly occurs in the range {$-2, 1$ eV} where Ti and Al DOS run similarly especially in the neighbourhood of E_F.

The features of chemical bonding can be further assessed by using the COOP given for total contribution within Fig. 2. The COOP have bonding character between -6 and -5 eV, i.e. above 1.5 eV^{-1} – to be taken as zero COOP in this presentation; they follow the skyline of C 2p DOS so that these COOP are dominated by Ti–C interactions. It should be noted however that Al–C bonding cannot be excluded but should be of a lesser importance from the weaker intensity of the Al DOS as with respect to Ti DOS. The anti-bonding counterpart can be seen in the energy range {$2, 6$ eV}. The large separation between the bonding and anti-bonding massifs is indicative of a strong bonding mainly driven by the Ti–C interaction which stabilises the lattice. Were it not for the interactions in the neighbourhood of E_F, i.e. in the DOS region where Ti–Al interactions predominate, the COOP would resemble those of TiC which have vanishingly small values around E_F – calculated for the pupose of this work but not shown here for lack of space. From this, Ti–Al interaction present in the neighbourhood of E_F plays a role in reducing the brittleness of the ternary carbide where Ti–C interactions are reminiscent of TiC precipitates formed between the carbon fibre and Ti$_2$AlC.

References

[1] D. Vujic, Z. Li, S.H. Wang. Metall. Trans. 19a (1988) 2445.

[2] M. Morinaga, J. Saito, N. Yukawa, H. Adach. Acta Metall. Mater. 38 (1990) 25.

[3] S.R. Chubb, D.A. Papaconstantopoulos, B.M. Klein, Phys. Rev. B 8 (1988) 12 120.

[4] L. Clochefert, Ph.D. Thesis, Université Bordeaux 1, 1995.

[5] A.R. Williams, J. Kübler, C.D. Gelatt Jr., Phys. Rev. B 19 (1979) 6094.

[6] V. Eyert, Electronic structure calculations for crystalline materials, in: M. Springborg (Ed.), Density Functional Methods: Applications in Chemistry and Materials Science, Wiley, Chichester, 1997, pp. 233–304.

[7] V. Eyert, S.F. Matar, Results to be published.

COMPUTATIONAL
MATERIALS
SCIENCE

ELSEVIER

Computational Materials Science 10 (1998) 319–324

Electronic structure of stannous oxide

M. Meyer [a,*], G. Onida [b], A. Ponchel [a], L. Reining [a]

[a] *CNRS-CEA/DSM, Laboratoire des Solides Irradiés, Ecole Polytechnique, 91128 Palaiseau Cedex, France*
[b] *Dipartimento di Fisica, Università di Roma "Tor Vergata", I-00133 Rome, Italy*

Abstract

We present an ab initio study of the electronic structure of SnO. Density functional theory in the local density approximation (DFT-LDA) is used in conjunction with carefully tested smooth pseudopotentials. Total energies and charge densities are calculated and analysed as a function of the atomic geometry, with a particular emphasis on the importance of low-charge-density contributions to the interlayer cohesion. SnO_2 has already been studied in the past and is used for comparison. Copyright © 1998 Elsevier Science B.V.

Keywords: DFT-LDA; Oxide; Cohesion; Pseudo-potential

1. Introduction

Tin oxides are used in many fields of technological importance such as catalysis, chemical gas sensing, heat reflection and microelectronics. This is the reason for the numerous studies of their physical and chemical properties (including their electronic structure). Most of these investigations concern tin dioxide. However, stannous and stannic oxide coexist frequently due either to an oxygen loss associated with the reduction of SnO_2, or to the oxidation of SnO.

The crystalline structures of SnO and SnO_2 are tetragonal at room temperature and normal pressure (Figs. 1(a) and (b)) and belong to space groups D_{4h}^7 (P4/nmm) [1–3] and D_{4h}^{14} (P42/nmm), respectively. The oxidation process results mainly in the insertion of an oxygen plane between two tin planes in the layered SnO crystalline structure (see Figs. 1(c) and (d)).

As a result SnO_2 is a more densely packed crystal where each tin atom is surrounded by a slightly distorted oxygen octahedron while in SnO the tin atoms sit on the vertices of pyramids with an oxygen square basis (Fig. 1(b)). These edge sharing pyramids form the layers of the SnO structure with tin vertices lying alternately above and below them. The layers are stacked perpendicularly to the **c** crystallographic axis with tin atoms facing each other (Fig. 1(d)). In order to well understand the stability of this structure and the cohesion between its layers, it is important to have an accurate description of the atomic and electronic structure of SnO and of the relation between electronic and geometric configurations.

2. Methodology

Ab initio calculations using density functional theory (DFT) with the local density approximation

---

* Corresponding author. Tel.: 33 1 69 45 01, fax: 33 1 69 33 30 22; e-Mail: Madeleine.Meyer@polytechnique.fr.

SnO₂   SnO

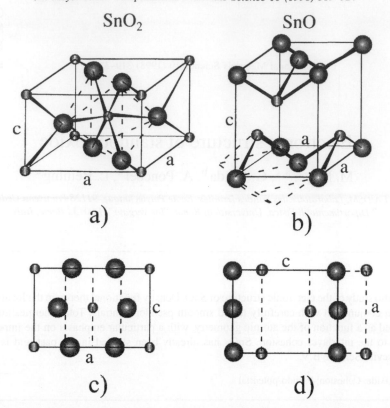

a)   b)

c)   d)

Fig. 1. Atomic configurations of SnO₂ and SnO, the large spheres correspond to oxygen atoms and the smaller ones to tin atoms: (a) Structure of SnO₂: unit cell. The dashed lines link the O atoms forming an octahedron surrounding a tin atom. (b) Structure of SnO: unit cell. The dashed lines link the O atoms and the tin atom forming square-based pyramids. (c) Projection of the SnO₂ unit cell onto a (1 0 0) plane showing the traces of the alternate O and Sn atomic planes. (d) Projection of the SnO unit cell onto a (0 1 0) plane showing the traces of the O and Sn atomic planes. An extra Sn plane has been added for the sake of comparison with (c).

(LDA) are performed in order to investigate the density of charge in SnO and its evolution with the atomic configuration. The Kohn Sham equations [4] are solved using the Car–Parrinello (CP) scheme [5] with normconserving pseudopotentials and a plane wave basis. This approach has been used with great success for various materials, but, in the case of oxides, the localised O 2p states require the use of a largely extended plane wave basis and consequently an important computational effort is necessary. The number of plane waves can be significantly reduced by the use of soft pseudopotentials [6] which have already proved to be efficient for oxides such as TiO₂ [7], SnO₂ [8] and Li₂O [9]. The oxygen soft pseudopotential created with a core cut-off radius of 1.45 a.u. allows us to get a good convergence of the total energy with

a kinetic energy cut-off equal to 70 Ry. A standard hard core norm conserving pseudopotential is used for the tin atoms [10] and the relative core contribution due to the 4d electrons is taken into account via a nonlinear core correction [11]. As usual with the CP method the Kleinman–Bylander (KB) separable form of the pseudopotential [12] is used with the p and d components as reference states for oxygen and tin, respectively. Eight k points of the irreducible Brillouin zone are used to calculate the total energy of the electronic ground state associated with the atomic positions of tin and oxygen.

The unit cell of SnO contains two molecular units with atoms located as follows: O(0 0 0; 1/2 1/2 0), Sn(0 1/2 u; 1/2 0 ū). There are some discrepancies between the lattice parameters obtained by powder

X-ray data measurements ($a = 3.796$ A, $c = 4.816$ A, $u = 0.2356$) [1] or neutron diffraction measurements ($a = 3.799$ A, $c = 4.827$, $u =?$) [3]. The main incertitude concerns the tin location since there is a significant lack of accuracy in the u determination. In our calculation the total energy is minimised, for each selected cell volume, with respect to the tin atom location (u). The CP code allows us to take into account the symmetry of the forces and we apply for SnO a C_4 symmetry around the **c** axis. This reduces the length of the calculations but keeps the atoms free to move in the [0 0 1] direction and allows for the determination of u. In order to determine the lattice constants that minimise the total energy, the calculations are performed for different values of the lattice parameters, in a range $c = 4.0 - 4.827$ A and $a = 3.61 - 3.99$ A.

3. Results and discussion

The 3D densities of charge calculated for SnO_2 and SnO at the experimental atomic configuration are plotted in Figs. 2 and 3(a). A simple look at the charge density distribution shows that the cohesion of SnO_2 is easy to understand. For SnO the explanation of the

cohesion is not straightforward in terms of a description only involving the Madelung energy due to nominal charges. In fact, within the range of variation of c, the distance between tin atoms belonging to neighbouring layers is smaller than the Sn–O distance of atoms in second neighbour position (the O atoms first neighbours of Sn are located in the same layer). The layers should repel each other. The interatomic distances plotted in Fig. 4 show that when the structure is compressed in the **c** direction the Sn–Sn distance decreases less rapidly than the Sn–O second neighbour distance. Thus, the Sn–Sn repulsive interaction becomes more and more counterbalanced by the attractive Sn–O interlayers contribution. However, in this simple picture, the repulsive component remains dominant. A close look to the graphs of the densities of charge, plotted in Fig. 3, is necessary to understand the cohesion when c decreases. "Hats" of charge covering the Sn atoms appear which screen the Sn ions and decrease the repulsive forces, thus allowing for cohesion. This screening is hardly visible for the experimental value of c/a (Fig. 3(a)) and becomes more important when this ratio tends towards the value minimising the total energy (Fig. 3(d)).

A quantitative evaluation of the results shows however that our description of SnO contains some uncertainty. Our calculated values of the equilibrium lattice constants are $a = 3.799$ A, in very good agreement with experiment, and $c = 4.286$ A, which underestimates the most quoted experimental value of 4.827 A by as much as 11%. No reliable experimental values are available for a comparison of u. The error in c is hence significantly higher than the few percent discrepancies which are typically found in LDA calculations.

We have checked that neither the particular choice of the pseudopotentials nor their KB form are responsible for this discrepancy. One might wonder about possible contributions of the 4d states of Sn, which have been treated as core states. They are in fact relatively close in energy to the oxygen 2s level. But first, the most important contribution of the Sn 4d electrons is the core-valence exchange, which is already taken into account by using a nonlinear core correction and second, the core relaxation effects are

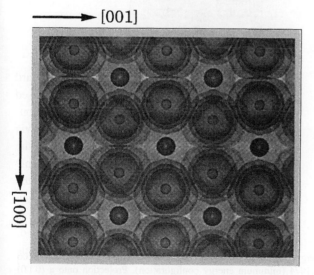

[001]

[100]

Fig. 2. 3D plot of the charge density calculated for the experimental configuration of SnO_2 ($a = 4.74$ A, $c = 3.18$ A, $u = 0.307$), projection onto a (1 0 0) plane. Four isodensity surfaces are plotted. They correspond to the following values: 5×10^{-5}, 5×10^{-4}, 2×10^{-4}, 4×10^{-4} electrons/(a.u.)^3.

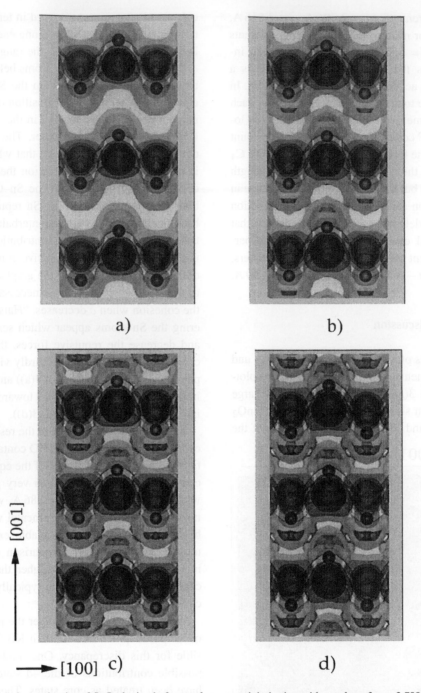

Fig. 3. 3D plot of the charge density of SnO obtained after total energy minimisation with a value of $a = 3.799$ A: (a) $c/a = 1.2706$ (experimental value); (b) $c/a = 1.2$; (c) $c/a = 1.15$; (d) $c/a = 1.1074$ (minimum energy configuration). Projection onto a $(0\,1\,0)$ plane. Four isodensity surfaces are plotted. They correspond to the following values: 5×10^{-5}, 5×10^{-4}, 2×10^{-4} and 4×10^{-4} electrons/(a.u.)^3.

generally very small. It remains that the hybridisation of the Sn 4d with the O 2s levels could in principle give non-negligible modifications. However, calculations on SnO using localised basis sets and including the Sn 4d levels have explicitly shown that the Sn 4d states form a separate, narrow band and do not mix into the true valence states [13–14].

Since, as pointed out above by our qualitative discussion, the interlayer cohesion is due to a very delicate balance between ionic repulsion and screening, one could suspect particular difficulties linked to the LDA. They could be related to the fact that the LDA does not correctly cancel the electron self-interaction of the Hartree term, and hence underestimates the electron localisation. A smeared-out charge density between the layers would then be the consequence, with enhanced screening and hence overbinding. Although this hypothesis cannot be completely excluded, and may at least partially contribute to the observed behaviour, it is somehow contradicted by our charge density plots: the screening of the Sn atoms for small c/a ratios is in fact mostly due to the relatively localised "hats" on each Sn atom, and not to delocalised charge.

There is another point, however, which merits reflection. As in most standard CP codes, we have, as a first approach, used the d component of the pseudopotential as the local reference component for the Sn atom. This is in principle not recommended, since this component has been created with an excited configuration of the atom, and bears hence a bigger arbitrariness than the s and p components, which are created from the ground state [10]. Such a choice is nevertheless widely used, and generally does not lead to problems, apart from those which are indirectly linked to the KB separation through the remaining s and p non-local components. Bulk silicon, for example, does not show any visible difference when calculated with the p instead of the d component as local potential. In the case of SnO, however, more care seems to be needed. The equilibrium geometry is indeed extremely sensitive to a change in the reference component. The naïve solution would be to switch from the d to an s or p reference component. Unfortunately, this choice only worsens the results. This can be explained intuitively: the p (or s) potentials are

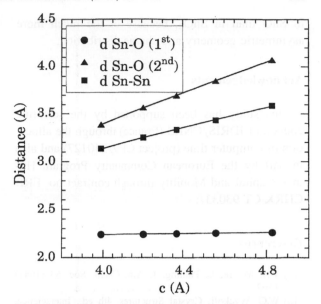

Fig. 4. Graph of the Sn–Sn and Sn–O distances plotted as a function of the c parameter. Two types of Sn–O distances are reported: the smaller ones (first neighbours) correspond to atoms belonging to the same layer, the larger ones correspond to atoms belonging to neighbouring layers.

more attractive than the d component. More charge is hence accumulated close to the Sn atoms, which means enhanced screening, which, in turn, leads to even smaller c/a values. This does actually happen; at constant $a = 3.799$ A, when using p as a reference state, c/a decreases again by more than 6% with respect to the value obtained with the d local potential. This implies that a reference component which is even less attractive than the d potential would be needed in order to get the results closer to experiment. In other words, the true potentials acting on higher angular components cannot be substituted in a satisfactory way by one of the lower components. This hypothesis is far from absurd, if one looks at the problem from the point of view of perturbation theory using a localised basis: in tin compounds, the Sn 4f states are more likely to contribute to the perturbed eigenstates than the 4f states of smaller atoms like silicon for example. One should hence expect that higher angular moments, in particular $l = 3$, are relatively important in tin oxides, and hardly allow a quick, approximate treatment. This problem is probably more pronounced in SnO than in SnO_2, because of the much more asymmetric geometry and charge distribution.

Acknowledgements

This study has been supported by the scientific council of IDRIS/CNRS (France) through the allocation of computer time (project CP9/940127) and also in part by the European Community Program Human Capital and Mobility through contract no. ERB CHRX CT 930337.

References

[1] W.J. Moore, L. Pauling, J. Am. Chem. Soc. 63 (1941) 1392.
[2] W.G. Wyckoff, Crystal Structures, 4th ed., Interscience, New York, 1974, p. 136.
[3] D.M. Adams, A.G. Christy, J. Haines, Phys. Rev B 46 (1992) 11 358.
[4] W. Kohn, J.L. Sham, Phys. Rev. 140 (1965) A1133.
[5] R. Car, M. Parrinello, Phys. Rev. Lett. 55 (1985) 2471.
[6] N. Troullier, J.L. Martins, Phys. Rev. B 43 (1991) 1993.
[7] K.M. Glassford, J.R. Chelikowsky, Phys. Rev. B 46 (1992) 1284.
[8] M. Palummo, L. Reining, M. Meyer, C.M. Bertoni, in: D.J. Lockwood (Ed.), Proceedings of the 22nd ICPS, Vancouver, 1994, World Scientific, Singapore, 1995, p. 161.
[9] S. Albrecht, G. Onida, L. Reining, Phys. Rev. B 55 (1997) 10 278.
[10] G. Bachelet, D.R. Hamann, M. Schlüter, Phys. Rev. B 26 (1982) 4199.
[11] S. Louie, S. Froyen, M. Cohen, Phys. Rev. B 26 (1982) 1738.
[12] L. Kleinman, D.M. Bylander, Phys. Rev. Lett. 48 (1980) 1425.
[13] J.M. Themlin, M. Chtaïb, L. Henrard, P. Lambin, J. Darville, J.M. Gilles, Phys. Rev. B 46 (1992) 2460.
[14] J. Terra, D. Guenzburger, Phys. Rev. B 44 (1991) 8584.

ELSEVIER

Computational Materials Science 10 (1998) 325–329

COMPUTATIONAL
MATERIALS
SCIENCE

First principles investigations of a "quasi-one-dimensional" charge-transfer molecular crystal: TTF-2,5Cl$_2$BQ

C. Katan [a,*], C. Koenig [a], P.E. Blöchl [b]

[a] *Groupe Matière Condensée et Matériaux, UMR au CNRS 6626, Université Rennes-1, 35042 Rennes Cedex, France*
[b] *IBM Research Division, Zürich Research Laboratory, CH-8803 Rüschlikon, Switzerland*

Abstract

We performed first principles calculations using the projector augmented wave method in order to get a clear description at a microscopic level of the electronic distribution far from the transition for TTF-2,5Cl$_2$BQ. Our calculations predict for the first time a low-symmetry structure (ground state) of this compound. We relate them to a simple tight-binding scheme which allows us to discuss the pertinence of the one-dimensional models used to study the phase transitions in this class of materials. Copyright © 1998 Elsevier Science B.V.

Keywords: Molecular crystals; Electronic structure; Ab initio; Charge transfer; TTF; Benzoquinone

Some mixed-stack organic compounds exhibit uncommon electronic instabilities, which give rise to different types of reversible structural transitions under pressure, temperature or photo-irradiation [1–3]. A common feature of all these transitions is the loss of symmetry-inversion in the crystal below a critical temperature T_c or above a critical pressure P_c, resulting in a "dimerization" of donor–acceptor pairs of slightly deformed molecules along the chains. This symmetry breaking is accompanied by a more or less abrupt variation of the charge transfer ρ from the donor (D) to the acceptor (A). For example, TTF-2,5Cl$_2$BQ versus pressure presents a rather large ionicity variation from the neutral high-symmetry (HS) phase (0.2e$^-$) to the ionic low-symmetry (LS) phase (0.8e$^-$), without any noticeable discontinuity at the critical pressure ($\simeq$ 4 GPa), in contrast with TTF–Chloranil (or

TTF–CA), where the observed discontinuity at the transition is of about 0.4e$^-$ for temperature- (and less for pressure-) induced transitions [1,4]. In this latter case, a discontinuous shrinking of the unit cell [5] has also been measured at the transition. Despite considerable experimental and theoretical efforts, the microscopic mechanisms at the origin of these neutral to ionic (N–I) transitions and their differences are not yet fully understood.

The aim of this paper is to illustrate the new possibilities provided by ab initio simulation techniques on TTF-2,5Cl$_2$BQ (26 atoms per unit cell). The method used is the PAW method [6] within the gradient corrected local density approximation (LDA)[7] of the density functional theory (DFT). From the ab initio determination of the electronic structure of the HS and LS phases of this compound, we show how the charge transfer variation is coupled to the loss of inversion symmetry. Moreover, we can extract a coherent set of parameters for the one-dimensional models which are

* Corresponding author. Tel.: +33 02 99 28 26 04; fax: +33 02 99 28 67 17; e-mail: claudine.katan@univ-rennes1.fr.

currently used to analyse the thermodynamical aspects of these transitions.

The crystal structure of TTF-2,5Cl$_2$BQ [3] has been measured at room temperature and pressure, i.e. in its HS phase: the unit cell is triclinic and contains one TTF and one 2,5Cl$_2$BQ molecule, forming a mixed chain in the **b** direction. For this experimental structure, we have performed self-consistent electronic-structure calculations with three or four **k** points between Γ and $Y = \frac{1}{2}\mathbf{b}^*$ in the Brillouin zone (BZ), or with 12 **k** points in $\frac{1}{8}$ of this BZ. We have found no stable magnetic configuration for this compound, in agreement with its experimental properties. So all the results presented here are obtained for the paramagnetic ground state.

We have already shown [8] that the dominant feature of the band structure along the $\Gamma - Y$ direction in **k** space, for the valence band (VB) and the conduction band (CB), is a maximum dispersion at the BZ boundary and not at Γ. This is simply due to the fact that the molecular orbitals involved in these bands are of different symmetries: the HOMO (highest occupied molecular orbital) of TTF is inversion antisymmetric whereas the LUMO (lowest unoccupied molecular orbital) of 2,5Cl$_2$BQ is symmetric. The resulting dispersion curves can be reproduced in a tight-binding scheme by considering two alternating hopping integrals t and $-t$ along the chain. This symmetry argument is valid for the entire series of mixed-stack TTF-p-benzoquinones, and differs from the assumption of one single integral t underlying the one-dimensional models used up to now to analyse the N–I transition in these compounds.

Fig. 1 displays isodensity curves for the Bloch wave function of the VB at Γ and Y for the HS phase, The electronic density at Γ is concentrated on the

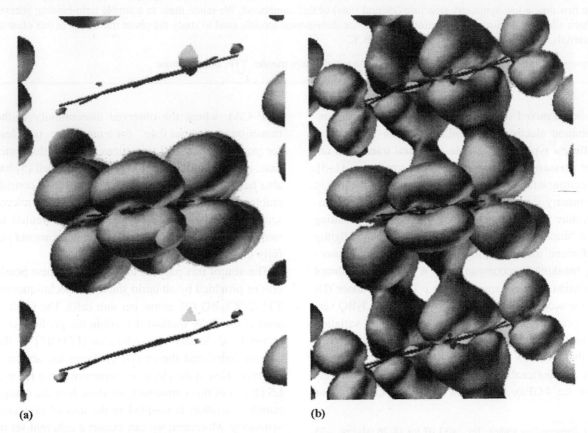

(a)   (b)

Fig. 1. Isodensity representation of the valence band states in the HS structure of TTF-2,5Cl$_2$BQ: three molecules (A–D–A) are shown along a vertical chain axis (a) for $\mathbf{k} = \Gamma$ and (b) for $\mathbf{k} = Y$.

TTF molecule and is nearly equivalent to the one of the HOMO of pure TTF [8]. At Y, it is mainly built from a linear combination of the HOMO of TTF and the LUMO of $2,5Cl_2BQ$. The charge transfer can be estimated as the mean value in the BZ of the weight of the $2,5Cl_2BQ$ states in this VB Bloch function. As the molecular orbitals cannot hybridize at Γ, the centre of the BZ brings no contribution to the charge transfer.

The LS phase of TTF-$2,5Cl_2BQ$ is obtained at high pressure and no crystallographic studies of this phase have been possible up to now. One can nevertheless get an idea of this structure by considering the LS phase of TTF–CA at low temperature [5]. This non-centro-symmetric phase is characterized by a slight deformation (or bending) of the molecules, with atomic displacements of the order of 0.1 Å. It is often described as a "dimerization" in D–A pairs along the chain.

To analyse the LS phase of TTF-$2,5Cl_2BQ$, we have proceeded in the following way: the PAW method allows straightforward atomic relaxations in a given unit cell. So we have determined the ground-state electronic configuration in the experimental lattice at normal pressure. The structure thus obtained is very similar to the LS structure of TTF–CA, and should be a good starting point to study the electronic redistribution driven by the loss of inversion symmetry. However, up to now there is no evidence for this LS phase for temperatures above 15 K.

The dispersion curves of this LS phase are very similar to those of the HS phase, apart from a global lowering of the energies. The comparison of the electronic distribution in real space is much more striking. Fig. 2 shows the isodensity curves in the LS phase. The distributions at Γ and Y are now both very similar to linear combinations of the HOMO

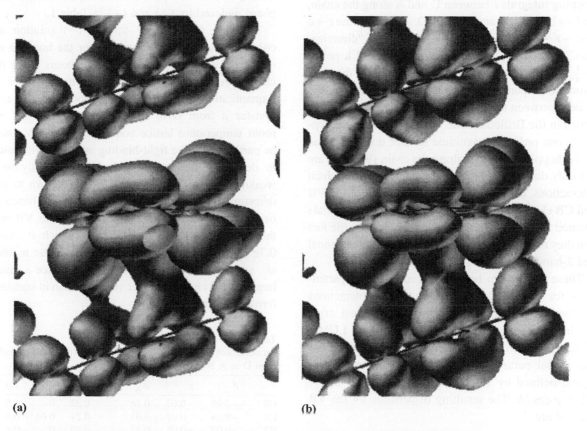

(a)   (b)

Fig. 2. Isodensity representation of the valence band states in the LS structure of TTF-$2,5Cl_2BQ$: three molecules (A–D–A) are shown along a vertical chain axis (a) for $\mathbf{k} = \Gamma$ and (b) for $\mathbf{k} = Y$.

of TTF and the LUMO of 2,5Cl$_2$BQ, with a slightly larger weight on TTF. The "dimerization" in D–A pairs, which is not at all conspicuous when looking at the molecular deformations, is manifested in this electronic distribution. The hybridization at Γ is no longer symmetry-forbidden, and induces an increase of the charge transfer.

So our self-consistent calculations lead to a very clear understanding of the relation between symmetry-breaking, pairing of the molecules and charge-transfer increase. It is due to the deformation of a rather delocalized electronic cloud along the chain, much more than to the simple image of the hopping of an electron from TTF to the acceptor. The simplest way to estimate this charge-transfer variation is to analyse these results in terms of a tight-binding scheme.

The dominant interactions in this system are the hopping integrals t between D and A along the chain, and $-t$ between A and the next D, which become $t'+\epsilon$ and $-t' + \epsilon$ in the LS phase, due to the "dimerization". However, preliminary studies with 12 **k** points in $\frac{1}{8}$ of the BZ revealed small dispersions in the band structure perpendicularly to the chain. Moreover, the middle between the energies E_+^k of CB and E_-^k of VB varies in the Brillouin zone. All these effects are typically one order of magnitude smaller than those due to the hopping integrals along the chain [9]. They are not only due to small additional intra- and inter-chain interactions, but also due to hybridizations of the VB and CB electronic states with other molecular orbitals situated higher or lower in energy, expressing the fact that they do not form a really isolated one-dimensional and 2-band system.

These effects can be reproduced by fitting parameters for all the D–A, D–D and A–A interactions between neighbours in the three-dimensional crystal, involving the HOMO of TTF and the LUMO of 2,5Cl$_2$BQ. In a one-dimensional model this adds two global parameters β and γ for the on-site energies, defined by $E_D = E_D^0 + \beta \cos kb$ and $E_A = E_A^0 + \gamma \cos kb$. The resulting dispersions for the VB and CB are

$$E_\pm^k = s \pm \sqrt{d^2 + 4t'^2 \sin^2 \tfrac{1}{2}kb + 4\epsilon^2 \cos^2 \tfrac{1}{2}kb},$$

where $s = \frac{1}{2}(E_A + E_D)$ and $d = \frac{1}{2}(E_A - E_D)$, whereas the weight of the LUMO of A in the VB for two spin directions is given by

$$|C_A^k|^2 = 1 - \frac{d}{\sqrt{d^2 + 4t'^2 \sin^2 \tfrac{1}{2}kb + 4\epsilon^2 \cos^2 \tfrac{1}{2}kb}}.$$

From the values of the ab initio energies and the weights at the three **k** points Γ, $\frac{1}{4}\mathbf{b}^*$ and Y, all the parameters of the model can in principle be determined. However, a precise estimate of the tight-binding weights from an ab initio calculation based on a completely different representation of the electronic states requires a proper numerical treatment. Here we give a first estimate of the parameters based on the sole value of $|C_A^\Gamma|^2$, which is strictly zero in the HS phase, and about 0.6 for the LS phase. Due to the non-linear character of the equations to be solved for the fit, two sets of parameters are possible for each phase, indexed by (+) and (−) in Table 1.

The sign of β and γ for the (−) solution are consistent with what is expected for the hopping integrals between second nearest neighbours along the chain. These two sets will be discriminated later by a quantitative estimate of $|C_A^Y|^2$. Both give a charge transfer ρ from D to A which is too high for the room temperature lattice considered here. This may be partly due to the tight-binding approximation used to determine this charge transfer, and partly to a weakness of the DFT–LDA, which is known to underestimate the gap of semi-conductors and hence to overestimate the hybridization between the VB and the CB. However, their difference, which is about 0.3e$^-$, shows that in this class of molecular materials, small atomic displacements may induce a much larger global charge transfer variation than in standard ferroelectric transitions.

Table 1
Parameters of the tight-binding model, in eV (the charge transfer from D to A is expressed in e$^-$)

	β	γ	$E_A^0 - E_D^0$	t	ϵ	ρ(e$^-$)
HS$^+$	−0.05	0.03	0.16	0.22	0	0.6
LS$^+$	−0.04	0.03	0.03	0.24	0.06	0.9
HS$^-$	0.03	−0.05	0.32	0.20	0	0.4
LS$^-$	0.03	−0.04	0.18	0.23	0.06	0.7

The global charge-transfer variation is due to several additional structural and electronic effects: variation of the lattice parameters versus temperature or pressure, distortion of the molecules in the LS phase, variation of the Madelung interactions in the crystal, due to the change in the geometry of the crystal and mainly to the change in ionicity itself. These effects are of course coupled to thermodynamical effects which are not taken into account in the present ab initio calculations.

A quantitative estimate of the corresponding parameters is needed to build models able to analyse the transition and explain the differences from one compound to the next. In this work, we have for the first time obtained a visualization of the effect of the loss of inversion symmetry in a fixed lattice on the electronic distribution. The most spectacular effect is a clear pairing of the electronic states on the D and A molecules, and an additional hybridization at the centre of the Brillouin zone. This leads to a significant charge-transfer enhancement, due both to an increase of t and to an additional parameter ϵ, which is characteristic of the symmetry breaking and is as large as $\frac{1}{4}$ of t. The induced variation of the Madelung interactions in the crystal is included in the variation of the on-site energy difference $E_A^0 - E_D^0$. In order to study the global effect of pressure on TTF-2,5Cl$_2$BQ, the sensitivity of these parameters to lattice contractions could be estimated by further ab initio calculations.

Acknowledgements

This work is part of the Joint Research Project (No. 82280512) between the two above-mentioned research partners. It benefited from collaborations within, and has been partially funded by, the Human Capital and Mobility Network on "Ab initio (from electronic structure) calculation of complex processes in materials" (Contract No. ERBCHRXTC930369). Part of the calculations has been supported by the "Centre National Universitaire Sud de Calcul" (France).

References

[1] C.S. Jacobsen and J.B. Torrance, J. Chem. Phys. 78 (1983) 112.
[2] A. Girlando, C. Pecile and J.B. Torrance, Solid State Commun. 54 (1985) 753.
[3] A. Girlando, A. Painelli, C. Pecile, G. Calestani, C. Rizzoli and R.M. Metzger, J. Chem. Phys. 98 (1993) 7692.
[4] M. Hanfland, A. Brillante, A. Girlando and K. Syassen, Phys. Rev. B 38 (1988) 1456.
[5] M. Le Cointe, M.H. Lemée-Cailleau, H. Cailleau, B. Toudic, L. Toupet, G. Heger, F. Moussa, P. Schweiss, K.H. Kraft and N. Karl, Phys. Rev. B 51 (1995) 3374.
[6] P.E. Blöchl, Phys. Rev. B 50 (1994) 17953.
[7] J.P. Perdew and A. Zunger, Phys. Rev. B 23 (1981) 5048; A.D. Becke, J. Chem. Phys. 96 (1992) 2155; J.P. Perdew, Phys. Rev. B 33 (1986) 8822.
[8] C. Katan, C. Koenig and P.E. Blöchl, Solid State Commun. 102 (1997) 589.
[9] C. Katan et al., unpublished.

ELSEVIER

Computational Materials Science 10 (1998) 330–333

COMPUTATIONAL
MATERIALS
SCIENCE

Ab initio calculation of electron affinities of diamond surfaces

M.J. Rutter *, J. Robertson

Department of Engineering, University of Cambridge, Trumpington Street, Cambridge CB2 1PZ, UK

Abstract

Diamond surfaces combine chemical inertness with, in some cases, a negative electron affinity. Such surfaces have great potential for use on cold cathodes in flat displays. We present ab initio plane wave electronic structure calculations which enable us to predict the electron affinities of many different diamond surfaces with various terminations and reconstructions. Such calculations give good accuracy and enable the study of perfect surfaces with a range of terminating species so that the effect of the passifying layer can be readily seen. Results for the (1 0 0) and (1 1 1) surfaces will be presented, giving a range of surfaces more comprehensive than previously published. We find that the electron affinity varies by over 5.5 V between oxygen and hydrogen coverings, and that this magnitude of variation can be understood as simply a rising from surface dipoles as polarised covalent bonds would be expected to produce. A brief discussion of some of the technical points of performing such a calculation is given, which combines ab initio LDA work with experimental results for the band gap for diamond in order to estimate accurately the position of the unoccupied levels. Copyright © 1998 Elsevier Science B.V.

1. Introduction

Diamond is a material of considerable technological interest, for it can exhibit a negative electron affinity (NEA) on a chemically inert surface. The term NEA means that the bulk conduction band minimum is above the vacuum level, and therefore that any electron excited to the conduction band is above the vacuum level, and can be emitted relatively readily. Such a material is therefore a good candidate for use as a cold cathode and for displays based on such technology.

An NEA is often obtained by the use of a highly dipolar surface coating such as caesium or caesium oxide [1]. However, caesium is readily contaminated and cathodes based on such coatings degrade rapidly. Diamond, however, has been shown to possess an NEA when terminated with hydrogen, a termination which is chemically inert. Experimental evidence for this NEA for the (1 1 1) surface has been known for almost two decades [2,3], and more recently the (1 0 0) surface has been shown to have an NEA [4]. Theoretical calculations also predict NEAs [5,6].

This work uses ab initio electronic structure calculations to predict the electron affinities of a range of diamond surfaces. By covering a wide range of terminations and configurations the large effect of the surface termination is clearly seen, and the use of computer modelling yields precise unambiguous knowledge of the atomic structure of the surface studied.

---

* Corresponding author. Address: TCM, The Cavendish Laboratory, Madingley Road, Cambridge CB3 0HE, UK. Tel.: +44 1223 33 7254; fax: +44 1223 33 7356; e-mail: mjr19@cam.ac.uk.

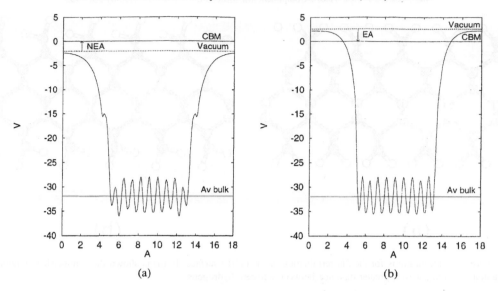

Fig. 1. Potentials across a diamond slab. In (a) is shown $(100)2 \times 1$:H which has an NEA, and in (b) $(100)1 \times 1$:O with a positive EA. The average bulk potentials have been aligned to place the conduction band minimum at 0 V.

2. Method

Ab initio calculations have a history stretching over more than two decades, and have been successful in many areas. A review of the general techniques is given in [7]. In calculating electron affinities it is necessary to overcome two restrictions inherent in plane wave codes, namely the inability of the local density approximation (LDA) [8] used for the exchange–correlation energy to give the energies of excited states, and the requirement for periodicity in all three dimensions. The use of an infinite periodic sandwich of bulk and vacuum regions successfully models an isolated surface whilst retaining periodicity if care is taken to ensure that both the vacuum and the bulk are sufficiently wide.

The method used to obtain the electron affinities is as follows [5]. The average Kohn–Sham potential and highest occupied electronic level are calculated for bulk diamond. The experimental value of the band gap is then added to find the position of the conduction band minimum (CBM) with respect to the bulk potential. A calculation is then performed on a cell containing both bulk and vacuum and the surface of interest. From this calculation is obtained the average value of the potential in the centre of the bulk and vacuum regions, and thus the difference between the average bulk potential far from the surface, and the vacuum potential. An averaging scheme similar to that of Baldereschi et al. [9] was used in the bulk region. The positions of both the CBM and vacuum level are then known with respect to a common reference, and hence with respect to each other. The electron affinity is simply the difference between them. This is shown in Fig. 1 where the bulk potentials have been aligned so as to place the CBM at 0 V.

3. Structures

There are many different possible species and arrangements for terminating diamond surfaces. A full review is beyond the scope of this paper, but some of the more interesting points are breifly mentioned here.

There has been much debate about the stability and structure of the (100) dihydride surface [10,11]. We have not attempted to investigate its stability, but we do find that a canted, but not twisted,

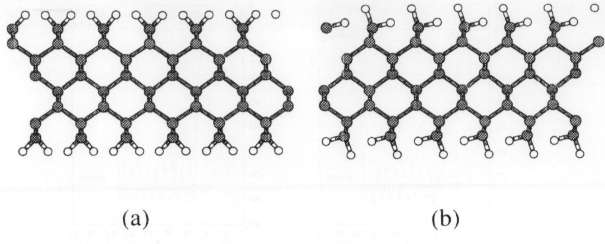

(a)   (b)

Fig. 2. Two possible configurations for the 2H termination of the (1 0 0) surface. In (a) is shown the symmetric configuration, whilst in (b) the canted structure, with a greater distance between adjacent hydrogens.

configuration, as shown in Fig. 2 is the most energetically favourable, being 0.4 eV per surface carbon atom lower in energy than the symmetric uncanted untwisted form. This canting of the CH_2 units by 27° increases the separation of adjacent hydrogens from 1.10 to 1.38 Å. A 2 × 1 reconstruction of this surface has been proposed [5], but we find it to be slightly higher in energy. These results for the dihydride surface structure agree with recent work by Hong and Chou [12].

The position of the oxygens on the (1 0 0):1 × 1 surface has also been a matter of debate. The obvious choices are between placing the oxygens directly above each carbon atom and double bonded to them, or placing them between each pair and single bonded to both in a bridge-like ether configuration. Thus one contains a double C = O bond, and the other two single C–O bonds. Standard chemical bond energies would favour the double bonded structure, and the calculations of Whitten et al. [13] also favour this structure. We find the bridge structure to be lower in energy, although we present results for both configurations.

4. Results and discussion

The results of these calculations are given in Table 1. It can be seen that good convergence with respect to the number of layers has been achieved. The vacuum separation between nuclei on opposing surfaces varied between 8 and 10 Å.

The ordering of the results from oxygen through bare suface to hydrogen termination is expected: the dipole moment set up by oxygen at the surface will hinder the escape of electrons, and conversely that set up by hydrogen will favour the escape. The magnitude of the variation, and its dependence on the precise surface geometry, are quite large. Both are explicable in terms of the dipole moment per unit cell. A shift in the EA of 7 eV can be produced by a change in the dipole moment of just 0.25 eÅ. This dipole moment would be expected to vary considerably as the precise geometry of the surface changes. For instance, the 'ketone' oxygen configuration places them 1.2 Å above the surface carbons, whereas the 'ether' configuration places them only 0.8 Å above, and hence produces less of a dipole moment, and thus less of an increase in the EA.

Table 1
Electron affinities of diamond surfaces

Surface	Reconstruction	Termination	Number of layers	EA (eV)
(1 0 0)	2 × 1	None	10	0.69
	1 × 1	2H (sym)	8	−3.19
	1 × 1	2H (canted)	8	−2.36
	2 × 1	H	8	−2.03
			12	−2.07
	1 × 1	O (ketone)	8	3.64
	1 × 1	O (ether)	8	2.61
			10	2.70
(1 1 1)	2 × 1	None	10	0.35
(1 1 1)	1 × 1	H	8	−2.03

5. Conclusions

Using ab initio computational techniques, the electron affinities (EAs) of a range of diamond surfaces have been calculated. The values obtained are shown to be converged with respect to slab thickness, and are in accord with the known experimental data, giving the clean surfaces a small positive EA, and the hydrogen terminated (1 0 0) and (1 1 1) surfaces a negative EA. A more detailed discussion of this work is given in [14].

Acknowledgements

The authors would like to thank Dr. Mike Payne for access to computer resources. The major calculations were performed on the Hitachi SR2201 at the High Performance Computing Facility of the University of Cambridge. Cambridge University is the recipient of a Motorola Corporated Research Laboratory research grant.

References

[1] R.L. Bell, Negative Electron Affinity Devices (Clarendon Press, Oxford, 1973).

[2] F.J. Himpsel, A. Knapp, J.A. van Vechten and D.E. Eastman, Phys. Rev. B 20 (1979) 624.

[3] C. Bandis and B.B Pate, Phys. Rev. B 52 (1995) 12 056.

[4] J. van der Weide and R.J. Nemanich, J. Vac. Sci. Technol. B 12 (1994) 2475.

[5] Z. Zhang, M. Wensell and J. Bernholc, Phys. Rev. B 51 (1995) 5291.

[6] J. Furthmüller, J. Hafner and G. Kresse, Phys. Rev. B 53 (1996) 7334.

[7] M.C. Payne, M.P. Teter, D.C. Allan and J.D. Joannopoulos, Rev. Modern Phys. 64 (1992) 1045.

[8] W. Kohn and L.J. Sham, Phys. Rev. A 140 (1965) 1133.

[9] A. Baldereschi, S. Baroni and R. Resta, Phys. Rev. Lett. 61 (1988) 734

[10] S.H. Yang, D.A. Drabold and J.B. Adams, Phys. Rev. B 48 (1993) 5261.

[11] A.V. Hamza, G.D. Kubiak and R.H. Stulen, Surf. Sci. 237 (1990) 35.

[12] S. Hong and M.Y. Chou, Phys. Rev. B 55 (1997) 9975.

[13] J.L. Whitten, P. Cremaschi, R.E. Thomas, R.A. Rudder and R.J. Markunas, Appl. Surf. Sci. 75 (1994) 45.

[14] M.J. Rutter and J. Robertson, Phys. Rev. B (1997), in preparation.

ELSEVIER

Computational Materials Science 10 (1998) 334–338

COMPUTATIONAL
MATERIALS
SCIENCE

Potential energy of two structures of $\Sigma = 11\langle 0\,1\,1\rangle$ tilt grain boundary in silicon and germanium with empirical potentials and tight-binding methods

J. Chen, B. Lebouvier, A. Hairie, G. Nouet *, E. Paumier

Laboratoire d'Etudes et de Recherche sur les Matériaux, UPRESA 6004 CNRS, Institut des Sciences de la Matière et du Rayonnement, 6 Bd du Maréchal Juin, F-14050 Caen Cedex, France

Abstract

Two atomic structures A and B of the $\Sigma = 11\langle 0\,1\,1\rangle$ grain boundary were observed in silicon and germanium. We have performed a complete study of the stability of these two grain boundaries using some empirical potentials and also the semi-empirical, tight-binding (TB) method. The TB method has confirmed the experimental observations at low temperatures. The A structure is more stable in silicon whereas for germanium the B structure is obtained. The empirical potentials, such as those of Keating (1966), Baraff et al. (1980) and of Stillinger and Weber (1985), give the A structure as the most stable for both germanium and silicon. The non-ability of these empirical potentials to make a difference between germanium and silicon and the advantage of TB method are discussed. Copyright © 1998 Elsevier Science B.V.

Keywords: Potential energy; Empirical potential; Tight-binding; Tilt grain boundary; Silicon; Germanium

1. Introduction

Atomic structure of symmetric tilt grain boundaries in silicon and germanium was extensively analysed by high resolution electron microscopy [1]. These grain boundaries are rather well-defined by using the concept of structural units [2]. However, this description may lead to several arrays of structural units for the same grain boundary. Thus, it was shown that the ⟨0 1 1⟩ tilt grain boundaries in silicon could exhibit different atomic structures A, B and C and some of them were actually observed [3]. Energetic calculations were performed on these atomic structures in order to calculate the potential energy by means of empirical potentials: Keating [4], Baraff et al. [5], Stillinger and Weber [6], Tersoff [7] and Vanderbilt et al. [8]. According to the potentials, agreement between the structure observed by high resolution electron microscopy and the structure with minimum potential energy was never perfect [9]. As some of these atomic structures were observed after deformation at high temperature of a silicon bicrystal, new calculations were made by introducing an entropic term to reach the free energy [10] or to determine the possible transformation from one atomic structure to an equivalent one on the basis of molecular dynamics [11–14].

* Corresponding author. Tel.: +33 2 31 45 26 47; fax: +33 2 31 45 26 60; e-mail: gerard@leriris1.ismra.fr.

Thus, it was shown that the main term in the calculation of the free energy is the potential energy [10] and that some atomic jumps could occur at the level of the grain boundary resulting in the modification of its local structure [11]. No significant progress on the correlation between the energetic calculations and the experimental observations was obtained. However, more recently, new results concerning the deformation of germanium bicrystals [15] showing an opposite behaviour for a special tilt grain boundary, $\Sigma = 11$, prompt us to improve our methods of calculation.

So, in a first part, empirical potentials are used again to calculate the potential energy of both atomic structures of $\Sigma = 11$ in silicon and germanium. Then, in a second part, calculations based on a tigh-binding (TB) approach are presented.

2. Atomic structure of the $\Sigma = 11$ tilt grain boundary

This orientation corresponds to a rotation $\theta = 50.48°$ around a $\langle 110 \rangle$ rotation axis and the indices of the symmetric grain boundary plane are $\{2\,3\,3\}$. The limiting structural units are the specific structural units M and T corresponding to $\Sigma = 9$ ($\theta = 38.94°$) and to $\Sigma = 3$ ($\theta = 70.51°$), respectively. The M unit is based on a 5–7 atom ring, whereas the T unit corresponds to the boat configuration of a six-atom ring. The period of $\Sigma = 11B$ is twice larger than that of $\Sigma = 11A$ and involves another structural unit, P, which results from the transformation of two T units into a new 5–7 atom ring. It is worth noting that the $\Sigma = 3$ twin could be built with this P structural unit but its potential energy would be much higher. A sketch of both atomic structures $\Sigma = 11A$ and $\Sigma = 11B$ is given in Fig. 1. The sequences of two atomic structures are, respectively:

$$\Sigma = 11A: M^-TM^+T$$
$$\Sigma = 11B: M^+TM^-P^+M^-TM^+P^-$$

where $-$ and $+$ correspond to structural units which are connected by a mirror symmetry with respect to the grain boundary plane.

Experimentally, these equivalent grain boundaries were obtained after deformation by compression of a $\Sigma = 9$ silicon bicrystal at two different temperatures: low temperature ($T < 950°C$) for $\Sigma = 11A$ and high temperature ($T = 1200°C$) for $\Sigma = 11B$ [3]. This second structure, B, was also observed in as-grown germanium bicrystal [16]. More recently, observation of a germaninum bicrystal has given an opposite result with respect to silicon with the B structure for the lowest temperatures and the A one for the highest temperatures [15].

3. Investigation with empirical potentials

The two atomic structures A and B of $\Sigma = 11$ grain boundary in silicon and germanium were relaxed with three empirical potentials: Keating [4], Baraff et al. [5], and Stillinger and Weber [6]. Keating potential is widely used in this type of calculation due to its simplicity. It depends on two coefficients α and β, the first one, α, for the bond length changes and the second one, β, mainly for the bond angle changes. Several values of the two parameters α and β have been proposed. Initially, for silicon these parameters have been fitted on the elastic constants C_{11}, C_{12} and C_{44} ($\alpha = 48.50\,J/m^2$, $\beta = 13.81\,J/m^2$). By modifying these values ($\alpha = 51.51\,J/m^2$, $\beta = 4.70\,J/m^2$), Baraff et al. [5] adjusted this potential on phonon frequencies to account for the short spatial extension deformations. For the analysis of more distorted structures such as amorphous ones, Stillinger and Weber potential [6] is well-adapted since it does not need tetracoordinance. All these potentials only include two and three body interactions. The results for the potential energy are given in Table 1. With Keating and Stillinger–Weber potentials the $\Sigma = 11A$ structure is the most stable one whereas with Baraff et al. potential, the two A and B structures are almost equivalent, even if the latter gives the B configuration to be the most stable. However, this difference seems to be too low to be significant. So, it is clear that the simplicity of these empirical potentials prevents them from giving an excellent agreement with the experimental observations.

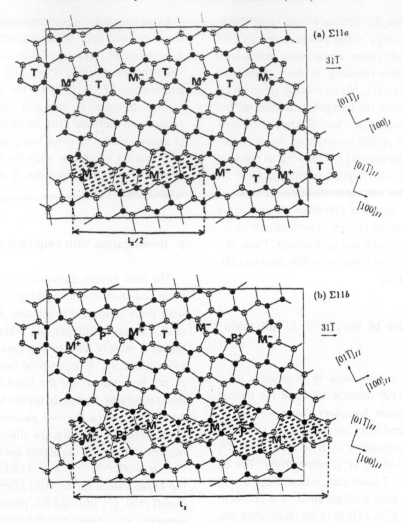

Fig. 1. Atomic structure of $\Sigma = 11A$ and $\Sigma = 11B$ of $\langle 110 \rangle$ tilt grain boundary (L_x defines the period of the B structure).

Table 1
Potential energies, E_p, for $\Sigma = 11A$ and 11B grain boundaries in Si and Ge by means of three empirical potentials: Keating, Baraff et al. and Stilling–Weber (SW)

Potential	Si			Ge		
	$\Sigma = 11A$	$\Sigma = 11B$	Δ	$\Sigma = 11A$	$\Sigma = 11B$	Δ
Keating	803.4	844.6	41.2	656.2	691.0	34.8
Baraff et al.	328.2	328.3	0.1	242.8	242.1	−0.7
SW	622.6	644.5	21.9	553.2	572.6	19.4

Δ is the difference of E_p between $\Sigma = 11B$ and $\Sigma = 11A$ ($\Sigma = 11B - \Sigma = 11A$) (unit mJ/m^2).

4. Tight-binding method

We have used an $O(N)$ semi-empirical TB scheme to compute the total energy at zero temperature.

The detail of the method is given in [17]. After obtaining a local tridiagonalized hamiltonian by the ways of the Lanczos–Haydock recursion method, we diagonalized it directly to obtain the energy and

Table 2
Potential energies, E_p, for $\Sigma = 11A$ and 11B grain boundaries in Si and Ge by means of tight-binding methods

Parameters	Si			Ge		
	$\Sigma = 11A$	$\Sigma = 11B$	Δ	$\Sigma = 11A$	$\Sigma = 11B$	Δ
Mercer	576.6	597.3	20.7	994.3	911.9	−82.4
Goodwin	765.6	828.2	62.6			

Two types of parameters were chosen: the set of Goodwin et al. and that of Mercer et al. Δ is the difference of E_p between $\Sigma = 11B$ and $\Sigma = 11A$ ($\Sigma = 11B - \Sigma = 11A$) (unit mJ/m^2).

the forces within a finite region in a cutoff radius R_c.

We have chosen the parameters of Mercer and Chos [18] which work for silicon and germanium. For the sake of comparison, the results obtained by the set parameters of Goodwin et al. [19] are also reported for silicon (Table 2). Mercer's transferable TB parameters come from the fit to the valence eingenvalues with ab initio calculation of band structure of silicon and germanium. The Ceperley–Alder and Wigner exchange–correlation functions were used for silicon and germanium, respectively. The repulsive term takes into account the energy values calculated in local density approximation of different structures and also the metallic nature of germanium.

5. Result and discussion

The empirical potentials always give the A structure as the stable one without any difference between silicon and germanium which is contrary to the experimental results. Balamane et al. [20] had pointed out that some empirical potentials, like Stillinger–Weber one, are not totally transferable in the case where the angular forces are important. Each of these empirical potentials may describe only one type of deformation: either the large spatial extension deformation (with Keating's potential) or the short spatial extension deformation (with Baraff one). Lebouvier et al. [9] have shown that the latter model favours the B structures.

Moreover, the classical empirical potentials are fitted on macroscopic quantities as elastic ones. A more detailed description is possible by using quantum-mechanical methods. On large systems as the present ones, totally ab initio methods are too much computing

and time consuming. TB approximation allows to reduce the treatment to valence electrons involved in bonding. Some parametrization is needed but based on ab initio or experimental data about the band structure. Thus, TB method is expected to better take into account the material properties as shown by the following results.

The results of the TB calculations are given in Table 2. The difference of potential energy between the two structures B and A is given by Δ. For silicon, Δ is positive with both sets of parameters (Mercer and Chou, and Goodwin et al.). It means that A structure in $\Sigma = 11$ is the most stable one. For germanium, the potential energy of $\Sigma = 11B$ is smaller and then B is the most stable one.

The differences between silicon and germanium are weak. The results of Table 2 are in agreement with experimental results [1]; in particular, the parameters of Goodwin et al. give the A structure as the stable one for silicon, as do the empirical potentials.

The results given by the empirical potentials are controlled by the ratio between angular and radial terms. The question is now to understand what kind of improvement is brought by the method of Mercer and Chou. In this method, an angular component is introduced in the repulsive term which is material dependent and this modification of the repulsive term could be responsible for the improvement.

6. Conclusion

We have performed a TB total energy calculation with a direct evaluation of local density matrix method for $\Sigma = 11$ tilt grain boundaries in silicon and germanium. For the first time, the result confirms the experimental observations [1]: the A structure of $\Sigma =$ 11 is more stable in silicon and the B structure can be found in germanium at low temperature. The empirical potentials were disqualified in this type of calculation in which the energy differences are too small.

References

[1] J. Thibault, J.L. Rouvière, A. Bourret, in: Schröter (Ed.), Mater. Sci. Technol. 4 (1991) 321.

[2] A.P. Sutton, V. Vitek, Trans. Roy. Soc. London A 309 (1983) 1, 37, 55.

[3] J. Thibault, J.L. Putaux, A. Jacques, A. George, H.M. Michaud, X. Baillin, Mater. Sci. Eng. A 164 (1993) 93.

[4] P.N. Keating, Phys. Rev. 145 (1966) 637.

[5] G.A. Baraff, E.O. Kane, M. Schluter, Phys. Rev. B 21 (1980) 5662.

[6] F.H. Stillinger, T.A. Weber, Phys. Rev. B 31 (1985) 5262.

[7] J. Tersoff, Phys. Rev. B 37 (1988) 6991.

[8] D. Vanderbilt, S.H. Taole, S. Narasimhan, Phys. Rev. B 40 (1989) 5657; 42 (1989) 11 373.

[9] B. Labouvier, A. Hairie, F. Hairie, G. Nouet, E. Paumier, Mat. Sci. For. 207/209 (1996) 277.

[10] B. Lebouvier, A. Hairie, F. Hairie, G. Nouet, E. Paumier, Interface Science and Materials Interconnection, Proceedings of JIMIS, vol. 8, edited by the Japan Institute of Metals, 1996, p. 299.

[11] O. Hardouin Duparc, Sol. State Phen. 37/38 (1994) 75.

[12] A. Hairie, F. Hairie, B. Lebouvier, G. Nouet, E. Paumier, N. Ralantoson, A.P. Sutton, Inter. Sci. 2 (1994) 17.

[13] O. Hardouin Duparc, M. Torrent, Inter. Sci. 2 (1994) 7.

[14] M. Hou, A. Hairie, B. Lebouvier, E. Paumier, N. Ralantoson, O. Hardouin Duparc, A.P. Sutton, Mat. Sci. For. 207/209 (1996) 249.

[15] M. Elkajbaji, J. Thibault, H.O.K. Kirchner, Phil. Mag. Lett. 73 (1996) 5.

[16] A. Bourret, J.J. Bacmann, Rev. Phys. Appl. 22 (1987) 563.

[17] J. Chen, A. Béré, A. Hairie, G. Nouet, E. Paumier, Comput. Mater. Sci. (1997), these proceedings.

[18] J.L. Mercer Jr., M.Y. Chou, Phys. Rev. B 47 (1993) 9366.

[19] L. Goodwin, A.J. Skinner, D.G. Pettifor, Europhys. Lett. 9 (1989) 701.

[20] H. Balamane, T. Halicioglu, W.A. Tiller, Phys. Rev. B 46 (1992) 2250.

A comparative study of the atomic and electronic structure of F centers in ferroelectric KNbO₃: Ab initio and semi-empirical calculations

E.A. Kotomin [a,b,*], N.E. Christensen [b], R.I. Eglitis [a,c], G. Borstel [c]

[a] *Institute of Solid State Physics, University of Latvia, 8 Kengaraga, Riga LV-1063, Latvia*
[b] *Institute of Physics and Astronomy, University of Aarhus, Aarhus C, DK-8000, Denmark*
[c] *Universität Osnabrück – Fachbereich Physik, D-49069 Osnabrück, Germany*

Abstract

The linear muffin-tin-orbital method combined with density functional theory (in a local density approximation) and the semi-empirical method of the intermediate neglect of the differential overlap (INDO) based on the Hartree–Fock formalism are used for the supercell study of the F centers (O vacancy with two electrons) in cubic and orthorhombic ferroelectric KNbO₃ crystals. The two electrons are found to be considerably delocalized even in the ground state of the defect. Their wave functions extend over the two Nb atoms closest to the O vacancy and over other nearby atoms. Thus, the F center in KNbO₃ resembles much more electron defects in the partly covalent SiO₂ crystal (the so-called E_1' center) rather than usual F centers in ionic crystals like MgO and alkali halides. This covalency is confirmed by the analysis of the electronic density distribution. The absorption energies were calculated by means of the INDO method using the ΔSCF scheme after a relaxation of atoms surrounding the F center. For the orthorhombic phase three absoprtion bands are predicted, the first one is close to that observed experimentally under electron irradiaton. Copyright © 1998 Elsevier Science B.V.

Keywords: Ferroelectrics; KNbO₃; F center; Defects; Ab initio; Quantum chemistry

1. Introduction

Potassium niobate, KNbO₃, a perovskite-type ferroelectric material, has lately been subject to numerous ab initio electronic structure calculations stimulated by its technological importance and applications. Many of the calculations were based on the *local density approximation* (LDA) [1,2] combined either with the linearized muffin-tin-orbital (LMTO) [3] or with the pseudopotential method [4,5] as well as with the linearized augmented plane wave (LAPW [6–8]) scheme.

Complementary to this approach is the Hartree–Fock (HF) formalism. Compared to the LDA the HF scheme has the advantage of the exact treatment of exchange interactions. Recent implementations have no restrictions on the spatial form of the potential, no potential effects due to use of muffin-tin boundary conditions and/or space-packing empty spheres. It gives the effective charges and suggests a bond-population analysis between pairs of atoms, lastly, it allows us easily to perform the calculation of excited states and optical absorption energies.

Since such calculations are quite time-consuming, there exist only a few HF studies for perovskite

---

* Corresponding author. Address. Institute of Solid State Physics, University of Latvia, 8 Kengaraga, Riga LV-1063, Latvia.

systems (e.g., the cluster calculations of [9]). Instead, a simplified (semi-empirical) version of the HF method widely known as intermediate neglect of the differential overlap (INDO) [10,11] has been applied successfully to calculations for many oxide crystals, including MgO [12], α-Al$_2$O$_3$ (corundum) [13], TiO$_2$ [14], SiO$_2$, etc. In recent studies of pure KNbO$_3$ and KTaO$_3$ crystals [15,16] their electronic structure, equilibrium ground state structure for several ferroelectric phases as well as Γ-phonon frequences were reproduced in surprisingly good agreement with both LDA calculations and available experimental data.

It is well understood now that *point defects* play an important role in the electro-optic and non-linear optical applications of KNbO$_3$ and related materials [17]. In particular, its use for light frequency doubling is seriously affected by presence of unidentified defects responsible for induced IR absorption [18]. The photorefractive effect, important in particular for holographic storage, is also well known to depend on the presence of impurities and defects.

One of the most common defects in oxide crystals is the so-called F center, an O vacancy (V$_O$) which traps two electrons [19]. In KNbO$_3$ structure each O atom is surrounded by four K atoms, two Nb atoms and eight next-nearest O atoms. In electron-irradiated KNbO$_3$ a broad absorption band is observed around 2.7 eV at room temperature and tentatively ascribed to F centers [20] (see also [21]). This defect is also of great theoretical interest for two reasons:

(A) Due to a low local symmetry of the O sites in the lattice, the three-fold degenerate $2p$-type excited state could be split into several levels responsible for *several* absorption bands. (This effect has been observed a long time ago for the F$^+$ centers in corundum, but theoretically it was examined [22] only very recently.) Upon cooling from a high temperature, KNbO$_3$ undergoes a sequence of phase transitions from a paraelectric cubic phase to ferroelectric tetragonal phase (at 708 K), then to the orthorhombic structure (at 498 K), and finally to the rhombohedral (at 263 K) phase. The atomic positions in all these phases have been determined experimentally [23]. Under these phase transitions the local symmetry of the O vacancy also changes, which can, in principle, affect

the optical properties of the F centers. This problem has never been addressed earlier.

(B) Qualitative theoretical analyses of the F centers in perovskites predict the effect of the symmetry breaking of one-electron orbitals associated with the *asymmetric* electron density delocalization over the two Nb atoms closest to the O vacancy: Nb$_1$–V$_O$–Nb$_2$ [24].

To answer these questions, as well as to check the assignment of the 2.7 eV absorption band, we study in the present paper the F center in KNbO$_3$ using the supercell model and two different theoretical techniques: full potential LMTO and INDO.

2. Methods used

2.1. Local density approximation

Even the crudest approximation, LDA, to the density functional theory has been successfully applied to predict structural and dynamical properties of a large variety of materials. Equilibrium volumes, elastic constants, phonon frequencies, surface reconstruction, magnetism are just some examples of properties which could be successfully calculated for systems without particularly strong electron correlations within the LDA (an LSDA, the local spin-density approximation). The LDA usually leads to some overbinding in solids (equilibrium volumes are typically 1–3% underestimated). Considerably larger errors are found in cases where the LDA is not sufficiently accurate; the ionic compounds like MgO serve as examples where the simple LDA fails [25].

The LDA calculations for KNbO$_3$ performed in [1] yield an equilibrium volume which is $\approx 5\%$ too small indicating that the LDA overbinding in this case is not considerably exceeding the 'acceptable' limits. This is why, in the present paper, we apply the LDA to the F center in KNbO$_3$. The LDA exchange–correlation contribution is accounted for by means of Perdew and Zunger's [26] parametrization of the calculations by Ceperley and Alder [27].

The self-consistent solution of the one-electron equation is performed by means of the LMTO method

[3]. We have used the 'atomic-spheres-approximation' (ASA) [3] as well as a 'full-potential formalism' [28] (LMTO-FP). Whereas LMTO-ASA uses potentials and charge densities that are made spherically symmetric inside (slightly) overlapping atomic spheres, no shape approximations are made in LMTO-FP. The atomic relaxations around the F center cannot be calculated by means of the ASA. We therefore performed the structural optimization by minimizing the LMTO-FP total energy calculated for a supercell. A similar method was used in earlier LMTO-FP simulations of defects in KCl [29] and MgO [30]. The supercell used in the present work contains 40 atoms for the perfect $KNbO_3$ (eight formula units) and 39 atoms plus one empty sphere in the F center case.

2.2. INDO

The INDO calculation scheme and the computer code CLUSTERD were discussed in detail in [10,11,15]. With this code it is possible to perform both cluster and periodic system calculations containing hundreds of atoms as well as to carry out automated geometry optimization which is especially important in defect calculations. In the periodic calculations the so-called large unit cell (LUC) model is used [31]. Its idea is to perform the electronic structure calculations for an extended unit cell at the wave vector $\mathbf{k} = 0$ in the narrowed Brillouin zone which is equivalent to band calculations at several special points of the normal BZ, transforming to the narrow BZ center after the corresponding extension of the primitive unit cell. In the $KNbO_3$ case the unit cell contains five atoms whereas the $2 \times 2 \times 2$ extended (super)cell consists of 40 atoms. Detailed analysis of the $KNbO_3$ parametrization for the INDO method is presented in [15]. In that work [15] considerable covalency was found for the chemical bonding in pure $KNbO_3$. The effective charges found from Mulliken population analysis are (in units of $|e|$): +0.543 for K, +2.019 for Nb and −0.854 for O, which is very different from the expectation of the generally accepted *ionic* model: +1, +5 and −2, respectively. This is in agreement with the effective atomic charges found in an experimental study of $LiNbO_3$ [32]. Our

results emphasize a high degree of covalency of the Nb–O bond as may be expected from intuitive electronegativity considerations and the fact of a strong overlap between O 2p and Nb 4d orbitals and partial densities of states. We discuss below how covalency of the chemical bonding may have important consequences for the physics of defects in ferroelectrics, in particular for the F centers.

To simulate F centers, we started with a 40-atom supercell with one of the O atoms removed. In the cubic phase all O atoms are equivalent and have the local symmetry C_{4v} whereas in the orthorhombic phase there are *two* kinds of non-equivalent O atoms whose symmetry is lower, C_{2v} or C_s. After the O atom is removed, the atomic configuration of surrounding atoms is re-optimized via search of the total energy minimum as a function of the atomic displacements from regular lattice sites. Calculation of the adiabatic energy curves for the ground and excited states permits to find the optical absorption energy using the so-called Δ SCF procedure according to which the E_{abs} sought for is the difference of the total energies for the ground and excited state with the defect geometry of the ground state unchanged (vertical optical transition). To extend the basis set in the F center calculation, additional 1s, 2p atomic orbitals (AO) were centered on the O vacancy. Their parameters were chosen close to those used in the F center calculations in MgO crystal [33]. During the defect geometry optimization, we make no a priori assumptions on the electron density distribution.

3. Ground state properties

3.1. LDA calculations

The band structure as derived from the straight LDA underestimates the gap between occupied and empty states. Since the supercell which we use is rather small (40 atomic sites) the defect states form a band of a finite width (≈ 0.8 eV). These two effects cause the defect band to overlap with the conduction band, and the supercell calculations within the LDA predicts $KNbO_3$ with the F centers to be a metal. This affects the charge distribution; the number of electrons in the vacancy 'atomic sphere' is 0.24, much smaller than that obtained from the INDO (0.6). When we artificially increase the optical gap by applying an upshift in each iteration to the Nb-d states, the defect band lies completely within the optical gap and disperses over ≈ 0.8 eV due to the small supercell size. After adjustment of the gap the (self-consistent) calculation yields 0.6 electrons inside the O vacancy sphere, i.e. more than twice the amount found before and very close to the INDO calculation for the *relaxed* structure. When atomic relaxations (see below) are included, the LMTO calculation yields a lower electron number in the vacancy sphere. This is simply caused by the outward motion of the nearest neighbors (Nb).

The relaxation of atoms surrounding the F center was first calculated within the LDA without any attempt to correct for the effect of the overlap between the defect state band and the conduction band. First, the nearest neighbor Nb atoms were relaxed, and the result is illustrated in Fig. 1(a) which shows the total energy as a function of the outward displacement, Δ_z, of Nb atom from its equilibrium position in the bulk crystal. The relaxed atomic positions correspond to $\Delta_z = 3.5\%$ of the lattice constant of $a_0 = 4.016\,\text{Å}$. This is about half the relaxation found in the INDO calculation. Further, the relaxation energy found here, ≈ 0.5 eV, is much smaller than the value of 3.7 eV obtained in the INDO.

We do not wish to rely on total energy calculations where we applied 2.5 eV upshift to the Nb-d bands. However, self-consistent calculations using supercells large enough for obtaining a small width of the defect band non-overlapping with the LDA conduction band are impractical. An approximate calculation was instead made of using the same supercell size as before but sampling only the Γ point of the BZ in the **k**-space integration. This is the point where the defect band has its minimum energy and lies inside the gap, even in the LDA calculation. Further, this sampling is the same as used in the INDO-LUC calculations. As expected, this changes the charge distribution and the atomic relaxations. The value of $\Delta_z = 4.8$ (Fig. 1(b)) is closer to that obtained in the INDO calculation. The relaxation energy 1.2 eV is also somewhat closer to the INDO result.

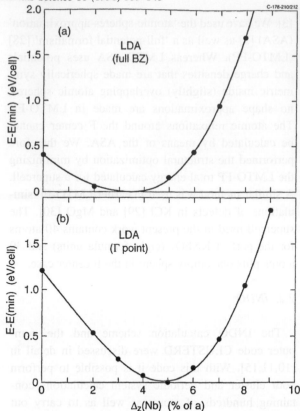

Fig. 1. (a). LMTO-FP calculation within the LDA of the total energy vs. the outward displacement, δ_z, of the two Nb atoms closest to the O vacancy. Note that no corrections were made in this case to insure that the defect band lies entirely inside the optical gap. Integration in **k** space used 40 points in the irreducible part of the BZ. (b) Same as (a) but only a single Γ point was here included in the **k**-space sampling, in order to simulate the LUC used in the INDO calculations. The curve is a high-order polynomial fit.

3.2. INDO

3.2.1. Cubic phase

The positions of 14 atoms surrounding the F center in a cubic phase after lattice relaxation to the miminum of the total energy were calculated for the two cases: with additional AO on vacancy site, and without the AO. The conclusion is that the largest relaxation is exibited by the two nearest Nb atoms which are strongly (by 6.5% of a_0) displaced *outwards* the O vacancy along the z axis. This is accompanied by much smaller, 0.9% outward displacement of K atoms and by 1.9% inward displacement of O atoms. The two Nb atoms give the largest ($\approx 80\%$) contribution to the lattice relaxation energy (3.7 eV) whereas O atoms give most of the rest energy gain of 1 eV.

The analysis of the effective charges of atoms surrounding the F center shows that of the two electrons associated with the removed O atom only $\approx -0.6|e|$ is localized inside V_O and a similar amount of the electron density is localized on the two nearest Nb atoms. The F center produces a local energy level, which lies ≈ 0.6 eV above the top of the valence band. Its molecular orbital contains primarily contribution from the atomic orbitals of the two nearest Nb atoms.

3.2.2. Orthorhombic phase

The orthorhombic phase of $KNbO_3$ is important since it is stable in a broad temperature range around room temperature and thus is subject to most studies and practical applications. The displacements of Nb atoms nearest to V_O were calculated for the both kinds of F centers existing in this phase, and again, with and without atomic orbitals centered at the O vacancy. In fact, Nb displacements are very similar in magnitude (6.6%) and also close to those found for the cubic phase. The relevant relaxation energies are considerable (3.6 eV) and nearly the same as that for Nb relaxation found in the cubic phase.

4. Optical properties

Because of the C_{4v} local symmetry of the F centers in the $KNbO_3$ cubic phase, its excited state splits into two levels, one of which remains two-fold degenerate. Our ΔSCF calculations predict the two relevant absorption bands: at 2.73 and 2.97 eV (Table 1). Neglect of the additional orbitals centered on the V_0 slightly affect this result, the relevant absorption energies turn out to be 2.67 and 3.02 eV, respectively.

Around the room temperature, in the orthorhombic phase there exist two kinds of F centers associated with two non-equivalent O atoms revealing the C_{2v} and C_s symmetry. The corresponding three absorption bands for each of them are shown in Table 1. Their difference is the largest for the lowest energy

Table 1
INDO-calculated absorption (E_{abs}, eV) and Nb-atom relaxation (E_{rel}, eV) energies for the F center for the cubic and the orthorhombic phases

Symmetry, phase	E_{abs}			E_{rel}
C_{4v}, cubic	2.73	2.97	—	3.7
C_s, orthorhombic	2.56	3.03	3.10	3.6
C_{2v}, orthorhombic	2.72	3.04	3.11	3.6

bands (0.16 eV) and very small for the other two bands.

5. Summary

Our results are in a sharp contrast with what is known for F centers in ionic crystals (in particular, in MgO and alkali halides [13,19]) where the two electrons are well localized by the V_O in the ground state of the F center. Evidently, this discrepancy arises from a considerable degree of *covalency of the chemical bonding in* $KNbO_3$ which is neglected in all previous models of defects in this material (as well as similar ABO_3 perovskites, e.g., [35,36]); the only exception known to us is an X_α cluster calculation of F centers in $LiNbO_3$ [34].

Electron defects similar to what we have observed here are known, in particular, in partly covalent SiO_2 crystals (e.g., in the so-called E'_1 center an electron is also not localized inside V_0 but its wave function mainly overlaps with the sp^3 orbital centered on the neighboring Si atom [37]).

We found that the ground state of the F center is associated with a strong *symmetrical* relaxation of the two nearest Nb atoms outwards relative to the O vacancy. These Nb atoms remain to be identical, i.e., we did not see formation of dipole moments of the Nb_1–V_0–Nb_2 type, as suggested in [24]. Note that the relevant relaxation energy is several eV which is typical for many point defects in ionic and partly ionic solids. Its magnitude is by several orders of magnitude larger than the tiny energy gain due to the phase transitions (meV per cell).

We presented a strong argument that the 2.7 eV absorption band observed in electron-irradiated crystals

[20] could be due to the F-type centers, and predicted existence of two additional absorption bands (at 3.04 and 3.10 eV) for the same defect in the orthorhombic phase of $KNbO_3$ (see also discussion in [21]). At higher temperatures where the cubic phase is stable, the latter two energies which hardly could be separated experimentally because of the large half-width of absorption bands, degenerate into a single, double degenerate level at 2.97 eV.

Our results also suggest that the blue-light-induced-IR absorption effect [18] mentioned in Section 1 could be triggered by the F center absorption which may lead to its subsequent ionization where an electron is transferred to the conduction band. The re-trapping of this electron by another defect (shallow state) is then responsible for the IR absorption when the re-trapped electron is excited to the conduction band. We recall that the UV excitation (blue light) energy used in these laser-light-frequency doubling experiments is very close to our calculated absorption energy of the F centers.

The reason for the discrepancy in LDA and INDO relaxation energies for the F center needs further study; probably it arises due to strong electron correlation effects which are crudely approximated in the two methods in quite different ways.

Acknowledgements

RE and GB have been supported by the Niedersächsische Ministerium für Wissenschaft und Kultur and the Deutsche Forschungsgemeinschaft (SFB 225), respectively. EK greatly appreciates the financial support from Danish Natural Science Research Council (contract 9600998) and Latvian National Program on *New Materials for Microelectronics*. The authors are grateful to A.I. Popov, A.V. Postnikov and E. Polzik for fruitful discussions.

References

[1] A.V. Postnikov, T. Neumann, G. Borstel and M. Methfessel, Phys. Rev. B 48 (1993) 5910; A.V. Postnikov and G. Borstel, Phys. Rev. B 50 (1994) 16 403.

[2] A.V. Postnikov, T. Neumann and G. Borstel, Phys. Rev. B 50 (1994) 758.

[3] O.K. Andersen, Phys. Rev. B 12 (1975) 3060.

[4] R.D. King-Smith and D. Vanderbilt, Phys. Rev. B 49 (1994) 5828.

[5] W. Zhong, R.D. King-Smith and D. Vanderbilt, Phys. Rev. Lett. 72 (1994) 3618.

[6] D.J. Singh and L.L. Boyer, Ferroelectrics 136 (1992) 95; D.J. Singh, Phys. Rev. B 53 (1996) 176.

[7] R. Yu and H. Krakauer, Phys. Rev. Lett. 74 (1995) 4067.

[8] D.J. Singh, Ferroelectrics 164 (1995) 143.

[9] H. Donnerberg and M. Exner, Phys. Rev. B 49 (1994) 3746.

[10] E. Stefanovich, E. Shidlovskaya, A. Shluger and M. Zakharov, Phys. Stat. Sol. B 160 (1990) 529.

[11] A. Shluger and E. Stefanovich, Phys. Rev. B 42 (1990) 9664.

[12] E.A. Kotomin, M.M. Kuklja, R.I. Eglitis and A.I. Popov, Mater. Sci. and Eng. B 37 (1996) 212.

[13] E.A. Kotomin, A. Stashans, L.N. Kantorovich, A.I. Lifshitz, A.I. Popov, I.A. Tale and J.-L.Calais, Phys. Rev. B 51 (1995) 8770.

[14] A. Stashans, S. Lunell, R. Bergström, A. Hagfeldt and S.-E. Lindqvist, Phys. Rev. B 53 (1996) 159.

[15] R.I. Eglitis, A.V. Postnikov and G. Borstel, Phys. Rev. B 54 (1996) 2421.

[16] R.I. Eglitis, A.V. Postnikov and G. Borstel, Proc. SPIE 2867 (1997) 150.

[17] P. Günter and J.-P. Huignard, eds., Photorefractive Materials and Their Application, Topics in Applied Physics, Vols. 61, 62 (Springer, Berlin, 1988).

[18] L. Shiv, J.L. Sørensen, E.S. Polzik and G. Mizell, Optics Lett. 20 (1995) 2271.

[19] J.H. Crawford Jr., Nucl. Inst. Meth. B 1 (1984) 159; J.-M. Spaeth, J.R. Niklas and R.H. Bartram, Structural Analysis of Point Defects in Solids, Springer Series in Solid State Sciences, Vol. 43 (Springer, Berlin, 1993).

[20] E.R. Hodgson, C. Zaldo and F. Agullo-López, Sol. State Comm. 75 (1990) 351.

[21] E.A. Kotomin, R.I. Eglitis and A.I. Popov, J. Phys.: Cond. Matt. 9 (1997) 315; R.I. Eglitis, N.E. Christensen, E.A. Kotomin, A.V. Postnikov and G. Borstel, Phys. Rev. B 56 (1997), in press.

[22] A. Stashans, E.A. Kotomin and J.-L. Calais, Phys. Rev. B 49 (1994) 14 854.

[23] A.W. Hewat, J. Phys. C 6 (1973) 2559.

[24] S.A. Prosandeyev, A.V. Fisenko, A.I. Riabchinski, A.I. Osipenko, I.P. Raevski and N. Safontseva, J. Phys.: Cond. Matter 8 (1996) 6705; S.A. Prosandeyev and I.A. Osipenko, Phys. Stat. Sol. B 192 (1995) 37; S.A. Prosandeyev, M.N. Teslenko and A.V. Fisenko, J. Phys.: Cond. Matter 5 (1993) 9327.

[25] K. Doll, M. Dalg and P. Fulde, Phys. Rev. B 52 (1995) 4842.

[26] J.P. Perdew and A. Zunger, Phys. Rev. B 23 (1981) 5048.

[27] D.M. Ceperley and B.J. Alder, Phys. Rev. Lett. 45 (1980) 566.

[28] M. Methfessel, Phys. Rev. B 38 (1988) 1537.

[29] A. Svane, E.A. Kotomin and N.E. Christensen, Phys. Rev. B 53 (1996) 24.

[30] T. Brudevoll, E.A. Kotomin and N.E. Christensen, Phys. Rev. B 53 (1996) 7731.

[31] R.A. Evarestov and L.A. Lovchikov, Phys. Stat. Sol. B 93 (1977) 469.

[32] M.E. Lines, Phys. Rev. B 2 (1970) 698.

[33] R.I. Eglitis, M.M. Kuklja, E.A. Kotomin, A. Stashans and A.I. Popov, Comput. Mater. Sci. 5 (1996) 298.

[34] G.G. DeLeo, J.L. Dobson, M.F. Masters and L.H. Bonjack, Phys. Rev. B 37 (1988) 8394.

[35] K.L. Sweeney, L.E. Halliburton, D.A. Bryan, R.R. Rice, R. Gerson and H.E. Tomaschke, J. Appl. Phys. 57 (1985) 1036.

[36] T. Varnhorst, O.F. Schirmer, H. Kröse, R. Scharfschwerdt and Th. W. Kool, Phys. Rev. B 53 (1996) 116.

[37] K.L. Yip and W.B. Fowler, Phys. Rev. B 11 (1975) 2327.

ELSEVIER

COMPUTATIONAL
MATERIALS
SCIENCE

Computational Materials Science 10 (1998) 346–350

Local density calculation of structural and electronic properties for $Ca_{10}(PO_4)_6F_2$

V. Louis-Achille *, L. De Windt, M. Defranceschi

Institute for Protection and Nuclear Safety, DPRE/SERGD/LMVT, BP6-92265 Fontenay-aux-Roses Cedex, France

Abstract

The local-density-functional pseudopotential approach is applied to the calculation of the lattice properties and electronic structure of the fluoroapatite $Ca_{10}(PO_4)_6F_2$ which is one of the simplest apatites. The calculated lattice parameters agree well with the experiment. Valence charge density and Mulliken population are analysed to understand the nature of the bond between the different atoms. Copyright © 1998 Elsevier Science B.V.

Keywords: Local-density-functional; Pseudopotential; Charge density; Fluoroapatite

1. Introduction

Recently, ceramics such as apatites $(M_t(PO_4)_u (SiO_4)_{6-u}X_2$ where M is a cation, X is an anion) have been considered as alternative nuclear waste disposal materials [1]. It has been shown that it is feasible to incorporate large amounts of radioactive elements in their crystal lattice, and some of these minerals have survived hundred million years of weathering keeping their crystalline structure even after exposure to radiations [2]. In this paper, results of investigation on the fluoroapatite $Ca_{10}(PO_4)_6F_2$ are presented, which is one of the simplest apatites. Calculations on its structural and electronic properties are performed as a reference point for further studies on apatite retention properties.

$Ca_{10}(PO_4)_6F_2$ crystallizes in the hexagonal structure according to the space group $P6_3/m$ (see Fig. 1)

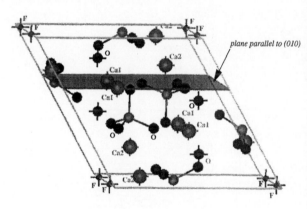

Fig. 1. Fluoroapatite primitive cell.

with $a = b = 9.398\,\text{Å}$ and $c = 6.878\,\text{Å}$. There is one $Ca_{10}(PO_4)_6F_2$ formula (42 atoms) per unit cell. The calcium are arranged on two non-equivalent sites. Ca in site one, denoted Ca(1), are surrounded by nine oxygens with six shorter bonds that defined a trigonal prism and three longer bonds that emerge through the prism faces. The mean interatomic distance is 2.55 Å

* Corresponding author. Tel.: 33 1 46 54 85 11; fax: 33 1 46 54 71 35; e-mail: vanina@sofia.far.cea.fr.

347

while the volume of the polyhedron is 30 Å³. Six oxygens plus one fluorine forming an irregular pentagonal pyramid surround the Ca in site 2 denoted Ca(2). The mean interatomic distance in these positions is 2.45 Å, and the volume of the polyhedron is 22 Å³[3]. The F are disposed in columns parallel to the sixfold helix axis and are at the centre of Ca triangles, whereas the 24 oxygens are distributed among three distinct sites, denoted O(1 to 3) [4].

2. Structural properties

The calculations are based on the local density functional theory (DFT) with the valence orbitals expanded in plane wave basis sets. Exchange and correlation are included through the local density approximation (LDA), using the Ceperley and Alder potential as parametrized by Teter [5]. The full ionic potential of the solid is replaced with a much softer pseudopotential generated using the procedure proposed by Troullier and Martins [6]. The minimization of the Kohn–Sham energy functional is realized with a conjugate-gradients technique [7]. A single k-point was used for the fluoroapatite. An energy cutoff equal to 1400 eV was fixed for all the calculations due to computer limitations, even if the absolute total energy is not fully converged. It has been shown that this value leads to correct physical properties [8].

Two parameters are sufficient to determine an hexagonal structure: the two lattice parameters a and c, or equivalently the cell volume V and the ratio c/a. The ground state of the system was obtained by a two-step procedure. The total energy was first minimized with respect to the cell volume V and the atomic coordinates were relaxed using Hellmann–Feynman forces as a guide. The c/a ratio was then calculated under the constraint that the stress tensor was isotropic. The number of plane waves changes discontinuously with respect to the size of the unit cell leading to discontinuancies in the energy and in the stress curves. Consequently, energy [9] and stress [10] correction factors were applied throughout the calculations. Hence, accurate energy differences and physical properties are computed using significantly smaller basis sets than those needed without applying the corrections.

The computed values are respectively $a = b = 9.375$ Å and $c = 6.844$ Å. They are within 0.3% of the experimental structures. The calculated structures are slightly more contracted than the experimental one. Underestimation of lattice parameters is inherent to the LDA approximation.

3. Electronic structure

In this section, electronic features of the fluoroapatite are studied qualitatively using valence charge density maps and semi-quantitatively from Mulliken population analysis.

3.1. Charge density maps

The valence electron distribution was calculated for the fluoroapatite. Fig. 2 shows contours of valence densities for two different planes. The first one is a (0 1 0) plane and contains fluorines and Ca(2) type atoms. The second plane is perpendicular to the y axis and contains Ca(1) type calciums.

The observation of this figure leads to some general features. First, the extremely localized nature of the electronic charge around fluorine shows its strong ionic character. This is a direct consequence of the high electronegativity of fluorine. Secondly, the PO_4 tetrahedrons are well individualized. The non-spherical form of the contour density around the oxygens reveals a slightly covalent character of the bond between O and P. On the other hand, the valence electron distributions around Ca(2) and Ca(1) appear to be isotropic as shown in Figs. 2(a) and (b), respectively. Comparison of the two maps does not exhibit any difference in charge repartition between the two types of calcium.

Owing to this qualitative observation, the apatite structure can be seen as a skeleton of PO_4 tetrahedrons with calcium and fluorine included in this lattice. However, in order to analyse in more details the electronic distribution, a Mulliken population analysis is carried out in the next section to give a semi-quantitative insight of these properties.

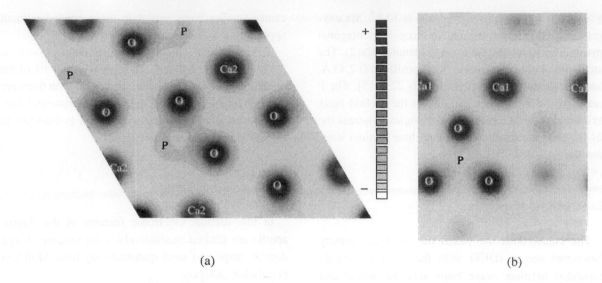

Fig. 2. Contour plot for the valence electron density (a) of the plane parallel to (0 0 1) containing Ca(2), F, and P sites and, (b) of the plane parallel to (0 1 0) containing Ca(1)sites.

Table 1
Mulliken population analysis for $Ca_{10}(PO_4)_6F_2$

	Atomic charge		AO	AO population
O(1)	−0.960	O(1)	2s	1.840
O(2)	−0.980		2p	5.120
O(3)	−0.930		3d	0.020
F	−0.820			
P	+1.400	F	2s	1.970
Ca(1)	+1.550		2p	5.850
Ca(2)	+1.640		3d	0.005
		P	3s	0.665
	Bond overlap population		2p	6.000
			3p	2.055
Ca(1)–O [a]	0.035		3d	0.880
Ca(2)–O [a]	0.020			
Ca(1,2)–F	0.000	Ca(1)	3s	2.000
P–O(1)	0.425		4s	0.175
P–O(2)	0.440		3p	6.005
P–O(3)	0.500		4p	0.270

[a] Mean value.

3.2. Mulliken population analysis

Even though the Mulliken analysis is an arbitrary representation without physical meaning, it provides helpful information on charge transfers occurring in a structure. Mulliken population analysis was performed on the previously optimized geometry with a DFT atomic orbitals code [11]. Because Mulliken atomic charges and populations are strongly dependent on the quality of the basis set, a double numerical basis including polarization functions (DNP) was used [11]. In Table 1 are reported the atomic charges of the various atoms, the bond overlap populations and the atomic orbital (AO) populations.

A general overlook of the results clearly shows an iono-covalent type structure for $Ca_{10}(PO_4)_6F_2$ with significant ionic charges. The crystallographic sites are perfectly identified by the Mulliken atomic charges distribution. Atoms of the same crystallographic site share the same electronic structure. The PO_4 tetrahedrons are well identified with important covalent contributions for the P–O bonds. Furthermore, there is a good correlation between P–O populations and the partial charges on the three atoms as well as with the experimental bond lengths: $PO(3)(1.532 Å) < PO(1)(1.537 Å) < PO(2)(1.539 Å)$ [4]. Much more important ionic characters are found for the Ca–O and Ca–F bonds.

Fluorine appears as being mainly ionic with an atomic charge equal to -0.820. A bond overlap population between adjacent atoms equal zero, and a very low 3d population reveal that there is no polarization of the atom.

Comparison of the atomic charges and Ca–O bond overlap data seems to demonstrate that the Ca(1) atoms are slightly more covalent than the Ca(2) atoms. This result is in accordance with previous experimental considerations [3]. Moreover, keeping in mind that Ca(1) sites are larger than Ca(2), the sum of the covalent radius for Ca and O is greater than the sum of the ionic radii. On the one hand, experimentalists suggested that the Ca(1) and Ca(2) site occupancies are determined very largely by bond type in subtitution mecanism [3]. On the other hand, the Ca(1) is sterically favoured. Such thumb rules failed to predict relative stabilities of various compounds [12]. Extensive quantum calculations (as on Mulliken population, charge density analysis or even DOS studies) can help quantitatively this prediction.

For both Ca(1) and Ca(2), a low 4s population in conjunction with a significant p polarization orbital population argue for important transfers of the 4s valence electrons. In contrast, the 3s and 3p electrons remain mainly undisturbed compared to the atomic state. A similar trend occurs for P atoms with unchanged 2p-3p electrons and transferred valence 3s electrons in parallel with population of the polarization orbital.

Mulliken analysis performed with the same DFT code using DNP basis set for BPO_4 and YPO_4 is reported in Table 2 for comparison with the results on $Ca_{10}(PO_4)_6F_2$. In these phosphate crystals, the

Table 2
Comparative Mulliken population analysis for some phosphate crystals

	BPO_4	YPO_4	$Ca_{10}(PO_4)_6F_2$
$\varepsilon(X)$ [a]	2.0	1.3	1.0
X-O population [a]	0.360	0.080	0.020 and 0.035
P-O population	0.375	0.425	0.425 to 0.500
X charge/nb e- [a,b]	0.06	0.54	0.78 to 0.82
O charge	-0.480	-0.845	-0.930 to -0.980
P charge	1.735	1.760	1.400

[a] X = B, Y or Ca.
[b] Valence electrons.

associated cations are of rather different natures. A good correlation is shown between the decrease of the atom electronegativity and the increase of the ionic character in the cation/oxygen bond. More valence electrons are transferred from cations on adjacent oxygens along the series. A parallel augmentation of the covalent character of the PO bonds occurs.

4. Conclusion

Modelling of mineral matrices as complex as apatites are very seldom found in the literature. It is quite a challenge, with nowadays computers, to reconcile good quality results and CPU. The present study, although being preliminary, has shown that DFT-LDA calculations can be useful for the determination of structural parameters, and also for the analysis of the electronic distributions. Work in progress along these lines on structures more directly connected with nuclear waste disposal materials gives promising prospective guidelines for the formulation of good candidates for the retention of a given species.

References

[1] J. Carpena and J.L. Lacout, Brevet 93 08676 (1993).
[2] D. Langmuir and M.J. Apted, Mat. Res. Soc. Symp. Proc. 257 (1992).
[3] V.S. Urusov and V.O. Khudolozhin, Geochem. Int. 11 (1975) 1048.
[4] J.M. Hughoo, M. Cameron and K.D. Crowley, Amer. Miner. 74 (1989) 870.
[5] Plane_Wave 3.0.0, MSI, San Diego (1995).
[6] N. Troullier and J.L. Martins, Phys. Rev. B 43 (1991) 1993.

[7] M.C. Payne, M.P. Teter, D.C. Allan, T.A. Arias and J.D. Joannopoulos, Rev. Mod. Phys. 64 (1992) 1045.

[8] V. Louis-Achille, L. De Windt and M. Defranceschi, J. Mol. Struct. (Theochem) (1997), in press.

[9] G.P. Francis and M.C. Payne, J. Phys.: Condens. Matter 2 (1990) 4395.

[10] P. Gomes Dacosta, O.H. Nielsen and K. Kunc, J. Phys. C 19 (1986) 3163.

[11] DSolid 4.0.0, MSI, San Diego (1996).

[12] M.E. Fleet, Y. Pan, J. Solid State Chem. 112 (1994) 78.

COMPUTATIONAL
MATERIALS
SCIENCE

ELSEVIER

Computational Materials Science 10 (1998) 351–355

Possible n-type dopants in diamond and amorphous carbon

S. Pöykkö *, M. Kaukonen, M.J. Puska, R.M. Nieminen

Laboratory of Physics, Helsinki University of Technology, FIN-02015 Espoo, Finland

Abstract

It has been extremely difficult to produce n-type conducting diamond, whereas it seems that n-type conducting tetrahedrally bonded amorphous carbon (ta-C) can be obtained using phosphorous or nitrogen as dopant. In this work we have studied substitutional group V–VII impurities (N, O and Cl) in diamond and ta-C using first-principles electronic structure methods. Electronic structure calculations reveal the differences between ta-C and diamond in the microscopic level and ease, therefore, the experimental search for producing n-type conducting diamond-like materials. Copyright © 1998 Elsevier Science B.V.

Diamond is a promising material for devices operating in extreme environmental conditions. This stems from its outstanding material properties such as the extreme hardness and the wide electron band gap. Usually, both n- and p-type conducting materials are needed in manufacturing electronic components. Diamond shows normally intrinsic p-type conductivity due to a small fraction of boron impurities. Also in laboratory environment p-type conducting diamond can be produced by using B-doping methods. In contrast, n-type conducting diamond has been extremely hard to produce and only very little success has been achieved so far. Tetrahedrally bonded amorphous carbon [1], which has many materials properties similar to diamond, can be n-doped using nitrogen or phosphorous. Tetrahedrally bonded amorphous carbon (ta-C) has been studied theoretically by various groups [2–7]. From the studies only those reported in [6,7] are based on fully self-consistent first-principles (SCFP) calculations.

Three-member carbon rings recently found in ab initio calculations for bulk ta-C [6,7] and for graphite surfaces [8] are absent in tight-binding (TB) studies of ta-C [3–5]. Also four-member rings are less probable in non-self-consistent studies than in fully self-consistent ab initio ones. Therefore, the TB method is not capable to describe ta-C accurately and fully SCFP methods have to be applied. To our knowledge, no fully SCFP studies for doping of ta-C exist. The works by Stumm et al. [2] and Stich et al. [3] are the only existing electronic structure calculations for the doping of ta-C.

Nitrogen doping has most often been performed by the chemical vapor deposition (CVD) method. The addition of N_2 molecules in the CVD reaction gas results in a narrowing of the band gap as the amount of nitrogen increases. The ensuing bulk resistivity is reported to reduce by 2–4 orders in magnitude [9].

Oxygen doping improves the film quality and reduces the number of lattice defects in CVD diamond. Oxygen is found to be incorporated into the diamond

---

* Corresponding author.

lattice [10]. It can remove the non-diamond-like hybridized carbon atoms from the film.

To our knowledge, there exists no experimental evidence for chlorine dopant atoms in diamond or in ta-C. Chlorine has been used as a reactant gas in the CVD growth, and therefore it may play a role in surface growth processes [11].

We have performed fully self-consistent plane-wave-pseudopotential calculations for carbon networks at the density $3.0 \, \text{g/cm}^3$. Our calculations employ the density-functional theory [12] within the local density approximation (LDA) for electron exchange and correlation effects [13]. We have used a 64-atom simple cubic simulation cell, and the supercell Brillouin-zone sampling consists of the $2 \times 2 \times 2$ Chadi–Cohen k-point mesh [14]. The use of the Vanderbilt ultrasoft pseudopotentials [15] reduces significantly the number of plane waves needed to treat sharply peaked electronic states of carbon, nitrogen, oxygen and chlorine accurately. A kinetic energy cut-off of only 20 Ry was needed in calculations for generating ta-C networks. Calculations involving chlorine or oxygen were performed with a kinetic energy cut-off of 27 Ry and for calculations involving nitrogen the cut-off energy of 25 Ry was sufficient.

We have used several methods to generate good non-doped ta-C structures to start with. These methods range from the direct total energy minimization starting from randomized atomic positions with several randomizing amplitudes to the molecular dynamics calculation in which a (liquid) sample at 5000 K was followed for 0.36 ps and thereafter instantly cooled to 0 K. All these calculations give very similar results. The sp^3 fraction in the simulated carbon networks varies from 55% to 68%. Three- and four-membered rings were present in all structures obtained. The radial distribution functions are almost identical and the total energies of these structures at 0 K nearly coincide. From these structures we chose the one with the lowest total energy to serve as our ta-C network in the subsequent doping studies. The final ta-C structure is visualized in Fig. 1 and the structure analysis is given in Table 1.

When calculating dopants in the diamond structure a 32-atom supercell and the $2 \times 2 \times 2$ Monkhorst–

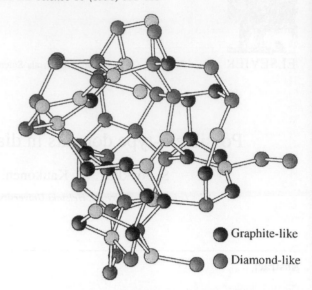

Graphite-like

Diamond-like

Fig. 1. 64-atom ta-C supercell. Carbon atoms in diamond- and graphite-like sites are indicated by gray and black spheres, respectively. We define that a carbon atom is 'diamond-like' if it has four neighbors within the distance of 1.53 ± 0.1 Å and the bond angles are in the range of $109.28° \pm 10°$. Similarly, an atom is considered to be 'graphite-like', if it has three neighbors, its bond lengths are 1.3–1.5 Å and the bond angles are 110°–130°. The white spheres correspond to carbon atoms which cannot be considered as 'diamond-like' or 'graphite-like'. Three-membered rings present in structure are visible eg. at the top of the figure and four-membered rings at the center.

Pack k-point mesh [16] has been employed. The calculated lattice constant of 3.531 Å is slightly smaller than the experimental value of 3.567 Å [17]. This is expected to be due to the well-known LDA tendency for overbinding. In all of our dopant calculations for the diamond as well as for the ta-C structure all the atoms in the supercells have been allowed to move to their minimum-energy positions without any symmetry assumptions.

We have studied substitutional nitrogen in diamond within the tetrahedrally and trigonally symmetric structures. The tetrahedral structure is stable for the positive charge state, while the trigonal structure is unstable. For the neutral and negative charge states the trigonal structure, for which one of the C–N bonds is broken (the C–N distance is 2.13 Å), was found to be the stablest structure. This is in good agreement with experiments and other theoretical studies [18]. The tetrahedral configurations are metastable in the

Table 1
Details of the simulated ta-C structure (the sp^3 fraction and the ring statistics are compared with other published data for densities close to $3.0 \, \text{g/cm}^3$)

Reference	$\rho(\text{g/cm}^3)$	N_{at}	sp^3(%)	3	4	5
This work	3.0	64	59	2	9	14
Clark et al. [7]	2.905	64	47	6	4	17
	3.147	64	53	4	5	21
Marks et al. [6]	2.9	64	65	3	3	21
Frauenheim et al. [4]	3.0	64	53	0	0	20
	3.0	128	64	0	3	18
Wang and Ho [5]	3.0	216	33	0	0	10

Table 2
Formation energies for the different nitrogen impurities in diamond and amorphous carbon

Defect	Formation energy (eV)
N_C^{1+} (diamond, tetrahedral)	$-0.01 + \mu_e$
N_C^0 (diamond, trigonal)	3.53
N_C^{1-} (diamond, trigonal)	$6.34 - \mu_e$
N_C^{1+} (ta-C sp^3)	$1.45 + \mu_e$
N_C^0 (ta-C sp^3)	1.62
N_C^{1-} (ta-C sp^3)	$1.83 - \mu_e$
N_C^{1+} (ta-C sp^2)	$-1.47 + \mu_e$
N_C^0 (ta-C sp^2)	-1.40
N_C^{1-} (ta-C sp^2)	$-1.21 - \mu_e$
N_C^{1+} (ta-C 3-ring)	$0.08 + \mu_e$
N_C^0 (ta-C 3-ring)	0.02
N_C^{1-} (ta-C 3-ring)	$0.05 - \mu_e$
N_C^{1+} (ta-C 4-ring))	$-0.53 + \mu_e$
N_C^0 (ta-C 4-ring)	-0.48
N_C^{1-} (ta-C 4-ring)	$-0.40 - \mu_e$

The reference chemical potential used for the nitrogen atom corresponds to the dimer N_2. For the carbon atom in ta-C the chemical potential is chosen as 1/64 of the total energy of the undoped supercell, even though the chemical potential of carbon atoms occupying diamond- and graphite-like sites should slightly deviate from each other, introducing a small inaccuracy into the formation energies given. The chemical potential for the carbon atom in diamond is 1/32 of the bulk diamond supercell total energy.

neutral and negative charge states. The total energy differences between the metastable tetragonal and the stable trigonal configurations are 0.2 and 1.1 eV for the neutral and negative charge states, respectively. The calculated donor level (2.3 eV below the conduction band minimum) is 0.6 eV lower in the

band gap than the experimental value (1.7 eV below the conduction band minimum [19]).

In the case of ta-C we studied the nitrogen doping by substituting an sp^3- or an sp^2-hybridized carbon atom or by substituting a carbon atom in a three- or four-membered ring by a N-atom. The nitrogen atom substituting an sp^2-bonded carbon atom was found to be the most probable one: the formation energy for the substitution of an sp^3-bonded carbon atom is in all charge states about 3 eV larger than that for the sp^2-bonded carbon atom. The energetics for the substitutional nitrogen in diamond and ta-C are collected in Table 2.

Group VI atoms such as oxygen are expected to be double donors in diamond and ta-C. The induced donor levels in diamond are too low in the band gap ($\sim$1.8 eV above valence band maximum) in order to contribute enough electrons to the conduction band and make the material n-type. In diamond the lattice relaxations lower the symmetry around the oxygen atom in the charge states 1+, 0, and 1−. In the charge states 2+, for which the gap states are empty, the tetrahedral symmetry is preserved. In ta-C the oxygen impurity prefers, similar to the nitrogen impurity, to substitute an sp^2-hybridized carbon atom. For the oxygen impurity this preference is calculated to be even stronger than for the nitrogen impurity. The calculated formation energies for the substitutional oxygen in diamond and ta-C are collected in Table 3.

The chlorine atom has seven valence electrons and can thus promote up to three electrons to the conduction band in diamond or ta-C. The chlorine-related electronic states are, however, so low (2.4 and 3.2 eV above valence band maximum) in the band gap that they became occupied immediately in n-type

Table 3

Formation energies for the different oxygen impurities in diamond and in amorphous carbon (the reference chemical potential used for the oxygen atom corresponds to the dimer O_2)

Defect	Formation energy (eV)
O_C^{2+} (diamond, tetrahedral)	$0.99 + 2\mu_e$
O_C^{1+} (diamond, trigonal)	$2.62 + \mu_e$
O_C^0 (diamond, trigonal)	4.45
O_C^{1-} (diamond, trigonal)	$7.41 - \mu_e$
O_C^{2+} (ta-C sp^3)	$1.50 + 2\mu_e$
O_C^{1+} (ta-C sp^3)	$1.56 + \mu_e$
O_C^0 (ta-C sp^3)	0.38
O_C^{1-} (ta-C sp^3)	$1.59 - \mu_e$
O_C^{2+} (ta-C sp^2)	$-1.85 + 2\mu_e$
O_C^{1+} (ta-C sp^2)	$-3.12 + \mu_e$
O_C^0 (ta-C sp^2)	-3.26
O_C^{1-} (ta-C sp^2)	$-2.62 - \mu_e$

Table 4

Formation energies for the different chlorine impurities in diamond and in amorphous carbon (the reference chemical potential used for the chlorine atom corresponds to the dimer Cl_2)

Defect	Formation energy (eV)
Cl_C^{3+} (diamond, tetrahedral)	$6.41 + 3\mu_e$
Cl_C^{2+} (diamond, trigonal)	$8.91 + 2\mu_e$
Cl_C^{1+} (diamond, trigonal)	$11.33 + \mu_e$
Cl_C^0 (diamond, trigonal)	14.47
Cl_C^{1-} (diamond, trigonal)	$17.70 - \mu_e$
Cl_C^{3+} (ta-C sp^3)	$8.01 + 3\mu_e$
Cl_C^{2+} (ta-C sp^3)	$8.12 + 2\mu_e$
Cl_C^{1+} (ta-C sp^3)	$8.25 + \mu_e$
Cl_C^0 (ta-C sp^3)	8.75
Cl_C^{1-} (ta-C sp^3)	$9.38 - \mu_e$
Cl_C^{3+} (ta-C sp^2)	$1.98 + 3\mu_e$
Cl_C^{2+} (ta-C sp^2)	$1.89 + 2\mu_e$
Cl_C^{1+} (ta-C sp^2)	$1.91 + \mu_e$
Cl_C^0 (ta-C sp^2)	2.17
Cl_C^{1-} (ta-C sp^2)	$2.50 - \mu_e$
Cl_C^{3+} (ta-C 3-ring)	$3.04 + 3\mu_e$
Cl_C^{2+} (ta-C 3-ring)	$2.85 + 2\mu_e$
Cl_C^{1+} (ta-C 3-ring)	$2.29 + \mu_e$
Cl_C^0 (ta-C 3-ring)	2.43
Cl_C^{1-} (ta-C 3-ring)	$2.73 - \mu_e$

materials. In diamond the lattice relaxation lowers the symmetry in the charge states 2+, 1+, 0, 1−. For the 3+ charge state there is no occupancy of the gap states and the symmetry is tetrahedral. In ta-C, the chlorine atom prefers also to substitute an sp^2-hybridized carbon atom. The formation energies for the substitutional chlorine in diamond and in ta-C are collected in Table 4.

In conclusion, all the substitutional impurity atoms in diamond studied in this work induce a symmetry lowering lattice relaxation in the charge states which have occupied levels in the band gap. Nitrogen is expected to be the most effective donor among the atomic species studied. The double (O) and triple (Cl) donors are shown to have too deep donor states in order to promote a useful concentration of electrons into the conduction band.

In ta-C the most probable site hosting an impurity atom is shown to be a graphite-like carbon atom site. The preference of the sp^2 site over sp^3 site is shown to increase as the number of the impurity valence electrons increases.

We would like to thank Dr. S. Clark and Dr. T. Mattila for useful discussions. We also acknowledge the generous computing resources of the Center for the Scientific Computing (CSC), Espoo, Finland.

References

[1] D.R. McKenzie, D. Muller and B.A. Pailthorpe, Phys. Rev. Lett. 67 (1991) 773.

[2] P. Stumm, D.A. Drabold and P.A. Fedders, J. Appl. Phys. 81 (1997) 1289.

[3] P.K. Stich, Th. Köhler, G. Jungnickel, D. Porezag and Th. Frauenheim, Solid State Comm. 100 (1996) 549.

[4] Th. Frauenheim, P. Blaudeck, U. Stephan and G. Jungnickel, Phys. Rev. B 48 (1993) 4823; Phys. Rev. B 50 (1994) 1489.

[5] C.Z. Wang and K.M. Ho, Phys. Rev. Lett. 71 (1993) 1184; J. Phys. Cond. Mat. 6 (1994) L239.

[6] N.A. Marks, D.R. McKenzie, B.A. Pailthorpe, M. Bernasconi and M. Parrinello, Phys. Rev. Lett. 76 (1996) 768; Phys. Rev. B 54 (1996) 9703.

[7] S.J. Clark, J. Grain and G.J. Ackland, Phys. Rev. B 55 (1997) 14 059.

[8] K. Nordlund, J. Keinonen and T. Mattila, Phys. Rev. Lett. 77 (1996) 699.

[9] G. Sreenivas, S.S. Ang and W.D. Brown, J. Electronic Mater. 23 (1994) 569.

[10] J. Ruan, W.J. Choyke and K. Kobashi, Appl. Phys. Lett 62 (1993) 1379.

[11] C. Pan, C.J. Chu, J.L. Margrave and R.H. Hauge, J. Electronic Mater. 141 (1994) 3246.

[12] W. Kohn and L.J. Sham, Phys. Rev. A 140 (1965) 1133.

[13] D.M. Ceperley and B.J. Alder, Phys. Rev. Lett. 45 (1980) 566; J. Perdew and A. Zunger, Phys. Rev. B 23 (1981) 5048.

[14] D.J. Chadi and M.L. Cohen, Phys. Rev. B 8 (1973) 5747.

[15] D. Vanderbilt, Phys. Rev. B 41 (1990) 7892; K. Laasonen, A. Pasquarello, R. Car, C. Lee and D. Vanderbilt, Phys. Rev. B 47 (1993) 10 142.

[16] H.J. Monkhorst and J.D. Pack, Phys. Rev. B 13 (1976) 5188.

[17] W. Kaiser and W.L. Bond, Phys. Rev. 115 (1959) 857.

[18] S.A. Kajihara, A. Antonelli, J. Bernholc and R. Car, Phys. Rev. Lett. 66 (1991) 2010.

[19] E. Rohrer, C.F.O. Graeff, R. Janssen, C.E. Nebel, M. Stutzmann, H. Güttler and R. Zachai, Phys. Rev. B 54 (1996) 7874.

ELSEVIER

Computational Materials Science 10 (1998) 356–361

COMPUTATIONAL
MATERIALS
SCIENCE

Ab initio calculation of excitonic effects in realistic materials

Stefan Albrecht [a],*, Giovanni Onïda [b], Lucia Reining [a], Rodolfo Del Sole [b]

[a] *Laboratoire des Solides Irradiés, URA 1380 CNRS-CEA/CEREM, École Polytechnique, F-91128 Palaiseau, France*
[b] *Istituto Nazionale per la Fisica della Materia, Dipartimento di Fisica dell'Università di Roma Tor Vergata, Via della Ricerca Scientifica, I-00133 Rome, Italy*

Abstract

Ab initio calculations, based on the density functional theory (DFT) in the local density approximation (LDA), allow for the description of the ground state properties of a wide class of materials. Also one-quasiparticle excitations can be obtained with good precision by adding self-energy corrections to the DFT–LDA eigenvalues. A realistic description of two-particle excitations, like the creation of electron–hole pairs in absorption experiments, is hardly feasible for systems where the electron and the hole interact. In this work we show how such excitonic effects can be included in ab initio electronic structure calculations, via the solution of an effective two-particle equation. Results for different systems are presented. Copyright © 1998 Elsevier Science B.V.

Keywords: Exciton; Optical absorption; Ab initio calculation; Condensed matter

Recent major advances in ab initio calculations, mostly density functional theory–local density approximation (DFT–LDA) applications, allow to determine the ground state properties and the Kohn–Sham (KS) electronic structure [1] for even complicated systems. However, excited electronic states, which correspond to electron addition and removal energies, are not correctly described by the KS eigenvalues. These quasiparticle (QP) energies should be obtained by using the true electron self-energy Σ (instead of the DFT-LDA exchange-correlation potential) in an equation similar to the KS one. Using Hedin's [2] GW approximation for Σ, the resulting band structure has been evaluated for many materials [3,4] in excellent agreement with experiment.

At the simplest level in the ab initio DFT–LDA framework, optical properties are calculated from one-electron transitions between KS states [5]. The computed absorption spectrum is not correct, since, e.g., the direct gap is generally wrong by 50% up to 100%. However, even when realistic QP energies are substituted for the KS eigenvalues, there is no quantitative agreement between theory and experiment [6]. Further, going beyond the one-particle picture by including local-field and exchange–correlation effects within the DFT does not improve the situation very much [7]. In fact, these approaches still suffer from the neglect of excitonic effects. Up to now, these effects have rarely been calculated for realistic systems, mostly in a semi-empirical way [8,9], or with an LDA-based ΔSCF approach [10].

Recently, the inclusion of excitonic effects starting from the QP energies in the calculation of the absorption spectrum of a small sodium cluster [11] greatly

---

* Corresponding author. Tel.: +33 1 69 33 40 90; fax: +33 1 69 33 30 22; e-mail: albrecht@seurat.polytechnique.fr.

improved the former rather poor agreement between calculated and experimental absorption spectra. In this paper we will give an overview of the physical basis and of the performance of that approach, with applications to different systems. Our procedure starts with a DFT–LDA calculation of the ground state properties and the KS electronic structure. We next determine the QP energies to obtain the occupied and the first empty bands of the considered material. Finally, we calculate the optical transition energies, including excitonic effects, by diagonalizing a two-particle equation containing the unperturbed band to band transitions and an interaction Hamiltonian.

The absorption spectrum is given by the macroscopic dielectric function ε_M

$$\varepsilon_M(\omega) = 1 - \lim_{q \to 0} v(\mathbf{q}) \chi_{\mathbf{G}=0,\mathbf{G}'=0}(\mathbf{q}; \omega), \tag{1}$$

where $\chi(1,2) = S(1,1;2,2)$ obeys the Bethe–Salpeter equation [8]

$$S(1, 1'; 2, 2') = S_0(1, 1'; 2, 2') + S_0(1, 1'; 3, 3')$$
$$\times \Xi(3, 3'; 4, 4') S(4, 4'; 2, 2'). \tag{2}$$

The term $S_0(1, 1'; 2, 2') = G(1', 2')G(2, 1)$ is directly related to the polarization function of independent particles χ_0, G being the one-particle Green's function. The electron–hole interaction is derived from the electron self-energy in the GW approximation and given by the kernel Ξ:

$$\Xi(3, 3'; 4, 4') = -i\delta(3, 3')\delta(4, 4')v(3, 4')$$
$$+i\delta(3, 4)\delta(3', 4')W(3, 3'). \tag{3}$$

The first term is the bare exchange part, which is responsible for local-field effects. v is the bare Coulomb interaction. The second term involving W is the screened Coulomb attraction between electron and hole.

We take the screened Coulomb interaction W in Eq. (3) to be static, which is well justified when the excitonic binding energies are small compared with the band gap. We invert Eq. (2) to get S, and transform the equation in the basis of LDA eigenstates. We avoid to invert the matrix $[1 - \chi_0(\omega)\Xi]$ for each absorption frequency ω by applying the identity

$$[H - \delta_{ij}\omega]^{-1}_{\alpha\beta} = \sum_\lambda \frac{\varphi_\lambda(\alpha)\varphi^*_\lambda(\beta)}{E_\lambda - \omega} \tag{4}$$

for a system of eigenvectors and eigenvalues defined by $H\varphi_\lambda = E_\lambda\varphi_\lambda$.

This leads to an effective two-particle Schrödinger equation given by

$$H^{(n_1 n_2)(n_3 n_4)}_{\text{exc}} A^{(n_3 n_4)}_\lambda = E_\lambda A^{(n_1 n_2)}_\lambda, \tag{5}$$

with the Hamiltonian

$$H^{(n_1 n_2)(n_3 n_4)}_{\text{exc}} = [E_{n_2} - E_{n_1}] \cdot \delta_{n_1 n_3}\delta_{n_2 n_4}$$
$$- (f_{n_2} - f_{n_1})\Xi^{(n_1 n_2)(n_3 n_4)}. \tag{6}$$

The energies E_n are the QP energy levels, while the f_n are Fermi–Dirac occupation numbers. [1] We find the exciton eigenvalues and eigenfunctions by diagonalizing the effective Hamiltonian matrix. The explicit knowledge of the coupling of the various two-particle channels, given by the coefficients $A^{(n_1 n_2)}_\lambda$, allows to identify the character of each transition. In particular, the exciton eigenstate is then described by

$$|N^*\rangle = \sum_{n_1 n_2} A^{(n_1 n_2)} c_{n_1} c^+_{n_2} |N\rangle, \tag{7}$$

where $|N\rangle$ is the ground state and c^+_n creates an electron in the Bloch state $|n\rangle$.

Furthermore, the macroscopic dielectric function in Eq. (1) is obtained as

$$\varepsilon_M(\omega) = 1 - \lim_{q \to 0} \sum_\lambda \left\{ v(\mathbf{q}) \sum_{n_1 n_2} \langle n_1|e^{-i\mathbf{q}\cdot\mathbf{r}}|n_2\rangle A^{(n_1 n_2)}_\lambda \right.$$
$$\left. \times \sum_{n_3 n_4} \langle n_4|e^{i\mathbf{q}\cdot\mathbf{r}}|n_3\rangle A^{*(n_3 n_4)}_\lambda \frac{f_{n_4} - f_{n_3}}{E_\lambda - \omega} \right\}. \tag{8}$$

In practice, the KS eigenvalues and eigenfunctions from a DFT–LDA calculation serve as input to the evaluation of the RPA screened Coulomb interaction, W, and the GW self-energy Σ. The KS eigenfunctions,

---

[1] We actually use $n_1 n_2 = (vc)$, where v stands for valence band and c for conduction band. Eq. (4) takes a more complicated form when also the small antiresonant, non-hermitian contributions to the Hamiltonian are taken into account.

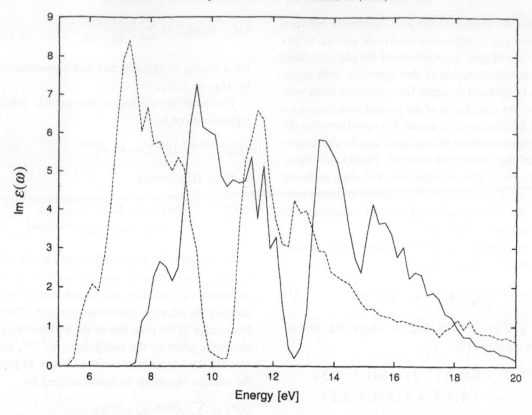

Fig. 1. Absorption spectrum of Li$_2$O. Dashed line: LDA calculation; continuous line: LDA + GW calculation.

together with the QP energies and W, are then used in the exciton calculation.

The first application of this scheme for a solid was done for lithium oxide in order to determine the optical absorption onset. The details of the calculation can be found in [12]. We will here show the absorption spectrum of Li$_2$O in order to illustrate the typical discrepancies which can be found with respect to experiment, when the standard approaches are applied. Li$_2$O is highly ionic and has anti-fluorite structure with an fcc cell consisting of one oxygen and two lithium atoms. Measured absorption spectra [13] show an onset above 6.6 eV, while our LDA spectrum (Fig. 1, dashed line) has its onset at the calculated direct gap of 5.3 eV. The GW corrections open the KS minimum direct band gap by 2.1 eV, while only slightly modifying the band dispersion. The resulting QP band gap of 7.4 eV is above the measured "optical gap". The absorption spectrum including GW corrections (continuous line) is hence

essentially shifted to higher energies, with a too high onset at 7.4 eV. Instead, application of Eq. (5) leads to a lowest allowed transition energy of 6.6 eV, consistent with experiment.

The calculation of the absorption spectrum, Eq. (8), is in principle straightforward, but demands for a very large number of points in the Brillouin zone sampling. Here we show preliminary results for bulk silicon, which is a much cited test case for optical absorption calculations. On one side, the energetic position of calculated LDA spectra are in reasonable agreement with experiment [14], but the E1 peak on the low energy side of the spectrum has a much too low intensity. On the other hand, a direct evaluation of Eq. (2) in a tight binding basis by Hanke and Sham [8] has shown that in fact excitonic effects are responsible for the discrepancies. An ab initio calculation of absorption including excitonic effects should hence be able to reproduce the observed features. The long-dashed

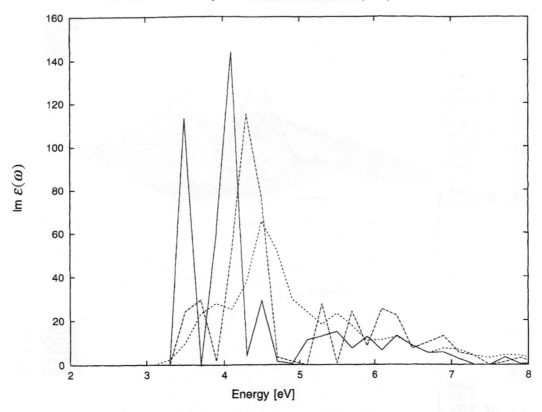

Fig. 2. Absorption spectrum of bulk silicon. Long-dashed line: $LDA + GW$ calculation using 10 special **k**-points in the IBZ; short-dashed line: same, but with 913 points in the IBZ. Continuous line: calculated LDA + GW + exciton spectrum, using 10 **k** points in the IBZ.

line in Fig. 2 shows the LDA+QP absorption spectrum obtained with a limited number of **k**-points (10 special points in the irreducible wedge of the Brillouin zone (IBZ)), and for comparison (short-dashed line) the converged spectrum with 913 **k**-points in the IBZ. The continuous line is our exciton spectrum obtained from Eq. (8) with the small set of **k**-points. Clearly, the desired effect of enhancing the E1 peak shows up. In order to obtain a well-converged spectrum, we have to use a number of **k**-points of the same order of magnitude as those used for the LDA spectrum. This is in principle feasible, since up to now we did not make use of the symmetry properties of the crystal. We are at present implementing the symmetries in our code, which will allow us to increase the number of **k**-points by an order of magnitude.

In order to illustrate the effect of excitonic interactions on the charge densities of excited states, we have

chosen the example of a small cluster, where (i) excitonic effects are dramatic, due to reduced screening and localized charges, and (ii) the effect on the charge distributions are clearly visible and relatively easy to understand, since the states are discrete. We have performed LDA and GW calculations on the monosilane SiH_4, using norm-conserving pseudopotentials for silicon [15] and hydrogen [16], with a cut-off of 20 Ry and a supercell of 30 a.u. Here we concentrate on the lowest dipole allowed transition which, in the LDA calculation, occurs between the 3-fold degenerate HOMO and the LUMO, at an energy of 7.5 eV. GW corrections shift this transition to about 12 eV, whereas the exciton binding energy of 4 eV gives again an onset energy close to the LDA one. However, the nature of the excited states has changed. looking at the lowest allowed transition, we find coefficients $A^{(n_1 n_2)}$ with absolute values of about 0.2–0.6 for the HOMO–

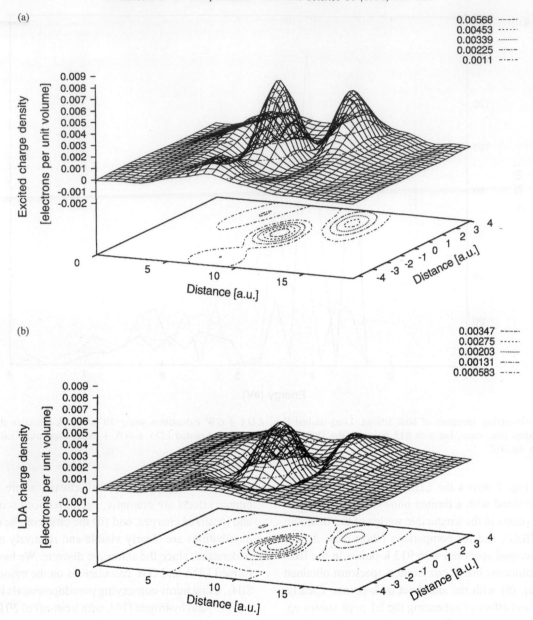

Fig. 3. Charge density of an electron in SiH₄, for the lowest allowed optical excitation. The charge is shown in a plane containing the silicon atom in the position $x = 8.59$ (horizontal axis), $y = 0$ (second in-plane axis), and the hydrogen atoms at $x = 5.88$, $y = -1.09$ and $x = 11.30$, $y = 1.09$, respectively. (a) Including excitonic mixing effects; (b) LDA result (charge density of the lowest unoccupied state).

LUMO transitions (which contribute of course 100% in each LDA case), and minor contributions from transitions from the HOMO to higher excited states, with a maximum for the fifth empty state with coefficients

$A^{(n_1 n_2)}$ up to 0.2. Fig. 3(a) shows the positive contributions to the resulting excitonic charge density ("electron") in a plane containing the silicon and two hydrogen atoms for one of the three degenerate transitions, calculated using (7). It can be compared to the charge density for the lowest unoccupied LDA (LUMO) state in Fig. 3(b). The electron is considerably perturbed with respect to the LUMO, which illustrates that excitonic effects are not limited to a pure shift of energies. The hole contains less effect of the mixing, since the lowest occupied state of SiH_4 is energetically far away and localized on the Si atom. Hence, the hole charge density is close to the HOMO one.

In conclusion, we have shown how excitonic effects can be treated within the ab initio framework, for a wide class of materials. Their inclusion is important in order to obtain correct optical gaps, absorption line shapes, and a realistic analysis of the nature of excited states.

This study has been supported by the scientific council of IDRIS/CNRS (France) through the allocation of computer time (projects 970906 and 970544).

References

[1] P. Hohenberg, W. Kohn, Phys. Rev. 136 (1964) B864; W. Kohn, L.J. Sham, Phys. Rev. 140 (1965) A1133.

[2] L. Hedin, Phys. Rev. A 139 (1965) 796.

[3] M.S. Hybertsen, S.G. Louie, Phys. Rev. Lett. 55 (1985) 1418; Phys. Rev. B 34 (1986) 5390.

[4] R.W. Godby, M. Schlüter, L.J. Sham, Phys. Rev. Lett. 56 (1986) 2415; Phys. Rev. B 37 (1988) 10 159.

[5] G.E. Engel, B. Farid, Phys. Rev. B 46 (1992) 15 812.

[6] R. Del Sole, R. Girlanda, Phys. Rev. B 48 (1993) 11 789.

[7] V.I. Gavrilenko, F. Bechstedt, Phys. Rev. B 54 (1996) 13 416.

[8] W. Hanke, L.J. Sham, Phys. Rev. B 21 (1980) 4656.

[9] N.A. Hill, K.B. Whaley, Phys. Rev. Lett. 75 (1995) 1130.

[10] F. Mauri, R. Car, Phys. Rev. Lett. 75 (1995) 3166.

[11] G. Onida, L. Reining, R.W. Godby, R. Del Sole, W. Andreoni, Phys. Rev. Lett. 75 (1995) 818.

[12] S. Albrecht, G. Onida, L. Reining, Phys. Rev. B 55 (1997) 10 278.

[13] W. Rauch, Z. Phys. 116 (1940) 652.

[14] H.R. Philipp, H. Ehrenreich, Phys. Rev. B 129 (1963) 1550.

[15] G.B. Bachelet, D.R. Hamann, M. Schlüter, Phys. Rev. B 26 (1982) 4199.

[16] N. Troullier, J.L. Martins, Phys. Rev. B 43 (1991) 1993.

[17] H.J. Monkhorst, J.D. Pack, Phys. Rev. B 13 (1976) 5188.

ELSEVIER

Computational Materials Science 10 (1998) 362–367

COMPUTATIONAL
MATERIALS
SCIENCE

LDA and tight-binding: Total energy calculations of polyparaphenylene

M.S. Miao [a],*, V.E. Van Doren [a], P.E. Van Camp [a], G. Straub [b]

[a] *Department of Physics, University of Antwerp (RUCA), B-2020 Antwerpen, Belgium*
[b] *Los Alamos National Laboratory, MS B221, Los Alamos, NM 87545, USA*

Abstract

A simple tight-binding model is presented and used to calculate the conformation and electronic structure of a single chain polyparaphenylene (PPP). A torsional angle of 52.5° is obtained by considering the conjugation of p_z orbitals on the carbon atoms and the Coulomb repulsion between neighboring hydrogen atoms. The characteristics of the filled and the unfilled π bands are calculated by solving the p_z-submatrix only at symmetry points in the Brillouin zone. The variations in these band characteristics versus the geometry parameters are discussed in detail and the results are confirmed by a full potential LDA calculation with a Ceperley–Alder term for the correlation energy. Copyright © 1998 Elsevier Science B.V.

Keywords: Polyparaphenylene (PPP); Tight binding; Torsion angle; Band structure

The discovery of blue light emission [1] from polyparaphenylene (PPP) has caused a new interest in this material. X-ray crystallographic measurements showed that the configuration of the crystalline oligomers of the phenyl ring was not planar but that there existed a twist torsional angle between neighboring phenyl rings [2]. Some evidence showed that PPP also crystallizes in a non-planar structure. The optical band gap has been determined by optical measurements and the ionization energy by X-ray (XPS) [3] and ultraviolet photo-emission spectroscopy (UPS) [4]. These experimental results were confirmed by theoretical calculations, ranging from Hartree–Fock (HF) [5] to density functional (DFT) [6] methods for both oligomers and polymer. In a recent article [7], we presented an LDA calculation of the total energy of a single chain of PPP for several different inter-ring torsional angles using a Ceperley–Alder term for the correlation energy. The band structure, especially the energy gap and band widths as well as the effect of torsion upon them, were investigated and compared to experimental data. Furthermore, we investigated in detail how the band characteristics vary as a function of the conformation parameters.

In this article, a tight-binding model will be constructed by using only nearest neighbor interactions and universal coupling constants as determined by Harrison [9]. The band characteristics and torsion angle are calculated and compared with our LDA results.

---

* Corresponding author. Tel.: +32 3 2180312; e-mail: miao@ruca.ua.ac.be.

0927-0256/98/$19.00 Copyright © 1998 Elsevier Science B.V. All rights reserved
PII S0927-0256(97)00188-2

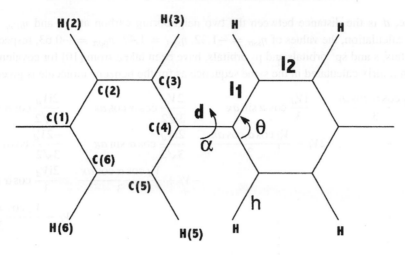

Fig. 1. Definition of structural parameters in the PPP chain.

Table 1
Geometry parameters used in TBA, in comparison with our LDA optimization, another LDF pseudo-potential calculation and HF calculated geometry parameters on phenyl oligomers as well as experimental results (for the definition of parameters, see Fig. 1)

	TBA PPP	LDA PPP	LDA[6] PPP	HF[5] Oligomer	exp[8][a]
l_1 (Å)	1.40	1.394	1.378	1.382	1.356–1.409
l_2 (Å)	1.40	1.394	1.399	1.388	1.371–1.425
d (Å)	1.48	1.473	1.456	1.492	1.469–1.505
h (Å)	1.10	1.098	1.100	1.074	0.960–1.129
θ (degrees)	120.0	120.95	121.2	120.3	104–128
α (degrees)	—	34.8	27.4	53.22	20–27

[a] For a complete lists of references, see [6].

The geometry of the PPP chain is presented in Fig. 1(a). The parameters used in the tight-binding approximation (TBA) are listed in Table 1 and defined in Fig. 1, in comparison with our LDA optimized values and other LDA and HF as well as experimental results.

As is well known, the TBA model has been developed as an effective method for calculating the properties of insulators, compounds and semiconductors [9]. Only nearest neighbor interactions are considered since the electrons are considered localized in these systems. Instead of original atomic s and p orbitals, the three sp^2 hybrids, in which the electron charge density is greatest in the direction of neighboring atoms in the plane of the phenyl rings, and a single p orbital perpendicular to the plane, are considered. The transfer matrix elements between the different orbitals in one atom are rather small and can be omitted in this approximate model.

The Hamiltonian can be expressed in an interaction matrix form using the Bloch orbitals constructed by the benzene molecular orbitals. Since the highest three filled and the lowest three unfilled bands are π symmetric and formed by the parallel p_z orbitals of carbon atoms, only the p_z orbitals will be considered now. The benzene molecule has six π symmetric orbitals Ψ_i with energies $\varepsilon_p - 2V_{pp\pi}$, $\varepsilon_p + 2V_{pp\pi}$, $\varepsilon_p - V_{pp\pi}$, $\varepsilon_p + V_{pp\pi}$, $\varepsilon_p - V_{pp\pi}$, $\varepsilon_p + V_{pp\pi}$, where the ε_p, with the value of -8.97 eV [9], is the energy of carbon p_z orbitals and the $V_{pp\pi}$ is the coupling coefficient between the neighboring p_z orbitals. These coupling coefficients are expressed by Harrison [10] as

$$V_{ll'm} = \eta_{ll'm} \frac{\hbar^2}{md^2}. \tag{1}$$

In this special case, d is the distance between the two neighboring carbon atoms and $\eta_{ll'm}$ is a dimensionless coefficient. In this calculation, the values of $\eta_{ss\sigma} = -1.32$, $\eta_{sp\sigma} = 1.42$, $\eta_{pp\pi} = -0.63$, respectively, for σ and π couplings of s orbitals, s and sp^2 orbitals and p_z orbitals, have been taken from [10] for covalent solids.

The Hamiltonian matrix calculated in the same sequence as for the benzene molecule is given by

$$
\begin{pmatrix}
-2V_l + \dfrac{V_d \cos\alpha\cos a\kappa}{3} & \dfrac{iV_d}{3}\cos\alpha\sin a\kappa & \dfrac{2V_d}{3\sqrt{2}}\cos\alpha\cos a\kappa & \dfrac{2iV_d}{3\sqrt{2}}\cos\alpha\sin a\kappa \\
 & 2V_l - \dfrac{V_d\cos\alpha\cos a\kappa}{3} & \dfrac{2iV_d}{3\sqrt{2}}\cos\alpha\sin a\kappa & \dfrac{-2V_d}{3\sqrt{2}}\cos\alpha\cos a\kappa \\
 & & -V_l + \dfrac{V_d\cos\alpha\cos a\kappa}{3} & \dfrac{2iV_d}{3}\cos\alpha\sin a\kappa \\
 & & & V_l - \dfrac{V_d\cos\alpha\cos a\kappa}{3} \\
 & & & -V_l \\
 & & & V_l
\end{pmatrix}
$$

where V_l and V_d are the coupling coefficients $V_{pp\pi}$ within the phenyl rings and between the phenyl rings. In this calculation, $l_1 = l_2 = l$, and ε_p is omitted in the matrix for the purpose of brevity.

Since all the π bands are monotonic, the band characteristics like the band widths, the ionization energy and the band gap can be obtained by solving the interaction matrix only at the high symmetry points in the Brillouin zone $\kappa = 0$ and $\kappa = \pi$. After some algebra, we get the eigenvalues at $\kappa = 0$:

$$
E_{3,1}(\kappa = 0) = \varepsilon_p + \frac{V_l}{2} + \frac{V_d}{3}\cos\alpha \mp \frac{3V_l}{2}\sqrt{1 + \frac{8}{81}\left(\frac{V_d}{V_l}\right)^2}\cos\alpha,
$$

$$
E_{4,2}(\kappa = 0) = \varepsilon_p - \frac{V_l}{2} - \frac{V_d}{3}\cos\alpha \mp \frac{3V_l}{2}\sqrt{1 + \frac{8}{81}\left(\frac{V_d}{V_l}\right)^2}\cos\alpha. \tag{2}
$$

For $\kappa = \pi$, the results can be obtained by just reversing the sign of the third term in the above equations. These eigenvalues are different from the ones given by formula (4.3) in [6]. The present results reproduce the molecular eigenvalues for $V_d = 0$ and are not degenerate at $\kappa = \pi$ and $V_d = V_l$. On the other hand, the results of [6] do not reproduce the molecular eigenvalues and are degenerate at $\kappa = \pi$ and $V_d = V_l$, i.e. $t_{22} = t_{11}$.

By taking the Tailor expansion of $\sqrt{1 + x}$ into account and keeping only the first two terms, we obtain for:

(1) the band gap:

$$
E_g = E_3(\kappa = 0) - E_2(\kappa = 0) \simeq -2V_l + \frac{2}{3}V_d\cos\alpha - \frac{4V_d^2}{27V_l}\cos\alpha, \tag{3}
$$

(2) the ionization energy:

$$
I = -E_2(k = 0) \simeq -\varepsilon_p - V_l + \frac{V_d}{3}\cos\alpha - \frac{2V_d^2}{27V_l}\cos\alpha, \tag{4}
$$

(3) the band widths:

$$
W_i = |E_i(\kappa = 0) - E_i(\kappa = \pi)| = -\frac{2}{3}V_d\cos\alpha, \tag{5}
$$

notice that V_d and V_l are less than zero.

Table 2

The TBA band characteristics compared with our LDA calculation using Ceperley–Alder correlation and the HF as well as the experimental results

	α	Valence band			Conduction band			Gap
		Min	Max	Width	Min	Max	Width	
TBA	0.0	−12.29	−10.83	1.46	−7.11	−5.65	1.46	3.72
TBA	16.0	−12.27	−10.86	1.40	−7.08	−5.67	1.40	3.79
TBA	34.8	−12.16	−10.96	1.20	−6.98	−5.78	1.20	3.97
TBA	90.0	−11.42	−11.42	0	−6.52	−6.52	0	4.90
LDA(PZ)[7]	0.0	−8.09	−4.71	3.38	−2.72	1.05	3.77	1.99
LDA(PZ)[7]	34.8	−7.51	−4.93	2.58	−2.39	0.47	2.86	2.54
LDA(PZ)[7]	90.0	−6.41	−6.00	0.41	−1.70	−0.54	1.16	4.30
HF[5]	22.7	−13.76	−8.85	4.91	−0.61	5.09	5.60	8.24
HF+MP2[5]	22.7	−11.83	−7.71	4.12	−2.83	2.39	5.22	4.88
EXP(UPS solid)[3]		−9.15	−5.65	3.5	—	—	—	2.8
EXP(UPS gas)[4]		−9.85	−6.95	2.9	—	—	—	3.4

All the values are expressed in units of eV. α is the torsional angle between the neighboring phenyl rings.

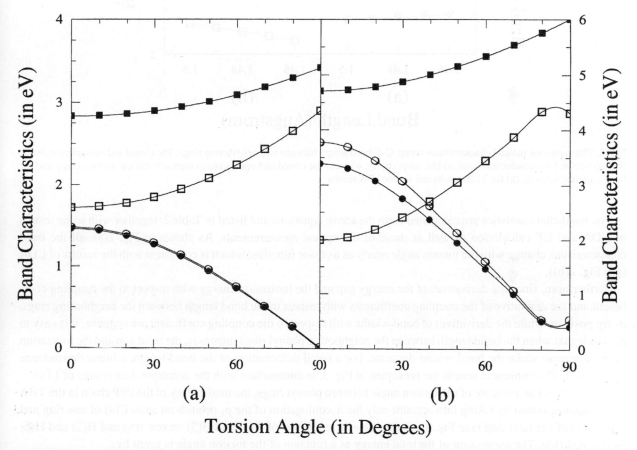

Fig. 2. Variations in band widths of the valence and the conduction bands, in the band gap as well as in the ionization energy as a function of the torsion angle. The closed and open circles denote the valence and the conduction band widths, respctively, whereas the closed and open squares represent the ionization energy and the band gap, respectively, (a) for TBA results, (b) for LDA results. For clarity, the band gap and the ionization energy are plotted with consistent shifts of −2 eV and −8 eV, respectively for TBA results.

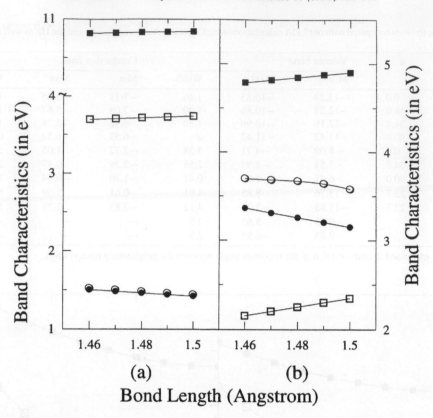

Fig. 3. The variations in band characteristics versus C–C bond length between two neighboring rings. The closed and open circles denote the valence and the conduction band widths, respectively, whereas the closed and open squares represent the ionization energy and the band gap, respectively, (a) for TBA results and (b) for LDA results.

The band characteristics are calculated from the above equations and listed in Table 2, together with some results of LDA and HF calculation as well as those of the optical measurements. As shown in Fig. 2(a), all the band characteristics change with the torsion angle nearly as a cosine function, which is consistent with the results of LDA (see Fig. 2(b)).

Furthermore, since the derivatives of the energy gap and the ionization energy with respect to the coupling coefficient and the derivative of the coupling coefficients with respect to the bond length between the neighboring rings, d, are positive, while the derivatives of band widths with respect to the coupling coefficient are negative, it is easy to conclude that when the bond length between the neighboring phenyl rings increases, the band gap and the ionization energy increase while the band widths decrease. For a small deformation of the bond length, a linear dependence is expected. The numerical results are presented in Fig. 3, in comparison with the corresponding results of LDA.

In order to get an estimate of the torsion angle between phenyl rings, the total energy of the PPP chain in the TBA can be approximated by taking into account only the π conjugation of the p_z orbitals on atom C(4) of one ring and atom C(1) of the next ring (see Fig. 1) and the charge repulsions of H(3) and H(5) on one ring and H(2) and H(6) on the next ring. The expression of the total energy as a function of the torsion angle is given by

$$E_T = \frac{2(Z^*e)^2}{\sqrt{(d+l-h)^2 + \frac{3}{2}(h+l)^2(1-\cos\alpha)}} + V_{pp\pi}\cos\alpha, \qquad (6)$$

where d, l and h are explained in Table 1 and Fig. 1. Z^* is the probability of an electron appearing on a hydrogen atom with lowest energy. By solving the carbon-hydrogen interaction submatrix, we obtain a Z^* value of 0.54 and therefore a torsion angle of 52.5° for the minimum energy. As was pointed out before, in this article the values of the coupling coefficients are the same as those for a covalent solid. Of course it would be better to fit $\eta_{ss\sigma}$, $\eta_{sp\sigma}$ and $\eta_{pp\pi}$ to the ground state properties of the LDA results of PPP.

In conclusion, it should be emphasized that the method presented here is very general and applicable to most of the polymer systems. The TBA results are comparable with the first principle calculation and experimental measurements.

This work is supported partly under grants no. 2.0131.94 and 6.0347.97 of the Belgian National Science Foundation (NFWO).

References

[1] G. Grem, G. Leditzky, B. Ullrich, G. Leising, Adv. Mater. 4 (1992) 36.
[2] J.L. Baudour, Y. Delugeard, P. Rivet, Acta Crystallogr, Sect. B 34 (1978) 625.
[3] J. Riga, J.J. Pireaux, J.P. Boutique, R. Caudano, J.J. Verbist, Y. Gobilon, Synth. Met. 4 (1981) 99.
[4] K. Seki, U.O. Karlsson, R. Engelhardt, E.E. Koch, W. Schimidt, Chem. Phys. 91 (1984) 459.
[5] L. Cuff, M. Kertesz, Macromolecules 27 (1994) 762; F. Bogár, W. Förner, E. Kapuy, J. Ladik, Theochem., in press; B. Champagne, D.H. Mosley, J.G. Fripiat, J. André, Phys. Rev. B 54 (1996) 2381.
[6] C. Ambrosch-Draxl, J.A. Majewski, P. Vogl, G. Leising, Phys. Rev. B 51 (1995) 9668.
[7] M. S. Miao, P.E. Van Camp, V.E. Van Doren, J.J. Ladik, J.W. Mintmire, submitted.
[8] Y. Delugeard, J. Desuche, J. L. Baudour, Acta Crystallogr, Sect. B 32 (1976) 702.
[9] W.A. Harrison, Electronic Structure and the Properties of Solids, Dover, New York, 1989.
[10] W.A. Harrison, Phys. Rev. B 24 (1981) 5835.

ELSEVIER

Computational Materials Science 10 (1998) 368–372

First-principles investigations of the electronic, optical and chemical bonding properties of SnO$_2$

Ph. Barbarat [a,*], S.F. Matar [b]

[a] CEA Le Ripault, Département Matériaux, 37260 Monts, France
[b] Institut de Chimie de la Matière Condensée de Bordeaux, CNRS Chateau Brivazac, Avenue du Docteur Schweitzer,
F-33608 Pessac, France

Abstract

The electronic structure of the rutile-type oxide SnO$_2$ is examined self-consistently using the augmented-spherical-wave (ASW) method within the density-functional theory (DFT). The influence of hybridization between the different l-states on the chemical bonding is discussed from the density-of-states (DOS) and the crystal-orbital-overlap-population (COOP) results. A description of the nature of chemical bonding in SnO$_2$ is provided along with the investigation of the optical properties. An overall agreement was found between the calculated and the experimental optical properties in the ultraviolet spectrum. Copyright © 1998 Elsevier Science B.V.

Keywords: Tin oxide; Density functional theory; Bonding; Optical properties

1. Introduction

The properties of metal oxides have received a great deal of interest for many years, due to many fields of application such as solar cells, optical devices and oxidation catalysts, in the particular case of SnO$_2$. Even though these applications are often related to the extrinsic behaviour of the oxide, a detailed understanding of the fundamental electronic structure and properties is required to obtain high-quality materials. Theoretical studies and density functional theory (DFT) calculations especially are expected to provide valuable information about the electronic or the ultraviolet optical properties of metal oxides. Therefore, we present

here a theoretical study of the electronic structure of SnO$_2$ that leads to a description of both the nature of chemical bonding and the character of the optical behaviour in the ultraviolet spectrum.

2. Calculational method

As in our earlier studies of oxide systems [1,2] we use the augmented-spherical-wave (ASW) method within the framework of the DFT. The chosen valence partial wave functions are taken to be (5s,5p) and (2s,2p) for Sn and O, respectively. The effects of exchange and correlation are treated within the local-density approximation (LDA) using the parametrization scheme of von Barth and Hedin [3] and Janak [4]. Since the rutile structure is not closely packed, empty

---

* Corresponding author. Present address: L' OREAL Recherchee Avancee, BP No. 22, 1 avenue Eugene, Schueller, 93601 Aulnay-Sous-Bois, Cedex, France.

spheres (ES1 and ES2) are introduced at appropriate sites within the SnO_2 unit cell to represent the interstitial space. Our non-unique choice of the atomic sphere radii which minimize the overlap between the atomic spheres was subjected to the following ratios: $r_{Sn}/r_O = 0.875$, $r_O/r_{ES1} = 1.333$, $r_{ES1}/r_{ES2} = 1.089$. Furthermore in this work chemical bonding features are discussed based on the crystal-orbital-overlap-population (COOP) initially developed by Hoffmann [5] from the quantum chemistry standpoint (extended Hückel calculations). This allows for the density-of-states (DOS) features to be assessed on the basis of *chemical bonding criteria* which in a lax notation consists of weighting the DOS with the sign and magnitude of the overlap integral between the relevant orbitals. The optical absorption spectrum is calculated as the real part of the frequency-dependent interband optical conductivity in the dipole approximation evaluated with the Kubo formula [6,7]. A detailed description of the calculational method is provided elsewhere [8].

3. Results and discussion

The calculated band structure of SnO_2 along the symmetry lines of the simple tetragonal Bravais lattice is shown in Fig. 1, where the Fermi level is chosen to be the zero of the energy scale. Moreover the projection of the DOS for the l-states for each one of the two constituent species is presented in Figs. 2(a) and (b). A detailed analysis of the band structure and DOS results is provided elsewhere [8]. However, it is of interest to note that the calculated minimum band gap is found to be 3.6 eV whereas the experimental value is within the [3.6 eV; 3.9 eV] range [9,10]. Furthermore, the SnO_2 valence band features obtained from our calculations are seen to be in agreement with ultraviolet photoelectron spectroscopy (UPS) measurements [11] as presented in Fig. 2(b).

3.1. Crystal-orbital-overlap-population results

The COOP results are then used to investigate the inter-atomic interaction. In the rutile structure the

metal is in a non-regular octahedron of oxygen atoms and SnO_6 octahedra are edge sharing. Because of this feature of stacking octahedra there could exist non-negligible O–O interactions beside the Sn–O ones. This we illustrate by showing the relevant COOP in following figures where positive, negative and zero values along the y-axis are relative to bonding, anti-bonding and non-bonding interactions, respectively. Beside the COOP given for two-body interactions for all the valence basis set, we give a projection as well for p- and d-orbitals for each atomic species in order to address their respective roles. Fig. 3(a) shows the COOP for the Sn–O and O–O interactions for one atom of each species. They are clearly dominated by both kinds of interactions. The energy range where Sn(p)-states predominate ([−4 eV; −2 eV]) shows bonding features of Sn–O interaction which then becomes antibonding. There are concomitantly bonding and antibonding O–O interactions. Thus the upper part of the valence band is formed of antibonding Sn–O and O–O interactions. This can be more clearly observed when l-projected COOP are plotted. In Fig. 3(b), the Sn–O interaction is resolved for Sn(p) and Sn(d) contributions. This clearly shows that Sn(p)–O interaction is bonding up to the top of the valence band, this is because Sn(p)-states are filled with less than one electron so that all electrons coming from oxygen will be bonding. On the contrary because Sn(d) band is nearly full all interactions with oxygen will be antibonding. The resulting COOP of Sn–O interaction is that the lower part of the valence band is bonding and the upper part antibonding as shown in Fig. 3(a).

3.2. Ultraviolet absorption spectrum

The ultraviolet absorption spectrum of SnO_2 is calculated for light polarized parallel to the tetragonal axis. Results are presented in Fig. 4. A shoulder is observed for photon energies right above the direct gap energy, i.e. 4.2 eV. This feature is due to transitions, from the O(p)-states in the valence band to Sn(s)-states in the conduction band, occurring at points located near the Γ-point in the Γ–M and Γ–Z directions of the BZ. A large number of direct transitions with a

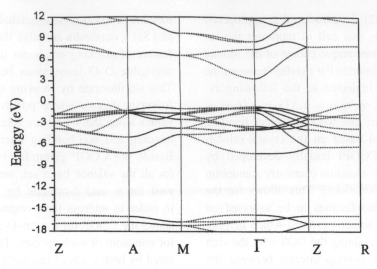

Fig. 1. Band structure of SnO$_2$.

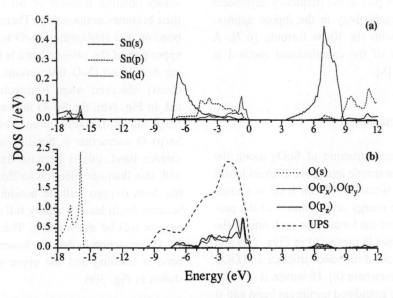

Fig. 2. Partial density-of-states (DOS) for SnO$_2$: l-projected DOS for tin (a); l- and cartesian-projected DOS for oxygen (b); also shown in (b) are the ultraviolet photoelectron spectroscopy (UPS) measurements from [11].

magnitude of about 8 eV are found to occur from O(p)-states to Sn(s)-states where the corresponding energy bands are relatively parallel, i.e. at the A-point and the M-point and near these points in the A–Z, A–M and M–Γ-directions of the BZ. These transitions give rise to the peak observed at 8.0 eV and their initial states are lying in the strong O–O antibonding interaction region plotted in Fig. 3(a). The peak at 9.6 eV

is probably due to transitions from the O(p)-states to Sn(s)-states occurring at the Z-point and the Γ-point and near these points in the Z–A, Γ–M and Γ–Z directions where numerous direct transitions of that magnitude are found. Another critical feature of the calculated absorption spectrum is a peak at 11.7 eV whose interpretation remains difficult due to a large number of transitions. However, that peak seems to

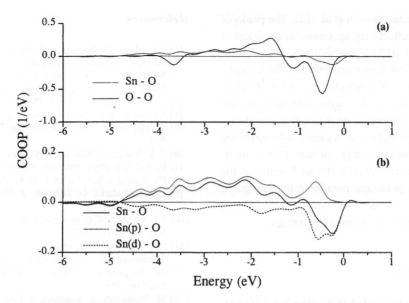

Fig. 3. Crystal-orbital-overlap-population (COOP) for Sn–O and O–O interactions (a); tin l-projected COOP for Sn–O interactions (b).

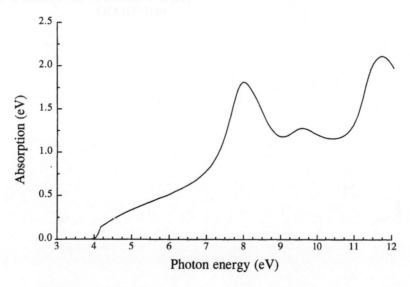

Fig. 4. Ultraviolet absorption spectrum of SnO_2 for light polarization direction parallel to the tetragonal axis.

be mainly caused by transitions occurring all over the volume of the BZ investigated except the M–Γ zone where relatively few transitions of this magnitude are observed. Most of these transitions take place from the O(p) states in the valence band to the Sn(p) states in the conduction band but a significant amount of transitions with a magnitude of about 11.7 eV is also found to occur from O(p)-states to Sn(s)-states. In contrast to the character of the SnO_2 optical gap which has been intensively studied over the past decades, there are very few experimental data about the optical properties of SnO_2 for photon energies larger than the minimum band gap energy. Our results are found to agree with the shape of the reflectivity spectrum of

SnO$_2$ measured by Jacquemin et al. [12]. The peaks of the experimental reflectivity spectrum were observed at 4.3, 7.4, 8.9 and 10.9 eV, respectively, whereas our calculated absorption spectrum exhibits three peaks at 8.0, 9.6 and 11.7 eV respectively, and a shoulder extending from 4.2 eV. The agreement between the two sets of results decreases as the photon energy increases, showing the degree of accuracy to be expected in the approach proposed here. At last, it is of interest to note that experimentally it was found that the absorption of light polarized parallel to the tetragonal axis exhibits a diffuse edge around 3.95 eV [13,14] which is in agreement with our calculations.

Acknowledgements

The calculations presented in Section 3.2 were carried out with the ESOCS® program from Molecular Simulations Inc.

References

[1] S.F. Matar, G. Demazeau, J. Sticht, V. Eyert, J. Kübler, J. de Phys. I France 2 (1992) 315.

[2] S.F. Matar, G. Demazeau, P. Mohn, V. Eyert, S. Najm, Eur. J. Solid State Inorg. Chem. 31 (1994) 615.

[3] J. von Barth, D. Hedin, J. Phys. C 5 (1972) 1629.

[4] J.F. Janak, Solid State Commun. 25 (1978) 53.

[5] R. Hoffmann, Angew. Chem. Int. Ed. Engl. 26 (1987) 846.

[6] C.S. Wang, J. Callaway, Phys. Rev. B 9 (1974) 4897.

[7] R. Kubo, J. Phys. Soc. Jpn. 12 (1957) 570.

[8] Ph. Barbarat, S.F. Matar, J. Mater. Chem., submitted.

[9] J.L. Jacquemin, G. Bordure, J. Phys. Chem. Solids 36 (1975) 1081.

[10] T.L. Credelle, C.G. Fonstad, R.H. Rediker, Bull. Am. Phys. Soc. 16 (1971) 519.

[11] P.L. Gobby, G.J. Lapeyre, in: F. G. Fumi (Ed.), Physics of Semiconductors, North-Holland, Amsterdam, 1976.

[12] J.L. Jacquemin, C. Raisin, S. Robin–Kandare, J. Phys. C 9 (1976) 593.

[13] M. Nagasawa, S. Shionoya, J. Phys. Soc. Jpn. 30 (1971) 158.

[14] R.D. McRoberts, C.G. Fonstad, D. Hubert, Phys. Rev. B 10 (1974) 5213.

Computational Materials Science 10 (1998) 373–380

COMPUTATIONAL
MATERIALS
SCIENCE

Conceptual and computational advances in multiple-scattering electronic-structure calculations

Rudolf Zeller [1]

Institut für Festkörperforschung, Forschungszentrum Jülich GmbH, D-52425 Jülich, Federal Republic of Germany

Abstract

A physically transparent transformation of the Korringa–Kohn–Rostoker (KKR or multiple-scattering) method into a tight-binding form is described. The transformation replaces the complicated, slowly decaying, traditional KKR structure constants by exponentially decaying "tight-binding" parameters. The main computational effort consists in the inversion of sparse matrices and scales for surfaces and interfaces, i.e. for systems with two-dimensional periodicity, linearly with the number of layers. This gives the opportunity to treat high-indexed surfaces as an approximation for almost isolated surface steps. Additional adatoms on surfaces and at steps can also be treated and it is discussed that reliable atomic forces and geometric arrangements can be obtained. Copyright © 1998 Elsevier Science B.V.

PACS: 70.15.−m; 71.20.−b

1. Introduction

Since ab initio electronic-structure calculations require too much computing power except for rather simple modeling systems, the method of empirical potentials is widely used for the computational simulation and analysis of materials. One step forward to bridge the gap between quantum-mechanical electronic-structure calculations and the method of empirical potentials is to increase the efficiency of density-functional calculations for larger systems. The search for efficient methods, particularly for ones, which need a computational effort scaling linearly with the system size [1], has recently received much attention. Most of these methods gain their efficiency by the assumption that the electronic wave functions decay fast enough in space so that a local region around each individual atom can be treated separately. This allows to treat insulating or semiconducting materials, but probably not metallic systems, where long-ranged electronic states at the Fermi energy are decisive for many physical properties.

Close-packed metallic systems, on the other hand, are well treated by the multiple-scattering (KKR) method, which originated in the works of Lord Rayleigh [2], Kasterin [3] and Ewald [4] and was formulated for the Schrödinger equation by Korringa, Kohn and Rostoker [5]. The main historical disadvantages of the KKR method are the

---

[1] Tel.: +49 2461-615192, fax: +49 2461-612620, e-mail: ru.zeller@fz-juelich.de.

numerical difficulties connected with the long-range and complicated energy and wave vector dependencies of the so called KKR structure constants, and the original restriction to spherically symmetric, nonoverlapping potentials around each atom, the so-called muffin-tin approximation. The controversy whether space-filling potentials can be used has now been settled and it is commonly assumed that the KKR method can generally be applied [6,7]. This general applicability is also substantiated by calculations for atomic forces and displacements [8], which will be discussed in Section 6.

The main issue of this paper concerns a recent development which bypasses the numerical complexity of the traditional multiple-scattering equations and allows to determine, for instance, the electronic structure of surfaces with an effort, which increases only linearly with the number of surface layers. This opens the way to treat high-indexed surfaces and in this approximation isolated surface steps. An important feature of the KKR method is that it allows to determine the Green function for the considered material. Thus the treatment of local perturbations like impurities, adatoms on surfaces, at steps or even at kinks is straightforward [9,10].

The recent progress which drastically reduces the computational effort and algorithmic complexity of the traditional KKR method has its origin and motivation in the success of the tight-binding linear-muffin-tin-orbital (TB-LMTO) method of Andersen and Jepsen [11]. The concept of the TB-LMTO screening constants was first extended to energy-dependent screening parameters by Andersen et al. [12] as a tool to obtain a TB form of the KKR method. Since the original determination of these screening parameters was numerically difficult, a simplified procedure was suggested by Szunyogh et al. [13], but the real breakthrough came by the discovery that the screening parameters can be obtained in a physically transparent and easy way.

For the transparent determination and the subsequent use of the screening parameters, two different, but related techniques have been developed. One technique [14] is based on the concept of unitary spherical waves, which are defined as solutions for a hard sphere solid, and is described in a wave function language. The other technique [15] is based on a repulsive *reference* system, which consists of constant, finite potentials inside nonoverlapping spheres, and uses a Green-function language. The Green-function technique will be presented here and was used to obtain the results discussed in Section 4.

2. Theory

The theory has been described in previous publications [15–17] and only a short account is given here. The Green functions $G(\mathbf{r}, \mathbf{r}', E)$ and $G^r(\mathbf{r}, \mathbf{r}', E)$ for two different Kohn–Sham effective potentials $V(\mathbf{r})$ and $V^r(\mathbf{r})$ are connected by a Dyson equation:

$$G(\mathbf{r}, \mathbf{r}', E) = G^r(\mathbf{r}, \mathbf{r}', E) + \int G^r(\mathbf{r}, \mathbf{r}'', E)[V(\mathbf{r}'') - V^r(\mathbf{r}'')]G(\mathbf{r}'', \mathbf{r}', E)\, d\mathbf{r}'', \tag{1}$$

which by the multiple-scattering expression:

$$G(\mathbf{r} + \mathbf{R}^n, \mathbf{r}' + \mathbf{R}^{n'}, E) = \delta^{nn'} G_s(\mathbf{r} + \mathbf{R}^n, \mathbf{r}' + \mathbf{R}^n, E) + \sum_{LL'} R_L^n(\mathbf{r}, E) G_{LL'}^{nn'}(E) R_{L'}^{n'}(\mathbf{r}', E) \tag{2}$$

can be transformed [6] into a system of linear equations:

$$G_{LL'}^{nn'}(E) = G_{LL'}^{r,nn'}(E) + \sum_{n''} \sum_{L''} G_{LL''}^{r,nn''}(E) \sum_{L'''} [t_{L''L'''}^{n''}(E) - t_{L''L'''}^{r,n''}(E)] G_{L'''L'}^{n''n'}(E). \tag{3}$$

Cell-centered coordinates $\mathbf{r}$ and $\mathbf{r}'$ are used in (2) and E and $L = (\ell, m)$ denote energy and angular-momentum numbers. R_L^n and G_s are wave functions and Green function [6] for a single potential restricted to the Voronoi cell around the atomic center $\mathbf{R}^n$. The t-matrices $t_{LL'}^n$ and $t_{LL'}^{r,n}$ are also single-site quantities [6] determined by the potentials V and V^r restricted to the cell at $\mathbf{R}^n$.

The traditional reference system is empty space, denoted by 0 in the following. In empty space the potential V^0 and consequently the t-matrix $t_{LL'}^{0,n}$ vanish and the so-called structural Green-function matrix elements $G_{LL'}^{0,nn'}$ are analytically known as

$$G_{LL'}^{0,nn'}(E) = -4\pi i\sqrt{E}\sum_{L''}i^{l+l''-l'}C_{LL'L''}h_{l''}\left(\sqrt{E}|\Delta\mathbf{R}|\right)Y_{L''}(\Delta\mathbf{R}) \tag{4}$$

with $\Delta\mathbf{R} = \mathbf{R}^n - \mathbf{R}^{n'}$ and $C_{LL'L''} = \int_{4\pi}d\Omega_{\mathbf{r}}Y_L(\mathbf{r})Y_{L'}(\mathbf{r})Y_{L''}(\mathbf{r})$, where h_l denotes spherical Hankel functions and Y_L spherical harmonics.

Empty space seems to be the easiest reference system, but this is not true because it makes the calculations difficult and time consuming. For instance, in periodic geometries one needs Fourier sums like

$$G_{LL'}^0(\mathbf{k}, E) = \sum_{n'}\exp(i\mathbf{k}\mathbf{R}^n - i\mathbf{k}\mathbf{R}^{n'})G_{LL'}^{0,nn'}(E), \tag{5}$$

which converge only conditionally and require complicated Ewald procedures [4,18] for their evaluation as a consequence of the slow decaying Hankel functions in (4). This problem is connected with the well-known free-electron singularities in the traditional KKR structure constants $G_{LL'}^0(\mathbf{k}, E)$, which occur if the free-electron-band structure condition $E = |\mathbf{k} + \mathbf{K}^h|^2$ is satisfied, where $\mathbf{k}$ denotes the wave vector and $\mathbf{K}^h$ a reciprocal lattice vector. The problem of the traditional reference system thus arises from the eigenstates of empty space. When the basic equations (1) and (3) are solved, these eigenstates are eliminated in the solution process and replaced by the eigenstates of the material, for which the Green function is calculated. It would be much more advantageous to start with a reference system, which has no eigenstates in the energy range, which is used by density-functional electronic-structure calculations and typically covers energies below 1 Ry.

Such a reference system consists of an infinite array of repulsive, constant, finite potentials within nonoverlapping spheres around each center $\mathbf{R}^n$ and a zero potential in the interstitial region between the spheres [15]. From first-order perturbation theory one expects that the eigenstates in the repulsive reference system have higher energies than in empty space and that the energy shift is given by the product of the potential strength and the volume filling. Calculated density-of-states for the reference system are shown in Fig. 1 and confirm the shift to higher energies. Contrary to first-order perturbation theory the shift is not uniform and also shows a saturation effect. The lowest eigenstate is at 1.35 Ry for potentials of 2 Ry height, at 2.25 Ry for potentials of 4 Ry height and just below 3 Ry for potentials of 20 Ry height. In Fig. 1 it is important to notice that the density of states vanishes below the energy of the lowest eigenstate. The consequences of this fact will be discussed in the next section. It is also important to notice that the vanishing density of states has been achieved by using an angular-momentum cut-off $l_{\max} = 3$ in all equations. This means that the repulsive reference system already works if only low angular-momentum components are used in the calculations. Thus no convergence problems with respect to angular-momentum sums occur in the TB form of the KKR method.

3. Basic facts

Fig. 1 shows that for lower energies no eigenstates exist in the reference system and thus any local perturbation of the reference system must decay exponentially. This means that the Green function $G^r(\mathbf{r}, \mathbf{r}', E)$ and the structural Green-function matrix elements $G_{LL'}^{r,nn'}(E)$ decrease exponentially with increasing distance $\Delta r = |\mathbf{r} - \mathbf{r}'|$ or $\Delta R = |\mathbf{R}^n - \mathbf{R}^{n'}|$. The exponential decay was verified by numerical calculations [12,15] which showed that the matrix

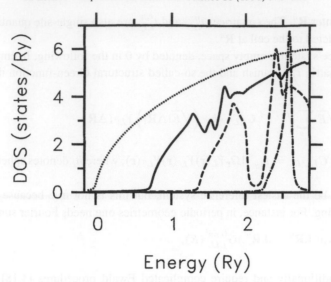

Fig. 1. Density-of-states (DOS) for a fcc reference system. The height of the repulsive potentials is chosen as 0, 1, 2, 4 Ry and the corresponding DOS are plotted with dotted, solid, dashed and dash dotted curves. The Brillouin sampling was done with 11 726 special points in the irreducible part and smoothed by a temperatute broadening with $T = 800$ K.

elements $G_{LL'}^{r,nn'}(E)$ decrease by a factor of about 10 000, if the distance ΔR is increased by the amount of a lattice constant. Because of this fast decrease it was speculated [12,15] that the matrix elements can be neglected beyond a rather short coupling range. This speculation was verified by density-of-states calculations for empty space [16,17], by total-energy calculations for the fcc metals Al, Cu and Pd [16] and by density-of-states calculations for the Cu surface [17]. Some of these results will be discussed in Section 4.

The short coupling range has important consequences for the numerical effort, both for the evaluation of the structural Green function matrix elements $G_{LL'}^{r,nn'}(E)$ of the reference system and for their use in (3). The matrix elements can be evaluated from the ones of empty space, defined in (4), by an equation, which similar to (3) is given by

$$G_{LL'}^{r,nn'}(E) = G_{LL'}^{0,nn'}(E) + \sum_{n''}\sum_{L''} G_{LL''}^{0,nn''}(E) t_{l''}^{r,n''}(E) G_{L''L'}^{r,n''n'}(E). \tag{6}$$

Here G^0 plays the role of the reference Green function and G^r is to be calculated. Because of the exponential decay of $G_{LL'}^{r,nn'}(E)$ the sum over n'' can be restricted to a finite number of centers around n'. Thus (6) can be solved in real space for each center separately. This task is ideally suited for parallel computing since the double loop over the individual atoms and over the different values of E can be completely distributed over the computing nodes.

The subsequent evaluation of (3) is simplified by the sparse matrix structure of $G_{LL'}^{r,nn'}(E)$, which is schematically given by

$$\begin{bmatrix} x & x & 0 & \cdot & 0 & 0 & x \\ x & x & x & \cdot & 0 & 0 & 0 \\ 0 & x & x & \cdot & 0 & 0 & 0 \\ \cdot & \cdot & \cdot & \cdot & \cdot & \cdot & \cdot \\ 0 & 0 & 0 & \cdot & x & x & 0 \\ 0 & 0 & 0 & \cdot & x & x & x \\ x & 0 & 0 & \cdot & 0 & x & x \end{bmatrix}, \tag{7}$$

where each x denotes a nonvanishing matrix block. For systems, which are periodic in two dimensions like surfaces, interfaces or multilayers, a two-dimensional lattice Fourier transform can be applied to (3) and the blocks in (7) are matrices in the angular-momentum indices. For general three-dimensional systems, each block x has again the structure shown in (7) with blocks x', which again have the same structure with blocks x'', which are matrices in the angular-momentum indices.

For surfaces, interfaces and multilayers the almost block tridiagonal structure (7) can efficiently be treated [17]. The general case is more difficult, existing algorithms for general sparse matrices usually do not exploit the regular structure of (7) and are not tailored to obtain the blocks on the diagonal of the inverse of the sparse matrix. These blocks, essentially the on-site ($n = n'$) Green function matrix elements $G_{LL'}^{nn}(E)$, are needed to determine the electronic density, the basic quantity of density-functional calculations. For that, a contour integration with about 20–40 complex energies is applied [21].

In this respect the TB-KKR method also differs from standard TB electronic-structure methods, where a TB Hamiltonian is used and the eigenvalue problem $H\Psi = E\Psi$ must be solved. In the TB-KKR method the diagonal blocks of the inverse of the matrix $M = 1 - G^r(t - t^r)$ are obtained by solving systems of linear equations. Another important difference concerns the coupling range. In usual TB methods a short coupling range is assumed in the considered material, whereas in the TB-KKR method the short range of TB parameters $G_{LL'}^{r,nn'}(E)$ is an exact property of the reference system and *not* of the material, for which the electronic-structure calculations are intended. This makes the TB-KKR method equally well suited for insulating, semiconducting and metallic systems.

4. Accuracy

For the numerical accuracy of the TB-KKR method, it must be investigated how many TB parameters $G_{LL'}^{r,nn'}(E)$ must evaluated in (6) and used in (3). Such investigations are described in previous publications [16,17] and showed that finite real space clusters of repulsive potentials can be used to determine the TB parameters. The clusters were constructed around the central site n with a repulsive potential on this site and repulsive potentials on several shells of neighboring sites. Both indices n' and n'' in (6) were restricted to the sites in the cluster and the calculated TB parameters $G_{LL'}^{r,nn'}(E)$ from the central site n to neighboring sites n' were used in (3).

If enough repulsive potentials were used, densities of states could be calculated very well almost up to the energy of the lowest eigenstates of the reference system. Already only 0.1 Ry below this energy the plotted densities of states [16,17] cannot be distinguished from the exact ones obtained by standard KKR calculations. Results for equilibrium total energies, lattice constants and bulk moduli [16] for the fcc metals Al, Cu and Pd are shown in Fig. 2 as functions of the number of repulsive potentials used to determine the TB parameters $G_{LL'}^{r,nn'}(E)$. The convergence with respect to the number of repulsive potentials is always rapid. Typical errors for total energies, lattice constants and bulk moduli are 1 mRy, 0.2 pm and 1 GPa, if 19 potentials on central, nearest and next-nearest neighbor sites are used, and 5 μ Ry, 0.01 pm and 0.05 GPa, if 79 potentials are used. The calculations applied full-potential KKR codes [19] and used an angular-momentum expansion for single-site charge densities and potentials up to $l_{max} = 6$. They were nonrelativistic and the exchange–correlation potential as given by Vosko et al. [20] was used. The $\underline{k}$ point Brillouin sampling was facilitated by a temperature of $T = 800$ K as described in [16,21].

5. Efficiency for large-scale calculations

The main computational tasks of multiple-scattering theory consist in the calculation of single-site quantities like $R_L^n(\mathbf{r}, E)$ and $t_{LL'}^n(E)$ and of the structural Green-function matrix elements $G_{LL'}^{nn'}(E)$. All quantities can be

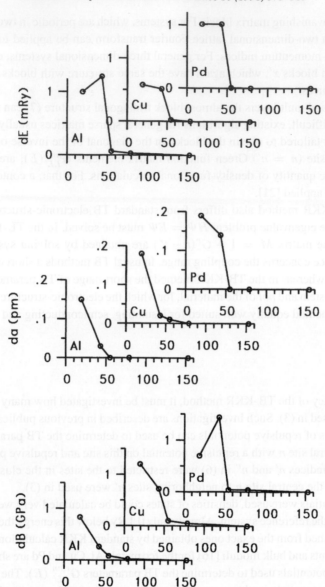

Fig. 2. Deviations dE, da, dB from standard KKR results for total energies (top), lattice constants (middle) and bulk moduli (bottom) for the fcc metals Al, Cu and Pd as function of the number of repulsive potentials used to determine the TB parameters.

calculated in parallel for each energy E. A second parallelization strategy concerns the calculation of the single-site quantities independently for each atom and the distribution of the inversion of the KKR matrix M onto the computing nodes by standard software [24]. The TB-KKR method uses matrices of economic size. With dimensions $16N$, where N denotes the number of inequivalent *atoms*, good quantitative all-electron results are obtained even for transition metals. This matrix size should be compared to the one used in $O(N)$ methods [1], where N denotes the number of *electrons* and where the prefactor is usually larger.

Contrary to the traditional KKR matrix, the TB-KKR matrix is sparse for large systems and the TB-KKR matrix elements can be calculated much easier. Fourier transformations similar to (5) are straightforward, thus bypassing any complicated Ewald sums of the traditional KKR method. The sparsity of the TB-KKR matrix can directly be exploited for systems with two-dimensional periodicity like surfaces. By combining several layers into a principal layer it is achieved that only neighboring principal layers interact in the reference system. As a consequence, the numerical complexity is O(N) for N different and O(log N) for N identical principal layers. Such fast calculations allow to treat complicated surfaces, e.g. high-indexed surfaces with a periodic repetition of surface steps.

6. Usefulness for perturbed systems

Since multiple-scattering method in its standard or TB form provides the full Green function for the considered material, the electronic structure of locally perturbed materials can straightforwardly be calculated by the KKR-Green function technique [9]. Thus the electronic structure for impurities in the bulk, in the surface layers or for adatoms on surfaces can be determined. This is also true for adatoms and substitutional atoms at steps and even at kinks, which can be approximated by missing or additional chains of a few adatoms. For perturbed materials the atoms are displaced from the ideal positions of the underlying periodic lattice, and it is satisfying that the displacements can now be calculated by the KKR method [8]. Reliable forces and displacements around the perturbing atoms are obtained by an extension of the full-potential formalism and by the application of a modified, "ionic" Hellmann–Feynman theorem, which allows to treat the core electrons in spherical approximation. The treatment of the displaced atoms is based on a reexpansion of the structural Green function matrix around the shifted positions by a matrix transformation $\overline{G} = UGU^{-1}$. The transformation matrix U depends on the shifts $\mathbf{s}^n$ and is given by

$$U_{LL'}^{nn'}(E) = -4\pi\delta^{nn'}\sum_{L''}i^{l+l''-l}C_{LL'L''}j_{l''}\left(\sqrt{E}|\mathbf{s}^n|\right)Y_{L''}(\mathbf{s}^n),\tag{8}$$

where j_l denotes spherical Bessel functions. The reliability of the calculated forces has been checked by calculating them in two different ways, either by the ionic Hellmann–Feynman theorem or by numerically differentiating total energies. It was found that these forces agreed very well, if the high lying core states are carefully treated. For instance, for a Ti impurity in Cu, a spherical, atomic-like treatment of the 3p core states overestimated the forces by 50%. Both the full potential anisotropy and the hybridization with the valence states were found to be important effects for calculating reliable forces. A spherical-potential approximation around the atoms must completely fail as numerical tests have shown [22], which overestimated the forces typically by a factor of 10. The reliability of the calculated forces and resulting displacements, obtained by zero-force condition, can nicely be checked against the experimental information available from extended-X-ray-absorption-fine-structure (EXAFS) measurements for impurities in metals [23]. The measured nearest-neighbor displacements agree very well with the calculated ones [8]. This agreement and the sensitivity of the calculated forces represent a stringent test for the accuracy of the used full-potential KKR formalism.

7. Summary

The traditional multiple-scattering (KKR) method can be transformed into a TB form by the concept of a reference system with repulsive, nonoverlapping, spherical potentials. The TB parameters can be calculated with finite clusters of repulsive potentials and used for accurate density-functional calculations. Contrary to the traditional KKR matrix, the TB-KKR matrix is sparse for large systems and can be easily calculated. The sparsity can be exploited and leads to O(N) algorithms for surface calculations which are expected to be of great use for complicated surfaces and surface steps. The TB-KKR method allows to determine the Green function of the considered materials and thus the treatment of local perturbations. Recent improvements in full-potential multiple-scattering calculations show that reliable atomic forces and resulting geometric atomic arrangements can be calculated.

Acknowledgements

The work has benefited from collaborations within the HCM Network Contract No. ERBCHRXCT930369 and the TMR Network Contract No. EMRX-CT96-0089.

References

[1] M. Krajci and J. Hafner, Phys. Rev. Lett. 74 (1995) 5100; Y. Wang, G.M. Stocks, W.A. Shelton, D.M.C. Nicholson, Z. Szotek, and W. M. Temmerman, Phys. Rev. Lett. 75 (1995) 2867; I.A. Abrikosov, A.M.N. Niklasson, S.I. Simak, B. Johansson, A.V. Ruban and H.L. Skriver, Phys. Rev. Lett. 76 (1996) 4203; S. Wei and M.Y. Chou, Phys. Rev. Lett. 76 (1996) 2650. T. Zhu, W. Pan and W. Yang, Phys. Rev. B 53 (1996) 12 713; S. Goedecker and L. Colombo, Phys. Rev. Lett. 73 (1994) 122; A.F. Voter, J.D. Kress and R.N. Silver, Phys. Rev. B 53 (1996) 12 733; A.P. Horsfield, A.M. Bratkovsky, M. Fearn, D.G. Pettifor and M. Aoki, Phys. Rev. B 53 (1996) 12 694; S. Itoh, P. Ordejon, D.A. Drabold and R.M. Martin, Phys. Rev. B 53 (1996) 2132; E. Hernandez, M.J. Gillan and C.M. Goringe, Phys. Rev. B 53 (1996) 7147; A. Canning, G. Galli, F. Mauri, A. De Vita and R. Car, Comp. Phys. Commun. 94 (1996) 89.
[2] Lord Rayleigh, Philos. Mag. 34 (1892) 481.
[3] N.P. Kasterin, Diss. Moscow (1903).
[4] P.P. Ewald, Ann. Physik 49 (1916) 1; 49 (1916) 117; 64 (1921) 253;
[5] J. Korringa, Physica 13 (1947) 392; W. Kohn and N. Rostoker, Phys. Rev. 94 (1954) 1111.
[6] R. Zeller, J. Phys. C 20 (1987) 2347.
[7] R.G. Newton, Phys. Rev. Lett. 65 (1990) 2031; J. Math. Phys. 33 (1992) 44; W.H. Butler, A. Gonis and X.-G. Zhang, Phys. Rev. B 45 (1992) 11 527; 48 (1993) 2118; S. Bei der Kellen, Y. Oh, E. Badralexe and A.J. Freeman, Phys. Rev. B 51 (1995) 9560.
[8] N. Papanikolaou, R. Zeller, P.H. Dederichs and N. Stefanou, Phys. Rev. B 55 (1997) 4157.
[9] R. Zeller and P. H. Dederichs, Phys. Rev. Lett. 42 (1979) 1713; P.J. Braspenning, R. Zeller, A. Lodder and P.H. Dederichs, Phys. Rev. B 29 (1984) 703.
[10] K. Wildberger, V.S. Stepanyuk, P. Lang, R. Zeller and P.H. Dederichs, Phys. Rev. Lett. 75 (1995) 509.
[11] O.K. Andersen and O. Jepsen, Phys. Rev. Lett. 53 (1984) 2571.
[12] O.K. Andersen, A.V. Postnikov and S. Yu Savrasov, in: Application of Multiple Scattering Theory to Materials Science, eds. W.H. Butler, P.H. Dederichs, A. Gonis and R.L. Weaver, MRS Symposia Proceedings No. 253 (Materials Research Society, Pittsburgh, 1992).
[13] L. Szunyogh, B. Újfalussy, P. Weinberger and J. Kollár, Phys. Rev. B 49 (1994) 2721.
[14] O.K. Andersen, O. Jepsen and G. Krier, in: Lectures on Methods of Electronic Structure Calculations, eds. V. Kumar, O.K. Andersen and A. Mookerjee (World Scientific, Singapore, 1994).
[15] R. Zeller, P.H. Dederichs, B. Újfalussy, L. Szunyogh and P. Weinberger, Phys. Rev. B 52 (1995) 8807.
[16] R. Zeller, Phys. Rev. B 55 (1997) 9400.
[17] K. Wildberger, R. Zeller and P.H. Dederichs, Phys. Rev. B 55 (1997) 10 057.
[18] F.S. Ham and B. Segall, Phys. Rev. 124 (1961) 1786.
[19] B. Drittler, M. Weinert, R. Zeller and P.H. Dederichs, Solid State Comm. 79 (1991) 31.
[20] S.H. Vosko, L. Wilk and M. Nusair, Can. J. Phys. 58 (1980) 1200.
[21] K. Wildberger, P. Lang, R. Zeller and P.H. Dederichs, Phys. Rev. B 52 (1995) 11 502.
[22] K. Abraham, B. Drittler, R. Zeller and P.H. Dederichs, KFA-JÜL-Report, Jül-2451 (1991).
[23] U. Scheuer and B. Lengeler, Phys. Rev. B 44 (1991) 9883.
[24] J. Choi, J. Demmel, I. Dhillon, J. Dongarra, S. Ostrouchov, A. Petitet, K. Stanley, D. Walker and R.C. Whaley, Comp. Phys. Comm. 97 (1996) 1.

ELSEVIER

Computational Materials Science 10 (1998) 381–387

COMPUTATIONAL
MATERIALS
SCIENCE

LDA modeling of the electronic structure of the halogen-bridged transition-metal chain compounds

M. Alouani [a,b,*], R.C. Albers [c], J.M. Wills [c], J.W. Wilkins [b]

[a] IPCMS, Université Louis Pasteur, 23 rue du Loess, Strasbourg, France
[b] Department of Physics, The Ohio State University, Columbus, OH 43210-1368, USA
[c] Theoretical Division, Los Alamos National Laboratory, Los Alamos, NM 87545, USA

Abstract

The electronic-structure for the halogen-bridged transition-metal (MX) linear chain compound $Pt_2X_6(NH_3)_4$ (X = Cl, Br, I) under uniaxial stress is studied using the full-potential linear muffin-tin method within the local density approximation. We have found that the ground state of the full structure (with the ammonia ligands) is insulating with a strong lattice distortion in good agreement with experiment. This highlights the importance of the ligand structure to the stability of MX systems and to its ground state nature. The amplitude of the charge density wave in these systems as a function of the metal–metal distance produced a linear shift of the core level energies of the two non-equivalent metal sites. This core level energy shift could be checked experimentally using X-ray absorption spectroscopy. The band structure along the chain is dominated by a strong coupling ($pd\sigma$) between $Pt(d_{3z^2-r^2})$ and $X(p_z)$ orbitals, which causes the system to be of a quasi-one-dimensional character. The relevant band structure as a function of the metal–metal distance is nicely fitted to a two band tight-binding Su–Schrieffer–Heeger Hamiltonian with a single set of parameters. This model revealed that the mechanism of the dimerization is a consequence of a competition between a strong electron–phonon coupling along the chain and a repulsive anharmonic potential. © 1998 Elsevier Science B.V.

1. Introduction

The linear halogen-bridged transition-metal chains are quasi-one-dimensional insulating compounds [1,2] and have a very rich variety of physical properties. In particular, they exhibit (i) mixed valence, and strong electron–electron and electron–phonon interactions; (ii) competing charge–density–wave, and spin–density–wave, which are sensitive to tuning by chemical substitutions, pressure, or doping [1–11]; (iii) interesting nonlinear excitations such as po-

larons, kinks, and excitons which are produced by introducing carriers into the lattice [8,9].

All of these fascinating properties make these compounds a fruitful class of materials for detailed study. Most of the theoretical effort to date [4,8,9] has involved one-band or two-band Peierls–Hubbard model Hamiltonian calculations to understand their properties. The parameters used in these models are usually extracted from experiment data. In this paper, we present electronic-structure calculations in the local density approximation (LDA) to attempt to unravel some of the details relating electronic structure and physical structure, which we hope will

---

* Corresponding author.

provide some guidance to the more phenomenological modeling. Because of strong electron–phonon coupling and chemical substitution effects, this connection between the electronic and physical structure is particularly strong for this class of materials. We have also looked at the mechanism of the lattice dimerization and the insulating ground state in terms of the basic interactions in the MX chains. We will focus on the specific MX chain compound: $Pt_2Br_6(NH_3)_4$ (called PtBr for simplification). This is the simplest material with the smallest number of atoms in the unit cell, which makes it easiest to tackle from a first-principles point of view. A short version of this paper was published elsewhere [12–14]. In Section 2 we briefly introduce our method of calculation; in Section 3 we discuss the physical structure of the neutral MX chain compounds. In Section 4 we present and analyze the band structure of PtBr and discuss the interplay between the physical structure and the electronic structure, whereas in Section 5 we discuss the dimerization effect as a function of the metal–metal distance and analyze it using a two band Su–Schrieffer–Heeger model. We will also calculate the phonon energies involved in PtBr bonds, and in Section 6 we present our summaries and conclusions.

2. Method of calculation

The electronic structure of the MX systems has been produced by an all-electron self-consistent, scalar-relativistic full-potential linear-muffin-tin orbitals (FP-LMTO) method [15]. In the FP-LMTO method the charge density in the interstitial region is evaluated correctly by transforming the wave functions in the interstitial region into a plane wave expansion using fast Fourier-transforms. To make the FP-LMTO basis set complete we used three basis function for each angular momentum. Each basis function has a different kinetic energy κ^2 in the interstitial region. The three kinetic energies are $\kappa^2 = -27.2, -13.6,$ and -0.14 eV. The Br 4s state is also used as a valence state because it is found to extend beyond the Br muffin-tin sphere. The core-electron charge density is allowed to relax, i.e., it is recalculated at each iteration in the self-consistent loop. The Brillouin zone integration uses a Gaussian sampling with 16 **k** points in the irreducible part of

the Brillouin (IBZ) for PtBr with the ammonia ligands. We use the exchange and correlation potential and energy in the Von Barth and Hedin approximation [16].

3. Physical structure

The most characteristic feature of this class of materials is their linear chain structure. Indeed, these materials are often called MX materials because of their backbone of alternating transition-metal atoms (M) and halogen atoms (X) along the chains; we will not consider the related MMX class of materials. Surrounding the linear chains is a ligand structure of various complexity. The whole matrix forms a three-dimensional crystal with the chains aligned parallel to each other. Typical transition-metal atoms (M) are Ni, Pd, and Pt; typical halogen atoms (X), Cl, Br, and I. Depending upon M, X, and the ligand structure (and pressure or doping), these materials have a variety of ground states. For example, PtX compounds with X = I or Br have a weak charge disproportionation and lattice distortion (Peierls distortion) [2]. For X = Cl the ground state exhibits a large-amplitude dimerization of the Cl sublattice, producing a strong charge-density wave distortion that increases the optical gap [8,9]. If we substitute Ni for Pt, then these compounds have an insulating and antiferromagnetic or spin-density wave ground state. Depending on the Ni–Ni distance the compound dimerizes or not [5–7].

Besides the ground state properties, another useful classification category for the MX compounds is the distinction between charged and neutral chains. Most of the experimental work has been done on the charged chains, which have a net charge on each chain and, to maintain overall charge neutrality, have oppositely charged counter ions in between the chains. The PtBr system that we explore in this paper is the other type of chain: a neutral chain. Each isolated chain is charge neutral, and no counter ions are required or present.

The currently available crystallographic information for the system that we are directly interested with in this paper, i.e., $Pt_2Br_6(NH_3)_4$, is incomplete, since the relative phases between the different chains is not available [3]. In this superimposed crystal structure, the X-ray data do not distinguish between the two valence states of the Pt atoms, which form in a body-centered orthorhombic structure with lattice parameters ($a = 8.23$ Å, $b = 7.76$ Å, $c = 5.55$ Å). For the calculations presented in this paper, we have left this Pt crystal structure as given by experiment, and have then simply moved the Br atom along the chain in order to find its stable position between Pt atoms. We have also used the experimentally determined distances between the Pt atoms and the *ligands* in the dimerized structure: Pt–Br = 2.43 Å and Pt–N = 2.06 Å. The H atoms in the NH_3 ligands are too light to show up in the X-ray data; we have assumed that they are the usual N–H = 1.11 Å of the ammonia molecule. In the geometry optimization studies to determine the optimum dimerization, the Pt-ligand distances were frozen at their experimental values.

4. Electronic structure

In this section we present the semi-relativistic band structure of PtBr. Because Pt is a fairly heavy

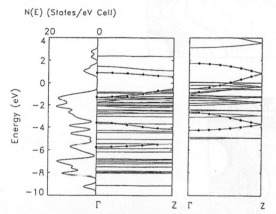

N(E) (States/eV Cell)

Fig. 1. Band structure along the chain direction ΓZ of $Pt_2Br_6(NH_3)_4$ (left plot) and of Pt_2Br_6 (right plot). The Fermi level is at the zero of the energy scale. The total density of states, $N(E)$, of $Pt_2Br_6(NH_3)_4$ is also included in the left plot; it shows an energy band gap is about 1.0 eV as compared to 1.5 eV given by experiment. The bands with most Pt $d_{3z^2-r^2}$ and Br p_z character are presented with full dots. Without the ammonia ligands (right plot) non-bonding bands appear at Fermi level, and destroy the insulating ground state and the dimerization (the extra bands in the left plot are due to the N–H bonding not included in the right plot). The least-squares fit of the two-band SSH Hamiltonian (full-thick curve in the left plot) to the full structure Pt $d_{3z^2-r^2}$ and Br p_z is excellent. This suggests that the dimerization is caused by an electron-phonon coupling along the chain.

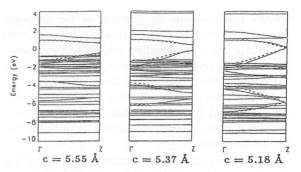

Fig. 2. Calculated LDA band structure along the chain direction ΓZ of $Pt_2Br_6(NH_3)_4$ for different metal–metal distances of 5.55, 5.37, and 5.18 Å. The most relevant bands with Pt $d_{3z^2-r^2}$ and Br p_z character for different Pt–Pt distances are nicely fitted by a two-band tight-binding SSH model using a single set of parameters. The Fermi level is at the zero of the energy scale.

atom with a spin–orbit splitting of the d states of about 1 eV in the MX chains [17], we would ideally like to have fully relativistic (Dirac equation) calculations for the PtBr system. Because of the large number of atoms in the unit cell, however, this is very expensive in terms of computer time (the number of basis functions must be doubled, which leads to an approximately eight-fold increase in running time due to diagonalization costs), and we have restricted the current calculations to be semi-relativistic, i.e., with all relativistic effects except for spin–orbit coupling.

Fig. 1 shows the band structure along the chain direction (ΓZ) for the complete PtBr compound (left plot) and for PtBr without ammonia ligands (right plot). The Fermi level is at the zero of energy scale. The total density of states $N(E)$ is also included; it shows a band gap of 1.0 eV as compared to 1.5 eV given by the experiment. The bands with the most Pt $d_{3z^2-r^2}$ and Br p_z character are presented with full dots. Without the ammonia ligands (right plot) nonbonding bands appear at Fermi level, and destroy the insulating ground state and the dimerization (the extra bands in the left plot, due to the N–H bonding, are not included in the right plot).

We have found that the width of the important one-dimensional bands (Pt $d_{3z^2-r^2}$ and Br p_z) is unchanged by the presence of the ammonia ligands, which suggests that the ammonia ligands are not essential in determining the conducting properties of these important bands. Instead, their main role seems to be limited to removing the non-bonding Pt d orbitals that point in the direction of the ammonia ligands. The ammonia-ligand lone pair interacts with the Pt($d_{x^2-y^2}$, s) nonbonding orbitals to form bonding and anti-bonding molecular orbitals that both move away from the Fermi level. We have also found that the others bands along the chain direction are unaffected by the interchain expansion, which suggest that the interchain coupling is not affecting the dimerization.

Fig. 2 shows the band structure of PtBr along ΓZ direction for three Pt–Pt distances, 5.55, 5.37, and 5.18 Å. The most relevant bands are also fitted to a two-band SSH Hamiltonian (dashed bands) using a single set of parameters $2e_0 = 2.32 \pm 0.03$ eV, $t_0 =$

1.5 ± 0.02 eV, $\alpha = 1.2 \pm 0.1$ eV, $\beta_M = \beta_X = 0$ (see end of Section 5 for details). This good fit together with the total energy fit indicates that the dimerization is caused by an electron–phonon coupling along the chain.

To better visualize the interchain coupling and the role of the ammonia ligands, we have plotted in Fig. 3 the total charge-density contours, in units of milli-electron/(a.u.)^3 in the (001) and (011) planes. These figures show a very small interchain coupling along the (110) plane, but a more pronounced interchain coupling in the (001) plane. The chemical bonding is such that the PtBr is better described as a strongly bonded planes with a relatively weak interplane

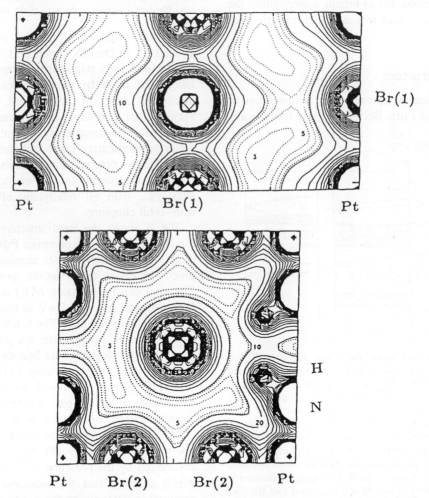

Fig. 3. Calculated total charge-density contours of $Pt_2Br_6(NH_3)_4$ compound along the (001) and (110) planes, in units of millielectrons/(a.u.)^3. The chemical bonding is such that the material is better described as a set of strongly bonded planes with a relatively weak inter-plane bonding through the ammonia molecules.

bonding through the ammonia molecule. This is because the hydrogen atoms repel each other due to the polarizability of the ammonia molecule. Despite the strongly bonded planes, the band structure along the chain seems to be mostly governed by two bonding and antibonding bands formed from the Pt $d_{3z^2-r^2}$ and the Br(1) p_z orbitals. The change of dispersion of these two bands with the ligand Br(2) position is due mainly to a coupling between the $d_{3z^2-r^2}$ and the Br(2) p_x orbitals. Thus the band crossing the Fermi level, which determine the physical properties of the system, is the result of a complex molecular orbital composed of Pt($d_{3z^2-r^2}$)–Br(1)(p_z)–Br(2)(p_x)–N(p_y) character.

5. Dimerization effect and phonon involving Pt–Br bonds

We have studied the dimerization effect in PtBr by minimizing the total energy of the system as a function of the degree of dimerization. In principle, should the system dimerize, the curvature of this function would provide the Pt–Br electron–phonon coupling constant needed for the many-body modeling. Because the relative phases between the chains is not known for the dimerized structure, we have chosen that set of phases which minimizes the number of atoms in the unit cell (c.f., the discussion at the end of Section 3).

We found that *all the ligands were necessary* to generate the insulating dimerized ground state with the large lattice distortion that is observed experimentally. Figure Fig. 4 shows a surface plot of the LDA calculated total energy as a function of Br dimerization and the Pt–Pt distance. The dimerization ζ is defined as the ratio of the off-center distance of the Br atom away from its symmetric position (midway between the Pt atoms) to the Pt–Pt distance. The minimum of the total energy occurs at a dimerization of $\sim 4\%$ which is close to the values of 4.8% and 5.2% found experimentally [2,3]. The Pt–Pt lattice parameter of PtBr without ammonia ligand is found to be 10% smaller than the experimental value. It is surprising that the LDA calculated Pt–Pt lattice parameter with all the ligands is 3% larger than the experimental value despite the fact that LDA has an overbinding tendency. If the experimental lattice parameter is correct, this suggests

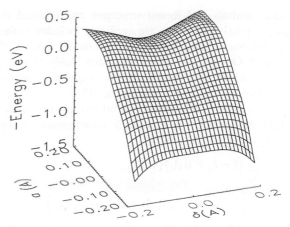

Fig. 4. Three dimensional plot of the LDA total energy of PtBr (the energy surface is actually an excellent cubic spline fit to the LDA data) as a function of dimerization ratio in (%) and Pt–Pt a distance (the experimental Pt–Pt distance 5.55 Å is the zero of the lattice scale). The Br dimerization is about 4% and the condensation energy is 0.1 eV. The lattice parameter a is 3% larger than the experimental value.

either that the MX compounds have a more localized electronic structure than the one predicted by LDA or that the atomic relaxations and the optimization of the transversal lattice parameters are important for the determination of the structural properties.

We have used the curvature of the total energy versus lattice dimerization at the minimum energy to calculate the effective phonon energy of a Raman-active breathing mode as a function of the Pt–Pt lattice parameter [12–14]. We have also calculated the symmetric optical phonon which involves the Pt and Br(2) ligand and have found a frequency of 21.3 meV (the experimental value of this phonon frequency is not yet known) and a Pt–Br(2) equilibrium ligand distance of about 2.35 Å, which is 4% smaller than the experimental value.

To analyze the mechanism of the lattice distortion and the origin of the band gap we have folded down the relevant band structure in terms of a two-band single-chain SSH model used to describe the electronic structure of defects in the charged MX chains [8,9]. In this model, both the hopping parameter and the on-site energy levels depend linearly on the degree of dimerization through the electron–phonon couplings α and β. Only a single $d_{3z^2-r^2}$ orbital on the Pt sites and a single p_z orbital on the Br sites along the chain are considered important. All the other orbitals and ligand structures are assumed to play a passive role, and are implicitly included in the anharmonic elastic energy of the model (see $K^{(j)}$ below). On-site and nearest-neighbor Hubbard terms can be added as needed for a more explicit treatment of electron–electron correlation. We have fit our LDA results to a slightly generalized version of the SSH model for the electronic contribution to the energy that is given by

$$H = \sum_{l\sigma} \{(-t_0 + \alpha\Delta_l)(c_{l\sigma}^+ c_{l+1\sigma} + h.c.)$$

$$+ \left[(-1)^l e_0 - \beta_l(\Delta_l + \Delta_{l-1})\right] c_{l\sigma}^+ c_{l\sigma}\}$$

$$+ \sum_{l,j=0,4} K^{(j)} \Delta_l^{(j)}. \tag{1}$$

Here $c_{l,\sigma}^+$ ($c_{l,\sigma}$) creates (annihilates) an electron at site l with spin σ, and where M and X occupy even and odd sites, respectively. The parameters α, $\beta_{2l} = \beta_M$, and $\beta_{2l+1} = \beta_X$ are electron–phonon parameters that couple the atomic positions to the electronic structure through $\Delta_l = x_{l+1} - x_l$, where x_l is the deviation of the lth atom from its positions in the symmetric structure. The zero of energy is chosen to be midway between on-site energy levels of the metal and halogen atoms for the undimerized structure. The M–X spring coefficients $K^{(j)}$ model the anharmonic repulsive potential between a halogen atom and a metal atom.

We have done a least-squares fit of Eq. (1) to the four LDA bands that have the most Pt $d_{3z^2-r^2}$ and Br p_z character along the chain direction (ΓZ), and we have found that nearest-neighbor interactions provide a reasonably good representation of these LDA bands. The fit provides values for e_0, t_0, α, β, and we have found that a single set of parameters fits nicely the relevant band structure along the chain for all the M–M distances (see Fig. 2). To extract the elastic coefficients $K^{(j)}$, we fitted the total energy of the SSH model total energy to the LDA band structure. Again a single set of $K^{(j)}$ is required to fit the total energy of PtBr under uniaxial stress. We have found $K^{(0)} = 0.05$ eV, $K^{(1)} = \_10.13$ eV/Å, $K^{(2)} = 8.10$ eV/Å^2, $K^{(3)} = \_5.35$ eV/Å^3, $K^{(4)} = 34.65$ eV/Å^4 [12–14].

Fig. 5 shows the energy shift of all the core states of Pt(II) with respect to Pt(IV) as a function of the Pt–Pt distance (dot points). The thick line is a linear

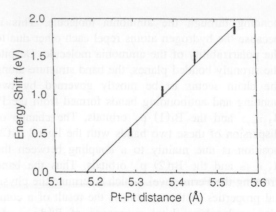

Fig. 5. Energy shift of the core states of Pt(II) with respect to Pt(IV) as a function of the Pt–Pt distance (thick dots). The linear curve is a least squares fit to the data. Pt(IV) core levels move downwards because Pt(IV) loses electrons and Pt(II) core levels move upwards because Pt(II) gains electrons. The change in the energy shift with respect to the Pt–Pt distance is linear. The shift is zero for a metal–metal distance where the CDW vanishes. This energy core shift could be observed experimentally by means of X-ray absorption spectroscopy.

least-squares fit to the data. Pt(IV) core levels move downwards because Pt(IV) loses electrons and Pt(II) core levels move upwards because Pt(II) gains electrons. The energy shift is proportional to the charge disproportionation between Pt(II) and Pt(IV) and vary linearly with respect to the Pt–Pt distance. The shift is zero for a metal–metal distance where the CDW vanishes. This energy core shift is an indirect signature of the CDW phase in PtBr and could in principle be observed experimentally by means of X-ray absorption at the Pt(II) and Pt(IV) sites.

6. Conclusions

We have found that LDA correctly predicts dimerization for the MX chain compound $Pt_2Br_6(NH_3)_4$ and produces the right dimerization. The bands at the Fermi energy dominates the physics of this system and allows the electron–phonon coupling to be effective in causing a dimerized and insulating ground state. The LDA electronic structure folds down nicely into a modified two-band SSH model which allows us to conclude that the mechanism of the dimerization is dominated by an electron–phonon coupling along the chain.

We have found that the ligand structure surrounding the chains play a crucial role in determining the electronic properties: The dimerization and the insulating ground state occur only when all the ligand structure is included. Without the ligands our calculations show the presence of nonbonding Pt orbitals at the Fermi energy that prevents dimerization and produces the wrong ground state.

Acknowledgements

We thank A.R. Bishop and his group for interesting discussions. This research was supported in part by the U.S. Department of Energy Basic Energy Sciences, Division of Materials Sciences and by NSF, grant number DMR-9520319. This research was conducted using the resources of the Cornell Theory Center, which receives major funding from the National Science Foundation (NSF) and New York State, with additional support from the Advanced Research Projects (ARPA), the National Center for Research Resources at the National Institutes of Health (NIH), IBM Corporation, and other members of the center's Corporate Partnership Program. We also acknowledge computer time from the Ohio State Supercomputer.

References

[1] H.J. Keller, in: J.S. Miller (Ed.), Extended Linear Chain Compounds, vol. 1, Plenum Press, New York, 1982, p. 357.
[2] R.J.H. Clark, R.E. Hester, in: Infrared and Raman Spectroscopy, vol. 11, Wiley Heyden, NY, 1984, p. 95.
[3] H.J. Keller et al., Acta Cryst. B 37 (1981) 674.
[4] D. Baeriswyl, A.R. Bishop, J. Phys. C 21 (1988) 339.
[5] K. Toriumi, Y. Wada, T. Mitani, S. Bandow, M. Yamashita, Y. Fujii, J. Am. Chem. Soc. 111 (1989) 2341.
[6] K. Toriumi et al., Mol. Cryst. Liq. Cryst. 181 (1990) 333.
[7] H. Okamoto, K. Toriumi, T. Mitami, M. Yamashita, Phys. Rev. B 42 (1990) 10381.
[8] J.T. Gammel et al., Phys. Rev. B 45 (1992) 6408, and references therein.
[9] X.Z. Huang et al., unpublished.
[10] M.H. Wangbo, M.J. Foshee, Inorg. Chem. 20 (1981) 113.
[11] M. Tanaka et al., Chem. Phys. 91 (1984) 257.
[12] M. Alouani, J.W. Wilkins, R.C. Albers, J.W. Wills, Phys. Rev. Lett. 71 (1993) 1415.
[13] M. Alouani, J.W. Wilkins, R.C. Albers, J.W. Wills, Phys. Rev. Lett. 73 (1994) 3599.
[14] T. Gammel, Phys. Rev. Lett. 73 (1994) 3598.
[15] J.M. Wills, unpublished.
[16] U. Von Barth, L. Hedin, J. Phys. C 5 (1988) 1629.
[17] R.C. Albers, Synthetic Metals 29 (1989) F169.

ELSEVIER

Computational Materials Science 10 (1998) 388–391

COMPUTATIONAL
MATERIALS
SCIENCE

Total-energy tight-binding modelisations of silicon

O.B.M. Hardouin Duparc *, M. Torrent

Laboratoire des solides irradiés, CEA-CEREM, CNRS URA 1380, Ecole Polytechnique, 91128 Palaiseau, France

Abstract

We present a systematic comparison of several tight-binding total-energy modelisations of silicon. These modelisations include the tight-binding electronic cohesive energy and a phenomenological description of a repulsive term. We examine both the electronic band gap and various structural properties such as structural stability, bulk modulus, elastic constants, four typical vibrational modes and their Grüneisen constants. We then apply these models to the calculations of the minimised energies of some multivariant grain boundaries. We discuss the influence of the choice of cut radii. Copyright © 1998 Elsevier Science B.V.

Keywords: Total energy; Tight binding; Silicon; Atomic simulations

1. Modelisations and preliminary checks

The improvement of the modelisation techniques goes parallel with the increase in computer powers. In the field of atomic simulations, the total-energy tight-binding approximation comes in between ab initio calculations and semi-empirical potential modelisations. The recent development of order-N techniques now allows one to study systems containing a few hundred atoms on a work station, at least for semi-conductors. [1] From that latter point of view, silicon has the tremendous advantage of being both a model material for physicists and the still most widely used element in the semiconducting device industry.

The tight-binding (TB) approximation has a long tradition in the calculation of electronic band structure.

When properly used, it also usually allows for a physically intuitive interpretation of some electronic effects in cases where such an interpretation is not obvious from ab initio calculations.

Strictly speaking, the TB approximations only deal with the way one calculates the electronic cohesive energy. In order to perform total-energy calculations, one must obviously be able to add to the TB band energy all that it misses with respect to a proper total energy. This is usually done in a rather empirical manner and the additional energy then goes under the accepted name of "repulsive energy".

TB band energy calculations may range from orthogonal, valence orbitals, with simple power-law exchange integrals limited to first-nearest neighbours only, to non-orthogonal orbitals, valence and conduction orbitals, with complicated and orbital-dependant exchange integrals slowly cut off by Fermi-like functions. The exchange integrals, and even the "atomic" orbital energies, can also be considered as environment dependant. When charge transfer

---

* Corresponding author. Tel.: +33 1 69 33 36 92; fax: +33 1 69 33 30 22; e-mail: ohd@corot.polytechnique.fr.

[1] See, e.g. [1]. Since then, several other N-order techniques have been proposed.

Table 1
Some model features (these models are described in details (although sometimes not completely, see text) in their original articles)

	Conduction orbitals	Neighbours cutting off	Repulsive energy
Cha-M (1979)	No	First	Simple power expansion, pairwise
GSP1-M (1983)	No	First	Functional form, pairwise
SawK-M (1990)	No	Third smooth	Functional form, environment-dependant
THD-M (1997)	Yes (s*)	Second smooth	Functional form, environment-dependant

s* refers to Vogl et al.'s way to describe at least the bottom edge of the conduction bands [8].

occurs, in alloys for instance, it must be treated self-consistently.

Various ranges of sophistication also exist to treat the additional "repulsive energy" term, from simple pairwise power-law functions limited to first neighbours, to complicated more long ranged and environment functionals.

There is a risk of inflation in the number of adjustable parameters and in the complexity of the functional forms. If not for aesthetical reasons, it is highly desirable that these latters remain analytically derivable if one wants to carry out molecular dynamics simulations.

But complexity is probably a part of the price to pay for realism in order to be able to treat real materials containing extended defects like grain boundaries for instance.

With these criteria in mind, we have selected four total-energy tight-binding models. The first two, namely the Chadi model "Cha-M" (1979) [2,3] and the Goodwin–Skinner–Pettifor model "GSP-M" (1983), [4] have been chosen for their pioneering values. The last two are more modern and realistic. These are the Sawada–Kohyama model "SawK-M" (1990,1991), [5,6] and the Torrent–Hardouin Duparc model "THD-M" (1997), [7]. The THD-M model has been designed to improve on the SawK model which still exhibit some weaknesses as will be seen below.

We cannot describe here in detail the four models and we have simply summarised in Table 1 a few features which we consider as specially important. They all consider the atomic orbitals as orthogonal. Only THD-M considers a "conduction band orbital" s* [8]. Except for Cha-M, they enforce local charge neutrality self-consistently to some extent via a Hubbard term the ponderation of which varies from $1\,eV/e^-$ (GSP-

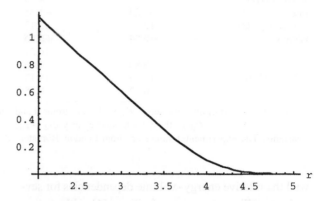

Fig. 1. GSP scaling function $f(r_0/r) = (r_0/r) \exp[-(r/r_c)^{n_c}]$. $r_c = 3.67\,\text{Å}$, $n_c = 6.48$, $r_0 =$ first n–n distance in diamond Si, $\sim 2.3517\,\text{Å}$.

M) to $10\,eV/e^-$ on the ϵ_s "atomic" level (THD-M). Chadi's model is restricted to the four nearest neighbour models of the diamond structure. Although the interactions of the GSP model have been devised with a rescaling "smooth step function, with the step positioned midway between first and second nearest neighbours in diamond silicon", the actual choice does not look like a step function as can be seen in Fig. 1. The repusive terms for SawK-M and THD-M are both distance- and coordinence-dependant, an idea which is akin to Tersoff's [9] phenomenological description of silicon.

We cannot either describe here all the physical properties one can get with these models and we restrict ourselves to just a few which are given in Table 2.

We shall now comment the results with respect to the various assumptions of the models. The property values indicated in Table 2 for GSP-M have been obtained with a sharp cut-off taken between the first and the second neighbours. Within this limit the GSP model proves rather successful. It also describes pretty

Table 2
Some physical properties

	Chadi-M	GSP1-M	SawK-M	THD-M	Exptal
a (Å)	5.44	5.43	5.43	5.43	5.43
E_g (eV)	1.2	0.8	2.1	1.15	1.16
B (GPa)	98	101	92.5	98	98
c_{11} (GPa)	146	145.1	164.5	167	165.7
c_{12} (GPa)	74	79.5	56.5	63.8	63.9
c_{44} (GPa)	64	67.2	104	80.4	79.6
C_d (GPa)	+10	+12.3	−47.6	−16.6	−15.7
TA(X) (THz)	5	4.49	6.94	6.1	4.49
$\gamma_{TA(X)}$	−1.25	−1.35	+0.01	−1.04	−1.4
LOA(X) (THz)	12.7	13.17	10.93	10.72	12.32
$\gamma_{LOA(X)}$	−0.54	+0.95	1.27	1.1	0.9
TO(X) (THz)	15.24	16.37	15.52	13.7	13.9
$\gamma_{TO(X)}$	−0.80	1.16	1.7	1.98	1.5
TO(Γ) (THz)	17.15	18.13	16.66	15.95	15.53
$\gamma_{TO(\Gamma)}$	−0.72	0.95	1.37	1.48	0.98

a is the equilibrium parameter, E_g is the electronic band gap, B the bulk modulus and c_{ij} are the cubic elastic constants, $B = (c_{11} + 2c_{12})/3$, C_d is the Cauchy discrepancy $c_{12} - c_{44}$, then comes four vibrational modes together with their Grüneisen constants. The experimental values are from Landolt–Börnstein III/17a [10].

well the relative energy–volume dependencies for several crystalline structures of silicon [4] with respect to the ab initio calculations by Yin and Cohen [11]. It constitutes a pioneering work from that point of view with respect to Cha-M. However, simply going up to the third neighbours raises the equilibrium lattice parameter for the diamond structure by 2%, to 5.54 Å, and the bulk modulus by 11%, to 111 GPa (not to mention the band structure which becomes semimetallic). This had already been noted by several authors, e.g. [6]. Note that the recent model of Mercer and Chou [12] is also limited to first neighbours. Although the concept of first neighbours can be defined for a perfect crystalline structure, it fails in the case of disordered structures like complex grain boundaries or liquids. The choice of an abrupt cut-off function is also problematic for such systems since it gives rise to artificially strong forces. This unfortunately limits the actual transferability of such potentials to noncrystalline structures. Even the sharpness and the centring of a Fermi cut-off function is a delicate matter and constitues a part of the fitting procedure.

The SawK-M does not give too good a c_{12} or c_{44} and fails to reproduce a negative Grüneisen parameter associated to the transverse acoustic mode but it performs rather well in the energy–volume dependencies

of the crystalline non-diamond silicon structures [6]. Its main drawback is that its electronic band gap, although correctly indirect, is almost twice too large. The THD-M still fails to correctly reproduce the low value of the TA(X) mode.

2. Application to grain-boundary calculations

We have also applied these techniques to study several symmetric tilt grain boundaries in silicon [7]. We use the real space order-N density matrix technique [1] to optimise the structures and we eventually carry out one standard reciprocal space calculation to get the excess interfacial energies and check band structures. We use a cut-off of (at least) 6 Å for the density matrix (is this a hint that tight-binding interactions are not so short ranged?). Möbius (anti-periodic) boundary conditions [13] are directly used in real space and indirectly used in reciprocal space to half the sampling of the Brillouin zone. Depending on the complexity of the structure, these Möbius boxes contain from 60 to 175 atoms so that the (anti-) periodic grain boundaries are separated by almost twice their structure-period length. We present in Table 3 some comparative results where we have also included the

Table 3
Excess interfacial energies of various symmetric tilt grain boundaries (see text) in mJ/m^2

	SW-M	T(C)-M	Cha-M	GSP-M		SawK-M	THD-M
				3.5 Å	5 Å		
$\Sigma13$ IH$^+$	885	906	660	825	940	1115	985
$\Sigma13$ IH$^\pm$	927	956	875	1005	990	1200	1055
$\Sigma13$ IV$^+$	967	975	650	895	1045	1165	1020
$\Sigma25$ IH$^+$	836	854	660	765	925	1090	980

classical Stillinger–Weber model "SW-M" [14] and the Tersoff Si(C) model "T(C)-M" [15]. The grain boundaries are the $\Sigma = 13$ [0 0 1] (5 1 0) $\theta = 22.62°$ and the $\Sigma = 25$ [0 0 1] (7 1 0) $\theta = 16.26°$ which have been observed in high resolution electron microscopy by Rouvière and Bourret [16]. These authors have built several model structures and have proposed that the real GBs are combinations of these structures. This has been substantiated by classical molecular dynamics simulations [17]. We only present in Table 3 the results for some model structures for the sake of simplicity and since we only aim at comparing our various models.

Besides the fact that the Cha-M model gives results rather at odds with the other models, it is interesting to note the drastic influence of the cut-off radius choice for the GSP model. The two most recent models give the same ranking but still disagree on the absolute values by about 10%. This reflects the empirical nature of the parametrisation of the tight binding.

3. Conclusion

Tight-binding modelisations are hybrid in the sense that they are both quantum mechanical and empirical. It is clear however that "ab initio" (by the way, is there really such a divine knowledge?) minimisations to get realistic excess interfacial energies, on not too small periodic boxes, for all the possible model structures of the complex grain boundaries considered for instance in this article is still out of reach. Realistic tight-binding modelisations thus still remain a worth

way to study complex systems both for their atomic and electronic structure.

Acknowledgements

One of us (OHD) spent two months at the MML (Oxford) to learn about the Vanderbilt density matrix technique.

References

[1] X.-P. Li, R.W. Nunes and D. Vanderbilt, Phys. Rev. B 47 (1994) 10891.
[2] D.J. Chadi, J. Vac. Sci. Technol. 16 (1979) 1290.
[3] D.J. Chadi, Phys. Rev. B 29 (1984) 785.
[4] L. Goodwin, A.J. Skinner and D.G. Pettifor, Europhys. Lett. 9 (1989) 701.
[5] S. Sawada, Vacuum 41 (1990) 612.
[6] M. Kohyama, J. Phys.: Condens. Matter 3 (1991) 2193.
[7] M. Torrent, Thesis, University of Paris VI (1996).
[8] P. Vogl, H.P. Hjalmarson and D. Dow, J. Phys. Chem. Solids 44 (1983) 365.
[9] J. Tersoff, Phys. Rev. Lett. 59 (1986) 632.
[10] Landolt-Börnstein, III/17a (Springer, Berlin, 1984).
[11] M.T. Yin and M.L. Cohen, Phys. Rev. B 26 (1982) 5668.
[12] J.L. Mercer and Chou, Phys. Rev. B 47 (1993) 9366.
[13] O.B.M. Hardouin Duparc and M. Torrent, Interface Sci. 2 (1994) 7.
[14] F.H Stillinger and T.A. Weber, Phys. Rev. B 31 (1985) 5262.
[15] J. Tersoff, Phys. Rev. B 38 (1988) 6991.
[16] A. Bourret and J.L. Rouvière, in: Polycrystalline Semiconductors: Grain Boundaries and Interfaces, eds. H.J. Möller, H.P. Strunk and J.H. Werner (Springer, Berlin, 1989) pp. 8 and 19.
[17] O.B.M. Hardouin Duparc and M. Torrent, Mater. Sci. Forum 207–209 (1996) 221.

COMPUTATIONAL
MATERIALS
SCIENCE

ELSEVIER

Computational Materials Science 10 (1998) 392–394

Comparison of two O(N) methods for total-energy semi-empirical tight-binding calculation

J. Chen, A. Béré, A. Hairie, G. Nouet *, E. Paumier

Laboratoire d'Etudes et de Recherche sur les Matériaux, UPRESA 6004 CNRS, Institut des Sciences de la Matière et du Rayonnement, 6 Bd du Maréchal Juin, F-14050 Caen Cedex, France

Abstract

We compare two O(N) methods, to calculate the total energy of a system, based on a local evaluation of the density matrix (DM) via the Lanczos–Haydock recursion scheme. We have recently introduced the first method, in which an approximated DM is directly computed from the local tridiagonalized hamiltonian (LTH). In the second method, the DM is obtained from the diagonalized LTH without any further approximation. The use of the first method is restricted to temperatures higher than 1000 K, but there is no limitation for the second one. The methods are compared when applied to a grain boundary in silicon and germanium. Copyright © 1998 Elsevier Science B.V.

Keywords: Tight-binding; Density matrix; O(N) method; Electronic energy; Forces

1. Introduction

Atomic scale modelling in materials is increasingly used, thanks to its ability to predict any relevant property. The main problem is the computation cost of interatomic forces, which has a direct effect on the maximum number of atoms of tractable systems. Interatomic forces are derived from the total potential energy of the system, for which three different computational approaches can be used.

In the first one, the forces are derived from empirical or semi-empirical potentials, with simple analytical form, leading to fast computations. The accuracy is good for some materials, such as noble metals [1], and

not too bad for group IV semiconductors [2] and III–V compounds [3].

The second approach for total energy and force computation is the semi-empirical tight-binding method (SETBM), which is the frame of the present work, and gives a good compromise between computing cost and accuracy [4].

The third method is the local density functional (LDF) method which has a high computing cost and a good accuracy and is hardly usable to study kinetics in large defects such as grain boundaries.

2. Linear scaling methods in the SETBM scheme

In the SETBM scheme, the total energy of the system is the sum of two terms: the electronic energy and repulsive energy. The electronic energy is computed

* Corresponding author. Tel.: +33 2 31 45 26 47; fax: +33 2 31 45 26 60; e-mail: gerard@leriris1.ismra.fr.

via the tight-binding (TB) theory, and the repulsive energy is a simple analytical pair potential. The main computing cost comes from the TB computation, for which efficient algorithms are needed. In the TB theory a one-electron hamiltonian H is written using a basis of atomic orbitals from which the density matrix is deduced:

$$\rho = 2 \left(1 + \exp \frac{H - E_F}{k_B T} \right)^{-1}$$

for a given temperature T and Fermi level E_F.

Then, the electronic energy is calculated:

$$E = \mathrm{tr}(\rho \cdot H)$$

and the electronic forces are obtained via the Hellman–Feynman theorem:

$$F = -\mathrm{grad}\, E = -\mathrm{tr}(\rho \cdot \mathrm{grad}\, H).$$

The heavy part of the problem is the computation of the matrix formula giving ρ from H, because H is generally a large nondiagonal matrix. To improve the situation, a possibility is to make H either small or diagonal or both. We can also transform or approximate the formula for faster calculation. Historical attempts have, of course, tried to explore these ways.

The first strategy (S1) has been to keep the exact formula and a large matrix but to make it diagonal. This was the classical method for many years, with a negligible cost for the formula computation, but a high one for the diagonalization.

The second strategy (S2) has been proposed by Li et al. [5], with improvement of two points. On one part, the matrix size is reduced by supposing ρ local, on the other part, the formula is simplified with $T = 0$ and approached through an iterative procedure.

We have proposed [6] a third strategy (S3), analogous to that of Vanderbilt supposing a finite range for ρ and we have successfully applied it to the diffusion of iron, cobalt and nickel in silicon [7]. However, our approximation for the $\rho(H)$ formula is different, valid at high temperature, calculated in one step and without any need of iterations.

We now propose a fourth strategy (S4) to improve the last one, by eliminating the approximation made in the $\rho(H)$ formula. The question is then to see whether this last improvement in accuracy has to be payed by some extra computing cost. This is the aim of the present work, but before the S3/S4 comparison, we must give more details about our new strategy S4.

3. Our new O(N) methods for density matrix computing

The basic assumption to achieve a linear scaling computing cost is the finite range of ρ, just as in S2 and S3 methods. The consequence is that when we want to calculate the electronic energy contribution and forces on a given atom, we just consider a finite region around it, within a cutoff radius R_c.

In this region, starting from the central atom, we apply the Lanczos–Haydock recursion method [8] to build a local basis on which H is approximated as a local tridiagonal hamiltonian (LTH). This ensures that we obtain a reduced but pertinent representation of the local properties. At this point, this method is identical to our S3 scheme, but the improvement of S4 with respect to S3 is that, in the following, we do not need other approximations.

We diagonalize the local tridiagonal hamiltonian and, then, the $\rho(H)$ formula leads to scalar and not matrix computations. The counterpart is that we have now to operate two basis transformations to come back to the original atomic orbitals basis set and we must now evaluate the effective cost of these computations.

4. Comparison of our "S3" and "S4" O(N) methods

The recursion procedure is the common first step of our two methods. The convergence of the results is related to the size of the region which is treated and to the number of iterations to be performed in the recursion operation. We have already studied the effect of these two parameters in a previous publication [6]. The conclusion was that a cutoff radius equal to a few interatomic distances and a recursion level equal to 20 ensure satisfactory results.

Table 1
Total energy (mJ m^{-2}) for the two structures A and B of the $\Sigma = 11$ tilt grain boundary in silicon, computed with our two methods S3 and S4 (an approximation used in S3 is eliminated in S4)

Method	$\Sigma = 11$ A	$\Sigma = 11$ B
S3	659.3	673.7
S4	709.2	779.4

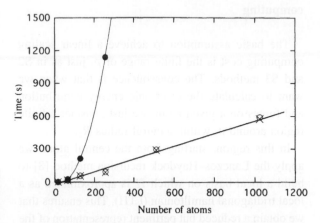

Fig. 1. Computing time as a function of the number of atoms of the system: classical diagonalization method (●), our S3 (○) and S4 (×) methods.

The two parameters being fixed, our first step produces a local tridiagonal hamiltonian, from which we deduce the density matrix in two ways. The calculation was an approximation in our S3 method, and is exact in the present S4 method. The difference can be seen in Table 1, giving the total energy of two possible structures of the $\Sigma = 11$ grain boundary in silicon [9], using the parameters of Goodwin et al. [10] for the SETB hamiltonian. The relative stability of the structures is not modified, but with our new method, we eliminate the question of the quality of the approximation used in the previous method.

Another question was to know whether the progress in accuracy had to be payed in computing time. We have treated systems of various sizes with our two methods and also with the classical diagonalization method. The results are shown in Fig. 1. The computing cost of the two O(N) methods is almost the same, and, of course, linear and much lower than the cost of the classical diagonalization method.

5. Conclusion

Atomic scale simulations of dynamical processes are greatly needed and more and more practised. Efficient and accurate methods must be developed in this field. We have created and adapted our own methods to solve the problems we encountered in the interpretation of HREM observations. We routinely use these methods because they give a good compromise between accuracy and computing cost. We have shown that the last one, the S4 method, is more accurate than the previous S3 one, without noticeable extra cost in computing time. To extend the application field of these methods, we think that the next effort is to improve the SETB method itself to apply it to a larger set of chemical species.

References

[1] M.W. Finnis and J.E. Sinclair, Phil. Mag. A 50 (1984) 45.
[2] F.H. Stillinger and T.A. Weber, Phys. Rev. B 31 (1985) 5262.
[3] M. Ichimura, Phys. Stat. Sol. A 153 (1996) 431.
[4] A.P. Sutton, M.W. Finnis, D.G. Pettifor and Y. Ohta, J. Phys. C 21 (1988) 35.
[5] X.P. Li, R.W. Nunes and D. Vanderbilt, Phys. Rev. B 47 (1993) 10891.
[6] A. Hairie, F. Hairie, B. Lebouvier and E. Paumier, Mat. Sci. Forum 207/209 (1996) 105.
[7] A. Hairie, F. Hairie, B. Lebouvier, J. Chen, E. Paumier and G. Nouet, Sol. Stat. Phen. 51/52 (1996) 99.
[8] R. Haydock, Solid State Physics, eds. F. Seitz and D. Turnbull, Vol. 35 (1980) p. 215.
[9] J. Chen, B. Lebouvier, A. Hairie, G. Nouet and E. Paumier, Comput. Mater. Sci. (1997), these proceeding.
[10] L. Goodwin, A.J. Skinner and D.G. Pettifor, Europhys. Lett. 9 (1989) 701.

Computational Materials Science 10 (1998) 395–400

Spin-flip contribution to the "in-plane" conductivity of magnetic multilayers

Ricardo Gomez Abal [1,*], Ana María Llois [1], Mariana Weissmann

Departamento de Física, Comisión Nacional de Energía Atómica,
Avda. del Libertador 8250, 1429 Buenos Aires, Argentina

Abstract

The "in-plane" conductivity of magnetic multilayers is calculated within the semiclassical approximation. The band structure is obtained with a tight-binding Hubbard Hamiltonian solved in the Hartree–Fock approximation. Once self-consistency is reached, the spin–orbit coupling term is added and a further diagonalization is performed. As it is well-known, quantum well states appear in the band structure of superlattices. By calculating the conductivity as a function of the Fermi energy we find that there is a considerable influence of these quantum well states if their energy is close to the Fermi level when the spin–orbit coupling is taken into account. Copyright © 1998 Elsevier Science B.V.

Keywords: Conductivity; Magnetic multilayers; Spin–orbit coupling

1. Introduction

The conductivity and the magnetoresistance of transition metal alloys and superlattices are usually interpreted within the two-current model, one for each spin component. However, the recently observed oscillations of the "in-plane" conductivity of Ni/Co superlattices as a function of layer thickness [1] are not explained within that model [2]. For this reason we attempt here a calculation of the spin-flip contribution to the conductivity.

The simplest way to perform conductivity calculations is within the semiclassical approximation (Boltzmann equation) [3]. In this paper, as in our previous one [2], from now on called Paper I, we study only the band structure contribution to the conductivity, but now adding a spin–orbit coupling term to the Hamiltonian. For this purpose we assume that the relaxation time τ is a constant, independent of wave number, band index and number of layers of each element in the superlattice. In this approximation the conductivity for each band n along direction i is given by

$$\sigma_n^{ii} \propto \int |v_n^i(k)|^2 \delta(\varepsilon_n(k) - \varepsilon_F) \, \mathrm{d}^3k \tag{1}$$

* Corresponding author. Tel.: 754-7099; fax: 754-7121; e-mail: ricgomez@cnea.edu.ar.
[1] Also at: Departamento de Fisica, Facultad de Ciencias Exactas y Naturales, Universidad de Buenos Aires.

as we have taken the relaxation time out of the integral sign. $v_n^i(k)$ is the derivative of the band energy with respect to k_i. The total conductivity is obtained by summing over all bands.

In Paper I we calculated the conductivity for Co/Ni superlattices as a function of the number of atomic layers of each material, within the two-channel model. Exceptions to this model have been analyzed in the past [4,5] but recently some cases in which the spin flip seems to be more important have been reported [6,7]. In this contribution we introduce the spin–orbit interaction (SOI) in the intrasite elements of the Hamiltonian, thus giving rise to hybridization of spin-up and spin-down bands (spin-flip). We are therefore left with only one conduction channel. A simple model that is able to show the important factors of the effect of SOI in the conductivity is studied first. Afterwards we introduce it in a realistic calculation for Ni/Co and Co/Cu superlattices.

2. Model calculation

From the results of Paper I we know that conduction is mainly due to the majority spin channel, with few bands that cross the Fermi level with a large slope. There are more minority bands near the Fermi energy but with a small dispersion, and also some pure d levels that are really non-dispersive along the superlattice growth direction. We have called these last ones quantum well states (QWSs) and found that they shift in energy as a function of the number of layers of each element. We believe that they are going to be the source for a conductivity reduction due to spin flip, if their energy is very close to E_F. The simple model will therefore contain three bands: two are dispersive majority bands, one of sp type and another of d type, and they will hybridize. The resulting band will be able to interact, via the spin–orbit term that connects only states of the same orbital quantum number, with the non-dispersive minority (d only) QWS.

To calculate the "in plane" conductivity (σ_{xx}) we consider only one value of k_z. The superlattice unit cell is large along the growth direction (z) and therefore the reciprocal cell is short along k_z making this approximation very good for the calculation of "in-plane" conductivity. Therefore, our model Hamiltonian will be that of a two-dimensional square lattice whose matrix given in a local basis is as follows:

$$H = \begin{bmatrix} \varepsilon_{s\uparrow} - t_{ss} \cdot (\cos(k_x) + \cos(k_y)) & -t_{sd} \cdot (\cos(k_x) + \cos(k_y)) & 0 \\ -t_{sd} \cdot (\cos(k_x) + \cos(k_y)) & \varepsilon_{d\uparrow} - t_{dd} \cdot (\cos(k_x) + \cos(k_y)) & \xi \\ 0 & \xi & \varepsilon_{d\downarrow} \end{bmatrix}, \tag{2}$$

where $\uparrow$ indicates majority and $\downarrow$ indicates minority spin. The parameter values: $t_{ss}/t_{dd} = 31$, $t_{sd}/t_{dd} = 1$, $(\varepsilon_{s\uparrow} - \varepsilon_{d\uparrow})/t_{dd} = 63$, $(\varepsilon_{d\downarrow} - \varepsilon_{d\uparrow})/t_{dd} = 8.6$ were chosen so as to simulate the relevant bands in the Ni/Co multilayers. The SOI parameter ξ for Ni and Co is of the order of 0.1 eV. In Fig. 1 we show the conductivities as a function of energy and we see a very important reduction due to spin-flip when the QWS coincides with the Fermi energy. To model the different superlattices, instead of changing the energy position of the QW ($\epsilon_{d\downarrow}$) and recalculating the bands, we have moved the Fermi energy slightly. This is possible because the majority bands do not depend on layer thickness [2] The SOI produces a gap and changes the slope of the bands locally, modifying the conductivity only if this happens at E_F. When a small dispersion is considered in the $d_\downarrow$ band the effect of spin-flip decreases, but still remains large.

3. Preliminary calculations for superlattices

We have performed calculations for two different superlattices Ni_4/Co_4 grown in the (1 1 1) direction and Cu_3/Co_3 grown in the (0 0 1) direction. The band structure is calculated using a spin-polarized tight-binding (TB) spd

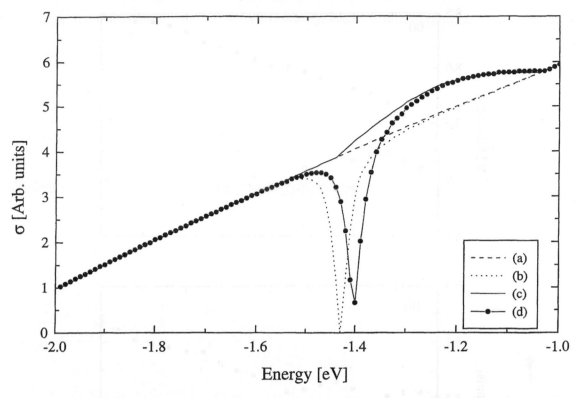

Fig. 1. Conductivity of the three-band model system as a function of the position of the Fermi energy: (a) $\xi = 0$, QWS exactly non-dispersive; (b) $\xi = 0.1$, QWS exactly non-dispersive; (c) $\xi = 0$, QWS only slightly dispersive; (d) $\xi = 0.1$, QWS only slightly dispersive.

Hamiltonian with parameters taken from [8]. The charge transfer and magnetic terms are introduced in the on-site elements of the Hamiltonian as in Paper I. The SOI is given in the usual approximation by $H_{so} = \xi \widehat{L}.\widehat{S}$, where ξ is the SOI parameter for which we use the atomic value for p and d states given by Herman-Skillman. We first obtain a self-consistent solution of the Hamiltonian without SOI, and once self-consistency is achieved one further diagonalization including H_{so} is performed in order to obtain the eigenvalues needed to calculate the conductivity.

The evaluation of the integral in Eq. (1) needs a large number of k-points. The introduction of the SOI term demands a much denser mesh of points for the k-space integration than the two-channel calculation. As there is almost no dispersion in the k_z-direction and in this paper we are only interested in the "in-plane" transport, we use a mesh of 112 000 k-points in only one plane of the First Brillouin zone ($k_z = 0$).

The superlattice Co_4/Ni_4 has a Ni QWS of $d_\downarrow$ character at $E = -1.44$ eV in the Γ point and the Fermi energy of this system lies at -1.4 eV. In Fig. 2(a) we plot σ_{xx}, calculated with and without SOI, as a function of the energy for a small interval around E_F. For this system the SOI reduces the conductivity in the whole energy range and shows a local minimum at $E = -1.4$ eV just above the QWS energy as in the three-band model.

The QWS at $E = -1.44$ eV has no dispersion along k_z and only a small dispersion for $|k| < 0.2$ around Γ as can be seen in Fig. 3(a), where we show the energy bands along a symmetry direction in $k_\parallel$. There are three crossings between $sp_\uparrow$ conduction bands and non-dispersive $d_\downarrow$ states. The effect of the SOI in this case is to decrease the conductivity by about 20% of its two-channel value.

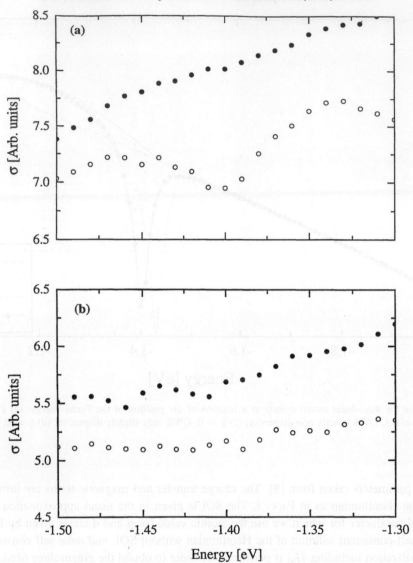

Fig. 2. Conductivity of the superlattices as a function of energy with SOI (hollow circles) and without it (solid circles): (a) Ni$_4$/Co$_4$; (b) Cu$_3$/Co$_3$.

To study the effect of QWSs on the conductivity for another real system, that is known to present QWSs [9] but with different dispersion characteristics in the $k_\parallel$ plane, we have chosen as an example the superlattice Cu$_3$/Co$_3$. This system has a QWS at $E = -1.46$ eV and the Fermi level lies at $E = -1.37$ eV. In Fig. 3(b) we show the band structure along a symmetry line in $k_\parallel$. It is seen that the QWS disperses strongly along $k_\parallel$ in this case and that there are no sp$_\uparrow$ – d$_\downarrow$ crossings as those appearing in Fig. 3(a). In Fig. 2(b) we show $\sigma_{xx}(E)$ for this system. Even if there is an overall decrease in the conductivity as the SOI term is included, there is no significant change in σ_{xx} as the QWS crosses E_F.

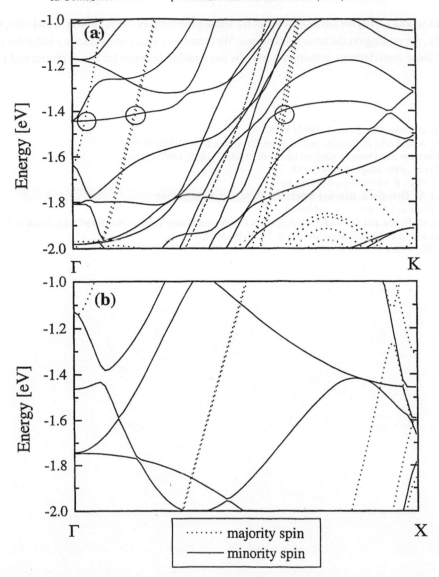

Fig. 3. Energy bands of the superlattices, along one of the $k_\parallel$ directions: (a) Ni_4/Co_4,; (b) Cu_3/Co_3. The crossings marked with circles, that are close to E_F, involve spin flip.

4. Conclusions

Spin flip reduces the "in-plane" conductivity of the superlattices studied and this reduction is enhanced considerably if the Fermi energy coincides with a QWS. Actually, this effect is present if there are: (a) one or more conducting states; (b) one or more non-dispersive states of the opposite spin with energy near the Fermi level; (c) both kinds of states are degenerate in the non-dispersive k-space region. We have shown that these conditions are fulfilled for the Ni_4/Co_4 superlattice while condition (c) is not given in the Cu_3/Co_3 superlattice. This last system shows thereafter no local minimum in $\sigma_{xx}(E)$.

In our previous works [2,10] we have shown that for some superlattices, such as Ni/Co and Ni/Cu, QWSs appear periodically at E_F, depending on the layer thicknesses. We therefore expect an oscillatory behavior of the "in-plane" conductivity for these superlattices, although it may be too small to account for the experimental results.

References

[1] J.M. Gallego et al., Phys. Rev. Lett. 74 (1995) 4515.

[2] M. Weissmann, A.M. Llois, R. Ramírez and M Kiwi, Phys. Rev. B 54 (1996) 15 335.

[3] J.M. Ziman, Electrons and Phonons (Oxford University Press, Oxford, 1967) Chap. 7.

[4] I.A. Campbell et al., Phil. Mag. 15 (1967) 977.

[5] A. Fert et al., J. Phys. F. Metal Phys. 6 (1976) 849.

[6] D. Lederman, J. M. Gallego, S. Kim and I. Schuller, Phys. Rev. B, Submitted.

[7] J. Banhart, H. Ebert and A. Vernes, unpublished.

[8] O.K. Andersen, O. Jepsen and D. Glötzel, Highlights of Condensed Matter Theory (North-Holland, Amsterdam, 1984).

[9] J.L. Pérez-Díaz and M.C. Muñoz, Phys. Rev. Lett. 76 (1996) 4967.

[10] M. Kiwi et al., Phys. Rev. B 55 (1997) 14 117.

ELSEVIER

Computational Materials Science 10 (1998) 401–405

COMPUTATIONAL
MATERIALS
SCIENCE

Atomistic study of dislocation cores in aluminium and copper

A. Aslanides *, V. Pontikis [1]

Laboratoire des Solides Irradiés, CEA-DRECAM, URA CNRS 1380, Ecole Polytechnique, F-91128 Palaiseau Cedex, France

Abstract

The core structure of an edge dislocation, dissociated or not, has been investigated through atomistic calculations. The simulation relies on two N-body phenomenological potentials for aluminium and copper. The various dislocation configurations are characterized in terms of strains and energies for both materials. The limit of validity of the elastic theory of dislocations is then discussed by comparing its predictions to those obtained by simulation. Copyright © 1998 Elsevier Science B.V.

1. Introduction

During plastic deformation, dislocations interact not only elastically but also through various highly non-linear elementary mechanisms such as cross-slip or dipole annihilation. The detailed understanding of such interactions that involve the overlap of dislocation cores is facilitated by resorting to an atomistic approach, the application of the elastic theory being no longer justified. Considering an edge dislocation line parallel to the $[1\,1\,\bar{2}]$ direction, present work demonstrates that both approaches are needed in order to reach a complete description of the dislocation. Two FCC metals are being considered: aluminium and copper, with respectively high and low stacking fault energy, i.e. a small and a large dissociation width. In the following, the numerical procedure as well as the different technical tools developed for this study are described. Then, using lattice relaxation, we investigate the dissociation mechanism in which a perfect edge

dislocation with Burgers vector $b = \frac{1}{2}[1\,\bar{1}\,0]$ dissociates into two Shockley partials separated by a stacking fault ribbon. Thereby, the dislocation core width for the two model potentials used is obtained in terms of strain and excess energy distributions. These show that two different regions exist on the graph displaying the excess energy as function of the radial dimension of a system containing a dislocation in its centre. An atomistic (resp. elastic) approach is required for small (resp. large) dimensions of the system considered.

2. Numerical procedure

Interactions in copper derived from a phenomenological N-body potential according to which the energy of an atom i in the crystal is expressed by [1]:

$$E_i = \sum_{j \neq i} A_{Cu} \exp\left[-p_{Cu}\left(\frac{r_{ij}}{r_0} - 1\right)\right]$$
$$+ \left\{ \sum_{j \neq i} \xi_{Cu}^2 \exp\left[-2q_{Cu}\left(\frac{r_{ij}}{r_0} - 1\right)\right] \right\}^{1/2},$$

* Corresponding author. Tel.: + 33 01 69 33 38 26; fax: + 33 01 69 33 30 22; e-mail: antoine.aslanides@polytechnique.fr.
[1] E-mail: vassilis.pontikis@polytechnique.fr.

Table 1
Values of the parameters for the copper and aluminium potentials used in the present work

	A	p	ξ	q	$C1$	$S1$	$C2$
Cu	0.07157	11.562	1.14850	2.02139			
Al	0.1780	6.5000	1.3831	2.0700	0.00947	0.00515	0.01664

$A\,\xi$, $C1$, $S1$ and $C2$ are in eV units. Cut-off radii in lattice parameter units: 1.581 (Cu), 2.236 (Al).

where j runs over all the particles, $r_{ij} = |r_i - r_j|$ denotes the euclidian distance between atoms i and j, r_0 is the first neighbour distance and A_{Cu}, p_{Cu}, ξ_{Cu}, q_{Cu} are adjustable parameters. The total energy of the crystal is obtained by summing all the atomic contributions.

The model used for aluminium yields the energy of an atom i as [2]:

$$E_i = \sum_{j\neq i} A_{Al} \exp\left[-p_{Al}\left(\frac{r_{ij}}{r_0} - 1\right)\right]$$
$$+ \left\{ \sum_{j\neq i} \xi_{Al}^2 \exp\left[-2q_{Al}\left(\frac{r_{ij}}{r_0} - 1\right)\right]\right\}^{1/2}$$
$$+ \sum_{j\neq i} \left\{ \left[\frac{C1\cos(2k_f r_{ij})}{(r_{ij}/r_0)^3}\right] + \left[\frac{S1\sin(2k_f r_{ij})}{(r_{ij}/r_0)^4}\right]\right.$$
$$\left. + \left[\frac{C2\cos(2k_f r_{ij})}{(r_{ij}/r_0)^5}\right]\right\}$$

where k_f stands for the Fermi wave vector of aluminium, and A_{Al}, p_{Al}, ξ_{Al}, q_{Al}, $C1$, $S1$ and $C2$ are adjustable parameters.

Parameters of both models are fitted on experimental values of the cohesive energy, elastic moduli, lattice parameter, vacancy formation energy [3] and intrinsic stacking fault energy [4]. Table 1 displays the corresponding optimal values.

The computational cell with X,Y and Z directions parallel to the $[1\,\bar{1}\,0]$, $[1\,1\,\bar{2}]$ and $[1\,1\,1]$ directions is generated by repeating this cell in the three space directions, so that the crystal contains $70 \times 3 \times 30$ cells. A perfect edge dislocation with line the Y axis and Burgers vector $b = \frac{1}{2}[1\,\bar{1}\,0]$ is then created. Its elastic displacement field [5], whose origin is arbitrarily taken in the middle of two consecutive atomic planes in the $[1\,\bar{1}\,0]$ as well as in the $[1\,1\,1]$ directions, is applied to all the atoms of the perfect crystal. Two extra $(1\,\bar{1}\,0)$ half-planes are then inserted in the cut that has appeared along the $[1\,1\,\bar{2}]$ direction. The initial configuration thereby obtained contains $N = 38070$ atoms. An infinite straight dislocation line is being simulated by using periodic boundary conditions along the $[1\,1\,\bar{2}]$ direction, whereas free surfaces in the $[1\,\bar{1}\,0]$ direction (Burgers vector direction) enable shear deformation of the crystal. Finally, atoms belonging to the outermost four $(1\,1\,1)$ atomic layers are kept fixed in the $[1\,1\,1]$ direction, while in plane motion is left unconstrained.

Energy minimization is performed using a damped newtonian algorithm [6] with a time step $\delta t = 10^{-15}$ s. The position of the dislocation is fixed and its behaviour e.g. dissociation or displacement is controlled by using the following relaxation scheme: (i) First, the atoms belonging to the extra half-planes are held in their initial positions while all other atoms of the simulation box are displaced towards the potential energy minimum. (ii) In a second stage, the rules are inverted by fixing all the atomic positions except those of the atoms pertaining to the extra half-planes. Both stages are repeated until a configuration of minimal energy is reached. Since simultaneous displacements of the atoms in the half-planes and their neighbourhood is forbidden, the dislocation is effectively immobile while the distance between the extra half-planes, i.e. the dissociation width, is kept fixed to its initial value. This procedure leads to configurations equivalent to those resulting from the application of an external stress. The validation of this procedure is achieved by checking that at the equilibrium dissociation distance the atomic configurations and the associated excess energy values obtained with or without making use of this scheme are identical.

The deformation profile close to the glide plane and along the $[1\,\bar{1}\,0]$ direction helps in localizing the

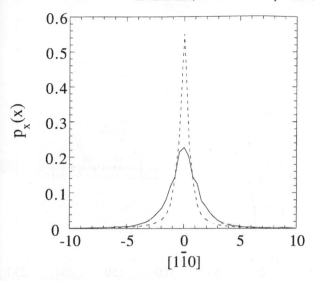

Fig. 1. Edge component of the Burgers vector density distribution for the perfect edge dislocation in aluminium. Dashed line: unrelaxed elastic configuration, solid line: relaxed configuration. Abscissas in lattice parameter units.

dislocation. Within a continuous description of the dislocation core by an infinitesimal dislocation distribution density, the profile displayed in Fig. 1 represents the edge component of the corresponding Burgers vector distribution [7]. The positions of the maxima on this graph identify the position of the extra half-planes.

3. Results and discussion

As in a real crystal, the energy of the model is lowered by splitting the dislocation into two Shockley partials. The above-mentioned procedure allows us to investigate the constrained undissociated state of the dislocation.

3.1. Perfect edge dislocation in aluminium

The deformation profiles along the $X//[1\,1\,0]$ direction in the initial and relaxed configurations are presented in Fig. 1. Both show that high strain values are obtained in the core region. The comparison of the two curves reveals that the elastic prediction overestimates strain in the core region while leading to a narrower core extension in the $[1\,\bar{1}\,0]$ direction. A satisfactory

fit of the relaxed Burgers vector density distribution is obtained by $f(x) = A/\cosh(x)$ with $A = 0.431$. The suggestion has been made in previous work [8] that the half-height width of the peak is a reasonable estimate of the core extension. The value we obtain amounts 3.36 b (b: Burgers vector modulus).

3.2. Dissociated state of the dislocation in aluminium and copper

Without applied stress, a spontaneous dissociation of the perfect edge dislocation (Burgers vector b) into two 60° Shockley partials (Burgers vectors b_1 and b_2) occurs: $\frac{1}{2}[1\,\bar{1}\,0] \rightarrow \frac{1}{6}[1\,\bar{2}\,1] + \frac{1}{6}[2\,\bar{1}\,\bar{1}]$. The resulting deformation profile exhibits two peaks, each identifying the position of the partials. Our model leads to a dissociation distance $d = 9.1$ Å (Fig. 2) while the elastic theory predicts $d = 7.7$ Å, a result that is obtained by using calculated values of the shear modulus C_{44} and stacking fault energy, γ, at $T = 0$ K. Both values are in good agreement with experimental observations suggesting the dissociation distance to be smaller than 8 Å [9,10]. Limitations inherent to both approaches might however explain the small difference between the above calculated values. Indeed, the narrow ribbon fault separating the two partials is very different from the perfect stacking fault considered in the elastic approach. Moreover, the limited extension of the simulation box in the $[1\,\bar{1}\,0]$ direction and the associated free surface boundary condition implies that image forces are far from being negligible. These increase the dissociation width but their influence is still limited as is shown from the small difference existing between the two evaluations, elastic or atomistic, of the dissociation distance.

Due to a low stacking fault energy in copper, the equilibrium dissociation distance is larger than that in aluminium ($d = 52.4$ Å). This result is in agreement with the elastic theory prediction (54 Å) and with experiments [11]. Fig. 2 illustrates differences existing between the density distribution of Burgers vectors in the two metals: the two partial dislocation cores are perfectly seperated in copper while they overlap in aluminium.

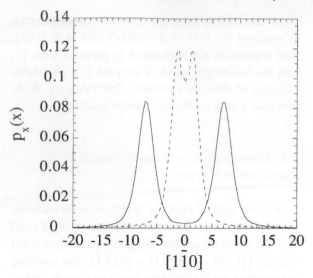

Fig. 2. Edge component of the Burgers vector density distribution for the dissociated dislocation in aluminium (dashed line) as well as in copper (solid line). Position in the [1 1 0] direction in lattice parameter units.

3.3. Comparison of the simulation and the elastic theory predictions

The two descriptions, atomistic and continuous, can be joined to provide the excess energy of a system containing a constrained perfect edge dislocation, obtained by using the procedure described in Section 2, in its centre. To this purpose, continuity conditions written for the excess energy and its gradient lead to a $R^{core} = 11.4$ Å value of the radial distance from the dislocation line. Below this distance, the stored energy should be calculated atomistically whereas above it, the elastic formula $(\mu b^2/4\pi(1-\nu))\ln(R/R^{core})$ (with $\mu = C_{44}$ and ν Poissons ratio) can be used. Values of the excess energy stored in a cylinder of radius R surrounding the dislocation line are displayed in Fig. 3. It is seen that the above choice of R^{core} ensures the continuity of excess energy calculated either atomistically or elastically. The inset figure shows that within the atomistic region i.e. $R < R^{core}$, potential energies of atoms significantly exceed the cohesion energy of the perfect crystal. Non-elastic effects associated to the dislocation core are therefore entirely contained in the cylindric region of radius R^{core}.

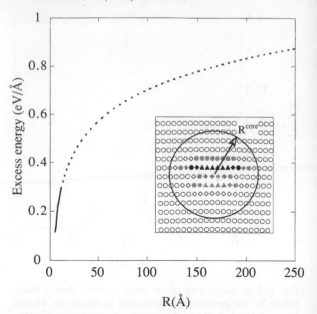

Fig. 3. Excess energy stored in a cylinder with axis the dislocation line and outer radius R. Excess energy in eV per unit length of dislocation, R in Å. The solid line corresponds to the non-elastic region where an atomistic treatment has to be applied, the dashed line corresponds to the elastic prediction. [1 1 $\bar{2}$] projection of the core region of the perfect edge dislocation in aluminium is shown in insert. Each symbol stands for an atomic site. Dimensions in Å; calculated excess energies in me V/atom. (**Black** – *Gray* – Open symbols: triangles $\approx$ **70** – *35* me V/at; diamonds $\approx$ **60** – $\overline{30}$ – <u>10</u> me V/at; circles $\approx$ **45** – *20* – <u>3</u> me V/at.)

4. Conclusion

Dislocation core structures in aluminium and copper have been investigated using two phenomenological potentials. This study has demonstrated the validity of the two models as well as it has served to develop appropriate numerical procedures. The continuity we have obtained between the atomic and the elastic approach allows a complete description of the dislocation. Work in progress focuses on the influence of dislocation cores on other elementary mechanisms involved in plastic deformation.

References

[1] C. Rey Losada, M. Hayoun and V. Pontikis, Mater. Res. Soc. Symp. Proc. 291 (1993) 549.

[2] A. Aslanides, A. Legris and V. Pontikis, in preparation.

[3] M.J. Gillan, J. Phys.: Condens. Mater 1 (1989) 689.

[4] B. Hammer, K.W. Jacobsen, V. Milman and M.C. Payne, J. Phys: Condens. Mater 4 (1992) 10 453.

[5] J.P. Hirth and J. Lothe, Theory of Dislocations (Wiley, New York, 1982).

[6] J.R. Beeler and G.L. Kulcinski, in: Interatomic Potentials and Simulation of Lattice Defects, eds. P.C. Gehlen, J.R. Beeler and R.I. Jaffee (Plenum, New York, 1972) p. 735.

[7] D.J. Bacon and J.W. Martin, Phil. Mag. A 43 (1981) 883.

[8] P. Beauchamp and J. Lépinoux, Phil. Mag. A 74 (1996) 919.

[9] M.J. Mills and P. Stadelmann, Phil. Mag. A 60 (1989) 355.

[10] J.M. Penisson and A. Bourret, Phil. Mag. A 40 (1979) 811.

[11] B. Weiler, W. Sigle and A. Seeger, Phys. Stat. Sol. A 150 (1995) 221.

COMPUTATIONAL
MATERIALS
SCIENCE

Computational Materials Science 10 (1998) 406–410

Mechanisms of formation and topological analysis of porous silicon – computational modeling

L.N. Aleksandrov *, P.L. Novikov

Institute of Semiconductor Physics, Academy of Science, Novosibirsk 630090, Russian Federation

Abstract

Porous silicon formation under anodization in HF solution is studied by means of computer simulation, using the model, which takes into account the Si dissolution, thermal generation, diffusion and drift of holes and quantum confinement. Structures of porous layer, obtained in computational experiments at various doping level of initial Si, temperatures, HF concentrations, anode current densities, are shown and analyzed. The porosity and fractal dimension of the obtained porous structures are also analyzed. Copyright © 1998 Elsevier Science B.V.

Keywords: Porous silicon; Quantum confinement; Fractal dimension

1. Introduction

Porous silicon, prepared by anodization, has become very promising material for device applications due to visible luminescence [1], large specific surface and wide technological opportunities [2–4]. It is also an interesting object for fundamental investigation, since fractal structure of porous silicon exhibits number of optical, diffusion and transport and thermodynamical properties [5], which are not fully understood. The mechanism of formation of PS is still a controversial subject [6–9]. In the present work computer simulation is carried out, which allows to control the process of porous silicon formation via doping level of p-Si, anodic current density, temperature and HF concentration. The simulation uses the model of porous silicon formation, which introduces the processes of thermal generation and recombination of holes as well as the effect of quantum confinement in nanoparticles. The modeling was performed on two- and three-dimensional lattices. The porosity and fractal dimension of the obtained porous structure were analyzed.

2. The model and method of calculation

In the model diffusion and drift, generation and recombination of holes in crystal, formation of space charge region at pore tips and near interface between crystal and electrolyte, as well as quantum confinement effects in crystal nanoparticles are considered as the basic processes, providing the mechanism of porous silicon formation. The combination of conditions determines, what processes are dominating.

Under anodization of p^+-Si ion charge is concentrated and focuses on the electric field lines mainly at pore tips, where the current density j is higher than an

* Corresponding author.

average one over volume. In the region near the pore tip holes concentrate at hillocks on crystal anode surface. Ions of electrolyte are attracted to the hillocks, leading to electropolishing at the pore tips. Computer simulation of p^+-Si anodization is performed on a two-dimensional (160 × 160) lattice. Each position in the lattice is defined as a cell. The step of the lattice corresponds to real size of surface charge fluctuations. N holes randomly walk on the lattice, jumping as far as a typical free path length. Once a hole occurs within Debye radius R_D from the nearest pore tip, it drifts toward this tip. After it reaches the interface at some point, the analysis of interface curvature is carried out within interval L from the point. The dissolution occurs in the most marked protuberance over the interval. The value of L (lattice units) depends upon the HF concentration c (wt%) and current density j (mA/cm^2) as

$$L \approx 48(j/c)^{(1/2)}. \tag{1}$$

In lightly doped crystal ($p < 10^{16}$ cm^{-3}) the space charge density and the electrical fields value are low. Therefore in p^--Si dominant mechanisms of the pore formation are diffusion, thermal generation and recombination of holes and – in nanocrystallites – quantum confinement effect. The quantum confinement is known to increase the band gap. Consequently a barrier appears between the bulk region and nanoparticles, which prevents hole penetration inside nanoparticles. The size, at which the quantum confinement becomes significant, is denoted R_Q and equal to 10 nm and less [10].

Computer simulation of the pore formation in p^--Si is performed on two- (160 × 160) and three-dimensional (160 × 160 × 80) lattices. The lattice unit corresponds to average distance between neighboring atoms of crystal lattice. The algorithm differs from the one, described above. Before jumping a hole may disappear from the current point and appear in another position of crystal volume. The probability G of this is $\sim 1/t_{rec}$, where t_{rec} is the time of hole recombination. After a hole reaches the interface, the analysis of surrounding space is carried out. If the crystal volume contains the sphere of radius R_Q, then the dissolution occurs. Otherwise the hole walks on.

In n-Si holes are formed by photogeneration. The positive space charge is formed by holes and by ionized acceptor impurities. Therefore with an increase of doping level the effect of focusing electric field lines at pore tips is amplified. This results in formation of column-like pores in n^+-Si, especially in the case of backside illumination.

The analysis of the obtained three-dimensional structures was carried out. The porosity was calculated as a function of the depth:

$$P(h) = N_{PS}(h)/N_{ij}(h), \tag{2}$$

where $N_{PS}(h)$ is the number of cells, contained in pores in the monolayer at depth h, and $N_{ij}(h)$ is the whole number of cells in this layer.

The fractal dimension of porous structure was calculated by formula of cellular fractal dimension [11]:

$$D(i, j, k) = 3 \log_{i \cdot j \cdot k / (i^* \cdot j^* \cdot k^*)}$$
$$\times (N_{PD}(i, j, k)/N_{PD}(i^*, j^*, k^*)), \tag{3}$$

where $N_{PD}(x, y, z)$ is the number of parallelopipeds $x \times y \times z$ ($x \geq 1, y \geq 1, z \geq 1$), which fill the lattice and contain at least one crystal cell. It was taken into account that an average size of parallelopiped $x \times y \times z$ is equal to $\sqrt[3]{x \cdot y \cdot z}$. The neighboring values (i, j, k) and (i^*, j^*, k^*) differ not more, than 1 unit.

3. Results and discussion

A matrix plot, obtained by computer simulation on a two-dimensional lattice (160 × 160), with $N = 80$ is shown at Fig. 1. Black color corresponds to pores. The pores grow downwards from the top of each plot. The effective parameters R_D and L are shown for each line and column of the matrix. The pore diameter is seen to increase with the increase of L, which corresponds to current density increase and/or HF concentration decrease. Within radius L pores are continuous in plan-parallel directions, which agrees with high-resolution and cross-sectional TEM photographs of a micro- and macroporous Si layers [12,13]. This result is due to including the factor L into modeling, since simulation, based on the conventional diffusion-limited model, leads only to ramified pores [6]. For

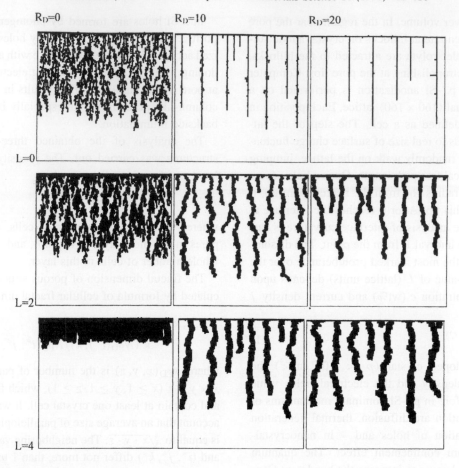

Fig. 1. A matrix of computer simulation plots with varying effective parameters L and R_D ($N = 80$).

$L = 4$ and $R_D = 0$ the current density is large enough to achieve polishing. Increasing R_D leads to straightening of the pores and the increasing distance between them. By the conditions $L = 0$ and $R_D = 10$ and 20 the direction of growth is practically predetermined, since the active zone for each pore is limited with its lowest point. The similar conditions were provided by Lehmann et al. [8] in n-Si, using inducing pits and backside illumination.

No dependence of porous structure upon N was found. This means that direct analogy between the number of holes in the Monte-Carlo process and the hole concentration in an actual crystal would be erroneous. In experiment the concentration gives rise to a number of effects (Fermi level shift, Debye screening, formation of space charge region, etc.), which are not

automatically included in modeling with parameter N.

A matrix of plan-parallel cross-sections of simulated three-dimensional structures ($160 \times 160 \times 80$) for $N = 3000$ was obtained. The calculations showed that the dP/dh becomes smaller with including hole thermal generation and recombination factor into simulation. The $P(h/h_{\max})$ curves are plotted in Fig. 2. The short slope near the initial surface results from difference between conditions of pore growth at initial surface and at some depth. The average porosity depends strongly on R_Q.

The fractal dimensions of crystal skeletons, formed by computer simulation on three-dimensional lattice, were calculated in accordance with Eq. (3). The dependence of fractal dimension upon the size of detail (defined as $\sqrt[3]{i \cdot j \cdot k}$) is shown in Fig. 3. It changes

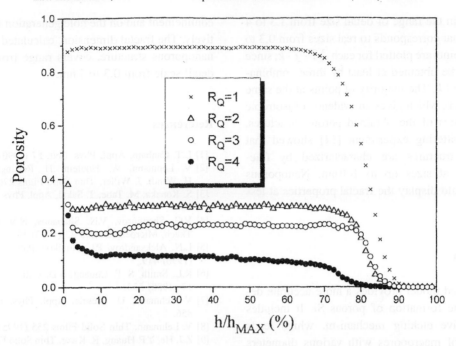

Fig. 2. Porosity of structures, simulated on three-dimensional lattice, vs. depth. (R_Q is given in lattice units).

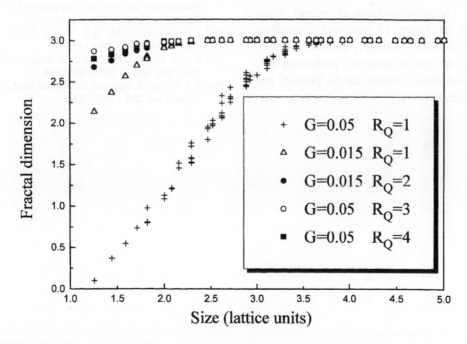

Fig. 3. Fractal dimension of structures, formed by computer simulation on three-dimensional lattice, vs. size of detail. (R_Q is given in lattice units).

from 0.1 to 3 in the range of detail size from 1.3 to 4 lattice units, that corresponds to real sizes from 0.3 to 1 nm. Three points are plotted for each $\sqrt[3]{i \cdot j \cdot k}$, since its value may be obtained at least by three combinations of i, j and k. The majority of points at the same abscissa overlap, which gives an evidence of isotropic fractal properties of the obtained porous structures. A neutron scattering experiment [14] showed that mesoporous structures are characterized by fractal dimension at sizes up to 100 nm. Nanoporous structures should display the fractal properties at less sizes.

4. Conclusion

The presented model provides a more accurate description of the formation of porous Si. It includes a relief-selective etching mechanism, which leads to formation of macropores with various diameters and shapes. In case of p^{-}-Si the model takes into account thermal generation and recombination of holes and quantum confinement effect in crystalline nanoparticles. The cross-sections of porous structures, obtained by means of computer simulation, exhibit visual similarity within experimental ones. The analysis of the porous structure, performed on a three-dimensional lattice, shows that the porosity and its gradient depend strongly on the scale of quantum confinement and on the hole generation factor, respectively. The fractal dimension, calculated for simulated nanoporous structure, covers range from 0.1 to 3 in detail scale from 0.3 to 1 nm.

References

[1] L.T. Canham, Appl. Phys. Lett. 57 (1990) 1046.
[2] V. Lehmann, W. Honlein, H. Reisinger, A. Spitzer, H. Wendt, J. Willer, Thin Solid Films 276 (1996) 138.
[3] S. Konaka, M. Tabe, T. Sakai, Appl. Phys. Lett. 41 (1982) 86.
[4] V.G. Shengurov, V.N. Shabanov, N.V. Gudkova, B.Ja. Tkach, Microelectronika 22 (1993) 19.
[5] L.N. Aleksandrov, P.L. Novikov, Phys. Stat. Sol. A 158 (1996) 419.
[6] R.L. Smith, S.-F. Chuang, S.D. Collins, J. Electr. Mater. 17 (1988) 533.
[7] V. Lehmann, U. Gæsele, Appl. Phys. Lett. 58 (1991) 856.
[8] V. Lehmann, Thin Solid Films 255 (1995) 1.
[9] Z.J. He, Y.P. Huang, R. Kwor, Thin Solid Films 265 (1995) 96.
[10] L.N. Aleksandrov, P.L. Novikov, Pis'ma v ZhETF 65 (1997) 685.
[11] B.B. Mandelbrot, The general properties of fractals, Proceedings of the 6th International Symposium on Fractals in Physics, Triest, Italy, 1985.
[12] C.H. Lee, C.C. Yeh, H.L. Hwang, Klaus Y.J. Hsu, Thin Solid Films 276 (1996) 147.
[13] E. Zur Muhlen, Da Chang, S. Rogashevski, H. Niehus, Phys. Stat. Sol. B 198 (1996) 673.
[14] B.J. Heuser, S. Spooner, G.J. Glinka et al., Mat. Res. Soc. Symp. Proc. 283 (1993) 209.

ELSEVIER

Computational Materials Science 10 (1998) 411–415

COMPUTATIONAL
MATERIALS
SCIENCE

Embedded atom method calculations of vibrational thermodynamic properties of ordered and disordered Ni₃Al

J.D. Althoff [a], Dane Morgan [a], D. de Fontaine [a,*], M.D. Asta [b], S.M. Foiles [b], D.D. Johnson [b]

[a] *Department of Materials Science, 577 Evans Hall, University of California, Berkeley, CA 94720-1760, USA*
[b] *Sandia National Laboratories, Livermore, CA, USA*

Abstract

Recent work had suggested that vibrational effects can play a significant role in determining alloy phase equilibria. In order to better understand these effects, we investigate the vibrational properties of disordered and ordered Ni₃Al using the embedded atom method and calculate vibrational thermodynamic quantities within the quasi-harmonic approximation. The vibrational entropy is found to be strongly dependent on volume. For fully relaxed structures the dependence on lattice decoration of the vibrational entropy is compared to that suggested by recent experimental results. Copyright © 1998 Elsevier Science B.V.

Keywords: Vibrational energy; Ni₃Al; Embedded atom method

1. Introduction

In ab initio calculations of alloy free energies, the difference of vibrational entropy between ordered and disordered phases at the same composition was often neglected, as it was considered to be small compared to the corresponding configurational entropy difference. It therefore came as somewhat of a surprise when Fultz and co-workers at Caltech [1,2] found experimental evidence for vibrational entropy differences of roughly the same order as the configurational. A well-studied example is the technologically important Ni₃Al compound, a superstructure of fcc, generally denoted by the symbol L1₂. The corresponding fcc disordered phase does not exist at equilibrium, but

has been prepared experimentally by laser quenching [3], vapor deposition onto cold substrates [1], and by mechanical alloying [2,4]. Specific heat measurements [1] up to 343 K indicated an entropy difference $\Delta S_{\text{vib}} = S_{\text{dis}} - S_{L1_2}$ of at least $+0.19\,k_{\text{B}}$ per atom, and an extrapolation based on the Debye model gave $\Delta S_{\text{vib}} = +0.27k_{\text{B}}$ per atom at 1700 K. Extended electron energy-loss fine structure spectroscopy measurements [1] gave a similarly large vibrational entropy excess for the disordered state. Inelastic neutron scattering measurements [2] likewise gave a large entropy difference: $\Delta S_{\text{vib}} = (0.27 \pm 0.1)k_{\text{B}}$ per atom.

Such large values pose several problems of theoretical and practical nature: are these effects real, or are they artifacts of the method of preparation of the disordered state; if real, what are they due to physically, and how might one take vibrational free energy into account in ab initio calculations? In an attempt to

---

* Corresponding author. Tel.: 510-642-8177; fax: 510-643-6629; e-mail: ddf@isis.berkeley.edu.

answer these questions, we have calculated the vibrational density of states (DOS) and corresponding thermodynamic functions of the ordered and disordered states of Ni_3Al in the quasi-harmonic approximation. A preliminary account of these calculations is given elsewhere [5]. In a paper actually preceding the experimental work of the Caltech group, and which received little attention, Ackland [6] performed similar calculations employing a Finnis and Sinclair potential, but without producing the all-important DOS which holds the key to the physics of the entropy effect. Results of the present calculations will depend to some extent on the choice of potentials, thereby casting some uncertainty on the conclusions drawn. However, a significant advantage of the calculations over experimental procedures is that unstable (or metastable) phases can be prepared very cleanly on a computer, hence avoiding such extended defects as dislocations, grain boundaries, free surfaces and other imperfections which are unavoidable by-products of rapid quenching, vapor deposition, and ball milling operations, and which may contribute to the measured vibrational entropy in an uncontrolled way.

2. Computations

Two extreme types of Ni_3Al configurations were considered: fully ordered ($L1_2$) and fully disordered, the latter obtained by occupying fcc sites by Ni and Al atoms at random in the proportion $3:1$ in a cubic 256-atom (occasionally extended to 500 atoms) supercell with periodic boundary conditions. Energies and atomic forces were calculated by the embedded atom method (EAM) [7] using the potentials developed for Ni_3Al by Voter and Chen [8]. The EAM functions were fit to the lattice constants, cohesive energies, bulk moduli, elastic constants and vacancy formation energies of elemental fcc Ni and Al and to the bond length and bond energy of the diatomic molecules Ni_2 and Al_2, as well as to the lattice constant, cohesive energy, elastic constants, vacancy formation energy, intrinsic stacking fault energy, and $(1\,0\,0)$ and $(1\,1\,1)$ antiphase boundary energies of Ni_3Al, and also to the lattice constant and cohesive energy of the B2 phase of NiAl. No

information concerning the temperature dependence of the parameters was used in the fit. Both volume and local atomic relaxations were performed, the former by a method to be described below, the latter by conjugate gradient energy minimization. At temperatures (T) of interest, and for a range of volumes (V), dynamical matrices were calculated and diagonalized at each k point to obtain normal mode (j) frequencies ω_j. Usually, $64\,k$ points were chosen, since calculations showed that the entropy differences were converged to within 1% with this choice.

In the quasi-harmonic approximation [9] the Helmholtz free energy may be written as

$$F = k_B T \sum_k \sum_j \ln\left[2\sinh\frac{x_j(\mathbf{k})}{2}\right], \tag{1}$$

where k_B is Boltzmann's constant and we have defined the volume-dependent argument $x_j(\mathbf{k}) = \hbar\omega_j(\mathbf{k}, V)/k_B T$. Other thermodynamic quantities of interest are obtained in the same approximation, such as the specific heats at constant volume and pressure

$$C_V = k_B \sum_k \sum_j \left[\frac{x_j(\mathbf{k})}{2\sinh(x_j(\mathbf{k})/2)}\right]^2$$

and

$$C_P = C_V + TVB_T\beta^2 \tag{2}$$

with bulk modulus and thermal expansivity given by

$$B_T = V\frac{\partial^2 E}{\partial V^2}\bigg]_T \quad \text{and} \quad \beta = \frac{1}{V}\frac{\partial V}{\partial T}\bigg]_P, \tag{3}$$

respectively. Finally, the vibrational entropy is expressed as

$$S = k_B \sum_k \sum_j \left\{ \frac{x_j(\mathbf{k})}{2\tanh(x_j(\mathbf{k})/2)} - \ln\left[2\sinh\frac{x_j(\mathbf{k})}{2}\right]\right\}. \tag{4}$$

No configurational entropy contribution is included in the present treatment. For each temperature, the equilibrium volume is found by minimizing the free energy (1) with respect to V numerically. In that way, the thermal expansivity can be calculated as a function of T, at essentially zero pressure. All quantities

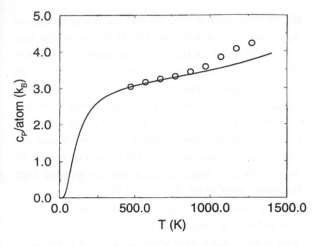

Fig. 1. Specific heat at constant pressure (per atom, in units of Boltzmann's constant) for the ordered $L1_2$ phase of Ni_3Al. The black line is the calculation, and the symbols are data from [10].

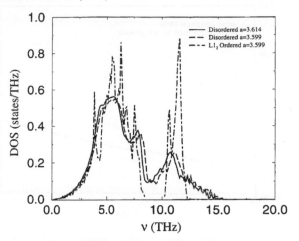

Fig. 2. Vibrational DOS for $L1_2$ ordered (dot–dash curve) and disordered Ni_3Al (full curve), both calculated at the equilibrium lattice constant for the $L1_2$ phase at 400 K and the DOS for the disordered state (dashed curve) calculated at the equilibrium lattice constant of the ordered state.

pertaining to the disordered state were calculated in the fully relaxed condition.

The isobaric specific heat (C_p) was calculated according to Eqs. (2) and (3) for the ordered state and plotted as a full line in Fig. 1. The open circles are experimental data [10]. Agreement is remarkably good over the temperature range for which vibrational effects alone are expected to contribute. Above approximately 800 K, atomic mobility allows configurational effects to come into play so that the experimentally measured total C_p curve begins to deviate from the calculated vibrational one. Indeed, calculations [11] suggest that the concentration of antisite defects in Ni_3Al increases from about 10^{-4} to 10^{-3} between 750 and 1000 K. Vibrational densities of state for both ordered and disordered states were obtained by calculating frequency spectra over roughly $1000 k$ points in the Brillouin zone and counting modes in 0.1 THz energy intervals. As is well known, calculations of the DOS requires a much larger number of k points than for such integrated quantities as the vibrational entropy, for example. Smooth DOS for the disordered state were obtained by averaging over 10 random atomic configurations. Resulting DOS at 400 K are shown in Fig. 2 for the $L1_2$ structure (dot–dash curve) and for the disordered state (full curve), both at their respective equilibrium volumes. Fig. 2 also shows (dashed

curve) the DOS of the disordered state at the volume of the fully ordered phase. The calculated ordered DOS agrees well with the one measured by Stassis et al. [12]. The effects of the disordering are striking: at same volume, the DOS of the disordered state appears as a broadened version of that of $L1_2$, in the sense that the band gap gets filled up, and sharp features of $L1_2$ become diffuse, although the main peak maxima appear roughly at the same frequency positions. The disordered state DOS at its equilibrium volume (full curve) is compressed to lower frequencies with respect to the DOS calculated at the equilibrium $L1_2$ volume. In short, the effect of disordering can be simply described as a broadening of the sharp features of the ordered-phase DOS and a downward shift in frequency with respect to what the DOS would be if the disordered state had the same lattice constant as the $L1_2$ phase.

Two entropy difference curves are plotted in Fig. 3 as a function of temperature, the full line (ΔS_{vib}) corresponding to the ordered and disordered states at their respective equilibrium lattice parameters, the dashed line corresponding to these states, both at the equilibrium lattice parameter of the $L1_2$ phase. Calculated values of ΔS_{vib} for equilibrium volumes are quite large, in fact of the same order of magnitude as those

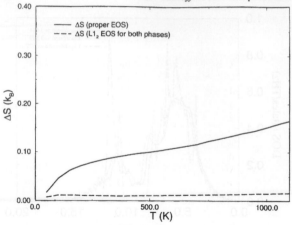

Fig. 3. Entropy differences (disordered minus ordered, in units of Boltzmann's constant) as a function of temperature; full curve: both phases at their equilibrium volumes, i.e using the equilibrium ("proper") equation of state (EOS); dashed curve: both phases at the equilibrium volume of the L1₂ phase.

observed experimentally: $0.11\,k_B$ at 600 K, $0.15\,k_B$ at 1000 K, and $0.27\,k_B$ at 1400 K [5]. It is seen, however, that the entropy difference at constant volume is much smaller. We therefore conclude that the large values of the "equilibrium" ΔS_{vib} are caused primarily by the fact that the volume of the disordered state is significantly larger than the one for the ordered state: the quasi-harmonic results for $V(T)$ for ordered and disordered states show that the equilibrium volumes of the two phases differ by about 1% at 0 K and by about 2% at 1000 K. This difference is apparently large enough to cause about 90% of the large ΔS_{vib} effect, as inferred form the two curves of Fig. 3 in the range from 500 to 1000 K. The other 10% must then be due to rearrangement of vibrational modes due to the disordering and relaxation processes. Our finding that the major cause of the entropy difference in this system is the volume increase on disordering confirms an earlier result of Ackland [6]. These issues will now be further discussed in the light of the DOS calculations.

3. Discussion

To investigate in more detail the nature of the broadening due to configurational disorder, we calculated

the DOS projected onto Ni and Al atoms in the ordered and disordered state [5]. In this way, the narrow double peak in the high frequencies of the DOS of the ordered structure has been identified as being due to the light aluminum atoms on their simple cubic sublattice. We find that the splitting of the high frequency peak from the rest of the spectrum is largely controlled by the mass difference between Ni and Al: a model calculation, in which the masses of Ni and Al were artificially made equal, shows no split-off high frequency peak at all. It was also found that the states of the narrow, high frequency peak in the L1₂ structure (which are predominantly of Al-character) are significantly more broadened upon disordering than the predominantly Ni states at lower frequencies. The fact that the Al states are more affected by the configurational disorder than the Ni states can be rationalized by a simple nearest neighbor bond counting argument [5].

Examination of Fig. 2 indicates that the disordered DOS (full line) is not well represented by a single-atom fcc DOS, thereby casting doubt on the virtual crystal approximation often used implicitly to analyze diffuse neutron diffraction data for disordered phases. In fact, as mentioned above, the disordered state DOS can be regarded as a broadened and shifted L1₂ DOS, the shift being responsible for the major portion of the entropy difference. These findings therefore do not support the interpretation given by Fultz et al. [2] whereby the large values of ΔS_{vib} found in this system are due to rearrangements of optic modes, quite independently of volume effects. Surprisingly, Fultz [13] claims that the lattice parameter of disordered Ni₃Al is less than that of the equilibrium ordered phase, contrary to the results of the present calculations, and to what was observed experimentally by Cardellini et al. [4]. How then do our calculated ΔS_{vib} values come to agree with the measured values obtained by the Caltech group [1,2]? Is the agreement fortuitous? If so, what is the source of the large entropy differences found experimentally in various alloy systems? These question will have to be resolved by further experimental and theoretical investigations.

Acknowledgements

We thank Daryl Chrzan for helpful and stimulating discussions and also Dr. I. Ansara for providing us with a copy of his paper prior to publication, thereby leading us to [10]. This work was supported by the US Department of Energy, Office of Basic Energy Sciences, Division of Materials Science under contract Nos. DE-AC04-94AL85000 (Sandia) and DE-AC03-76SF00098 (Berkeley).

References

[1] L. Anthony, J.K. Okamoto and B. Fultz, Phys. Rev. Lett. 70 (1993) 1128.

[2] B. Fultz et al., Phys. Rev. B 52 (1995) 3315.

[3] J.A. West, J.T. Manos and M.J. Aziz, Mater. Res. Soc. Symp. Proc. 213 (1991) 859.

[4] F. Cardellini et al., J. Mater. Res. 8 (1993) 2504.

[5] J.D. Althoff, D. Morgan, D. de Fontaine, M.D. Asta, S.M. Foiles and D.D. Johnson, Phys Rev. B 56 (1997) R5705.

[6] G.J. Ackland, in: Alloy Modeling and Design, eds. G. Stocks and P. Turchi (The Minerals, Metals, and Materials Society, 1994) p. 149.

[7] M.S. Daw and M.I. Baskes, Phys. Rev. B 29 (1984) 6443; M.S. Daw, S.M. Foiles and M.I. Baskes, Mater. Sci. Rep. 9 (1993) 251.

[8] A.F. Voter and S.P. Chen, Mater. Res. Soc. Symp. Proc. 82 (1988) 175.

[9] A.A. Maradudin, E.W. Montroll, G.H. Weiss and I.P. Ipatova, Theory of Lattice Dynamics in the Harmonic Approximation, 2nd Ed. (Academic Press, New York, 1971).

[10] A.I. Kovalev, M.B. Bronfin, Yu. V. Loshchinin and V.A. Vertogradskii, High Temp. High Pressures 8 (1976) 581.

[11] S.M. Foiles, J. Mater. Res. 2 (1987) 5.

[12] C. Stassis, F.X. Kayser, C.-K. Loong and D. Arch, Phys. Rev. B 24 (1981) 3048.

[13] B. Fultz, private communication.

COMPUTATIONAL
MATERIALS
SCIENCE

ELSEVIER

Computational Materials Science 10 (1998) 416–420

Structure of point defects in B2 Fe–Al alloys: An atomistic study by semi-empirical simulation

Rémy Besson [a,b,*], Joseph Morillo [a,1]

[a] *Laboratoire des Solides Irradiés, Commissariat à l'Energie Atomique, URA CNRS 1380, Ecole Polytechnique, 91128 Palaiseau Cedex, France*
[b] *Centre Science des Matériaux et des Structures, URA CNRS 1884, Ecole des Mines de Saint-Etienne, 42100 Saint-Etienne Cedex 9, France*

Abstract

We present the first results of a simulation study of point-defect properties in B2 ordered Fe–Al alloys ($34\% < x_{Al} < 52\%$). After examining the $T = 0\,K$ energetics of simple defects, we turn to complexes and discuss the usual hypothesis of independent elementary defects. Almost all complex defects involving an Al vacancy and an Al antisite atom in nearest neighbour position are shown to be unstable, confirming that Al vacancies are rare in these alloys. Small deviations from stoichiometry induce a change in the nature of defects: whereas on the Al-rich side, isolated Al antisite atoms dominate, Fe antisite atoms have a strong trend towards clustering on the Fe-rich side, leading to the formation of local D0₃ order. Copyright © 1998 Elsevier Science B.V.

Keywords: Point defects; B2 structure; Fe–Al alloys; Atomistic simulation; Semi-empirical potentials

1. Introduction

The major drawback of B2 ordered iron aluminides, which has strongly hindered their industrial use up to now, is their room-temperature intergranular brittleness, a property dependent on a wide variety of factors (for a review, see e.g. [1]). As a preliminary study of the atomic-scale mechanisms responsible for this weakness, we present here a detailed analysis of volume point defects simulated with a semi-empirical potential [2]. After a brief description of the compu-

tations (Section 2) we will first (Section 3) discuss the strengths and limitations of the widely used four-independent-simple-defect (4ISD) model [2–5], after which we will focus our attention on interactions between these defects (Section 4), possibly leading to the formation of complexes.

2. Computations

Most of the reported simulation results were performed with a 15*15*15 simulation cubic cell with periodic boundary conditions (for more technical details refer to [2,6]).

In the text, a defect D on the sublattice associated to atom A will be noted D_A, with A = Fe, Al and

---

* Corresponding author. Address: Laboratoire des Solides Irradiés, Ecole Polytechnique, 91128 Palaiseau Cedex, France. Tél.: 01 69 33 38 25; e-mail: besson@lyto.polytechnique.fr.
[1] Tél.: 01 69 33 46 82; fax: 01 69 33 30 22; e-mail: joseph.morillo@polytechnique.fr.

Table 1
Chemical potentials and formation energies of simple defects (eV) in B2 Fe–Al ($T = 0$ K) at and close to the stoichiometric composition

	Chemical potentials		Formation energies			
	μ_{Fe}	μ_{Al}	V_{Fe}	Al_{Fe}	V_{Al}	Fe_{Al}
Raw energy			5.48	1.92	6.42	−0.25
Fe-rich side	−4.20	−3.95	1.28	1.67	2.47	0.00
Stoichiometric	−4.58	−3.57	0.90	0.90	2.85	0.77
Al-rich side	−5.04	−3.11	0.44	0.00	3.31	1.68

D = Fe, Al or V (vacancy). A complex defect will be specified by mention of the simple defects forming it and of their level of neighbourhood: for example, a triple defect with components D1, D2 and D3 will be noted $(D1 + D2 + D3)_{N12, N13, N23}$, with Nij the number of neighbour shells between defects Di and Dj.

3. Four-independent-simple-defect approximation

3.1. General considerations

Within the framework of the usual grand canonical formalism, the concentrations of the four independent elementary defects in B2 compounds can be obtained as a function of temperature, chemical potentials (the latters being determined via the composition of the alloy and the zero pressure condition) and raw formation energies [2–6]. As usual, in the first-approximation level of description, the defect entropies and volumes are neglected. Defect formation energies appear to have a complex temperature and composition dependence via the chemical potentials. However, they have very well-defined values at and close to the stoichiometric composition in the $T = 0$ K limit (Table 1).

3.2. Low temperature defect structure

At stoichiometry, elementary B2 point defects can be classified (according to their formation energies) into two groups, one with the two antisite defects and the Fe vacancy, with formation energies close to 1 eV,

and the other with the Al vacancy, which appears to be a very energetic defect.

For small departures from stoichiometry, we recall [2–6] that (i) the chemical potentials of both species (and consequently the defect formation energies) are constant on each side of the equiatomic composition, at which they show a discontinuity and (ii) only one type of defect can accommodate deviations from stoichiometry. The nature of this defect (for the A-rich side for example) depends on the sign of the quantity $\Delta = E_f(A_B) - 2E_f(V_B)$. If this quantity is positive, the material possesses B (structural) vacancies at $T = 0$ K, otherwise its defect structure is constituted of A antisite atoms. Application of this criterion to Fe–Al leads to the values $\Delta(\text{Fe-rich}) = -4.93$ eV and $\Delta(\text{Al-rich}) = -0.90$ eV, which clearly indicates that, according to our model, B2 Fe–Al alloys should accommodate departures from stoichiometry by antisite defects of either type at low temperatures.

3.3. High temperature defect structure

The first point to be noted (in agreement with their high formation energy value) is that Al vacancies are present in very small quantities, even at relatively high temperatures and whatever the composition. Apart from this point, in the equiatomic compound, the three types of defects (both kinds of antisites and Fe vacancies) exist in very close proportions. This constitutes a further hint of the mixed behaviour of Fe–Al among B2 compounds. Most B2 alloys have been shown to belong either to the *substitutional defect* group (accommodating departures from stoichiometry as well as thermal disorder with various amounts of both antisite defects), or to the *triple defect* group (in which this is achieved with A vacancies and A antisite atoms). Due to the low formation energy of Fe vacancies, B2 Fe–Al alloys exhibit a change in their defect structure from antisite type at low temperatures to mixed type at high temperatures, which distinguishes them, in particular, from B2 Ni–Al alloys (of triple defect type whatever the temperature).

In Fe-rich alloys, Fe antisite atoms appear to be the predominant defects, whereas in Al-rich alloys, the defect structure is more complex, involving at the same time Al antisite atoms and Fe vacancies (this must be distinguished from the classical triple defect structure). Finally, with increasing Al concentration, the amounts of Fe antisite atoms and Fe vacancies respectively, decrease and increase, in agreement with Mössbauer [7] and dilatometry [8] experiments.

Several other works [4,5] have coupled the preceding independent-defect analysis to ab initio evaluations of formation quantities. The overall agreement between both approaches is satisfactory, the main difference lying in an underestimate of about 1 eV of the Al vacancy formation energy in our semi-empirical calculations. A possible explanation for this discrepancy could be a poor convergence of ab initio calculations with respect to the simulation cell size. Indeed, with our semi-empirical model we obtained an increase of the Al vacancy formation energy of about 1 eV when the cell size is comparable to those of the ab initio models.

Thus, the 4ISD approximation enables a qualitative description of the main trends experimentally detected in the defect structure of B2 Fe–Al alloys. However, it suffers from severe limitations. The first, and main one, is obviously that it leads to a strong underestimate of vacancies concentration (between one and two orders of magnitude) compared to that obtained from experiments [7,8]. Although the large dispersion of experimental results (reflecting probable non-equilibrium situations) may throw doubt on their reliability, this trend nevertheless remains clear. On the other hand, the remarkable agreement existing between the various approaches to estimate the defect formation energies (present work and [4,5])

suggests that the divergence is bound to come either from the existence of complex defects, or from the overlooking of the temperature dependance of defect formation free energies. This last point seems to be supported by the work of Würschum et al. [9], who obtained by positron annihilation measurements in Fe-rich alloys, values around $6 k_B$ for the formation entropies (much higher than the classical values of about $1 k_B$ associated with metals).

The second shortcoming of this 4ISD model lies in its inability to predict the probable existence of a maximum vacancy concentration with temperature (for T around 1250 K), detected by dilatometry measurements [10]. Strictly speaking, the 4ISD model is able to yield such a behaviour but, in the case of Fe–Al, the formation energy values are such that no maximum can exist. Thus, to account for this property, Paris and Lesbats [10] successfully used a mean-field model initially developed by Cheng et al. [11], based on the arbitrary introduction of complexes constituted of one vacancy surrounded by various numbers of antisite atoms.

To improve the 4ISD hypothesis, two ways are then available (either vibrational or complex defect contributions): whereas work is in progress to assess formation free energies more exactly, we will now discuss the influence of $T = 0$ K interactions between elementary defects.

4. Study of complex defects

The fundamental problem that leads to the concept of complex (or multiple) defect is related to the non-degeneracy of the energy of the system with respect to the relative positions of simple defects when these defects come close to each other (a few neighbour shells). In the 4ISD model, the energy was assumed to be degenerate in such a way that the relevant variables entering the expression of energy were simply the four numbers of simple defects. To go beyond this approximation, the usual procedure consists in regarding also the numbers of complex defects as additional new variables in the energy.

Table 2
Absolute binding and formation energies (eV) of the most stable double and triple defects in B2 Fe–Al ($T = 0\,K$) at and close to the stoichiometric composition

	Absolute binding energy	Formation energies		
		Fe-rich	Stoichiometric	Al-rich
$(Fe_{Al} + Al_{Fe})_1$	−0.31	1.36	1.36	1.36
$(2Fe_{Al})_3$	−0.18	−0.18	1.35	3.18
$(2V_{Fe})_2$	−0.21	2.35	1.58	0.67
$(3Fe_{Al})_{333}$	−0.55	−0.55	1.75	4.49
$(Al_{Fe} + 2V_{Fe})_{232}$	−0.31	3.92	2.39	0.56

4.1. Stoichiometric alloy

4.1.1. Double defects

From the systematic study of all double defects energetics as a function of the distance between component defects [6], it is found that many double defects have a significant binding energy (larger than 0.1 eV). Among these complexes, only three need to be retained (all the others having too high a formation energy to be present in non-negligible proportions): $(2V_{Fe})_2$, $(Fe_{Al} + Al_{Fe})_1$ and $(2Fe_{Al})_3$ (Table 2).

The second most striking feature concerning double defects relates to the instability of the complex $(Al_{Fe} + V_{Al})_1$, which spontaneously recombines at $T = 0\,K$ to give a single Fe vacancy. This feature, which proves to be quite independent of the presence of additional defects (see below the case of triple defects), partially rules out the model of Paris and Lesbats [10], since almost all the complexes involving an Al vacancy they considered are in fact unstable (on the whole, only three such complexes are found to exist: $(V_{Al} + 8Al_{Fe})$, $(V_{Al} + 7Al_{Fe})$ and $(V_{Al} + 6Al_{Fe})$, the latter being such that the two normal Fe atoms surrounding the Al vacancy are located along the $\langle 1\,1\,1 \rangle$ direction).

4.1.2. Triple defects

The study of triple defects is more intricate than that of double defects, because there exists an ambiguity upon the definition of binding energies. In fact, one can define several binding energies, the first of which (here called absolute), noted $E_b(T)$ for a triple defect T, uses as a reference the formation energies of the three component defects. But other binding

energies (here called relative and noted $\tilde{E}_b^1(T)$), that take as a reference the formation energy of a simple defect (labelled S_1) and that of a double defect (noted $D_{2,3}$), must also be considered. The following relation holds between absolute and relative binding energies: $\tilde{E}_b^1(T) = E_b(T) - E_b(D_{2,3})$. If $\tilde{E}_b^1(T)$ is close to zero, then all the states including triple defects T are properly described by single defects S_1 and double defects $D_{2,3}$. Thus, only those triple defects for which it is possible to find at least a decomposition $(S_1 + D_{2,3})$ leading to a relative or absolute non-zero binding energy value are really significant.

From the systematic study of compact triple defect energetics [6], it appears that (i) since the lowest triple defect formation energy is 1.75 eV, the equilibrium amount of such defects should be very low (compared to those of single and double defects), even at high temperatures, (ii) many triple defects (Table 2) present strongly negative absolute and relative binding energies and (iii) all triple defects in $(3\,3\,3)$ configuration (that is, in a $\{1\,1\,1\}$ plane) exhibit very negative binding energies, demonstrating a clear tendency of triple defects to arrange in $\{1\,1\,1\}$ planes. The case of $(3Fe_{Al})_{333}$ is particularly significant, since it can be regarded as the initial stage of a local $D0_3$ ordering (the low temperature structure of Fe_3Al).

Finally, as previously mentioned, almost all the triple defects including the complex $(Al_{Fe} + V_{Al})_1$ prove to be unstable. The only exception to this rule is the complex $(V_{Al} + Al_{Fe} + Fe_{Al})_{131}$, with a high formation energy of 3.27 eV. Conversely, the triple defect $(V_{Fe} + Al_{Fe} + Fe_{Al})_{151}$, not involving the complex $(Al_{Fe} + V_{Al})_1$, is found to recombine spontaneously into a single Fe vacancy by a collective movement of the pair (Al + Fe) along $\langle 1\,1\,1 \rangle$.

4.2. Influence of departure from stoichiometry

Exactly as in the case of single defects, the sharp variation of chemical potentials when shifting from the Fe-rich to the Al-rich side induces a strong variation of the formation energies of complexes (Table 2).

On the Al-rich side, the lowest formation energy is finally reached with the isolated Al_{Fe} defect (compare the last row of Table 1 and last column of Table 2), which is definitely the predominant defect. Secondary defects are single and double Fe vacancies and complexes involving both Al_{Fe} and V_{Fe}. The increase of formation energy when creating these double and triple defects is low enough to ensure that they are present in quantities almost equal to those of single Fe vacancies.

On the Fe-rich side, the main point concerns the negative formation energies of $(2Fe_{Al})_3$ and $(3Fe_{Al})_{333}$ (note the high absolute value of this last one: 0.55 eV), showing that the approximation of independent defects is unable to describe properly Fe-rich alloys. Moreover, according to this result, off-stoichiometry on the Fe-rich side is expected to be accommodated by small clusters of local DO_3 order.

5. Conclusion

This semi-empirical simulation study has shown that, whereas the four-independent-simple-defect model is accurate for the description of stoichiometric B2 FeAl (with a certainly important vibrational contribution to formation free energies), defect interactions are essential for the description of any, even small, departure from B2 stoichiometry. Moreover,

the defect structure is shown to be very sensitive to the composition, a point that might be critical for the reliability of measurements if samples are not fully homogeneous. In order to go beyond this simple static description of off-stoichiometry, Monte-Carlo simulations in the reduced grand canonical ensemble without vacancies are in progress. Although the latter approach is usually sufficient in common alloys, preliminary results seem to indicate that vacancies are needed to stabilize off-stoichiometric compositions, rendering the study more intricate.

Acknowledgements

We are indebted to M. Biscondi, A. Fraczkiewicz and V. Pontikis for helpful discussions. One of us (RB) also wishes to thank M. Biscondi and T. Magnin for providing him the opportunity of collaboration with the LSI.

References

[1] J.H. Westbrook and R.L. Fleischer, eds., Intermetallic Compounds: Principles and Practice (Wiley, New York, 1994).
[2] R. Besson and J. Morillo, Phys. Rev. B 55 (1997) 193.
[3] J. Mayer, C. Elsässer and M. Fähnle, Phys. Stat. Sol. B 191 (1995) 283.
[4] C.L. Fu, Y.Y. Ye, M.H. Yoo and K.M. Ho, Phys. Rev. B 48 (1993) 6712.
[5] Y. Mishin and D. Farkas, Philos. Mag. A 75 (1997) 169.
[6] R. Besson and J. Morillo, to be published.
[7] G.S. Collins and L.S.J. Peng, Il Nuovo Cimento D 18 (1996) 329.
[8] K. Ho and R.A. Dodd, Scripta Metall. 12 (1978) 1055.
[9] R. Würschum, C. Grupp and H.E. Schaefer, Phys. Rev. Lett. 75 (1995) 97.
[10] D. Paris and P. Lesbats, J. Nucl. Mater. 69–70 (1978) 628.
[11] C.Y. Cheng, P.P. Wynblatt and J.E. Dorn, Acta Metall. 15 (1967) 1045.

ELSEVIER

Computational Materials Science 10 (1998) 421–426

COMPUTATIONAL
MATERIALS
SCIENCE

Computational treatment of order–disorder processes by use of the cluster variation method

V.M. Matic [1]

Laboratory of Theoretical Physics and Solid State Physics, Institute of Nuclear Sciences – Vinca, PO Box 522,
11000 Belgrade, Serbia, Yugoslavia

Abstract

We present an implementation of the cluster variation method (CVM), applied to the Ising model, for computational characterization of order–disorder processes in binary systems. We show how the Newton–Raphson iteration scheme (NRIS) is used for numerical solving of system of nonlinear equations, obtained from the condition of the free energy minimum within the framework of the CVM approximation. An emphasis is made on the problem of the starting iteration point (NRIS being very sensitive to the choise of this point), for obtaining the low-temperature ordered phases. It was shown that an infinitesimally small breaking of symmetry of the high-temperature disordered phases supresses finding solutions which correspond to the metastable phases (saddle points) by NRIS (below the critical temperature T_c). This kind of problem is illustrated by an example of oxygen ordering in basal planes of $YBa_2Cu_3O_{6+x}$ system, modeled by the two-dimensional asymetric next-nearest neighbor Ising (ASYNNNI) model. Copyright © 1998 Elsevier Science B.V.

1. Introduction

At the present time, the cluster variation method (CVM) represents one of the most powerful theoretical methods for treatment of cooperative phenomena. It has been proposed by Kikuchi [1,2] as the method based on combinatorial counting of ensemble configurations which belong to a lattice macrostate defined through a set of cluster probabilities. Hijmans and de Boer, in their frequently cited paper [3], reformulated CVM on general theoretical grounds, while significant contribution to further development of the method was made by Morita [4], Kikuchi [5], van Baal [6] and Sanchez and de Fontaine [7,8].

[1] Tel.: 381 (11) 8363020 (direct), 458222/269, 716, 766; fax: 381 (11) 4440195; e-mail: vmatic@rt270.vin.bg.ac.yu.

The accuracy of the CVM depends on the size of the largest (basic) cluster included in the approximation – it has been well established that the accuracy of the CVM increases with the size of the basic cluster. An increase of size of the basic cluster causes an increase of the number of variational parameters, which makes the problem of the thermodynamic potential (free energy) minimization more complicated. In order to include all energy contributions in the total energy of the system, it is necessary to choose basic cluster sufficiently large to include all interactions that are contained in the Hamiltonian. In cases in which the characteristic structural features extend beyond the range of the interactions included in Hamiltonian, it is sometimes necessary to choose the basic cluster large enough to include these structural features. In this way, the correlation functions, which are convenient to describe the structural features, can be expressed through the quantities that determine the statistics of the basic cluster. Therefore, one has to find equilibrium between the quality of the approximation (which becomes improved by enlarging the size of the basic cluster) and the complexity of the calculations (which also rises with the size of the basic cluster).

Nowadays, when the high-speed computing machines have become widely available, numerical solving of the CVM equations depends on efficiency of the linear programming only, while the complexity of the calculations is related to the construction of CVM equations. Once the approximation is defined (i.e. the basic cluster, or a set of basic clusters), the system of nonlinear equations, expressing the condition for minimum of the thermodynamic potential Ω (or of the free energy F) for given values of the chemical potential μ (concentration c) and temperature T, must be solved numerically. The most widely used numerical method is the Newton–Raphson iteration scheme (NRIS), which is very sensitive (in the CVM calculations) to the location of starting iteration point in space of variational variables.

This paper is organized as follows. In Section 2, we present a general formalism of the cluster variation method, while in Section 3 we show an implementation of the NRIS technique for the numerical solution of CVM equations in case of ordered phases. In particular, we show that suitable starting iteration point for ordered structural phases (Section 3) can be obtained by a small breaking of symmetry of the high-temperature phase. This procedure is illustrated by an example of oxygen ordering in basal planes of the high-temperature superconducting material $YBa_2Cu_3O_{6+2c}$.

2. Brief rewiev of the cluster variation method

In the approximation of the cluster variation method the macrostate of the system is determined by the set of probabilities of the nonequivalent microstates of the basic cluster. The central idea is to express any extensive thermodynamic quantity in the form of the cluster expansion. For example, the thermodynamic potential per lattice site $\omega = \Omega/N$ is written as

$$\omega = \sum_l \gamma_l \Omega_l, \tag{1}$$

where the index l enumerates clusters of the Kikuchi's cluster family, while the coefficients γ_l (the so-called Kikuchi's coefficients) are introduced to take into account the cluster overlap. Ω_l is the thermodynamic potential of the l-cluster ensemble comprised of $\gamma_l N$ identical (and noninteracting) clusters of the type l, and it can be expressed in the form

$$\Omega_l = \sum_{i=1}^{b(l)} \alpha_{l,i} E_{l,i} x_{l,i} + k_B T \sum_{i=1}^{b(l)} \alpha_{l,i} x_{l,i} \ln x_{l,i}. \tag{2}$$

Here the index i denotes nonequivalent microstates of the cluster l (with respect to the symmetry of the structural phase considered), $b(l)$ is total number of microstates of the cluster l, $\alpha_{l,i}$ and $x_{l,i}$ are the factors of degeneracy and the probability of occurrence of the ith microstate of the cluster l, respectively, while $E_{l,i}$ is the total energy associated with the ith microstate of the cluster l. The cluster probabilities $x_{l,i}$ are related to the multisite correlation functions through the following set of equations

$$x_{l,i} = \frac{1}{2^{p(l)}} \left[1 + \sum_j v_{l,i,j} \xi_j \right], \tag{3}$$

where the index j denotes all nonequivalent subclusters of the cluster family $(l = 1, 2, \ldots)$, $p(l)$ is the number of spins in the cluster l and ξ_j is the multisite correlation function associated with the subcluster j. The coefficients $v_{l,i,j}$ are calculated as follows: for a given subcluster j of the cluster l, $j \subseteq l$ (the cluster l being in its microstate i), the product of its spins is calculated and these products (of all subclusters of the type j) are summed. The coefficients $v_{l,i,j}$ have few important properties:

$$\sum_{i=1}^{b(l)} \alpha_{l,i} v_{l,i,j} = 0, \tag{4a}$$

$$\sum_{i=1}^{b(l)} \alpha_{l,i} v_{l,i,j} v_{l,i,j'} = 2^{p(l)} v_{l,i=1,j} \delta_{j,j'}, \tag{4b}$$

$$n_j = \sum_l \gamma_l \nu_{l,i=1,j}. \qquad (4c)$$

In the above equations the first cluster microstate ($i = 1$) is chosen to be that in which all spins are oriented upwards – therefore $\nu_{l,i=1,j}$ simply denotes the number of subclusters j which are contained in the cluster l. The quantity n_j denotes the number of clusters of the type j per lattice site.

Inserting Eqs. (2) and (3) into Eq. (1) we obtain

$$\omega = \sum_j n_j V_j \xi_j + k_B T \sum_l \gamma_l \sum_{i=1}^{b(l)} \alpha_{l,i} x_{l,i} \ln x_{l,i}. \qquad (5)$$

The thermodynamic potential ω must be minimized with respect to the correlation functions ξ_j, yielding

$$\frac{\partial \omega}{\partial \xi_j} = 0$$

$$\Rightarrow n_j V_j = -k_B T \sum_l \frac{\gamma_l}{2^{p(l)}} \sum_{j=1}^{b(l)} \alpha_{l,i} \nu_{l,i,j} \ln x_{i,j}. \qquad (6)$$

In the above equations V_j denotes the effective interaction energy associated with the cluster j through the expression for the Ising model Hamiltonian

$$H = -\mu \sum_p \sigma(p) + \sum_{p \geq p'} V_{pp'} \sigma(p) \sigma(p')$$
$$+ \sum_{p,p',p''} V_{pp'p''} \sigma(p) \sigma(p') \sigma(p'') + \cdots, \qquad (7)$$

where $\sigma(p)$ denotes the Ising spin variable at the lattice site p. In Eq. (6), the interaction energy V_j associated with a single spin cluster equals the chemical potential μ (or, equivalently, the external field).

There is another approach to CVM, which uses effective fields as variational variables, instead the multisite correlation functions ξ_j (for some systems, on the other hand, the effective fields approach is more suitable). The fields are introduced in CVM via

$$x_{l,i} = \frac{1}{Z_l} e^{(-1/k_B T) \sum_j \nu_{l,i,j}(V_j + \Psi_{l,j})}, \qquad (8)$$

where Z_l denotes the statistical sum for the cluster l, $\Psi_{l,j}$ is an effective field of the cluster j (which is

considered as a subcluster of the cluster l), and k_B is the Boltzman constant. Inserting Eq. (8) in the system of Eqs. (6) we obtain

$$\sum_l \gamma_l \nu_{l,i=1,j} \Psi_{l,j} = 0. \qquad (9)$$

The above equation implies that not all effective cluster fields $\Psi_{l,j}$ are linearly independent. The thermodynamic equilibrium is obtained when the thermodynamic potential, which is determined by Eqs. (1) and (2), is minimized with respect to the subset of linearly independent cluster fields, i.e.

$$\frac{\partial \omega}{\partial \Psi_{l,j}} = 0. \qquad (10)$$

The systems of nonlinear equations (6), or (10), must be solved numerically for fixed values of the chemical potential μ and the temperature T.

3. Implementation of the Newton–Raphson iteration scheme

The Newton–Raphson iteration scheme (NRIS) is the most widely used technique to solve the systems of equations (6) and (10). For a given starting iteration point (SIP), in space of variational variables, NRIS creates the sequence of points in the direction of the smallest gradient until it reaches the minimum (MIN) of the thermodynamic potential Ω. If in the area B, which is located between the points SIP and MIN, the first and the second derivatives of the thermodynamic potential are not defined, the NRIS will clearly not be able to reach the thermodynamic equilibrium (i.e. the point MIN). Another kind of problem is detected when, in the area B, the thermodynamic potential Ω has values that are greater than that in the point SIP. Having in mind the fact that NRIS allways flows "downwards", it will form the sequence of points in the direction opposite to those in which the point MIN is located – consequently, it will never reach the equilibrium.

The best way to avoid these difficulties is to choose the SIP as close to the point MIN as possible. As a result, the area B would be small enough and the derivatives of the thermodynamic potential Ω would change monotonically. At temperatures which are high enough, the disordering temperature effects are so strong that the behavior of each spin is weakly influenced by the behavior of the neighboring spins. This means that correlation between spins is weak and consequently $\xi_j = 0$ (for all clusters j) is a good starting iteration point. When the first point is solved, we continue to decrease the temperature in small steps using the previously solved point as a good SIP for the new (decreased) temperature. As the critical temperature T_c is approached, which separates the high-temperature (disordered) structural phase from the ordered (low-temperature) phase, the equilibrium minimum of the thermodynamic potential Ω becomes more flat. The flatness is measured by the second increment $d^2\Omega$, which can be expressed by

$$d^2\Omega = \sum_{j,j'} \frac{\partial^2 \Omega}{\partial \xi_j \partial \xi_{j'}} \delta\xi_j \delta\xi_{j'}. \qquad (11)$$

If the minimal eigenvalue $\lambda_{\min}$ of the matrix of the second derivatives is positive, $d^2\Omega$ is positively definite quadratic form and the solution of the structural phase is stable. For $\lambda_{\min} = 0$, which occurs at $T = T_c$, the hypersurface of the thermodynamic potential Ω is flat, at the point which is the solution of the systems of equations (6) or (10).

When the temperature is further decreased below T_c, NRIS is not capable to avoid solutions for which $\lambda_{\min} < 0$. These solutions are the "saddle points", having the symmetry of high-temperature (disordered) phase. Physically, these solutions represent the metastable phase continuation (below T_c) of disordered phase. Thus, NRIS possesses a kind of inertial feature with respect to the change of symmetry of structural phases, i.e. it does not turn to a structural phase with reduced symmetry (ordered phase). Such behavior of NRIS is not strange since it always flows in the direction of the smallest gradient. The saddle point always has at least one direction in which the gradient is negative, which will force NRIS to flow to another saddle point, when the temperature is further decreased. $\lambda_{\min} = 0$ at $T = T_c$, and great number

of saddle points are located very close to that point (in space of variational variables) which corresponds to the critical temperature, and that is why the NRIS moves in the direction where the saddle point is located rather than in the direction where the minimum of ordered phase is located.

To avoid obtaining the metastable phase below T_c, we suggest the procedure of the symmetry breaking of high-temperature phase, at a point that corresponds to $T > T_c$, by an infinitesimally small amount. It is important to note that the symmetry breaking of high-temperature (disordered) phase above T_c must be performed in such a way to obtain the symmetry of the low-temperature (ordered) phase. The symmetry breaking is made by appropriate changes of the interaction energies $V_1, V_2, \ldots, V_i, \ldots$ which are included in Hamiltonian (7), i.e. $V_i \rightarrow V_i + \delta V_i$, $\delta V_i \cong 0$. This transformation will cause the order parameter to become different from zero at $T > T_c$ (although, it will acquire very small values) and no phase transition will take place at $T = T_c$. Instead, the simulation of the transition will occur at $T \cong T_c$ in the following way: as the temperature is decreased, from above T_c, the order parameter will start to increase at $T \cong T_c$ and continue to increase as the temperature is further decreased in small steps.

At the point, in the space of variational variables, in which the order parameter η gets a large enough value (say $\eta > 0.7$), we restore the original system by setting $\delta V_i = 0$. The values of variational variables (ξ_j's or $\Psi_{l,j}$'s) which correspond to this point, in the space of variational variables, define a very good starting iteration point for solution of ordered structural phase, for given values of the concentration (chemical potential) and temperature $T < T_c$.

The procedure of symmetry breaking of high-temperature phase is based on the fact that the mapping $\{V_1, V_2, \ldots, V_i, \ldots\} \rightarrow \zeta(c, T, V_1, V_2, \ldots, V_i, \ldots)$ is continious, i.e. $\delta V_i \rightarrow 0 \Rightarrow \delta\zeta \rightarrow 0$, where ζ denotes the equilibrium value of any thermodynamic quantity. The SIP, obtained in this way, is found to be located very close to the point in which thermodynamic potential Ω is minimal (since $\delta V_i \cong 0$).

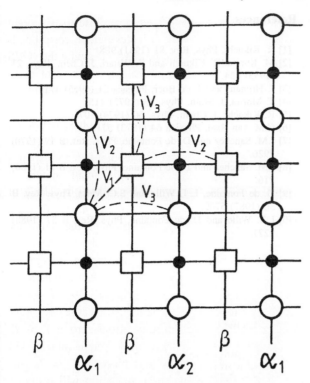

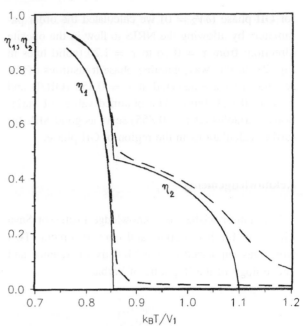

Fig. 1. The basal plane of YBa$_2$Cu$_3$O$_{6+x}$ material. Open circles and squares represent oxygen sites on α (α_1 and α_2) sublattices, respectively, while small black circles denote Cu(1) ions. The effective pair interactions $V_1 > 0$, $V_2 < 0$ and $V_3 > 0$ between oxygen atoms (i.e. the spins of Ising model) are also shown.

Fig. 2. Calculated values of the order parameters η_1 and η_2, as a function of reduced temperature $\tau = k_B T / V_1$, at $\mu = -7.9792 = $ const. The solid lines represent values obtained by use of the unchanged interactions ($\delta V_1 = 0$, $\delta V_2 = 0$ and $\delta V_3 = 0$); the temperature increases from $\tau = 0.76$ to $\tau = 1.20$. The dashed lines represent values obtained by use of modified interactions, as it was indicated in text (the temperature decreases from $\tau = 1.20$ to $\tau = 0.76$). Calculations were made by applying the NRIS to the cluster variation method in approximation of six 5/4 point basic clusters, for $V_2 = -0.83V_1$ and $V_3 = 0.11V_1$.

4. Oxygen ordering in basal planes of YBa$_2$Cu$_3$O$_{6+x}$. Ordered phases

Oxygen ordering in basal planes of YBa$_2$Cu$_3$O$_{6+x}$ material is the most succesfully described in terms of the well-known two-dimensional asymetric next-nearest neighbor Ising model (ASYNNNI) of de Fontaine et al. [9] and Wille and de Fontaine [10]. The basal plane lattice of YBa$_2$Cu$_3$O$_{6+x}$ is shown in Fig. 1 together with the effective pair interactions V_1, V_2, V_3, which define the ASYNNNI model. For $V_1 > 0$, $V_3 > 0$ and $V_2 < 0$, the model stabilizes two ordered structures [10]: OII in which all α_1 sites are occupied by oxygen atoms (while α_2 and β sites are unoccupied), and OI phase in which all α sites (α_1 and α_2) are occupied (while β sites are empty). In tetragonal phase (disordered) all sites are randomly occupied.

In order to reach OI and OII phases, we make the symmetry breaking of the tetragonal phase by the following transformations: $V_2^{(\alpha_1,\alpha_1)} \rightarrow V_2^{(\alpha_1,\alpha_1)} + 0.02V_1$, $V_2^{(\alpha_2,\alpha_2)} \rightarrow V_2^{(\alpha_2,\alpha_2)} - 0.02V_1$, $V_2^{(\beta,\beta)} \rightarrow V_2^{(\beta,\beta)} - 0.01V_1$ ($\delta V_1 = 0$, $\delta V_3 = 0$). Using the altered values of interactions, we calculated the order parameters $\eta_1 = c_{\alpha_1} - c_{\alpha 2}$ and $\eta_2 = c_{\alpha 1} - c_\beta$ as a function of reduced temperature $\tau = k_B T / V_1$, at constant value of chemical potential $\mu = -7.9792V_1 = $ const., decreasing temperature from $\tau = 1.20$ to $\tau = 0.76$ in small steps. The obtained values are presented by dashed lines in Fig. 2. We see that simulation of Tetra/OI and OI/OII transitions occur at $\tau \cong 1.12$ and $\tau \cong 0.86$, respectively. Using the calculated values of variational variables (in this case, the effective cluster fields $\Psi_{l,j}$'s) at $\tau = 0.76$ as a good first SIP

V.M. Matic / Computational Materials Science 10 (1998) 421–426

for OII phase ($\delta V_2 = 0$) we calculated the order parameters by allowing the NRIS to flow in the oposite direction: from $\tau = 0.76$ to $\tau = 1.20$ (solid lines in Fig. 2). In this way, genuine phase transitions of the second order are detected at $\tau = 0.855$ (OII/I) and $\tau = 1.10$ (OI/Tetra). The obtained values of variational variables (at $\tau < 0.855$) serve as good SIPs for further calculations in the region of OII phase.

Acknowledgements

The author wishes to acknowledge Professor Sava Milosevic for his support and encouragement. This work was supported by the Ministry of Science and Technology of the Republic of Serbia.

References

[1] R. Kikuchi, Phys. Rev. 81 (1951) 988.
[2] M. Kurata, R. Kikuchi and T. Watari, J. Chem. Phys. 21 (1953) 434.
[3] J. Hijmans and J. de Boer, Physica 21 (1955) 471.
[4] T. Morita, J. Math. Phys. 13 (1972) 115.
[5] R. Kikuchi, J. Chem. Phys. 60 (1974) 1071.
[6] C.M. van Baal, Physica 64 (1973) 571.
[7] J.M. Sanchez and D. de Fontaine, Phys. Rev. B 17 (1978) 2926.
[8] J.M. Sanchez and D. de Fontaine, Phys. Rev. B 21 (1980) 216.
[9] D. de Fontaine, L.T. Wille and S.C. Moss, Phys. Rev. B 36 (1987) 5709.
[10] L.T. Wille and D. de Fontaine, Phys. Rev. B 37 (1988) 2227.

Computational Materials Science 10 (1998) 427–431

COMPUTATIONAL
MATERIALS
SCIENCE

Angle dependence and defect production in metal-on-metal cluster deposition on surfaces

C. Félix [a,b], C. Massobrio [c,*], B. Nacer [d], T. Bekkay [d]

[a] *Institut de Physique Expérimentale, Ecole Polytechnique Fédérale, CH-1015 Lausanne, Switzerland*
[b] *Northwestern University, Department of Chemistry, Evanston, IL, 60208, USA*
[c] *Institut de Physique et de Chimie des Matériaux de Strasbourg, 23, rue du Loess, F-67037 Strasbourg, France*
[d] *Université Cady Ayyad, Faculté des Sciences Semlalia, Département de Physique, LPSCM, Marrakech, Morocco*

Abstract

We use molecular dynamics to analyze the dependence on the impact angle of the distribution of defects originated by the deposition of a Ag_{19} cluster on $Pd(1\,0\,0)$ at initial kinetic energies 0.1, 2, 20 and 95 eV. For increasing energy the cluster undergoes a transition from a multi-layered adsorbed structure to a two-dimensional one. Implantation of Ag atoms and promotion of Pd substrate atoms is common to all energies and angles and, for a given initial total kinetic energy, it increases with decreasing impact angle. Copyright © 1998 Elsevier Science B.V.

1. Introduction

The interest of the scientific community in the deposition and growth of homo- and hetero-structures on well-defined surfaces is triggered by both fundamental (how the individual atomic processes combine to give a very rich variety of structures, and how they grow [1,2]) and applied reasons (related to the growth of materials with controlled optical, catalytic, magnetic as well as mechanical properties [3,4]). The idea of using clusters as construction units instead of atoms is already fairly old [5,6]. However, the deposition process is more complicated than for atoms. In particular, contamination and possible fragmentation of the clusters need to be accounted for. If one is interested in using clusters as elementary units for the construc-

tion of controlled surface structures, it is necessary to know how a single deposited cluster looks like.

In a recent experimental and theoretical effort, the deposition of silver clusters (Ag_1, Ag_7 and Ag_{19}) on $Pd(1\,0\,0)$ has been investigated [7–11] via thermal energy atom scattering (TEAS) for impact energies E_{imp} equal to 20 and 95 eV. These experiments show that implantation and fragmentation are important in a way directly proportional to the impact energy per atom. Molecular dynamics simulations [10,12–15] provide an ideal complement to the very sensitive but global measurements of He scattering, which cannot yield direct information on the morphology of the adsorbate/substrate system created by deposition.

As part of a simulation study devoted to the collision of clusters on surfaces, we investigate in this paper the effect of an incidence angle on the impact of Ag_{19} clusters on $Pd(1\,0\,0)$ as a function of energy. Even though there is no direct experimental counterpart to

---

* Corresponding author. Tel.: 333 881 07040; fax: 333 881 07249; e-mail: carlo@marylou.u-strasbg.fr.

these simulations, the size of the cluster and the deposition energies have been chosen to match the one previously used in the experiments and simulations. For these reasons the results presented here can be highly useful to experimentalists willing to optimize innovative experimental conditions in the search of peculiar surface morphologies.

2. Model and calculations

Our simulations are based on the embedded atom (EAM) potentials of Foiles et al. [16] for Ag and Pd, with cutoff radii R_{cut} equal to 5.25 Å. For relevant details on the setting up of the calculations we refer the reader to a recent publication devoted to a full description of the simulation conditions [17].

The simulation slab consists of seven layers, modeling the (1 0 0) surface of Pd and submitted to periodic boundary conditions along the [0 0 1] and [0 1 0] directions. Each layer contains 200 Pd atoms. The structure for the Ag_{19} cluster is a highly distorted six-capped icosahedron selected among several local minima produced via annealing runs.

The temperature of the Ag_{19} cluster and the temperature of the Pd(1 0 0) surface are initially set to zero. The cluster is directed towards the substrate with an incidence angle $\theta = 15°$, $30°$ or $60°$ in the [0 0 1] direction and an initial kinetic energy E_{imp}. For each energy–angle combination 50 depositions are produced for statistical purposes, by varying randomly the initial cluster location with respect to the surface. The evolution of the system takes place in the microcanonical ensemble, with trajectories lasting up to 20 ps. By running simulations at $E_{imp} = 20$ eV we have checked that our results are unaffected by the consideration of a different, low ($T = 150$ K) substrate temperature. The issue of the dependence of the results on substrate temperature, system size and control temperature conditions have been thoroughly addressed in [17].

3. Results

The distribution of Ag and Pd atoms in the different layers and the number of vacancies are given in Table 1. It is worth noticing that implantation occurs for any initial deposition conditions. For low deposition energies one atom is implanted, while for bigger impact energies this number increases. Implantation of a Ag atom in the second substrate layer takes place only for $E_{imp} = 95$ eV. At low deposition energies, the adsorbates created by the impact have a clear three-dimensional character, which diminishes with increasing impact energy. Very few vacancies are created by the impacts and only for the higher impact energies. From Table 1 we deduce that the defect production and the concomitant change towards- two-dimensional clusters decrease with increasing incidence angle. It appears that the critical parameter is the energy perpendicular to the surface, while the incident energy in the direction parallel to the surface has a much smaller effect on the defect production. The differences between $E_{imp} = 0.1$ or 2 eV depositions are small and only the distributions of atoms above and in the adlayer differ slightly. This is not surprising since E_{imp} is negligible compared to the energy released during the adsorption process (~ 12 eV).

Fig. 1 shows the position of the silver and palladium atoms after a collision at $E_{imp} = 0.1$ eV projected onto the surface plane. For this initial energy the cluster remains three-dimensional after the collision. We notice that even at this very low deposition energy a Ag atom is implanted into the Pd(1 0 0) substrate while an atom from the substrate is pushed into the adlayer. Under similar conditions this never happens for Ag atoms on the same substrate [11]. As shown recently [11], the onset of chemical disorder is a direct result of the initial three-dimensionality of the cluster. We believe that this interesting phenomenon can be of high importance for the so-called "softlanding" deposition studies, since the implanted atom can act as a nucleation center and stabilize the deposited particle with respect to diffusion. However, at the same time, it prevents the occurring of a perfect defect-free adsorption of the cluster with no intermixing between the two species.

Figs. 2 and 3 show the positions of the atoms at the end of typical depositions with $E_{imp} = 95$ eV and impact angles $\theta = 30°$ and $\theta = 60°$, respectively. Here the cluster is no more three-dimensional. Some

Table 1
Averaged distribution of Ag and Pd atoms in different layers resulting from the deposition of Ag_{19} with an incident angle θ

Impact energy (eV)	θ (°)	Ag above adlayer	Ag in adlayer	Ag in 1st layer	Ag in 2nd layer	Pd in adlayer	Vacancies
0.1	30	8.9	9.8	1.0	0.0	1.0	0.0
	60	8.0	9.3	1.0	0.0	1.0	0.0
2	30	6.6	11.4	1.0	0.0	1.0	0.0
	60	7.6	10.4	1.0	0.0	1.0	0.0
20	15	3.9	10.9	4.0	0.1	4.1	0.0
	30	3.4	13.1	2.5	0.0	2.6	0.1
	60	6.8	11.2	1.0	0.0	1.0	0.0
95	15	0.1	7.5	7.3	4.1	12.0	0.6
	30	0.0	11.4	6.3	1.3	7.8	0.2
	60	0.5	14.5	3.9	0.1	4.1	0.1

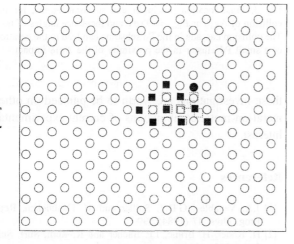

[010]

[001]

Fig. 1. Final location of cluster and substrate atoms, for the collision of Ag_{19} at $E_{imp} = 0.1$ eV and an incidence angle of 30°. Ag atoms are represented with squares while Pd atoms are represented with circles. White symbols refer to atoms in the first layer, black symbols to atoms in the adlayer and dotted contours to atoms above the adlayer. The cluster clearly stays three-dimensional and one Ag atom is implanted.

of the Ag atoms are implanted in a compact shape close to the point of impact, while the rest of them are positioned around the impact. In Fig. 2 one silver atom has escaped the impact point by diffusing along a channel in the [0 1 $\bar{1}$] direction and is no more bound to the other atoms of the initial cluster. Anytime a Ag atom is implanted in the substrate, a Pd

atom is promoted to the adlayer, as a consequence we do not observe the creation of interstitial defects. For low deposition energies or small incident angles the ejected Pd atoms are located in the proximity of the implanted Ag atoms. Figs. 2 and 3 show that the multiple collisions mechanism resulting in Pd atoms ejection into the adlayer can produce adatoms or clusters at ~ 5 sites away from the rest of the adatoms. This is a typical feature occurring in more than 50% of the depositions at $\theta = 60°$. Atoms, dimers, up to tetramers (see Fig. 3) are observed as result of this mechanism. Interestingly, we never observe two separate Pd adatoms away from the cluster impact region. Pd adatoms leaving the cluster impact region end up forming a small cluster, since the energy to initiate a second multiple collision event is likely to be higher than that required to move a second atom in an already existing cascade. It seems therefore that the system dissipates part of this energy in a long range multiple collision event which leads, if the impact energy parallel to the surface is high enough, to the appearance of isolated Pd adatoms or small clusters in the proximity of the impact of the initial particle.

In view of these indications experimentalists could choose to lower the amount of defects produced during the deposition by increasing the incidence angle of the in-coming particles, as the simulations discussed in this paper suggest. Long range collision effects could arise as a limitation. However, since they only appear

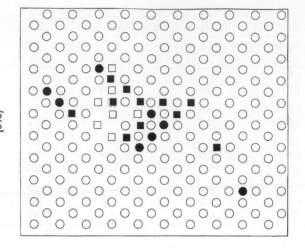

[001]

Fig. 2. Final location of cluster and substrate atoms, for the collision of Ag_{19} at $E_{imp} = 95$ eV and incidence angles of 30°. Ag atoms are represented with squares while Pd atoms are represented with circles. White symbols refer to atoms in the first layer, black symbols to atoms in the adlayer. The gray square (underneath a black square) represents a silver atom implanted in the second layer.

at $E_{imp} = 95$ eV and are not observed for smaller deposition energies we do not consider them as a serious limitation.

4. Conclusions

We have investigated by molecular dynamics simulations the angle dependence in the collision process of Ag_{19} on $Pd(1\,0\,0)$. The cluster is three-dimensional for low deposition energies but evolves towards a two-dimensional shape with increasing deposition energies. The amount of defect creation by an impact decreases with increasing incidence angle, thereby showing that the impact energy parallel to the surface is not responsible for implantation or interlayer mass transport. However, at 95 eV deposition energy and 60° incidence angle, adatoms or small Pd clusters appear due to a collision cascade in the near proximity of an impact. Due to the previous good agreement recorded between simulated depositions and experimental data, we believe that

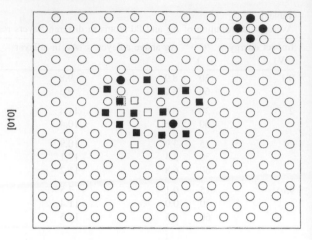

[001]

Fig. 3. Final location of cluster and substrate atoms, for the collision of Ag_{19} at $E_{imp} = 95$ eV and incidence angles of 60°. The collision cascade in the Pd substrate can lead to the creation of a small Pd cluster in the proximity of the Ag_{19} impact.

these investigations are worth to be pursued on other metal-on-metal combinations of current experimental interest.

References

[1] J.A. Venables, G.D.T. Spiller and M. Hanbücken, Rep. Prog. Phys. 47 (1984) 399.
[2] H. Röder, H. Brune, J.P. Bucher and K. Kern, Surf. Sci. 298 (1993) 121.
[3] X.P. Xu and D.W. Goodman, Catal. Lett. 24 (1994) 31.
[4] H. Haberland, M. Karrais, M. Mall and Y. Thurner, J. Vac. Sci. Technol. A 10 (1992) 3266.
[5] T. Takagi, I. Yamada and A. Sasaki, J. Vac. Sci. Technol. 12 (1975) 1128.
[6] S.B.D. Cenzo, S.D. Berry and E.H.H. Jr., Phys. Rev. B 38 (1988) 8465.
[7] G. Vandoni, C. Félix, R. Monot, J. Buttet and W. Harbich, Chem. Phys. Lett. 229 (1994) 51.
[8] C. Félix, G. Vandoni, C. Massobrio, R. Monot, J. Buttet and W. Harbich, to be published.
[9] K. Bromann et al., Science 274 (1996) 956.
[10] G. Vandoni, C. Félix and C. Massobrio, Phys. Rev. B 54 (1996) 1553.
[11] B. Nacer, C. Massobrio and C. Félix, to be published.
[12] H. Hsieh, R. Averback, H. Sellers and C. Flynn, Phys. Rev. B 45 (1992) 4417.
[13] F. Karetta and H. Urbassek, J. Appl. Phys. 71 (1992) 5410.

[14] P. Stoltze and J.K. Nørskov, Phys. Rev. B 48 (1993) 5607.

[15] H.-P. Cheng and U. Landman, J. Phys. Chem. 98 (1994) 3527.

[16] S.M. Foiles, M.I. Baskes and M.S. Daw, Phys. Rev. B 33 (1986) 7983.

[17] C. Massobrio, B. Nacer, T. Bekkay, G. Vandoni and C. Félix, Surf. Sci. (1997), to be published.

ELSEVIER

Computational Materials Science 10 (1998) 432–435

COMPUTATIONAL
MATERIALS
SCIENCE

Molecular dynamics simulations of spallation in metals and alloys

W.C. Morrey, L.T. Wille *

Department of Physics, Florida Atlantic University, 777 Glades Road, Boca Raton, FL 33431, USA

Abstract

We report on the results of large-scale molecular dynamics simulations of the mechanical behavior of two-dimensional metallic systems. The specific impact phenomenon studied is that in which a flyer of mass M moving with x-velocity v impacts a target of mass $2M$ moving with x-velocity $-v/2$. Simulations of such a spallation experiment have been performed for a generic metal, modelled with an embedded atom potential and also for a Cu–Ni alloy system, modelled with truncated Lennard–Jones potentials.

Our simulations indicate cold-welding upon impact, and shock wave generation, followed by rebound from the boundaries. The alloy was less ductile than the generic metal and consequently the system came apart due to the cooperative effect of the reflected shock waves. Copyright © 1998 Elsevier Science B.V.

Keywords: Fracture; Spallation; Molecular dynamics; Parallel computing

Fracture in metals has been a field of study for many years [1]. Interest in this area has been sustained by the pervasive economic impact of even the smallest advances. Understanding metallic fracture is at the heart of improvements in safety, reliability, and cost of a vast range of human endeavor. Tremendous benefits accrue to manufacturing, transportation, and construction, among others, by improving a metal's resistance to fracture. Understanding fracture mechanisms is the key to devising that improvement.

The path to illumination of mechanisms underlying fracture is microscopic simulations. Molecular dynamics (MD) simulations using realistic potentials are now performed on a scale that is approaching the actual dimensions of the microscopic features of interest [2]. This breakthrough has only become possible in recent years because of the availability of scalable parallel computers.

In general, parallel computers consist of a multiplicity of processors, which have interconnecting communications, and some overall controlling element. These computers are used to tackle a problem by having different elements of the processing section work on different parts of the problem. There are two basic architectural approaches to the design of parallel computers [3]. In order to break up the logic of the problem and dispense different tasks to different processors, each processor has to be powerful enough to handle its own task independently. This requirement gives rise to MIMD (multiple-instruction multiple-data) parallel computation. On the other hand, if the size and complexity of the problem arises from the number of objects in the problem, not by the variations of the actions of each object, the same logic can be used by each processor. This results in keeping the logic in the

---

* Corresponding author. Tel.: +1 561 297 3379; fax: +1 561 297 2662; e-mail: willel@acc.fau.edu.

controlling portion, leaving each processor as a simple calculating engine, with communications. This structure is SIMD (single-instruction multiple-data) parallel computation. Various strategies for parallelizing MD simulations of large systems have been described elsewhere [4].

The simulation of fracture in materials at the atomic level lends itself very nicely to an SIMD approach. Each particle is in a potential due to surrounding particles. These potentials result in applied forces which move the particle. Each step in the algorithms to calculate these potentials, forces, and motions can be applied in lock-step to a different particle by each processor. Several investigators have used SIMD machines to study fracture problems [5–8]. The present calculations were performed on one such computer, the MasPar MP-1.

A plane-wave impact experiment simulation using a square grid of ~200 000 atoms was conducted in a manner analogous to that employed in [5]. We took a two-dimensional array of generic atoms arranged on a hexagonal lattice. The atomic locations were laid out so that not quite all of the PEs on the MasPar were used. Sufficient free space was left around the grid to allow for expansion in all directions, the surfaces were relaxed and the velocities were quenched until the system was within a few Kelvin of absolute zero. The idea of the simulation was to have a segment of the particles, the flyer, impact the remainder of the particles, the target. The atoms were not physically separated between the flyer and the target: the start of the run represented the moment of impact. A velocity V was added to the left 1/3 of the atoms representing the flyer, while $-V/2$ was added to the right 2/3 of the atoms, which were the target. This provided a center-of-mass velocity of zero.

From the start of the simulation shock waves move out from the impact plane both forward toward the opposite end of the target and backward into the flyer. These waves reflect off the free ends and return, broadening due to nonlinear effects. Since the target is two flyer-widths (FWs), and assuming constant wave velocities, the waves travel three FWs and meet at a plane one FW from the end of the target. We chose the relative velocity to be $\sim 1.2\sqrt{\varepsilon/m}$, where ε represents

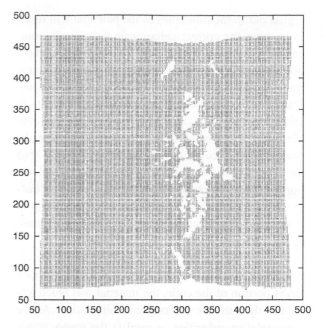

Fig. 1. Spallation in a generic metal using EAM potential.

the depth of the potential well and m is the atomic mass. This gave a high enough impact velocity that the superimposition of rarefaction wave reflections from the free surfaces led to spallation.

A snapshot of this generic metal spallation run is shown in Fig. 1. In particular, we note the necking at the top and bottom of the material centered around location 300. This is due to the bulk tensile strain in the X direction. There is also a concavity at the right and left around location 250. This is a rebound from the expansion caused when the shock waves hit the ends.

Next we took the step of expanding the applicability of the simulation to alloys. In doing this, it was decided to eliminate the EAM segment of the calculations, because we were eventually to use this model to study fracture and Holian and Ravelo [7] have shown that the LJ potential alone could be used in fracture studies. They discuss the differences between brittle and ductile materials as being characterized by the strength of the attractive portion of the potential at the point of critical strain. The higher the potential at that point, the more likely that the material will be brittle.

The LJ potential itself is inherently ductile, and retaining that shape of the potential in the repulsive region out to the minimum of the potential will result in keeping the correct anharmonic behavior in compression. Adding a cubic spline cutoff from the inflection point of the potential retains the ductile dynamic fracture behavior of the full LJ potential in tension. Starting the spline cutoff at the minimum results in brittle fracture behavior. Since both brittle and ductile behavior can be accommodated by a change in the shape of the LJ cutoff tail, it seems reasonable to start the study of alloy fracture with the LJ potential alone. Further, with the chosen material, Cu–Ni, being ductile, we will remain with the same cubic spline cutoff, starting at the inflection point (as used in the mono-atomic simulation).

We used this model to simulate spallation in an alloy. Since this model did not include the many-body EAM portion of the code, the material was much less ductile than in our mono-atomic simulation. At the same relative velocities as used in the mono-atomic spallation, this material exploded at the point of impact, as shown in Fig. 2. Particles were blown out of the lower edge of the PE range and disappeared from the simulation. What occurred at the point of impact can be seen at the top of the sample. Material

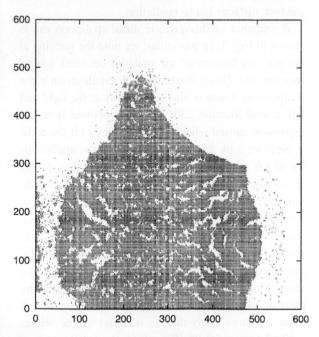

Fig. 2. Flyer/target impact in Cu–Ni using LJ potential. Relative velocity of impact same as mono-atomic EAM simulation.

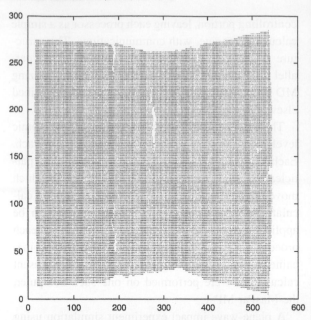

Fig. 3. Flyer/target impact in Cu–Ni using LJ potential. Relative velocity of impact one-fourth mono-atomic EAM simulation.

is extruded outward at high velocity. The sample is burst apart internally, and atoms can be seen boiling off from the surfaces.

Reduction of the impact velocity by a factor of 4 put the collision in the range where the elastic material response dominated the shock propagation. However, the pure LJ alloy does not spall at the 2/3 line as the mono-atomic simulation with EAM did. As can be seen in Fig. 3, this material eventually fails under tension. The forward and backward shock waves from the impact have both had time to reflect from the front and back surfaces, and the material is unable to sustain the subsequent elongation. Wagner et al. [5] attribute similar results for a pure metal in part to an expected decrease in total strain at the spall plane compared to that calculated for an EAM material.

The authors wish to thank Dr. B. Holian at Los Alamos National Laboratory for extremely helpful discussions and Dr. C. Halloy and the Joint Institute for Computational Science at the University of

Tennessee, Knoxville, for the generous use of time on their MasPar MP-2.

References

[1] L.B. Freund, Dynamical Fracture Mechanics (Cambridge University Press, Cambridge, 1990).

[2] N. Gronbech-Jensen, T. Germann, P.S. Lomdahl and D.M. Beazly, IEEE Comput. Sci. Eng. 2 (2) (1995) 4.

[3] K. Hwang, Advanced Computer Architecture – Parallelism, Scalability, Programmability (McGraw-Hill, New York, 1993).

[4] L.T. Wille, C.F. Cornwell and W.C. Morrey, Mater. Res. Soc. Symp. Proc. 409 (1996) 81.

[5] N.J. Wagner, B.L. Holian and A.F. Voter, Phys. Rev. A 45 (1992) 8457.

[6] F.F. Abraham, D. Brodbeck, R.A. Rafey and W.E. Rudge, Phys. Rev. Lett. 73 (1994) 272.

[7] B.L. Holian and R. Ravelo, Phys. Rev. B 51 (1995) 11 275.

[8] S.J. Zhou, P.S. Lomdahl, R. Thomson and B.L. Holian, Phys. Rev. Lett. 76 (1996) 13.

ELSEVIER

Computational Materials Science 10 (1998) 436–439

COMPUTATIONAL
MATERIALS
SCIENCE

Computer modeling of grain boundaries in Ni$_3$Al

M.D. Starostenkov, B.F. Demyanov *, S.L. Kustov, E.G. Sverdlova, E.L. Grakhov

General Physics Deptartment, Altai State Technical University, Lenin Str. 46, 656099, Barnaul, Russian Federation

Abstract

Computer modeling of symmetric tilt boundaries $\Sigma = 5\,[1\,0\,0]\,(0\,1\,2)$ and $\Sigma = 5\,[1\,0\,0]\,(0\,1\,3)$ in alloy Ni$_3$Al was carried out. The energy of grain boundaries (GBs) was calculated out by a method of construction of γ-surface by using Morse's empirical centralforce potentials. Researched GBs have four steady states: one is stable and three – metastable. These states of GBs differ by energy and atomic structure. It is shown that GBs in model of coincidence site lattice are unstable, the stabilization is achieved by additional displacement on some vector along plane of defect. Copyright © 1998 Elsevier Science B.V.

Keywords: Grain boundaries; Computer simulation; Coincidence site lattice; Superlattice L1$_2$; Ni$_3$Al

1. Introduction

The significant progress in study of atomic structure of grain boundaries (GBs) was achieved on the basis of geometrical model of coincidence site lattice (CSL) [1]. It became possible to research not only structure of boundaries of a special type within the framework of model CSL, but also boundary defects and processes of reorganization GBs. At the same time, geometrical criterion used by this model does not permit with confidence to speak about power stability of an atomic configuration GB, received in model CSL. The calculations of GB executed in number of works by methods of computer modeling showed features which were not predicted by geometrical models: availability of several metastable states, change of local volume, existence of additional shifts [2,3]. However, received data are not sufficient neither for determination of

common laws of thin GB structure, nor for revealing of restrictions of applicability of CSL model, therefore it is necessary to work out further researches with use of computer experiments.

In the present work computer modeling of symmetric tilt boundaries in ordered alloy Ni$_3$Al with superstructure L1$_2$ is carried out. Special boundaries $\Sigma = 5\,[1\,0\,0]\,(0\,1\,2)$ and $\Sigma = 5\,[1\,0\,0]\,(0\,1\,3)$ with angles misorientation 53.13° and 35.86° accordingly are investigated.

2. Method of calculation

The energy of GB was carried out using Morse's empirical central-force potentials. Radius of action of potentials was limited by the fourth coordination sphere, thus interaction of 54 atoms was taken into account. The energy of GB was determined as a difference of energies of a crystal, containing defect, and ideal crystal with identical quantity of atoms.

---

* Corresponding author. Tel.: +7 3852 368522; fax: +7 260516; e-mail: pva@agtu.altai.su.

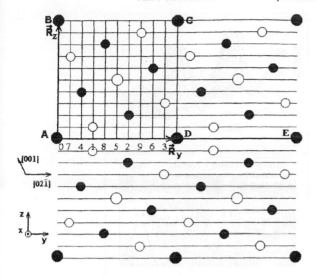

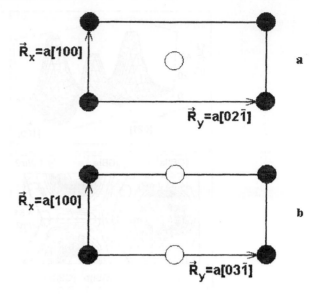

Fig. 1. The structure of GB [1 0 0] (0 1 2). (○) atoms Ni, (●) atoms Al, (●, ○) sites of CSL. The axis X is directed along axis of turn [1 0 0], the axis Y-along plane of defect, axis Z is normal to the plane of defect. ABCD – the unit cell of CSL.

Fig. 2. The unit cells in plane of GB: (a) GB (0 1 2), (b) GB (0 1 3).

The calculations of GB energy were carried out by a method of γ-surface construction [4]. The γ-surface represents the dependence of specific energy of GB on vector of shift of one grain relatively to the other along the plane of defect. The calculation was carried out in rigid model, in which the atoms at shift are in sites of corresponding lattices.

The structure of special boundaries is well investigated within the framework of model CSL. Therefore we chose the initial structure of researched GB coinciding with the boundary given by the model CSL in our calculations. In Fig. 1 structure GB (0 1 2) in projection to plane (1 0 0) is shown. The plane of symmetry AE is a plane GB. CSL is allocated by the spheres of large size. The unit cell of CSL for GB (0 1 2) has tetragonal structure with basic vectors $Rx = a$ [1 0 0], $Ry = a$ [0 2 1], $Rz = a$ [0 1 2]. For GB (0 1 3) $Rx = a$ [1 0 0], $Ry = a$ [0 3 1], $Rz = a$ [0 1 3]. Miller's indexes here and further are given using coordinate axes of the bottom crystal. Calculating γ-surface we used rigid translations along plane of GB, that means a kind of unit cell in plane of GB is of the main significance. The plane unit cells for GB (0 1 2) and GB (0 1 3) are shown in Fig. 2. There are two types of plane unit cells for any symmetric GB of a type $(0 k l)$, formed

by turn around of axis [1 0 0] in alloys with superstructure L1$_2$. The plane has an unit cell of the first type at odd k and l, if k or l is even then the second type. The calculation of γ-surface needs conducting in area of the corresponding plane unit cell.

3. Discussion

In Fig. 3 γ-surfaces and contour plots of their relief for GB (0 1 2) and (0 1 3) are indicated. The γ-surfaces have a complicated picture and essentially differ for considered boundaries, that reflect various symmetry of plane unit cells of defect (Fig. 2). The highest maxima correspond to positions of atoms in plane GB. Besides them, there is yet the number of maxima and minima, which determine formation stable or metastable defects and reorganization of one type GB in the other. We had conducted the analysis of all elements of γ-surface, arising as a result of displacement of the top grain on some vector f, and comparison of each extremum with atomic structure of GB was carried out.

In alloys with superstructure L1$_2$ takes place alternation of two types of planes: one-atomic consisting atoms of one sort and two-atomic consisting atoms

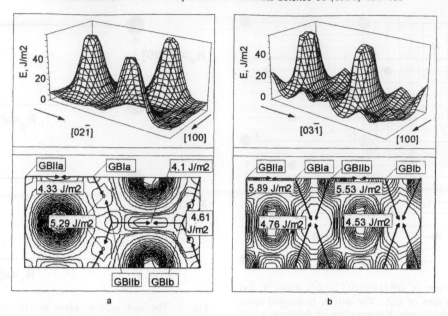

Fig. 3. The γ-surfaces and contour plots of GB: (a) GB (0 1 2), (b) GB (0 1 3). The arrows show directions of transitions of GB to metastable and stable states. The figures designate value of energy of minima of γ-surface.

of two sorts. Thus formation of two types of symmetric tilt boundaries is possible. In the alloy Ni₃Al GBI has the plane of symmetry consisting of atoms Ni and GBII has the plane of symmetry, consisting of the atoms Ni and Al. The formation of tilt GB is accompanied by closing atoms with one another, belonging to adjacent planes on both sides from the boundary, on a distance smaller than the radius of the first coordination sphere, that results in their overlapping. The overlapped atoms can give the essential contribution to increase of energy. The displacements along the plane GB of one grain relative to another separate these atoms and lower energy of defect. A plane GB and two planes adjacent to it form a GB core. The structure of cores of the main types of GB (0 1 2) in model CSL is shown in Fig. 4.

We shall conduct more detailed consideration of structure of defects on example GB (0 1 2).

GBI. For this type of boundary a plane of symmetry is a one-atomic plane, consisting of the atoms Ni. In this case two-atomic planes are overlapped, which permits to form two essentially different cores: with overlapping of the same name atoms Ni–Ni and Al–Al (Fig. 4(a)) and with overlapping of the different

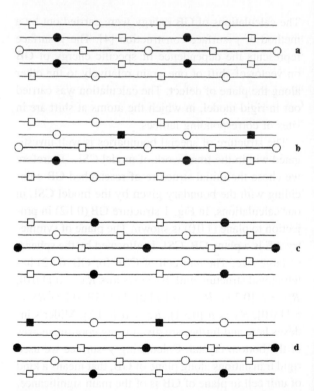

Fig. 4. The structure of core GB (0 1 2). It is shown two planes (1 0 0), crossing GB: (○) atoms Ni, (●) atoms Al in the top plane; (□) atoms Ni, (■)-atoms Al in the bottom plane.

atoms Ni–Al (Fig. 4(b)). Thus formed GBIa and GBIb differ by a vector of displacement $f = (Rx + Ry)/2$ and energy. Their positions on γ-surface as shown in Fig. 3(a) correspond to maxima with energy 6.98 and 5.81 J/m^2. The lower energy GBIa is connected with the closing of atoms Al in cores which gives potential of interaction near minimum.

GBII. This boundary has a two-atomic plane of symmetry. GBIIa is chosen by use as starting configuration for construction γ-surface, its location is characterized by a vector $f = 0$. GBIIb has a vector of displacement $f = (Rx + Ry)/2$. Locations of these boundaries also correspond to maxima on γ-surface with energies 7.32 and 7.83 J/m^2 accordingly. Overlapped planes are one-atomic, therefore at shift on $f = (Rx + Ry)/2$ the structure of a core is not changed (Figs. 4(b) and (c)). The small difference in energy is connected with the change of bonds only in the second planes from the boundary. The transition from GBII to GBI is realized by shift on vector $f = 4Ry/10$.

Consideration of structure GB (0 1 3), conducted under similar scheme, has also shown that multiplicity of core configurations GBIa, GBIb and GBIIa, GBIIb exists. The points, corresponding to these boundaries, are shown on a γ-surface (Fig. 3(b)). In contrast to GB (0 1 2), in which the plane unit cell contains atom in the center and includes reorganization of a core diagonal translations, GB (0 1 3) is reconstructed only at shift along axis Y. It results in that all maxima on a γ-surface GB (0 1 3) are lined up. Thus, all GB, given by model CSL, correspond to maxima on a γ-surface and are unstable. The stabilization can be achieved by transition in one of power minima by additional displacement of one grain to another. On a γ-surface of both boundaries there are four minima, with different values of energy, appropriate three metastable and one stable state. Each type of GB from state, given by model CSL, passes to one of these minima.

The transition from CSL structure to low energy state is not a barrier and as for the transition from metastable to stable state overcoming potential barriers are required. The stable state GB (0 1 2) has energy 4.33 J/m^2, GB (0 1 3) – 4.65 J/m^2. These values have reasonable significance for intermetallide Ni$_3$Al taking into account that the calculations were carried out in rigid model. Each GB type has three variants of transition to the low energy state.

4. Conclusion

Carried out calculations of γ-surface of tilt GB [1 0 0] (0 1 2) and [1 0 0] (0 1 3) in ordered alloy Ni$_3$Al have shown, that there is no simple conformity between atomic structure of GB and its geometrical parameters, such as an angle misorientation θ and reciprocal density of coincidence lattice sites. Being based on model CSL it is possible to allocate four GB types with various atomic configuration of cores. Calculations have shown that they are unstable, the stabilization is achieved at shift on some vector f, as a result GB passes in steady state. There are four steady states, appropriate on a γ-surface to positions of minima: one is stable and three – metastable. The transition of GB from one type to the other is connected with overcoming a potential barrier and is realized by shift of one grain relative to another on appropriate vector.

References

[1] H. Grimmer, W. Bollmann, D.H. Worrington, Acta Cryst. A 30 (1974) 197.
[2] T.I. Mazilova, I.M. Mikhailovski, Fiz. Tverd. Tela 37 (1995) 206.
[3] M. Kohyama, R. Yamamoto, Phis. Rev. B 49 (1994) 17102.
[4] M. Yamaguchi, V. Paidar, D.P. Pope,V. Vitek, Phil. Mag. A 45 (1982) 867.

COMPUTATIONAL
MATERIALS
SCIENCE

Computational Materials Science 10 (1998) 440–443

The effect of spillover in the electronic and magnetic properties of Ni, Co, and Fe clusters

Javier Guevara [a,b,*], Francisco Parisi [a,b], Ana Maria Llois [a,c], Mariana Weissmann [a]

[a] *Departamento de Física, Comisión Nacional de Energía Atómica, Avda del Libertador 8250, 1429 Buenos Aires, Argentina*
[b] *Escuela de Ciencia y Tecnología, Universidad Nacional de San Martin, Alem 3901, 1651 San Andrés, Buenos Aires, Argentina*
[c] *Departamento de Física "Juan José Giambiagi", Facultad de Ciencias Exactas y Naturales, Universidad de Buenos Aires,
Ciudad Universitaria, Pab. I., 1428 Buenos Aires, Argentina*

Abstract

We calculate the electronic structure of 3d transition-metal clusters with a model Hamiltonian that takes into account electron spillover at the cluster surface and uses bulk parameter values for the interactions. We perform calculations for fcc and bcc clusters of up to 260 atoms making use of symmetry properties. We obtain magnetic moments per shell and ionization potentials for Ni, Co, and Fe clusters starting from an spd-bulk parametrization. Copyright © 1998 Elsevier Science B.V.

Keywords: Clusters; Transition metals; Magnetism; Ionization potential; Spillover

1. Introduction

During the last decade there has been a great interest in studying small metallic clusters. Specifically, transition-metal clusters are very attractive because of their magnetic properties and their chemical reactivity in connection with catalytic processes.

In a recent contribution [1] we have shown that the use of bulk parameters together with a good treatment of the surface gives the necessary tools for cluster calculations within a parametrized Hubbard Hamiltonian in the unrestricted Hartree–Fock approximation. Transition-metal surfaces present two

interesting features: electron spillover and bulk-like d-orbital occupation. Electron spillover at the surface is considered by extending the spd basis outside the surface with empty orbitals with s symmetry, s', whose site energies are chosen in order to keep the d-band occupation similar to the bulk one. In this way the different electronic properties like magnetization and ionization potential can be calculated and compared with experiments.

In our previous work [1] we have calculated the average magnetic moment per atom and the ionization potential (IP) for Ni, Co, and Fe clusters. We have shown that both properties compare well with experimental results as a function of cluster size. In the present work we show the magnetic moment per atom and shell, for larger cluster sizes than in [1], and also the effect of improving the representation of the spillover.

---

* Corresponding author. Address: Departamento de Física, Comisión Nacional de Energía Atómica, Avda del Libertador 8250, 1429 Buenos Aires, Argentina. Fax: (0541) 754-7121; e-mail: guevara@cnea.edu.ar.

In a recent work Bouarab et al. [2], using a different criterium to parametrize the diagonal terms of the Hamiltonian, have calculated the magnetic structure of small Ni_n clusters obtaining a very good agreement with experiment by using optimized icosahedral geometries taken from [3]. Comparison with their results is included here.

2. Method of calculation

A Hubbard tight-binding Hamiltonian with spd orbitals and parameters from the corresponding bulk is used, magnetism is obtained by solving it in the unrestricted Hartree–Fock approximation. Only nearest-neighbor two-center parameters are used in the case of Ni and Co, and also second nearest-neighbor interactions for Fe clusters. In all cases bulk interatomic distances are considered. All many-body contributions appear in the diagonal spin dependent term $\epsilon_{im\sigma}$ given by

$$\epsilon_{im\sigma} = \epsilon_{im}^0 + \sum_{m'} U_{imm'} \Delta\eta_{im'}$$
$$-\sigma \sum_{m'} \frac{J_{imm'}}{2} \mu_{im'} + \Delta\epsilon_i^{MAD}, \qquad (1)$$

where $\Delta\eta_{im'}$ is the electronic occupation difference per orbital in the ith atom of the cluster with respect to the bulk paramagnetic values, $\mu_{im'}$ is the magnetization per orbital, and $\Delta\epsilon_i^{MAD}$ is the Madelung term, as already discussed in [1].

We add extra orbitals s' outside the clusters, parametrizing them in order to get adequate monolayer d-orbital occupations. The number of s' orbitals added is such that each cluster atom keeps its bulk coordination [1]. For the bcc clusters we discuss in this contribution which bulk coordination to choose.

The Hamiltonian is solved self-consistently diagonalizing at each step the spin-up and spin-down matrices and the electronic occupations per orbital, spin and atom are obtained. For $N \geq 27$ we calculate isomers of high symmetry by using techniques of group theory to block separate the Hamiltonian matrix. The matrix has as many blocks as the number of irreducible representations of the group, allowing us to deal with

clusters of up to 260 atoms (that is to say up to the 14th neighbors of the central atom or considering up to 14 shells).

3. Results

In Fig. 1 we present the calculated magnetic moment per atom and shell for some representative Ni, Co, and Fe clusters. In all the cases we take an atom with its first, second, third, etc neighbors, always as part of fcc or bcc structures. Horizontal lines mark

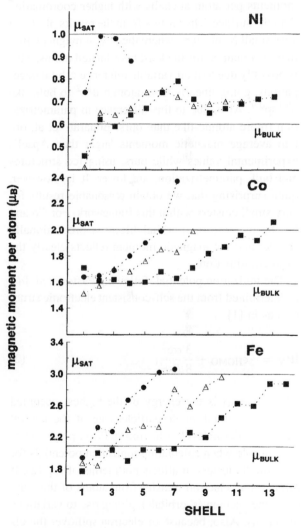

Fig. 1. Magnetic moment per atom vs. shell number of some selected Ni, Co, and Fe clusters (Ni_N, $N = 43$, 201, and 249; Co_N, $N = 87$, 135, and 249; Fe_N, $N = 65$, 137 and 259).

the experimental bulk magnetization and its saturation value. In the cases of Co and Fe, the magnetic moments of the surface atoms are close to saturation, going down to the bulk moment for inner shells. However, for Ni clusters the magnetic moment of the outer shells is not as large. There is a competition between low dimensionality (smaller number of nearest neighbors) and an increase of the d-band occupation at cluster surfaces with respect to the bulk. For Ni clusters this last effect prevails and therefore the magnetic moment of surface atoms does not saturate. It can also be seen that Ni clusters show in general larger magnetic moments per atom at shells with higher coordination than the surface. This differs from the results of [2] for very small Ni clusters, where the largest magnetic moment is obtained for the least coordinated atoms. This is possibly due to their variable reference orbital occupations, going linearly from atomic d^8s^2 to bulk like $d^{9.1}sp^{0.9}$. Also, due to this difference in parametrization (more atomic-like than ours) Bouarab et al. obtain average magnetic moments larger than Apsel's experimental values while ours, using fcc structures and bulk parametrization, are lower. It is, however, quite surprising that we obtain reasonable results for very small clusters within this framework. For Co and Fe the increase in d-orbital filling is proportionally smaller and the magnetic moment reflects clearly the local coordination.

The ionization potential, following Pastor et al. [4], is determined from the self-consistent electronic structure as in [1]:

$$IP_N = -\epsilon_{HOMO} + \frac{3}{8}\frac{\alpha e^2}{R_{eff}}, \tag{2}$$

where ϵ_{HOMO} is the energy of the highest occupied molecular orbital single-particle state of the neutral cluster (or Fermi level). Electron spillover plays an important role when calculating ionization potentials for very small clusters, it affects both terms of Eq. (2). It produces an electron transfer from atoms of the cluster to the vacuum (s' orbitals), giving rise to variations in ϵ_{HOMO}. Also, because of electron spillover the effective cluster radius R_{eff} is larger than the radius of a spherical cluster with N atoms and bulk solid density.

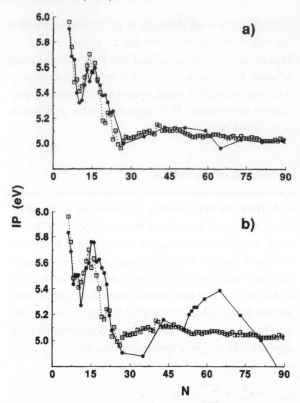

Fig. 2. Ionization potential as a function of N for Fe clusters. (□) experiments [7], (●) calculated values considering bcc structure. (a) calculated IP of Fe taking into account nearest and next nearest-neighbors in spillover; (b) with only nearest neighbors in the spillover.

The calculated IP's for Fe bcc clusters of size $6 \leq N \leq 90$ together with experimental results are given in Fig. 2. The zero of energies in our self-consistent calculations is fixed in this case by fitting the IP for $N = 90$ to the experimental value. We obtain a good agreement with the experimental data along the whole range of sizes. In Fig. 2(a) we choose the number of s' orbitals by taking the bulk coordination equal to 14, that is, nearest and next nearest neighbors, while in Fig. 2(b) we reduce it to bulk nearest-neighbor coordination of 8. In both cases the correspondence is good for the smallest clusters, but in the case of Fig. 2(a) the agreement is good along the whole curve, putting in evidence the importance of a good treatment of the spillover, as can be seen by comparing with [5] where no spillover has been taken into account, and the behavior of the IP for small bcc Fe clusters as a function of size is not obtained.

4. Discussion and conclusions

The magnetic moment per atom and shell was calculated for symmetric clusters and the results obtained show that there is a surface enhancement which is more important for those transition-metal systems with less 3d orbital occupation. The surface magnetic moment of Ni clusters is in general lower than the saturation value, while for Co and Fe clusters it is close to their μ_{SAT}. This seems to be in agreement with experimental results showing that Ni clusters of increasing size approach the bulk magnetization value per atom faster than Co and Fe ones [6].

From the IP calculation we show that it is very important to take the spillover correctly into account.

References

[1] J. Guevara, F. Parisi, A.M. Llois and M. Weissmann, Phys. Rev. B 55 (1997) 13 283 and references therein.

[2] S. Bouarab, A. Vega, M.J. López, M.P. Iñiguez and J.A. Alonso, Phys. Rev. B 55 (1997) 13 279.

[3] M.J. López and J. Jellinek, Phys. Rev. A 50 (1994) 1445.

[4] G.M. Pastor, J. Dorantes-Dávila and K.H. Bennemann, Chem. Phys. Lett. 148 (1988) 459.

[5] S. Bouarab, A. Vega, J.A. Alonso and M.P. Iñiguez, Phys. Rev. B 54 (1996) 3003.

[6] I.M.L. Billas, A. Châtelain and W.A. de Heer, Science 265 (1994) 1682.

[7] S. Yang and M.B. Knickelbein, J. Chem. Phys. 93 (1990) 1533.

ELSEVIER

Computational Materials Science 10 (1998) 444–447

COMPUTATIONAL
MATERIALS
SCIENCE

First-principles approach to the calculation of electronic spectra in clusters

Lucia Reining [a,*], Giovanni Onida [b], Stefan Albrecht [a]

[a] *CNRS-CEA, Laboratoire des Solides Irradiés, Ecole Polytechnique, 91128 Palaiseau, France*
[b] *Istituto Nazionale per la Fisica della Materia, Dipartimento di Fisica, Università di Roma "Tor Vergata" I-00133 Roma, Italy*

Abstract

We discuss a method for first-principles calculations of photoemission spectra in small clusters, going well beyond a standard density functional theory–local density approximation (DFT–LDA) approach. Starting with a DFT–LDA calculation, we evaluate self-energy contributions to the quasiparticle energies of an electron or hole in the GW scheme, where the self-energy $\Sigma = GW$ is constructed from the one-particle Green's function G and the RPA screened Coulomb interaction W. The contributions of structural relaxation are taken into account. We show the importance of these effects at the example of the photoemission spectrum of SiH_4. We also briefly discuss results for longer hydrogenated silicon chains, and address the problem of optical absorption. Copyright © 1998 Elsevier Science B.V.

Keywords: Clusters; Electron states; Spectroscopy; Ab initio calculations

Electronic–states spectroscopy represents one of the most valuable tools for the experimental investigation of structural and physico-chemical properties of clusters. Accurate and reliable theoretical calculations of photoelectron and absorption spectra are hence important, both for the interpretation of the experimental results and for the prediction of desired spectroscopical properties in the design of cluster-based or cluster-assembled materials. While highly accurate quantum-chemistry configuration interaction (CI) methods are available for the computation of spectroscopical properties of the smallest clusters [1], the excited state properties of aggregates containing more than 10 valence electrons can only be studied with approximate

or semi-empirical methods. In this paper we present an alternative, still parameter free approach for the study of cluster photoemission spectra, based on first-principles techniques which have been carried over from solid state physics. Due to the favourable scaling of the method with the number of valence electrons (N^3 at the worst), the present approach leads to less stringent size limitations than those of CI calculations.

Our paper is organised as follows: first, we give a brief summary of the state-of-the-art GW approach for band structure calculations in solid-state systems, which uses the density functional theory–local density approximation (DFT–LDA) results as a starting point. We then consider the additional problem of Jahn–Teller distortions, which in small clusters can produce dramatic effects not only on the energy positions of the spectral peaks, but also on their number. In

* Corresponding author. Tel.: +331 69333690; fax: +33 1 69333022; e-mail: Lucia.Reining@polytechnique.fr.

order to solve the problem raised by the geometry relaxation, we express the energy of the photoemission peaks as the sum of a "traditional" GW eigenvalue, plus a relaxation term, which is easily obtained within DFT–LDA. We illustrate this point with the results obtained for the paradigmatic case of SiH_4, showing that important features of the experimental photoelectron spectrum are well described by our approach.

DFT–LDA, being a ground-state theory, is not directly applicable as a tool for the determination of spectroscopic properties, which also require the determination of the excited states. In fact, the relevant energy in a photoemission experiment is the difference between the ground state of an N-electrons system and an excited state of the system with $(N - 1)$ electrons. This energy difference is the quasiparticle (QP) energy for a hole in the valence band, and it is formally given by the eigenvalue of a Schrödinger-like equation containing the classical Hartree potential and the self-energy Σ:

$$\left(-\frac{1}{2}\nabla^2 + v_{\text{ext}}(\mathbf{r}) + \int d\mathbf{r}' \frac{\rho(\mathbf{r}')}{|\mathbf{r} - \mathbf{r}'|} \right) \psi_i(\mathbf{r})$$
$$+ \int d\mathbf{r}' \Sigma(\mathbf{r}, \mathbf{r}'; \varepsilon_i^{\text{QP}}) \psi_i(\mathbf{r}) = \varepsilon_i^{\text{QP}} \psi_i(\mathbf{r}) \quad (1)$$

Σ is a non-local and energy-dependent operator, which takes into account the effects of exchange, correlation, and relaxation of the $(N - 1)$ electrons. It can be computed in the GW approximation [2], as the convolution of the one-particle Green's function G and the screened Coulomb interaction W. Formally, Eq. (1) is similar to the Kohn–Sham (KS) equations [3], in which Σ is replaced by the (local and energy independent) exchange–correlation potential $v_{xc}(\mathbf{r}, \rho)$. Generally, the solution of the KS equations within the LDA is considered as a good approximation for the QP energies and wave functions, and the correction to the KS eigenvalue can be evaluated to first order in the difference between Σ and v_{xc}. Those corrected eigenvalues, which we call in the following "GW eigenvalues", have been shown in many calculations on solids to describe well the relevant energies which are involved in photoemission and inverse photoemission spectra [4,5]. However, a naive application of this scheme "as it is" to small finite systems like

atomic or molecular clusters does not allow to obtain the correct results for the interpretation of measured photoemission spectra. In particular, the effects of geometrical relaxation must be taken into account: without their inclusion, even the number of peaks in the calculated photoemission spectrum can be different from the measured one. This effect can be nicely illustrated in the case of the simplest $Si_n H_m$ system, the silane molecule SiH_4. While neutral SiH_4 in its fully symmetric Td geometry has a non-degenerate ground-state, the electronic state for the SiH_4^+ cation in the T_d symmetry is degenerate. Hence, according to the Jahn–Teller theorem, its ground-state equilibrium geometry is a distorted, lower symmetry configuration. The situation, similar to that of methane, implies that the photoelectron spectrum does not only depend on the total energy difference between SiH_4 and SiH_4^+ (vertical ionisation energy). Instead, it is determined by the energy difference between the neutral symmetric cluster and the $N - 1$ electrons system in a distorted geometrical configuration:

$$E = E_{\text{ideal}}^N - E_{\text{dist.}}^{N-1} \quad (2)$$

This implies that it is no longer possible to simply identify the relevant energy difference as a GW eigenvalue of the cluster Hamiltonian, in either geometry. However, the energy difference in (2) can be rewritten adding and subtracting the total energy of the neutral cluster in the distorted geometry:

$$E = E_{\text{ideal}}^N - E_{\text{dist.}}^N + E_{\text{dist.}}^N - E_{\text{dist.}}^{N-1}$$
$$= \Delta_E^N + \varepsilon_i^{GW,\text{dist.}} \quad (3)$$

In this way, (3) can be calculated as the sum of an energy difference for the ground state of the neutral cluster in two different geometries, which is easily evaluated in DFT–LDA, plus a GW eigenvalue for the distorted cluster at fixed geometry.

We have applied this scheme to SiH_4, using a plane-wave basis in a large supercell geometry. We have performed the DFT–LDA calculations using the Car–Parrinello approach [6], and hard-core, norm-conserving pseudo potentials [7]. Optimised geometries for the neutral and ionised clusters have been found to possess the full T_d and the D_2 symmetries,

Table 1

Column 1 and 2: LDA eigenvalues of the occupied states for the neutral cluster in the ideal T_d geometry and in the distorted geometry, respectively. Column 3: Total energy difference for the neutral cluster in the two geometries. Column 4: GW eigenvalues for the neutral cluster in the distorted geometry. Note the large corrections with respect to LDA (column 2). Column 5: final photoemission energies according to Eq. (3). Values in eV

LDA (ideal)	LDA (dist.)	Geom. effect	GW eigenvalue	Final energy
−8.20	−6.99	−0.74	−10.49	−11.23
−8.20	−8.59	−0.74	−11.99	−12.73
−13.20	−13.18	−0.74	−16.88	−17.62

respectively. Si–H bond lengths are 1.52 Å for the tetrahedral geometry, while in the Jahn–Teller distorted configuration they become 1.55 Å, with H–Si–H angles of 98° and 136°. A further LDA calculation for the neutral SiH_4 in the equilibrium geometry of the cation was then necessary in order to evaluate (3). The total energy difference between the ideal and the distorted geometry for the neutral cluster has been found to be 0.7 eV, as reported in Table 1, where also the KS eigenvalues are summarised. We have checked the convergence of the result with respect to both the supercell size (fcc, with $a_0 = 30$ and 40 a.u.) and the kinetic energy cut-off (11 and 20 Ry). The 30 au supercell and 20 Ry cut-off are found to give well converged results.

The last ingredient of our calculation, i.e. the GW correction to the LDA eigenvalues of the neutral cluster in the distorted geometry, has been computed following the approach of Godby and Needs [8], with the inclusion of the cut-off in the Coulomb potential (see [9]) in order to eliminate the interaction between clusters in different supercells. The final GW corrected eigenvalues, together with our final results for the energy positions of the photoemission peaks evaluated according to (3), are summarised in Table 1. In Fig. 1, the calculated energies are superimposed to the experimental spectrum of Feher [10]. While the KS eigenvalues for the neutral cluster in its ground state clearly produce a spectrum with a wrong number and position of the peaks, the energies calculated according to (3) are in close agreement with the experimental data, with a remaining discrepancy of about 0.3–0.6 eV. An exact comparison is impossible, since the experimen-

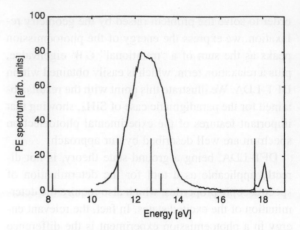

Fig. 1. Experimental photoemission spectrum for SiH_4, from [10]. Short and long bars represent the calculated energies corresponding to columns 1 and 5 of Table 1, respectively. In the experimental spectrum the two lower energy peaks are not resolved because of vibrational contributions.

tal spectrum is in fact given by a superposition of a large number of vibrational peaks ($\nu_n - \nu_m$ transitions), while in our calculation only the $\nu_0-\nu_0$ transition is considered, which should coincide with the onset of the peaks. However, some remaining discrepancy may be due to the fact that we evaluate the GW corrections perturbatively, using the LDA eigenvalues and eigenstates. The energy of the top-valence state should in principle be correctly given by a DFT calculation, but we find a considerable GW correction. This implies that the LDA is responsible for a relatively large error, which may not be completely cancelled by the perturbative approach.

Calculations for larger chains (Si_2H_6, Si_3H_8) using the same computational scheme are presently in progress, and first results further confirm the validity of our approach. [1] Also Si_2H_6 and Si_3H_8 undergo large Jahn–Teller distortions, with stretched bonds and flattened $Si–H_3$ dihedral angles. Similarly to the SiH_4 case, the calculated number of photoemission peaks gives a good account of the experimental results only when the geometrical relaxation effects are considered. Preliminary GW results indicate that the size of the correction to the LDA eigenvalues depends mainly

---

[1] This work will be the subject of a forthcoming paper.

on the type and localisation of the orbital considered, and is similar for the different clusters.

Finally, it should be noted that much experimental information on clusters comes from absorption measurements, where two quasiparticles (electron and hole) are created simultaneously, and may interact strongly. This problem has been recently studied in the case of the Na_4 cluster [9], where we have found that electron–hole interaction effects are of the same order of magnitude as the GW corrections.

Following that approach, which consists in solving an effective two-particle equation starting from the results of a GW calculation for the one-quasiparticle states, we have used the GW results calculated for SiH_4 in order to determine the exciton binding energy of the lowest dipole allowed transition. The exciton turns out to be bound by as much as 4 eV. This value almost completely cancels the self-energy correction to the gap (the gap is 7.5 eV in LDA, and 12 eV at the GW level) and yields a calculated absorption onset at about 8 eV.

In conclusion, we have computed the energy structure of the photoemission spectrum of SiH_4 from first principles, illustrating an approach which makes use of solid-state derived techniques for the evaluation of the QP energies starting from DFT–LDA results, and which allows the inclusion of the essential contributions from geometrical relaxation. The comparison of our results with the experimental spectrum shows that a good agreement can only be obtained if both the GW

corrections to the KS eigenvalues and the geometrical relaxation effects are taken into account.

Acknowledgements

We are grateful to Rodolfo Del Sole and Rex Godby for useful and stimulating discussions. This work has been partially supported by EEC contract CHRX-CT93-0337 Supercomputer time on a Cray C98 has been granted by IDRIS/CNRS (project no. 970906).

References

[1] V. Bonacic-Koutecky, J. Pittner, C. Scheuch, M.F. Guest, J. Koutecky, J. Chem. Phys. 96 (1992) 7938.
[2] L. Hedin, Phys. Rev. 139 (1965) 796.
[3] P. Hohenberg, W. Kohn, Phys. Rev. 136 (1964) B864; W. Kohn, L.J. Sham, Phys. Rev. 140 (1965) A1133.
[4] M.S. Hybertsen, S.G. Louie, Phys. Rev. Lett. 55 (1985) 1418.
[5] R.W. Godby, M. Schluter, L.J. Sham, Phys. Rev. Lett. 56 (1986) 2415.
[6] R. Car, M. Parrinello, Phys. Rev. Lett. 55 (1985) 2471.
[7] G. Bachelet, D.R. Hamann, M. Schlueter, Phys. Rev. B 26 (1982) 4199.
[8] R.W. Godby, R.J. Needs, Phys. Rev. Lett. 62 (1989) 1169.
[9] G. Onida, L. Reining, R.W. Godby, R. Del Sole, W. Andreoni, Phys. Rev. Lett. 75 (1995) 818.
[10] F. Feher, Molekülspektroskopische Untersuchungen auf dem Gebiet der Silane und der Heterocyclischen Sulfane, Forschungsbericht des Landes Nordrhein–Westfalen, Westdeutscher Verlag, Köln, 1977.

ELSEVIER

Computational Materials Science 10 (1998) 448–451

COMPUTATIONAL
MATERIALS
SCIENCE

Nonlinear electronic dynamics in free and deposited sodium clusters: Quantal and semi-classical approaches

F. Calvayrac [a],*, A. Domps [a], C. Kohl [b], P.G. Reinhard [b], E. Suraud [a,1]

[a] *Laboratoire de Physique Moléculaire et Chimie Quantique, Université Paul Sabatier, 118 route de Narbonne, F-31062 Toulouse Cedex, France*
[b] *Institut für Theoretische Physik, Universität Erlangen, Staudtst. 7, D-91077 Erlangen, Germany*

Abstract

The optical response of sodium clusters is studied using fully three-dimensional electronic dynamics, in real time, with an explicit ionic background. The obtained spectra are shown to come close to experimental results for free clusters. We use the approach also to compute the response of sodium clusters deposited on a NaCl(1 0 0) surface. The flattening of the clusters induced by the surface results in a shift and a large fragmentation of the plasmon resonance. Copyright © 1998 Elsevier Science B.V.

Keywords: Metallic clusters; Adsorbed clusters; Plasmons; TDLDA; TDKS; Sodium cluster

1. Introduction

The optical response is a key tool to explore the structure of metal clusters. Most theoretical models of it are based on the time-dependent local-density approximation (TDLDA), for reviews see [1]. Fully detailed ionic structure had been taken care of in very elaborate quantum chemical calculations [2]. On the side of the TDLDA, the development has been also carried forth to include ionic structure and to go beyond the linear regime as well, see e.g., [3] and work cited therein. We present here results of TDLDA calculations without any symmetry constraint, with ionic structure, at any excitation amplitude. Various semi-classical approximations to TDLDA are performed

consistently within the same numerical scheme. It is the aim of this contribution, to report new results for two typical examples, plasmon spectra in the free cluster Na_9^+ and for Na_8 on a NaCl(1 0 0) surface.

2. Model

The cluster is described as a system of valence electrons (one per Na atom) and ionic cores whose position is assumed to be frozen at the short timescale of the plasmon oscillations. To describe the interaction of the ions and electrons, we use a local pseudopotential adjusted to the local part of the pseudopotentials from [4]. In all calculations, the ionic ground state configurations are optimized with a Monte-Carlo annealing scheme. The forces from the NaCl surface on the cluster are incorporated via the effective interface potential of [5].

* Corresponding author. Fax: (33) (5) 61556065; e-mail: calvayra@irsamc2.ups-tlse.fr.
[1] Membre de l'Institut Universitaire de France.

We compute the electronic response in real time by solving directly the equations of motion for the single particle electronic wave functions in the time-dependent Kohn–Sham approximation [6]:

$$i\partial_t \psi_\alpha(\mathbf{r}, t)$$
$$= \left(-\frac{\hbar}{2m_e} \Delta V_{KS}(\mathbf{r}, \rho_\uparrow(\mathbf{r}, t), \rho_\downarrow(\mathbf{r}, t)) \right)$$
$$\times \, \psi_\alpha(\mathbf{r}, t), \tag{1}$$

where $\sigma = \uparrow, \downarrow$ denotes the spin. The Kohn–Sham potential V_{KS} depends on the density $\rho_\sigma(\mathbf{r}, t) = \sum_{\alpha \in \sigma} |\psi_\alpha(\mathbf{r}, t)|^2$, both in the direct (Hartree) term as well as in the exchange–correlation part where the local functional of [7] is used, for details see [3].

A simplified, semi-classical version is provided by the time-dependent Thomas–Fermi model (TDTF), where the total kinetic energy is evaluated as

$$T = \int d^3 \mathbf{r} \left\{ \frac{m \mathbf{j}^2}{2\rho} + (3\pi^2)^{2/3} \frac{3\hbar^2}{10m} \rho^{5/3} \right\}. \tag{2}$$

In Eq. (2), $\mathbf{j}$ represents the local density current and the second term is the usual Thomas–Fermi functional. This approximate form for the kinetic energy yields a simple hydrodynamical description of the moving electron cloud [8], at the price however, that single particle features are lost. The TDTF is solved by mapping it to a one-particle Schrödinger equation with the help of the Madelung transform [9].

Wave functions and fields are represented on a regular real-space grid in three dimensions spaced of 0.8 a_0. The Poisson equation is solved with the method of Lauritsch and Reinhard [10]. The Kohn–Sham ground state is computed with damped gradient iterations [11]. The electronic time step used employs an interlaced local and kinetic propagation in Fourier space (see [12,13] for details).

The spectral distribution is explored by initializing the dynamics with an instantaneous shift of the whole electron cloud out of its equilibrium position. The emerging dipole momentum $D(t)$ is Fourier transformed yielding $\tilde{D}(\omega)$. The photoabsorption strength is then simply $S(\omega) = \Im[\tilde{D}(\omega)]$. This is the appropriate observable for excitations in the linear regime. For details of the spectral analysis and its interpretation

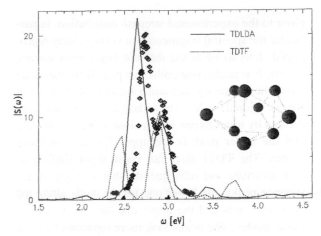

Fig. 1. Strength function for a Na_9^+ cluster with the configuration sketched on the right part of the figure, obtained by using the TDLDA method (continuous line) and TDTF (dashed line). Experimental data (dots) for the photoionization cross-section are also shown, with rescaled values to have the same height as the TDLDA results.

see [3]. In practice, we use here a small shift such that we stay in the linear regime with about 1 eV average excitation energy. The calculation run up to 10 000 time steps up to 48 fs.

3. Results

Vacuum cluster. For Na_9^+, we find the ground state configuration of Na_9^+ as illustrated in Fig. 1. This shape is similar to the one found in [14], but carries a larger octupole momentum. In an experimental setup, it is not possible to control the orientation of the clusters in a laser field. We thus have to average the results over all orientations. To this end, it suffices to average over three excitations along the three principle axes of the cluster. We consider here one-photon processes as they are induced by a faint laser pulse with extended time-profile (nanoseconds). This means that we drive TDLDA in the small amplitude regime. Fig. 1 shows the dipole strength function for Na_9^+ computed with TDLDA (solid line) in comparison with TDTF (dotted line) as well as with experimental data at low temperature (circles) from [15]. The result from the full TDLDA calculation comes close to the experimental strength distribution, in particular the spectral fragmentation is very nicely reproduced. It is to be noted that the larger peak covers, in fact, two peaks, one collective peak from the mode along the symmetry axis and a smaller side peak from the orthogonal modes. The orthogonal modes are fragmented by interference with a close particle-hole state and the upper peak is the upper fragment of these modes. The TDTF does not know about single particle structure and subsequently shows only pure resonance peaks, the lower one for the mode along the symmetry axis and the upper one for the two orthogonal modes. The peaks look more separated because there is no further spectral fragmentation broadening them. This splitting demonstrates the geometric effect

from the small global deformation of the configuration. It is also present, of course, in the full TDLDA, but washed out by fragmentation. The TDLDA result is thus closer to the data, but the result of TDTF is satisfying in view of the simplicity of the method. It will become a useful method for larger clusters where fully fledged TDLDA hits its limits.

Adsorbed clusters on a NaCl(1 0 0) surface. It is an interesting question whether clusters stay stable as slightly perturbed free clusters attached to the surface or if they are melting down to a planar cluster just a monolayer thick. The cluster Na_8 was found to show a competition of these extremes, a planar ground state and a stable drop-like isomer [16,17]. Both are illustrated in the upper panels of Fig. 2. The planar shape

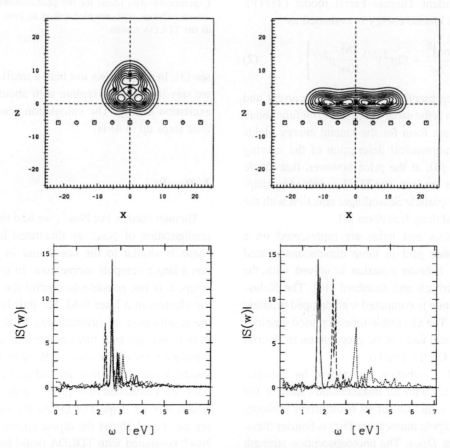

Fig. 2. Upper panel: Density contour plots in the x–z plane of the planar ground state (right column) and the droplet-like isomer (left column) of Na_8 absorbed to a NaCl(1 0 0) surface. The stars denote the ionic positions projected on the plane, the surface is indicated by squares (Cl^-), and circles (Na^+). Lower panel: Corresponding dipole strength in the TDLDA approach for the modes in x- (full line), y- (dashed) and z-direction (dotted line).

is shown in the right panel. It is, in fact, strongly tri-axial as the extension in the y-direction (not shown) is less than in the x-direction. The left panel shows the "formerly free" isomer where one sees how the attachment to the surface adds a slight octupole deformation to the cluster which was nearly spherical before. The TDLDA dynamics was followed here for 96 fs to enhance spectral resolution. The two lower panels of Fig. 2 show the resulting strength functions. They clearly demonstrate the effect of the two different geometries on the optical response. The much tri-axial planar cluster shows nicely three well separated peaks whereas the nearly free cluster has only an insignificant splitting which will show up in experiment as one single broad peak. Optical analysis of clusters on surfaces can thus provide useful information on the actual geometries. We note in passing that the surface interaction as such has little effect on the plasmon. It is exclusively the shape of the cluster which determines the spectra.

4. Conclusion

We have computed electronic dynamics in small Na clusters at the level of the time-dependent local-density approximation (TDLDA). The removal of any symmetry restriction allows one to deal with detailed ionic structure and to evaluate all possible modes of the system on top of a reliable ground state. The full TDLDA is capable of handling energetic excitations in the nonlinear regime, but the efficient numerical techniques used here make the code also competitive with three-dimensional linearized TDLDA models (which do not yet exist anyway). A semiclassical approximation to TDLDA is the time-dependent Thomas–Fermi (TDTF) method which describes the global pattern of the motion by sort of fluid dynamical approach.

We have compared the TDLDA and TDTF results in a realistic situation for the example of a free Na_9^+ cluster. Both methods give results close to the measured spectrum with a slight splitting due to a small deformation of the ground state configuration. Furthermore, we have applied TDLDA to Na_8 clusters

deposited on the insulating NaCl(1 0 0) substrate. The ground state is a planar cluster and the first isomer is related to the free cluster only slightly distorted by the surface. A clear triple splitting arises for a strongly triaxially deformed planar cluster, whereas the plasmon of the isomer, representing the adsorbed magic vacuum cluster, is just slightly affected and shows still one broad resonance peak.

Acknowledgements

We thank IDRIS and CNUSC for providing access to their parallel computers. The Institut Universitaire de France and the French–German exchange program PROCOPE (number 95073) are also acknowledged for financial support.

References

[1] W. de Heer, Rev. Mod. Phys. 65 (1993) 611; M. Brack, Rev. Mod. Phys. 65 (1993) 677.
[2] V. Bonačić-Koutecký, P. Fantucci and J. Koutecký, Chem. Rev. 91 (1991) 1035.
[3] F. Calvayrac, P.G. Reinhard and E. Suraud, Ann. Phys. (NY) 255 (1997) 125.
[4] G.B. Bachelet, D.R. Hamann and M. Schlüter, Phys. Rev. B 26 (1982) 4199.
[5] C. Kohl and P.-G. Reinhard, in preparation.
[6] R.M. Dreizler and E.K.U. Gross, Density Functional Theory (Springer, Berlin, 1990).
[7] O. Gunnarsson and B.I. Lundqvist, Phys. Rev. B 13 (1976) 4274.
[8] A. Domps, P.-G. Reinhard and E. Suraud, Phys. Rev. Lett., submitted.
[9] E. Madelung, Z. Phys. 40 (1927) 322.
[10] G. Lauritsch and P.G. Reinhard, Int. J. Modern Phys. C 5 (1994) 65.
[11] V. Blum, G. Lauritsch, J.A. Maruhn and P.G. Reinhard, J. Comput. Phys. 100 (1992) 364.
[12] F. Calvayrac, P.G. Reinhard, E. Suraud and C. Ullrich, in: Proc. PC'96 Conf. ed. CYFRONET KRAKOW ISBN 83-902363-3-8, pp. 163–166
[13] F. Calvayrac, E. Suraud and P.G. Reinhard, in preparation.
[14] B. Montag and P.-G. Reinhard, Z. Phys. D 33 (1995) 265.
[15] T. Reiners, C. Ellert, M. Schmidt and H. Haberland, Phys. Rev. Lett. 74 (1995) 1558.
[16] H. Häkkinen and M. Manninen, Europhys. Lett. 34 (1996) 177.
[17] C. Kohl and P.-G. Reinhard, Z. Phys. D 39 (1997) 225.

ELSEVIER

Computational Materials Science 10 (1998) 452–456

COMPUTATIONAL
MATERIALS
SCIENCE

The growth dynamics of energetic cluster impact films

Michael Moseler *, Oliver Rattunde, Johannes Nordiek, Hellmut Haberland

Freiburg Materials Research Center, Stefan–Meier-Str. 21, D-79104 Freiburg, Germany

Abstract

Dense and smooth thin films can be produced by the deposition of energetic clusters onto a solid surface. Roughnes, induced by random deposition of the clusters, is strongly suppressed by a downhill smoothing process. Molecular dynamic simulations of the impact of copper clusters onto tilted copper surfaces demonstrate that a downhill particle current proportional to the local slope of the surface is initiated. In the long wavelength limit the evolution of the surface profile is governed by the Edwards–Wilkinson equation. One parameter in this equation is related to the strength of the downhill movement due to the cluster impact and can be determined from the simulations. From the solution of the spatially Fourier transformed growth equation the power spectrum is calculated, which is in agreement with atomic force microscopy measurements for copper films. Copyright © 1998 Elsevier Science B.V.

1. Introduction

Thin films are usually grown by depositing atoms or molecules onto a solid surface. These particles may come from a vapour or a beam. In both cases, a random rain of particles hits the solid surface leading to spatial fluctuations in the amount of deposited matter and thus to a non-uniform film height [1]. If random deposition is the dominating growth process, the resulting film unavoidably will be rough.

In many cases, additional redistribution processes are present on the film surface causing a reduction of the film roughness. In general, this lateral transport of atoms needs an activation energy. Therefore, in conventional physical vapour deposition techniques the substrate is heated to several hundred Kelvin above

room temperature in order to get an enhanced surface diffusion.

Many materials, though, are not resistant to high temperatures. Therefore, instead of heating the substrate, the incoming particles can be accelerated to high kinetic energies in order to activate other lateral mass transport mechanisms. Unfortunately, energetic atoms can penetrate the solid causing radiation damage or sputtering [2]. On the other hand, molecular dynamic simulations reveal a penetration depth of energetic clusters of the order of some lattice constants at maximum [3,4], i.e. in contrast to atoms, clusters release their kinetic energy in the top layer of the growing film leading to a strong collective displacement of atoms. This is the reason for the use of energetic cluster impact (ECI) in thin film formation experiments. For instance, smooth and compact thin films can be produced by depositing metal clusters (10^3–10^4 atoms) with sufficiently high impact energy onto a room temperature substrate [5]. The surface of

* Corresponding author. Tel.: +49 761 203 4769; fax: +49 761 2034700; e-mail: mmos@frha06.physik.uni-freiburg.

500 nm thick copper films showed a rms-roughness of only 1 nm for cluster energies above 2 eV/atom [6]. In contrast, films deposited by low energy clusters exhibited an interface width of more than 10 nm, indicating that a strong lateral mass redistribution process must be active in energetic cluster film growth.

2. The evolution of the film height

The occurrence of overhangs on the film surface is very unlikely for compact films from energetic cluster impact. Therefore, the film profile can be described by the film height $h(\mathbf{x}, t)$, where $\mathbf{x}$ is a two-dimensional vector in the initially flat surface (i.e. $h(\mathbf{x}, 0) = 0$). Assuming a constant film density, particle conservation implies a continuity equation for the film height [1]

$$\partial h(\mathbf{x}, t)/\partial t = -\Omega \nabla \cdot \mathbf{j}(\mathbf{x}, t) + \eta(\mathbf{x}, t), \qquad (1)$$

where Ω is the atomic volume, $\mathbf{j}(\mathbf{x}, t)$ the two-dimensional particle current representing the lateral mass redistribution, and $\eta(\mathbf{x}, t)$ is the local height source due to the randomly deposited clusters.

To proceed further a knowledge of the particle current is necessary. Our macroscopic experience may provide an idea of the basic process. Assume that rain is falling onto a sand dune. Whenever the flank of the dune is hit by a rain drop some matter is pushed downhill. In this way the rain initiates a downhill mass current leading to the erosion of the dune and thus suppressing roughness. In order to answer the question, whether this macroscopic picture may be translated to a microscopic one the impact of an energetic cluster onto a sloped surface was studied by molecular dynamics (MD).

3. MD simulation of cluster impacts

Copper clusters impinging onto copper(1 0 0) surfaces were considered. In a so-called primary zone, where a lot of disorder occurs during the impact, the interaction between the atoms was represented by an embedded atom method (EAM) potential using

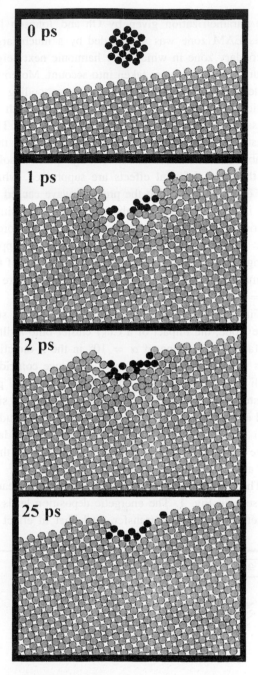

Fig. 1. The energetic impact of a Cu$_{87}$ cluster with 10 eV/atom onto a tilted Cu(1 0 0) surface. A cross section in the x–z plane is shown. The slope of 10° leads to a downhill asymmetry in the final distribution of atoms.

the parametrization given by Oh and Johnson [7]. This EAM zone was surrounded by a much larger secondary zone in which only harmonic next neighbour interactions were taken into account. Moreover, instead of a fcc-lattice only the corresponding sc-lattice was considered in the secondary zone with the mass of the atoms increased by a factor of 4. This drastically reduced the computational effort and thus enabled the simulation of a much larger piece of solid. In this way artificial effects are suppressed, which are due to the fact that the pressure wave caused by the impact is reflected at the boundary of the simulational zone [8]. The strength of the harmonic forces was chosen to fit the elastic constants of copper. Langevin-forces acting on the boundary atoms of the primary zone simulated the substrate temperature of $T = 300$ K [9].

First, in order to study the influence of the surface slope, the impact of a Cu_{87} cluster onto a tilted surface with a slope of $\alpha = 10°$ in the x-direction was simulated. The direction of incidence coincided with the z-axis. A typical impact energy of 10 eV per cluster atom was chosen. For a cluster of that size and energy roughly 28.000 atoms had to be included into the primary zone. The primary and secondary zone together corresponded to a system of more than 10^6 atoms.

The most important feature of the impact can clearly be seen in Fig. 1. The energetic deposition leads to a net downhill mass transfer, such that in effect at the end of the simulation, when all atoms are situated at lattice sites again, the local slope of the surface has been decreased in a region around the point of impact.

The downhill current is also present for larger clusters: The displacement of the atoms in the surface from the beginning to the end of the simulation were calculated. A cross section of the resulting vector field is depicted in Fig. 2 for a Cu_{1985} cluster impinging with 5 eV/atom onto a tilted piece of a Cu(1 0 0) surface representing roughly 10^7 atoms. Again a strong downhill displacement is perceptible, which results in a net downhill particle current.

4. Estimating the strength of the downhill current

Consider the deposition of clusters onto a surface with a slope α in x-direction. Let r be the number of clusters impinging onto a unit area in unit time. It is assumed that during each impact N atoms are displaced in x-direction by a distance u_i ($i = 1, \ldots, N$). The average displacement of an atom is given by δ/N, where $\delta = \sum_i u_i$ denotes the sum of all displacements caused by a single impact. Consider a straight line of length c parallel to the y-axis. All clusters impacting in the time Δt onto a strip of the width δ/N move $\delta cr \Delta t$ atoms downhill beyond the straight line leading

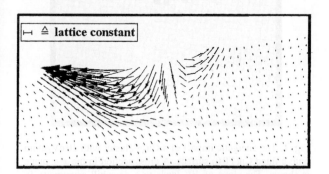

Fig. 2. The displacement of surface atoms by the impact of a Cu_{1985} cluster with 5 eV/atom onto a tilted Cu(1 0 0) surface. Only the displacement of surface atoms in a x–z cross section through the point of contact is shown. The asymmetry in the particle current causes a net downhill transfer of atoms.

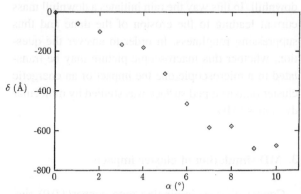

Fig. 3. The sum of displacements in x-direction δ versus tilt angle α of the Cu(1 0 0) surface for the impact of a Cu_{87} cluster with 10 eV/atom. The downhill particle current which is proportional to δ is roughly a linear function of the surface slope.

to a particle current $\mathbf{j} = (r\delta, 0)$. Therefore, the efficiency of this smoothing process can be characterized by the displacement sum δ. For Cu_{87} with $10\,eV/atom$ impacting onto $Cu(1\,0\,0)$ δ decreases with increasing slope (Fig. 3), which indicates that more tilted parts of the film are eroded more efficiently causing a smoothing of rough film features. Because for smooth films the small slope limit may be applied, δ can be assumed to be a linear function of $\tan(\alpha)$.

By inserting the downhill particle current

$$\Omega\mathbf{j}(\mathbf{x}, t) = -\nu\nabla h(\mathbf{x}, t) \tag{2}$$

into Eq. (1) one obtains the well-known Edwards–Wilkinson equation

$$\partial h(\mathbf{x}, t)/\partial t = \nu\nabla^2\mathbf{j}(\mathbf{x}, t) + \eta(\mathbf{x}, t), \tag{3}$$

which was discovered in the context of sedimentation [10]. The quantity

$$\nu = -r\delta\Omega/\tan\alpha \tag{4}$$

measures the strength of the downhill process.

Restricting oneself to the description of height fluctuations on length scales much larger than one cluster

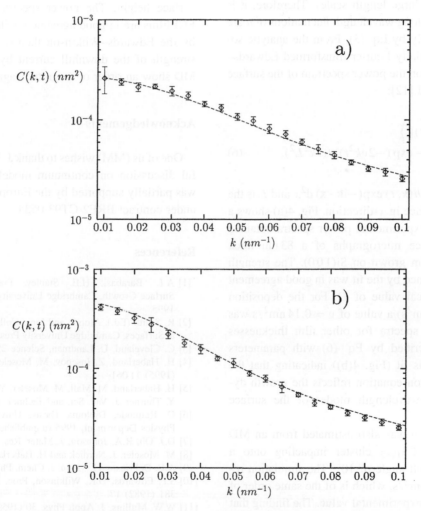

Fig. 4. Power spectra of Cu_{2000} (5 eV/atom) cluster films for different film thicknesses: (a) 83 nm, (b) 147 nm. Data points: experimental films studied by atomic force microscopy (scan length 3000 nm), dashed curves: the Edwards–Wilkinson power spectrum with parameters fitted to experiment (a).

diameter, one can approximate η by a temporally and spatially white noise [10] with covariance

$$\langle \eta(\mathbf{x}, t) \eta(\mathbf{x}', t') \rangle = 2D\delta(\mathbf{x} - \mathbf{x}')\delta(t - t'), \tag{5}$$

where D measures the strength of the noise and can be calculated from the deposition rate and details of the cluster size distribution. For instance, for a beam of clusters with N_c atoms/cluster the noise strength is given by $D = r N_c^2 \Omega^2 / 2$. Additional particle currents like surface or bulk diffusion are related to higher order spatial derivatives of the film height [11] and can be neglected for large length scales. Therefore, it is expected that the long wavelength fluctuations of h are described correctly by Eq. (3). From the analytic solution of the spatially Fourier transformed Edwards–Wilkinson equation the power spectrum of the surface height is obtained [12]:

$$\begin{aligned} C(\mathbf{k}, t) &:= \langle |h_\mathbf{k}(t)|^2 \rangle \\ &= D(1 - \exp(-2\nu k^2 t))/(\nu k^2 L^2), \end{aligned} \tag{6}$$

where $h_\mathbf{k}(t) = \int h(\mathbf{x}, t) \exp(-i\mathbf{k} \cdot \mathbf{x}) \, d^2 x$ and L is the length of the surface in x-direction. Fig. 4(a) shows a fit of Eq. (6) to experimental power spectra obtained from atomic force micrographs of a 83 nm thick Cu_{2000} cluster film grown on Si(1 0 0). The strength of the noise obtained by the fit was in good agreement with the theoretical value of D. For the deposition rate $r = 10^{-3}/(\text{nm}^2\text{s})$ a value of $\nu = 0.14 \, \text{nm}^2/\text{s}$ was obtained. Power spectra for other film thicknesses can also be described by Eq. (6) with parameters obtained from this fit (Fig. 4(b)) indicating that the Edwards–Wilkinson equation reflects the growth dynamics of long wavelength modes of the surface profile correctly.

The parameter ν was also estimated from an MD simulation of a Cu_{1985} cluster impacting onto a Cu(1 0 0) with a tilt angle of $10°$. The simulation revealed $\nu = 0.23 \, \text{nm}^2/\text{s}$, which is of the same order of magnitude as the experimental value. The finding that the experimental ν is 40% smaller may be explained by the fact that ECI films are polycristalline and thus the downhill current is suppressed by defects.

5. Summary

MD simulations of clusters impinging onto sloped surfaces suggest a new surface smoothing mechanism. The cluster causes a downhill particle current which transports hills on the film surface into valleys. The introduction of this concept into a continuum description of thin film growth reveals the Edwards–Wilkinson equation for the long wavelength components of the surface height. The power spectra of experimental ECI films are in good agreement with those predicted by the Edwards–Wilkinson theory. Estimates of the strength of the downhill current by experiment and MD show an order of magnitude agreement.

Acknowledgements

One of us (MM) wishes to thank J. Villain for a helpful discussion on continuum models. This research was partially supported by the European Commission under contract BRE2-CT93-0534.

References

[1] A.L. Barabasi, H.E. Stanley, Fractal Concepts in Surface Growth, Cambridge University Press, Cambridge, 1995.
[2] R. Smith (Ed.), Atomic and Ionic Collisions in Solids and at Surfaces, Cambridge University Press, Cambridge, 1997.
[3] C. Cleveland, U. Landman, Science 257 (1992) 355.
[4] H. Haberland, Z. Insepov, M. Moseler, Phys. Rev. B 51 (1995) 11 061.
[5] H. Haberland, M. Mall, M. Moseler, Y. Qiang, T. Reiners, Y. Thurner, J. Vac. Sci. and Technol. A 12 (1994) 2925.
[6] O. Rattunde, Diploma Thesis, University of Freiburg, Physics Department, 1995 (unpublished).
[7] D.J. Oh, R.A. Johnson, J. Mater. Res. 3 (1988) 471.
[8] M. Moseler, J. Nordiek and H. Haberland, PRB, accepted.
[9] A.E. DePriesto, H. Metiu, J. Chem. Phys. 90 (1989) 1229.
[10] S.F. Edwards, D.R. Wilkinson, Proc. Roy. Soc. Lond. A 381 (1982) 17.
[11] W.W. Mullins, J. Appl. Phys. 30 (1959) 77.
[12] J. Villain, J. Phys. I 1 (1991) 19.

ELSEVIER

Computational Materials Science 10 (1998) 457–462

COMPUTATIONAL
MATERIALS
SCIENCE

A combination of atomic and continuum models describing the evolution of nanoclusters

M. Strobel *, K.-H. Heinig, W. Möller

Research Center Rossendorf, Institute of Ion Beam Physics and Materials Research, PO Box 510 119, D-01314 Dresden, Germany

Abstract

Ion beam synthesis is a promising technique to form nanoclusters. However, the synthesis of an ensemble of nanoclusters having a specific size and depth distribution requires a comprehensive understanding of the physical processes. We present two models, a discrete and a continuous one, where each is suited to study special stages of ion implantation and annealing. On the atomic scale our kinetic 3D lattice Monte-Carlo (MC) model is used to study nucleation and growth of nanoclusters. On the mesoscopic scale, rate-equations are used to describe their coarsening. By a combination of both methods the use of data from the MC simulation provides for the first time realistic initial conditions for the rate-equation approach to Ostwald ripening. As an application, the combined atomic-scale and continuum simulation tools are used to study the evolution and self-organization of nanoclusters. Copyright © 1998 Elsevier Science B.V.

Keywords: Kinetic Monte-Carlo simulation; Nucleation; Nanocluster; Ostwald ripening; Self-organization

1. Introduction

Ion implantation is a powerful technique for manipulating the physical properties of solids or thin films. Depending on the dose, the modifications range from pure doping and generation of nanoclusters to formation of buried layers. Generally, ion beam synthesis (IBS) of nanostructures can be separated in five stages: supersaturation of the matrix by implanted impurity atoms, nucleation of nanoclusters, their growth, subsequent Ostwald ripening (OR) and, for high doses, coalescence. All stages have to be well understood in order to give reliable predictions.

As an atomic description the Monte-Carlo (MC) simulation allows to study, in principle, all basic

processes of IBS. However, due to its atomic nature, the size of simulation volume is too small for predictions of large scale and long time phenomena due to the restricted power of available computer systems.

On the other hand (equilibrium) thermodynamic concepts are the method of choice for quasi-equilibrium processes like OR of nanoclusters, which is usually observed during subsequent annealing of implanted samples. The first comprehensive description of OR was given by Lifshitz and Slyozov [1] and independently by Wagner [2] – known as LSW-theory. This theory has been successfully applied to describe coarsening in homogeneous systems. In inhomogeneous systems the self-organization of nanoclusters into layers of precipitates by OR, as it was observed in the SIMOX process [3], could be modeled by a direct integration of the reaction–diffusion equations [4].

---

* Corresponding author. E-mail: strobel@fz-rossendorf.de.

The main problem in simulations of OR of inhomogeneous systems remains the choice of the initial conditions, i.e. the positions $\{r_i\}$ and radii $\{R_i\}$ of the precipitates at the beginning of the simulation. This is much less a problem in infinite homogeneous systems, because all ensembles evolve towards an attractor state [5] regardless how unphysical the initial conditions were chosen. However, in finite systems small fluctuations may have a strong effect on the (spatial) evolution [4], so that a reliable knowledge of the initial conditions is crucial for the rate-equation approach to OR.

Here we present for the first time simulation results obtained by using a combination of atomic-scale and continuum modeling for the evolution of nanoclusters, which resolves the problem mentioned above. Within this model we demonstrate that self-organization of nanoclusters is an intrinsic property of OR and is not necessarily initiated by boundary effects.

2. Atomic and continuum simulation tools

2.1. Kinetic 3D Monte-Carlo method

Naturally the most appropriate method to study supersaturation, nucleation and growth phenomena during IBS is an atomic description. Therefore we have developed a kinetic MC code for atoms on a 3D lattice, where atomic interactions are simulated by a Cellular Automaton approach. Using the degree of freedom to gauge the binding energy of impurity atoms, we just consider effective nearest neighbor interaction among the impurity atoms, which includes interactions of the impurity atoms with the host matrix.

Using a fcc geometry in our particular lattice description, each atom is allowed to jump to any empty site of its 12 nearest neighbor positions. In the most simple implementation of configuration independent interactions, i.e. the Ising model, the binding energy kE_B is just proportional to the number k of nearest neighbor bonds, whereas the diffusion of monomers is determined by the activation energy E_A. Thus the cluster properties are defined by the effective binding energy per bond E_B. During an MC step the transition probability P_{if} for one atom to jump from the initial site i to the final site f is

$$P_{if} = \begin{cases} \exp\left\{-\dfrac{E_A}{k_B T}\right\}, & k_f \geq k_i, \\ \exp\left\{-\dfrac{(k_i - k_f)E_B + E_A}{k_B}\right\}, & k_f < k_i, \end{cases}$$

(1)

where $k_{(i,f)}$ represents the number of effective nearest neighbor bonds in the two states. In order to reduce simulation time it is convenient to allow a successful jump for each diffusion attempt, i.e. $E_A \to 0$, which means a renormalization of the jump probabilities and defines the timescale of an MC step. As common to MC simulations the Metropolis algorithm [6] is applied, i.e. transitions, which are energetically not favored, are allowed if a random number is less than P_{if}. In our simulations for each MC step all atoms with less than 12 effective nearest neighbor bonds have statistically one jump attempt.

2.2. Rate-equation approach

During annealing following ion implantation an ensemble of precipitates, which interact diffusionally, evolves in real and cluster size space due to OR. In the thermodynamic description the driving force of OR is the minimization of the energy associated with the precipitate/matrix interface. As a consequence large nanoclusters grow at the expense of smaller ones, which eventually dissolve. The pioneering work concerning the computational modeling of OR was done by Voorhees and Glicksman [7] using a local mean-field description. More general versions of this method allow the study of spatially inhomogeneous systems [4] and of systems having an arbitrary degree of diffusion and/or reaction control [8].

Our rate-equation approach solves the (stationary) diffusion equation $\Delta c(r) = 0$ in real space, where precipitates fix the boundary conditions c' on their interfaces (details of this method are discussed in [8]). In the case of diffusion controlled ripening the Gibbs–Thomson relation

$$c^{GT}(R) = c_\infty \exp\left\{\frac{R_c}{R}\right\} =: c', \quad R_c = \frac{2\sigma V_m}{k_B T} \quad (2)$$

determines this boundary condition, which is the equilibrium monomer concentration at the interface of a precipitate with radius R. Here, c_∞ is the solidus concentration and the capillarity length R_c is proportional to the interface energy σ and the atomic volume V_m.

A multipole expansion for each nanocluster is applied to construct the concentration field, and for volume fractions of the precipitated material of the order of a few percent, only the monopole terms are retained

$$c(r) = c_u + \frac{1}{4\pi D} \sum_i \frac{Q_i}{|r - r_i|}, \tag{3}$$

where c_u is the (global) mean concentration of impurity atoms. This leads to a linear system of equations for the monopole terms $\{Q_i\}$, which can be interpreted as source (sink) strengths for monomer detachment (attachment). In the case of material conservation, i.e. $\sum_i Q_i = 0$, c_u is determined self-consistently. However, in systems, where material conservation is not required, c_u is a free parameter and can be chosen to control the flux of impurity atoms through the boundaries of the system.

The evolution of the precipitates is calculated by the stepwise numerical integration of the reaction-diffusion equations: (i) one solves the linear system of equations to determine the source strengths $\{Q_i\}$; (ii) via the relation $(dR_i^3)/(dt) \propto -Q_i$ one calculates the individual evolution of the precipitates within a time interval Δt. With the new radii $\{R_i\}$ one repeats the two steps.

3. Combination of atomic and continuum description

Usually one starts the combined simulation by specifying the physical parameters like ion dose, flux or temperature of the system to be modeled. For example, for a given ion dose, the lateral dimensions of the simulation cell are chosen in a way to keep the total number of atoms below n upper limit given by the power of the computer system. The MC simulation describes implantation, nucleation and OR simultaneously and is usually terminated when the remaining number of precipitates drops below a limit, which can

be handled by a subsequent rate-equation approach. Fig. 1(a) demonstrates for a small system the result of an MC run after several 10^6 MC steps. In this example, at the beginning of the simulation, impurity atoms corresponding to a volume fraction of 3% were homogeneously distributed in a box with periodic boundary conditions. The MC simulation generates precipitates with the typical shape of octahedras. This can be understood in terms of the principle of minimization of surface energy, since for systems condensed on a fcc lattice, the $\langle 1\,1\,1 \rangle$ surfaces are energetically favored.

As the crucial parameters for OR are the radii of the precipitates, we need an approximation for the sizes. The radius characterizing a specific cluster is determined by the smallest sphere, which envelops the number of atoms the nanocluster contains (see Fig. 1(b)). Together with the coordinates of the center of mass of the simulated nanoclusters we have the appropriate initial conditions for the rate-equation simulation. Since two completely different descriptions of the evolution of nanoclusters are combined, we would like to make two comments on the data transfer from the MC to the rate-equation simulation:

- The basic thermodynamic relation, which governs in the late stage of annealing the evolution of nanoclusters, is the Gibbs–Thomson relation (2). Thus, a crucial test of the MC simulation is a consistency check with the Gibbs–Thomson relation. Therefore, a series of simulations have been performed, where at a given temperature a nanocluster of a certain size was put into a box and the number of dissolved monomers being in equilibrium with this cluster was determined [9]. The results for three different temperatures are shown in Fig. 2. Obviously, the MC simulation reproduces correctly the Gibbs–Thomson relation for clusters even on the nanometer scale.

- Another check for the reliability of the MC results for further usage in the rate-equation simulation is the particle radius distribution (PRD) of the precipitates. Fig. 3 shows the simulated PRD for an ensemble of precipitates in a box with periodic boundary conditions. In this example, the volume fraction of the precipitates is approximately 3%. The PRD obtained from the MC simulation is

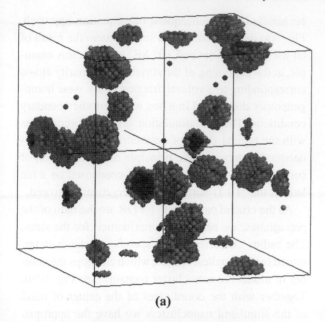

(a)

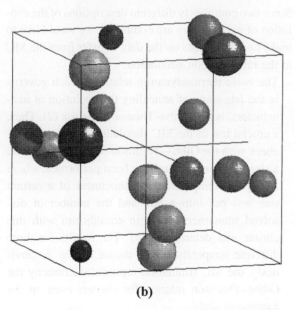

(b)

Fig. 1. Example of a combined simulation for a small system. The result of an MC simulation of impurity atoms in a matrix is presented in (a), where, for clarity, only the impurity atoms are shown. The impurity atoms form clusters, their diffusional interaction is mediated by monomer exchange. In (b) the nanoclusters are approximated by the smallest spheres, which envelope their atoms. The center of each sphere is located at the center of mass of its atomic configuration (with appropriate translations at the periodic boundaries).

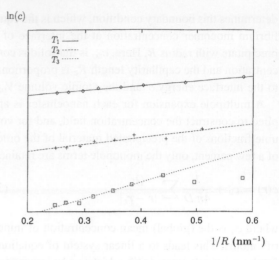

Fig. 2. For three different temperatures, $T_1 > T_2 > T_3$, the equilibrium concentrations of precipitates obtained from MC simulations are plotted versus the inverse radius. The least square fits for each temperature indicate the validity of the Gibbs–Thomson relation for nanocluster sizes of the order of a few nanometer.

in good agreement with the stationary PRD of the LSW-theory for the diffusion controlled limit, and is in accordance with PRDs obtained from the simulation of OR [8].

These aspects ensure that due to our combination of two theoretical approaches the predictability of the evolution of nanoclusters by OR is now no longer influenced by artificial or arbitrary choices of initial conditions. Thus, the origin for self-structuring of nanoclusters have to be traced back to the nucleation stage at the very beginning of the ion implantation.

4. Modeling of self-organization of nanoclusters

With the combination of both simulation tools, a discussion of the reasons for the self-organization of nanocluster into layers or other spatially periodic structures is now presented. Within the framework of OR we discuss the question of how boundary conditions and internal ripening interfere during the evolution of these structures.

Computer simulations based on the rate-equation approach successfully described the occurrence of

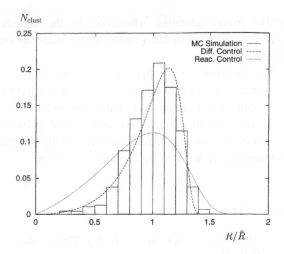

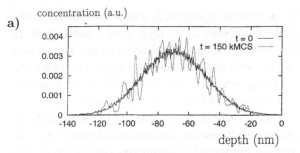

Fig. 3. PRD obtained from an MC simulation. For this ensemble the mean radius $\bar{R}$ of enveloping spheres is approximately 1.7 nm. For comparison, the stationary PRDs of the LSW-theory for the diffusion and reaction controlled cases are shown.

bands of nanoclusters [4,10] using a model, where an absorbing interface initiated the self-organization. Additionally the structure wavelength λ was determined as the characteristic interlayer distance, which is proportional to the diffusional screening length λ_D [11]:

$$\lambda(t) = \alpha \lambda_D(t) = \frac{\alpha}{\sqrt{4\pi n(t)\bar{R}(t)}}, \qquad \alpha \approx 3. \qquad (4)$$

Here, n and $\bar{R}$ represent the density and the mean radius of the precipitates, respectively. Generally, the diffusional screening length defines the critical distance between nanoclusters, beyond which their diffusional interaction is screened by the remaining ensemble of precipitates they are embedded in. The simplified discussions of Reiss and Heinig [4] and Borodin et al. [10] have assumed the rather artificial initial condition of equal-sized precipitates across the depth. Though experimental evidence supports this assumption [12], simulations beyond this approximation were needed to clarify the physics of this process.

Here we restrict our study to systems, where the physical conditions during implantation prevent a significant redistribution of the deposited atoms before annealing. This situation is encountered if during ion implantation the diffusion length l_D of impurity atoms is negligible with respect to the width of the

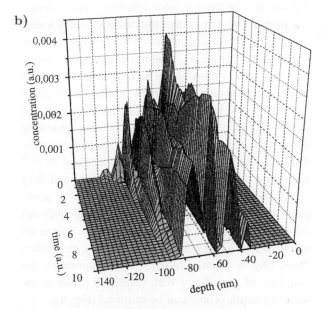

Fig. 4. Evolution of a depth profile of implanted impurity atoms due to nucleation and OR of nanoclusters. In (a) the result of an MC simulation of the nucleation stage is shown. The subsequent rate-equation simulation of OR starting with the final depth profile of (a) is shown in (b).

implantation profile, i.e. $l_D \ll \Delta R_p$. MC studies of nucleation phenomena in this case show a rather weak dependence of the mean precipitate radius across the profile [13]. However, right from the beginning local OR is initiated by these small size differences. This can clearly be seen in Fig. 4(a), where we plot the MC simulation results of the evolution of the depth distribution of implanted impurity atoms. For this example we simulate the redistribution of an as-implanted Gaussian-like depth profile of 1×10^{16} cm^{-2} impurity atoms. The projected range of the peak concentration is -70 nm and at a depth of -140 nm we set up an absorbing interface, which corresponds to the implantation into a thin layer A on a substrate B. In Fig. 4(b), the long time behavior of this system was studied with the rate-equation approach, where the absorbing interface was described by the boundary condition $c(-140\,\text{nm}) = c_\infty$ (as used in [4]).

Since according to the Gibbs–Thomson equation the equilibrium monomer concentration decreases with increasing cluster size, precipitates slightly larger than the average become a more effective sink to diffusing monomers. Thus small fluctuations initiate a self-amplifying coarsening process over the whole implantation profile (Fig. 4). Assuming an approximately constant mean radius of the nanoclusters after nucleation, the structure wavelength λ varies with depth and is smallest in the center of the profile, where the density of precipitates is highest. During annealing the diffusive screening length evolves (in the case of diffusion controlled coarsening) proportional to $t^{1/3}$ according to the time dependencies of $n(t)$ and $\bar{R}(t)$ (LSW-theory). This is the reason, why spatial structures, which are in phase with λ (i.e. self-amplifying) during a certain time, grow out of phase for later times and then start a competitive growth among each other. Thus, like in chaotic systems, during annealing the doubling of periodicity with respect to peaks in the impurity depth profile can be observed (Fig. 4).

Obviously, before a diffusional interaction between the impurity profile and the interfaces sets in, self-structuring has started within the profile. Generally, our observed self-organization towards sublayers does not result in equal-spaced layers of precipitates, because the self-structuring is triggered by random spatial inhomogeneities. Therefore, in the chaotic spatial distribution of nanoclusters some order can be found.

In conclusion for the first time the formation and self-organization of nanoclusters has been studied by a combination of atomic and continuum models. The self-strucuring of nanoclusters into spatial patterns was shown to be an intrinsic property of the system not necessarily initiated by boundary conditions.

References

[1] I.M. Lifshitz and V.V. Slyozov, J. Phys. Chem. Solids 19 (1961) 35.
[2] C. Wagner, Z. Elektrochem. 65 (1961) 581.
[3] P.L.F. Hemment, K.J. Reason, J.A. Kilner, R.J. Chater, C. Marsh, G.R. Booker, G.K. Celler and J. Stoemenos, Vacuum 36 (1986) 877.
[4] S. Reiss and K.-H. Heinig, Nucl. Instr. and Meth. B 102 (1995) 256; S. Reiss and K.-H. Heinig, Nucl. Instr. and Meth. B 112 (1996) 223.
[5] M.K. Chen and P.W. Voorhees, Model. Simul. Mater. Sci. Eng. 1 (1993) 591.
[6] N. Metropolis, A. Rosenbluth, M. Rosenbluth, A. Teller, and E. Teller, J. Chem. Phys. 21 (1953) 1087.
[7] P.W. Voorhees and M.E. Glicksman, Acta Metal. 32 (1984) 2001.
[8] M. Strobel, S. Reiss and K.-H. Heinig, Nucl. Instr. & Meth. B 120 (1996) 216; M. Strobel, S. Reiss, K.-H. Heinig and W. Möller, Rad. Def. Eff. Solids 141 (1997) 990.
[9] M. Palard and K.-H. Heinig, to be published.
[10] V.A. Borodin, K.-H. Heinig and S. Reiss, Phys. Rev. B 56 (1997) 5332.
[11] A.D. Brailsford, J. Nucl. Mat. 60 (1976) 257.
[12] R. Weber, R. Yankov, R. Müller, W. Skorupa, S. Reiss and K.-H. Heinig, Mater. Res. Soc. Symp. Proc. 316 (1994) 729.
[13] M. Strobel and K.-H. Heinig, to be published.

ELSEVIER

Computational Materials Science 10 (1998) 463–467

COMPUTATIONAL
MATERIALS
SCIENCE

A first principles study of small Cu_n clusters based on local-density and generalized-gradient approximations to density functional theory

Carlo Massobrio [a,*], Alfredo Pasquarello [b,c], Andrea Dal Corso [b]

[a] Institut de Physique et de Chimie des Matériaux de Strasbourg, 23 rue du Loess, F-67037 Strasbourg, France
[b] Institut Romand de Recherche Numérique en Physique des Matériaux (IRRMA), IN-Ecublens, CH-1015 Lausanne, Switzerland
[c] Department of Condensed Matter Physics, University of Geneva, CH-1211 Geneva, Switzerland

Abstract

Neutral and anionic Cu_n clusters (Cu_2, Cu_3, Cu_6 and Cu_7^-) are studied within density functional theory via (a) the local-density approximation (LDA) and (b) the generalized-gradient approximation (GGA) of Perdew and Wang (GGA-PW) for exchange and correlation. GGA reduces by ∼20% the binding energies, while the bond lengths are increased by ∼3–4%. The different levels of GGA approximation, involving optimization of the electronic density and/or of the geometry, are shown in detail. In the case of Cu_6 the GGA configurational ground state is a planar structure of D_{3h} symmetry. This result differs from the one obtained by LDA, where the three different isomers (one two-dimensional and two three-dimensional) were found to lie within 0.04 eV. Copyright © 1998 Elsevier Science B.V.

In recent years the electronic and structural properties of small charged and neutral Cu_n clusters have been the object of accurate first principles investigations [1–5]. In particular, calculations treating s and d electrons on an equal footing have been carried out within the local density approximation (LDA) of density functional theory by using the Car–Parrinello (CP) [6] scheme and a plane-wave basis set [2–4]. These studies have focused on clusters as large as Cu_{10} [3] by providing fully optimized equilibrium structures and the corresponding electronic density of states. Moreover a successful interpretation of photoelectron spectroscopy data on Cu_n^- clusters has been obtained by combining first principles molecular dynamics and a

simplified scheme to account for final-state relaxation effects [2].

The large amount of data collected in [2–4] makes Cu_n clusters well suited for a comparative study of their energetics and structural properties as obtained within the LDA and the generalized-gradient approximation (GGA) to density functional theory, in analogy with what has been achieved in the case of selected atoms and solids in [7].

In the case of Cu_n clusters two recent investigations, both limited to $n < 5$ have been performed by using the GGA scheme. In the first [1] the GGA is applied only as a perturbation to the self-consistent LDA solution for a given geometry and LDA electronic density, while in the second [5] fully optimized GGA geometries are obtained. The GGA approach appears to improve the binding energies as compared to LDA, while

* Corresponding author. Fax: +00333 388 107249; e-mail: carlo@marylou.u-strasbg.fr.

a moderate increase in bond lengths (~0.1 Å) is found.

This paper extends upon these previous investigations by considering larger cluster sizes. We first analyze the changes in binding energy of the different isomers when the GGA is applied at different levels of approximation. Then we consider the modifications introduced by the GGA in the relative stability of different isomers. This is of particular interest in the case of Cu_6 and Cu_7^-. The former cluster size marks the onset of the three-dimensional character in Cu_n clusters according to the LDA results of Massobrio et al. [3]. The latter is characterized by the presence of two different isomers very close in energy both of which were invoked to interpret spectral features [2].

We studied Cu_n clusters by first principles dynamical simulations [6] based on density functional theory. Two distinct approximations are adopted for exchange and correlation, the local-density approximation (LDA) [8] and the generalized-gradient approximation (GGA) proposed by Perdew and Wang [9].[1] We employ a plane-wave basis set with periodic boundary conditions in conjunction with Vanderbilt [10] ultrasoft pseudopotentials. The clusters are placed in a cubic cell of side equal to 25.9 a.u. The energy cutoffs are $R_1 = 20$ Ry for the orbitals expanded in plane waves at the Γ point and $R_2 = 150$ Ry for the augmented electron density. These values are significantly larger than those employed in [2–4] ($R_1 = 15$ Ry, $R_2 = 130$ Ry) and yield dimer binding energy and bond distance differing by less than 0.4% from the values obtained for $R_1 = 30$ Ry and $R_2 = 300$ Ry. The initial cluster geometries are those published in [3,4]. The electronic degrees of freedom are first relaxed at fixed ionic positions, and then the ground state and local minima are found by using a combination of steepest descent and damped CP molecular dynamics [11]. In Tables 1–3 we report the values of the exchange–correlation energies, total energies, average interatomic distance and binding energy for all the cluster sizes and isomers considered in this study. The binding energies are obtained by

subtracting from the total energy of the cluster the sum of the atomic energies given in Table 1. These latter include spin-polarization effects, as obtained with the empirical formula [12] $\Delta E = -0.18 \times n_p^2$ eV, where n_p^2 is the difference between the number of up and down spins. Four set of results are presented. The first is the fully optimized LDA result, in the second the LDA total energy is modified by replacing the value of the LDA exchange and correlation energy with the GGA one. Both LDA electronic density and ionic positions are left unchanged. In the third the electronic density is optimized at the GGA level by keeping the LDA positions and finally the results of a full GGA optimization with respect to both electronic density and ionic positions are given.

Overall GGA is seen to reduce by at least 20% the binding energies, while the bond lengths are increased by ~ 3–4%. We note that the reduction of binding energy is already appreciable at the first level of GGA approximation. The GGA binding energy of Cu_2 is in much better agreement with the experimental value ($E_b^{exp} = 1.97$ eV) than the LDA result. Dimer vibrational frequencies obtained within LDA and GGA are both close to the experimental value ($\omega_e^{exp} = 264$ cm^{-1}, $\omega_e^{LDA} = 272$ cm^{-1}, $\omega_e^{GGA} = 254$ cm^{-1}). In the case of Cu_3 both the LDA and the GGA results compare favorably with the data of Calaminici et al. [5]. We found an energy difference between the acute and the obtuse isomers[2] of Cu_3 equal to 0.02 eV for both LDA and GGA, while in [5] this value is slightly smaller within GGA (GGA, 0.006 eV, LDA, 0.023 eV). In Table 1 we note that at intermediate GGA levels, the two isomers of Cu_3 have total energies differing by less than 0.005 eV.

The equilibrium structures found for Cu_6 are displayed in Fig. 1. Within LDA three isomers are very close in energy, the fully three-dimensional (3D) C_{2v} structure being slightly lower in energy (by 0.02 and

---

[1] In the LDA case the exchange and correlation energy was included using the formulae given in [8].

[2] For the trimer the geometries are characterized by the value of the angle between two equivalent bond lengths, where 'acute' stands for $\alpha < 60°$ and 'obtuse' for $\alpha > 60°$. In the LDA case the distances are 4.58, 4.22 and 4.22 a.u. for Cu_3 'obtuse' and 4.18, 4.45 and 4.45 a.u. for Cu_3 'acute'. In the GGA case the distances are 4.87, 4.36 and 4.36 a.u. for Cu_3 'obtuse' and 4.34, 4.62 and 4.62 a.u. for Cu_3 'acute'.

Table 1
Exchange–correlation energy and total energy for the Cu atom and exchange–correlation energy, total energy, average equilibrium distance and binding energy for Cu_2 and Cu_3, the latter in the obtuse and acute equilibrium energy configurations

	R_I	ρ	XC	E_{xc} (H)	E_{tot} (H)	d (a.u.)	E_b (eV)
atom		LDA	LDA	−7.723	−50.086		
		LDA	PW	−7.924	−50.287		
		PW	PW	−7.926	−50.051		
dimer	LDA	LDA	LDA	−15.500	−100.271	4.11	2.71 (1.36)
	LDA	LDA	PW	−15.889	−100.659	4.11	2.29 (1.15)
	LDA	PW	PW	−15.894	−100.185	4.11	2.28 (1.14)
	PW	PW	PW	−15.886	−100.186	4.26	2.31 (1.16)
trimer C_{2v} (obtuse)	LDA	LDA	LDA	−23.251	−150.421	4.34	4.46 (1.49)
	LDA	LDA	PW	−23.824	−150.993	4.34	3.57 (1.19)
	LDA	PW	PW	−23.835	−150.282	4.34	3.53 (1.18)
	PW	PW	PW	−23.817	−150.283	4.53	3.59 (1.20)
trimer C_{2v} (acute)	LDA	LDA	LDA	−23.246	−150.420	4.36	4.44 (1.48)
	LDA	LDA	PW	−23.819	−150.993	4.36	3.57 (1.19)
	LDA	PW	PW	−23.830	−150.282	4.36	3.53 (1.18)
	PW	PW	PW	−23.813	−150.284	4.53	3.57 (1.19)

R_I refers to the level of optimization of the geometry (LDA or GGA), ρ to the level of optimization of the electronic density (LDA or GGA) and XC refers to the exchange–correlation functional (LDA or GGA) considered in the total energy. Energies are expressed in Hartree, distances in atomic units and binding energies in eV. The binding energy per atom is given in parenthesis. PW stands for the GGA scheme by Perdew and Wang (see [9]).

Table 2
Exchange–correlation energy, total energy, average equilibrium distance and binding energy for Cu_6 (for the meaning of the different symbols and quantities, see Table 1)

	R_I	ρ	XC	E_{xc} (H)	E_{tot} (H)	d (a.u.)	E_b (eV)
Cu_6 (D_{3h})	LDA	LDA	LDA	−46.600	−301.029	4.36	14.02 (2.34)
	LDA	LDA	PW	−47.718	−301.147	4.36	11.49 (1.92)
	LDA	PW	PW	−47.746	−300.722	4.36	11.38 (1.90)
	PW	PW	PW	−47.698	−300.728	4.51	11.54 (1.92)
Cu_6 (C_{5v})	LDA	LDA	LDA	−46.624	−301.030	4.40	14.04 (2.34)
	LDA	LDA	PW	−47.737	−302.143	4.40	11.40 (1.90)
	LDA	PW	PW	−47.767	−300.717	4.40	11.26 (1.88)
	PW	PW	PW	−47.712	−300.724	4.56	11.44 (1.91)
Cu_6 (C_{2v})	LDA	LDA	LDA	−46.624	−301.031	4.46	14.06 (2.34)
	LDA	LDA	PW	−47.730	−301.137	4.46	11.23 (1.87)
	LDA	PW	PW	−47.762	−300.711	4.46	11.10 (1.85)
	PW	PW	PW	−47.703	−300.718	4.61	11.27 (1.88)
Cu_6 (C_{2v}, planar)	LDA	LDA	LDA	−46.604	−301.019	4.36	13.75 (2.29)
	LDA	LDA	PW	−47.722	−302.137	4.36	11.23 (1.87)
	LDA	PW	PW	−47.751	−300.712	4.36	11.10 (1.85)
	PW	PW	PW	−47.705	−300.718	4.50	11.26 (1.88)
Cu_6 (C_{2h})	LDA	LDA	LDA	−46.556	−300.984	4.39	12.78 (2.13)
	LDA	LDA	PW	−47.676	−302.103	4.39	10.30 (1.72)
	LDA	PW	PW	−47.704	−300.679	4.39	10.22 (1.70)
	PW	PW	PW	−47.669	−300.685	4.51	10.38 (1.73)

Table 3
Exchange–correlation energy, total energy, average equilibrium distance and binding energy for Cu_7^- (for the meaning of the different symbols and quantities, see Table 1)

	R_I	ρ	XC	E_{xc} (H)	E_{tot} (H)	d (a.u.)	E_b (eV)
Cu_7^- (D_{5h})	LDA	LDA	LDA	−54.610	−351.335	4.50	20.01 (2.86)
	LDA	LDA	PW	−55.882	−352.608	4.50	16.22 (2.32)
	LDA	PW	PW	−55.919	−350.954	4.50	16.32 (2.33)
	PW	PW	PW	−55.843	−350.961	4.67	16.52 (2.36)
Cu_7^- (C_{3v})	LDA	LDA	LDA	−54.602	−351.336	4.47	20.03 (2.86)
	LDA	LDA	PW	−55.876	−352.610	4.47	16.27 (2.32)
	LDA	PW	PW	−55.911	−300.956	4.47	16.38 (2.34)
	PW	PW	PW	−55.835	−300.964	4.63	16.58 (2.37)

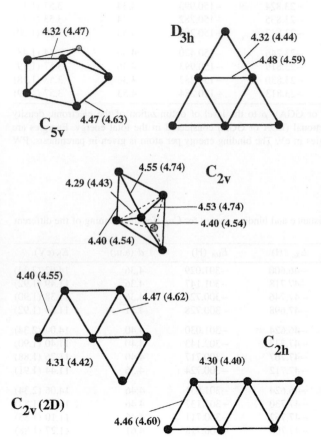

Fig. 1. Equilibrium geometries determined in the present work for Cu_6 by using the LDA and GGA approaches. Bond lengths are in atomic units, and are given in parenthesis for the GGA case.

0.04 eV, respectively) than the flat pentagonal pyramid of symmetry C_{5v} and the planar D_{3h} hexamer. Table 2 shows that the GGA has the effect of modifying the

relative stability of the different isomers obtained via LDA. The planar D_{3h} structure becomes the configurational ground state within GGA, and is lower by 0.1 eV than the C_{5v} structure and by as much as 0.27 eV than the C_{2v} isomers. It is worth noticing that within 1% these energy differences are essentially independent of the level of GGA approximation considered. As shown in Table 2 the results for Cu_6 given in [3] have been extended by considering two further local minima of the potential energy surface, corresponding to planar C_{2v} and C_{2h} symmetries. GGA makes the planar C_{2v} structure as stable as the three-dimensional C_{2v}, while the C_{2h} is definitely much higher in energy in both LDA and GGA approaches. These results suggest that planar structures of Cu_6 are more stable than what is found on the basis of the LDA results presented in [3], even though the differences in total energy remain quite small and close to the intrinsic precision of the calculation, which can be estimated to a few hundredths of eV.

Our data on two competing isomers (shown in Fig. 2) of Cu_7^- are displayed in Table 3. According to the analysis presented in [2,4] the predominant features of the photoelectron spectra of Cu_7^- can be associated with the C_{3v} isomer. This isomer was found to be essentially as stable as the D_{5h} one at zero temperature, but it becomes thermally favored at $T = 400$ K. In this respect it is interesting to note that our present calculations at the GGA level show a slightly more pronounced stability (by 0.06 eV against the 0.02 eV in the LDA case) of the C_{3v} isomer, by confirming the results and the

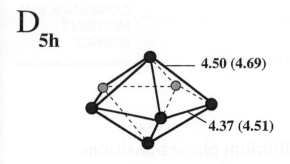

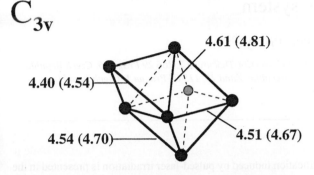

Fig. 2. Equilibrium geometries determined in the present work for Cu_7^- by using the LDA and GGA approaches. Bond lengths are in atomic units, and are given in parenthesis for the GGA case.

interpretation proposed on the basis of the LDA results.

In conclusion our analysis proved that in all the cases examined the first level of GGA approximation (the one that does not require any optimization of the electronic density at the GGA level) is already sufficient to improve the binding energies by an amount which does not differ significantly from the

value obtained via a full GGA optimization. Introduction of GGA changes the energy ordering among the isomers of Cu_6, although the differences in energies involved are not larger than 0.1 eV. More refined levels of GGA approximation do not modify the relative stability of the isomers considered.

Calculations have been performed on the computers of the IDRIS computing center of CNRS located in Orsay, France.

References

[1] K.A. Jackson, Phys. Rev. B 47 (1993) 9715.
[2] C. Massobrio, A. Pasquarello and R. Car, Phys. Rev. Lett. 75 (1995) 2104.
[3] C. Massobrio, A. Pasquarello and R. Car, Chem. Phys. Lett. 238 (1995) 215.
[4] C. Massobrio, A. Pasquarello and R. Car, Phys. Rev. B 54 (1996) 8913.
[5] P. Calaminici, A.M. Köster, N. Russo and D.R. Salahub, J. Chem. Phys. 105 (1996) 9546.
[6] R. Car and M. Parrinello, Phys. Rev. Lett. 55 (1985) 2471.
[7] A. Dal Corso, A. Pasquarello, A. Baldereschi and R. Car, Phys. Rev. B 53 1180 (1996).
[8] J.P. Perdew and A. Zunger, Phys. Rev. B 23 (1981) 5048.
[9] J.P. Perdew and Y. Wang unpublished, J.P. Perdew, in: Electronic Structure of Solids '91, eds. P. Ziesche and H. Eschrig (Akademie Verlag, Berlin, 1991).
[10] D. Vanderbilt, Phys. Rev. B 41 (1990) 7892; A. Pasquarello, K. Laasonen, R. Car, C. Lee and D. Vanderbilt, Phys. Rev. Lett. 69 (1992) 1982; K. Laasonen, A. Pasquarello, R. Car, C. Lee and D. Vanderbilt, Phys. Rev. B 47 (1993) 10 142.
[11] F. Tassone, F. Mauri and R. Car, Phys. Rev. B 50 (1994) 10 561.
[12] Y.M. Juan and E. Kaxiras, Phys. Rev. B 48 (1993) 6376.

ELSEVIER

Computational Materials Science 10 (1998) 468–474

COMPUTATIONAL
MATERIALS
SCIENCE

Computational model of nonequilibrium phase transitions in a Si–Ge system

Robert Černý [a], Petr Přikryl [b,*]

[a] *Department of Physics, Faculty of Civil Engineering, Czech Technical University, Thákurova 7, 166 29 Prague 6, Czech Republic*
[b] *Mathematical Institute of the Academy of Sciences of the Czech Republic, Žitná 25, 115 67 Prague 1, Czech Republic*

Abstract

A computational model of a Si–Ge system melting and solidification induced by pulsed-laser irradiation is presented in the paper. Phase transitions in the system are modeled using the theory of transition states so that undercooling or overheating of the interface and kinetic phase diagrams are taken into account. The computational solution of the mathematical model is performed using the Galerkin finite element method in a one-dimensional approximation and the moving boundary problem is solved by a front-fixing technique. In a practical application of the model, the melting and solidification of both Si–Ge alloys of various composition and thin Ge layers on the Si bulk induced by ArF excimer laser are simulated. The results of numerical simulations are compared with the available experimental data. Copyright © 1998 Elsevier Science B.V.

Keywords: Nonequilibrium phase transition; Laser irradiation; Computational model; Si–Ge materials

1. Introduction

Si–Ge materials are commonly applied in many microelectronic and optoelectronic devices. Using strained Si–Ge alloys, the application range for avalanche photodiodes can be extended [1], Si–Ge alloys can also be employed in devices requiring a heterojunction technology [2]. Si–Ge heterostructures consisting of only a few monolayers of Ge and a few monolayers of Si can create semiconductors with different electronic band structures [3].

In mathematical modeling of phase transitions in two-component systems, two approaches are commonly employed. The first of them consists of a generalization of the classical Stefan problem where the phase interface is modeled as a discontinuity surface [4]. The second one employs the mushy-zone concept [5] and determines the actual position of the phase interface a posteriori from the temperature and concentration fields.

The nonequilibrium phase transitions which can take place during fast processes such as laser induced melting and solidification [6] are modeled using an interface response function mostly. In one-component systems, the interface response function is formulated as a relation between the velocity of the interface and its temperature.

---

* Corresponding author. Tel.: +420 2 2221 1631; fax: +420 2 2221 1638; e-mail: prikryl@math.cas.cz.

In two-component systems, the interface response function is more complicated due to the effect of concentration changes. However, both experimental [7] and theoretical [8] treatments have been developed which make it possible to determine the segregation coefficient as a function of the course of the process – growth rate and undercooling. Kinetic phase diagrams which are obtained in this way are then employed in modeling phase transitions in these systems, and the velocity of the interface is a function of both temperature and concentration in the interface response functions. It should be noted that in modeling the nonequilibrium phase transitions the mushy-zone treatment is not feasible due to the difficulties in introducing the interface response functions. Therefore, the discontinuity-surface approach is favored.

2. Mathematical model of nonequilibrium phase transitions

We consider a binary alloy consisting of two components, A and B, and undergoing phase changes (melting and/or solidification). Supposing that heat conduction is the only mode of heat transfer and species diffusion is the only way of mass transfer, we can formulate the balance equations of mass of the component A and the balance equation of internal energy of the system only.

Supposing further that the problem can be treated in one space dimension, the species diffusion in the volume phases can be described by

$$\frac{\partial C_{A,i}}{\partial t} = \frac{\partial}{\partial x}\left(D_i \frac{\partial C_{A,i}}{\partial x}\right), \quad i = 1, s, \tag{1}$$

where C_A is the concentration of the component A defined as $C_A = \rho_A/\rho$, where ρ_A is the mass of the component A per unit volume of the mixture, ρ is the density of the system, D is the diffusion coefficient, and the subscripts l, s denote the liquid and the solid phase, respectively.

The balance of the internal energy reads

$$\rho c_i \frac{\partial T_i}{\partial t} = \frac{\partial}{\partial x}\left(K_i \frac{\partial T_i}{\partial x}\right) + S_i(x, t), \quad i = 1, s, \tag{2}$$

where T is the temperature, c the specific heat, K the thermal conductivity, and S is a volume heat source term.

In formulating the corresponding balance equations at the solid/liquid interface $Z(t)$, we employ the theory of discontinuity surfaces [4] and write

$$(C_{A,l} - C_{A,s})\frac{dZ}{dt} = -D_l\left(\frac{\partial C_{A,l}}{\partial x}\right)_{x=Z(t)-} + D_s\left(\frac{\partial C_{A,s}}{\partial x}\right)_{x=Z(t)+}, \tag{3}$$

$$\rho L_m \frac{dZ}{dt} = -K_l\left(\frac{\partial T_l}{\partial x}\right)_{x=Z(t)-} + K_s\left(\frac{\partial T_s}{\partial x}\right)_{x=Z(t)+}, \tag{4}$$

where L_m is the latent heat of fusion.

Assuming the nonequilibrium melting and solidification to take place in the system, the remaining condition at the solid/liquid interface has a form of an interface response function

$$\frac{dZ}{dt} = C_{A,l,Z}v_A + C_{B,l,Z}v_B, \tag{5}$$

where $C_B = 1 - C_A$ and (see [9])

$$v_j = X_j\left(1 - k_j \exp\left[\frac{L_{m,j}}{R}\left(\frac{1}{T_{E,j}} - \frac{1}{T_Z}\right)\right]\right), \quad j = A, B. \tag{6}$$

Here, T_Z, C_Z denote the temperature and the concentration at the interface, respectively, R is the universal gas constant, T_E is the equilibrium phase change temperature, X is a constant, depending on the parameters of the interface, such as the surface density of atoms, and on the probability factors, k is the nonequilibrium segregation coefficient which is defined as

$$k_j = \frac{C_{j,s,Z}}{C_{j,l,Z}}, \quad j = A, B. \tag{7}$$

The relation of k to the equilibrium segregation coefficient k_0 can be expressed, for instance, by the formula derived by Aziz [10]

$$k(T_Z) = k_0(T_Z) + [1 - k_0(T_Z)] \exp\left(\frac{\tilde{D}_i}{\lambda(dZ/dt)}\right), \tag{8}$$

where $\tilde{D}_i$ is the diffusion coefficient through the phase interface and λ is the interatomic distance.

In modeling the problem of pulsed-laser induced phase transitions in a Si–Ge system we suppose that a Si–Ge sample is irradiated by a laser pulse with the energy density

$$E = \int_0^\infty I_0(t)\, dt, \tag{9}$$

where $I_0(t)$ is the power density of the pulse (power per unit area). Initially, the sample is in the solid state and occupies the one-dimensional domain (interval) $[0, H]$, where H is the thickness of the sample. Due to the laser irradiation the sample begins to melt and we suppose that the solid phase occupies the interval $[Z(t), H]$, whereas the interval $[0, Z(t)]$ corresponds to the liquid phase.

The heat source terms in Eq. (2) have the form [6]

$$S_l(x, t) = (1 - R(t))\alpha_l(x) I_0(t) \exp\left(-\int_0^x \alpha_l(\eta)\, d\eta\right), \tag{10}$$

$$S_s(x, t) = (1 - R(t))\alpha_s(x) I_0(t) \exp\left(-\int_0^{Z(t)} \alpha_l(\eta)\, d\eta - \int_{Z(t)}^x \alpha_s(\eta)\, d\eta\right), \tag{11}$$

where $R(t)$ is the reflectivity, and $\alpha(x)$ is the optical absorption coefficient.

The initial and boundary conditions can be written in the form

$$C_A(x, 0) = C_0(x), \quad x \in [0, H], \tag{12}$$

$$T(x, 0) = T_0 = \text{const.}, \quad x \in [0, H], \tag{13}$$

$$\frac{\partial C_A(0, t)}{\partial x} = 0, \quad \frac{\partial C_A(H, t)}{\partial x} = 0, \qquad t > 0, \tag{14}$$

$$\frac{\partial T(0, t)}{\partial x} = 0, \quad T(H, t) = T_0, \qquad t > 0, \tag{15}$$

$$Z(0) = 0, \tag{16}$$

which is typical for a standard experimental setup.

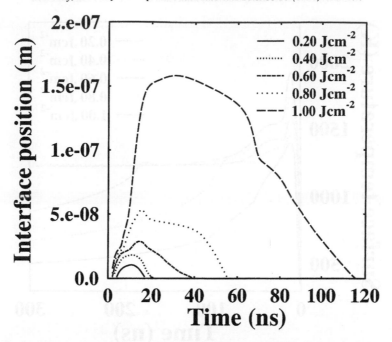

Fig. 1. Solid/liquid interface position for the Ge-layer thickness $A = 50$ nm, $E \in [0.2 \, \text{J/cm}^2, 1.0 \, \text{J/cm}^2]$.

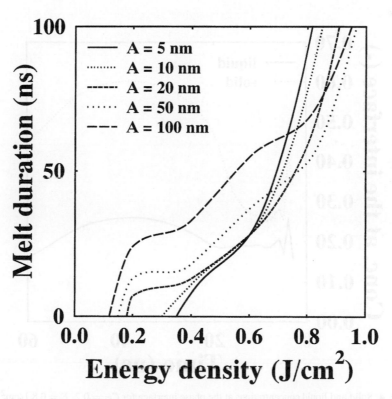

Fig. 2. Melt duration as a function of energy density for the Ge-layer thickness $A \in [5 \, \text{nm}, 100 \, \text{nm}]$.

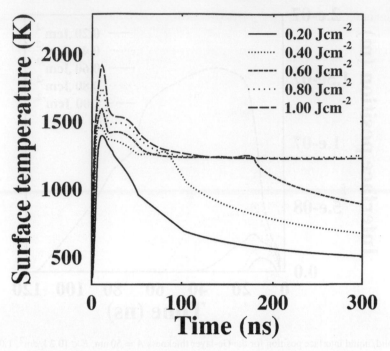

Fig. 3. Surface temperature for $C_{Si} = 0.2$, $E \in [0.2\ \text{J/cm}^2, 1.0\ \text{J/cm}^2]$.

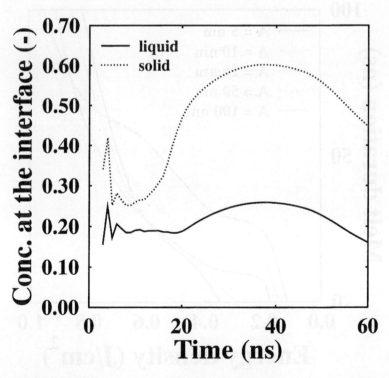

Fig. 4. Solid and liquid concentrations at the phase interface for $C_{Si} = 0.2$, $E = 0.8\ \text{J/cm}^2$.

3. Numerical results and discussion

In solving the problem (1)–(5), (12)–(16) numerically, we first employ the Landau transformation [11] to map both the liquid and the solid domain onto a fixed space interval $\xi \in [0, 1]$. To solve the fixed-domain initial-boundary value problem obtained in this way we apply the Galerkin finite element method. The space and time discretization performed in a common way with linear basis functions leads to a system of nonlinear algebraic equations for temperatures and concentrations to be solved at every time step of length Δt.

The iteration algorithm used to solve the nonlinear problem under consideration is based on the successive approximation approach with underrelaxation. The interface balance conditions (3) and (4) are incorporated in the right-hand side of the above system of equations, the interface values of temperature and concentrations being coupled through the phase diagram. The interface response function (5) is used as the convergence condition in the iteration procedure, i.e., the procedure is designed so that it tries to satisfy (5) at each time step.

Using this numerical model, a computer code in Fortran for the simulation of laser induced phase transitions in a Si–Ge system was developed. This code was employed in all numerical simulations described in what follows.

The following experimental situation was modeled: a Si–Ge sample in the form of either an a-Ge layer deposited on the c-Si substrate or a Si–Ge alloy of constant composition is irradiated by ArF excimer laser (193 nm, 13 ns FWHM). The energy of the laser beam ranged from 0.1 to 2.0 J/cm^2, the shape of the laser pulse $I_0(t)$ was obtained by experimental measurements [12]. The thickness A of the Ge layer in the first modeled case varied from 5 to 100 nm, the mass fraction of Si in the alloy in the second case was supposed to vary in the entire possible range, $C_{Si} \in [0, 1]$.

Tests of the numerical parameters of the model were performed first in order to identify their optimum values. The following parameters were considered: the lengths of the finite elements in the liquid and solid, the maximum value of the time step Δt, and the influence of the relaxation factor q used in the iterative procedure. For the optimum values of these parameters 3–4 iterations were sufficient in every time step.

In regular computational experiments, we calculated the temperature and concentration fields and the position and velocity of the phase interface. Selected typical results are shown in Figs. 1–4, where Figs. 1 and 2 refer to the case of an a-Ge layer on the c-Si substrate, and Figs. 3 and 4 to the case of a Si–Ge alloy.

4. Conclusions

In the previous sections we presented a model describing laser induced melting and solidification of Si–Ge systems. Its applicability to the solution of real experimental problems has been demonstrated on two typical experimental situations.

A direct comparison with experimentally determined data was performed using the results of Thompson et al. [13] obtained by ruby laser irradiation (694 nm, 30 ns FWHM) of a $Si_{0.5}Ge_{0.5}$ alloy. We have observed differences of $\sim 5\%$ in melt duration and of $\sim 25\%$ in the maximum melt thickness, which seems to be a reasonable agreement with respect to the uncertainties in calibrating the TRC data by Thompson et al. [13] and in the thermal conductivity of Si–Ge alloys used in our model [14].

Acknowledgements

This research was supported by the Grant Agency of the Academy of Sciences of the Czech Republic under grant A1010719 and by the Grant Agency of the Czech Republic under grant 201/97/0217.

References

[1] T.P. Pearsall, H. Tenkin, J.C. Bean and S. Luryi, IEEE Electron. Dev. Lett. 7 (1986) 330.
[2] C.A. King, J.L. Hoyt, C.M. Gronet, J.F. Gibbons, M.P. Scott and J. Turner, IEEE Electron. Dev. Lett. 10 (1989) 52.
[3] J. Bevk, A. Ourmazd, L.C. Feldman, T.P. Pearsall, J.M. Bonar, B.A. Davidson and J.P. Mannaerts, Appl. Phys. Lett. 50 (1987) 760.
[4] D. Bedeaux, A.M. Albano and P. Mazur, Physica A 82 (1976) 438.
[5] V. Alexiades, D.G. Wilson and A.D. Solomon, Q. Appl. Math. 41 (1985) 143.
[6] R. Černý, R. Šášik, I. Lukeš and V. Cháb, Phys. Rev. B 44 (1991) 4097.
[7] P. Baeri, G. Fotti, J.M. Poate, S.U. Campisano and A.G. Cullis, Appl. Phys. Lett. 38 (1981) 800.
[8] Z. Chvoj, Cryst. Res. Technol. 21 (1986) 1003.
[9] K.A. Jackson, Can. J. Phys. 36 (1958) 683.
[10] M.J. Aziz, J. Appl. Phys. 53 (1982) 1158.
[11] H.G. Landau, Quart. Appl. Math. 8 (1950) 81.
[12] V. Cháb, private communication.
[13] M.O. Thompson, P.S. Peercy, J.Y. Tsao and M.J. Aziz, Appl. Phys. Lett. 49 (1986) 558.
[14] B. Abeles, D.S. Beers, G.D. Cody and J.P. Dismukes, Phys. Rev. 125 (1962) 44.

ELSEVIER

Computational Materials Science 10 (1998) 475–480

COMPUTATIONAL
MATERIALS
SCIENCE

Author index to volume 10